BUSINESS STATISTICS
FUNDAMENTALS AND APPLICATIONS

ABOUT THE AUTHORS

Moshe Ben-Horim received his education in statistics, and B.A. and M.A. degrees in economics, from the Hebrew University of Jerusalem. He received his Ph.D. from the Graduate School of Business Administration at New York University.

Professor Ben-Horim has held research positions in statistics and finance at the National Bureau of Economic Research in New York, the Israeli Institute of Financial Research, the Hebrew University of Jerusalem, and a number of other institutions. He has also held teaching positions in statistics, decision-making, and finance at New York University, the University of Florida, the Hebrew University, and Montclair State College in New Jersey.

He has written a number of articles in professional journals, and has served as a consultant to public utilities and government agencies in the United States and elsewhere.

Haim Levy received B.A. and M.A. degrees in statistics and economics, and his Ph.D., from the Hebrew University of Jerusalem. He has taught statistics, finance, and decision-making at the Hebrew University, the University of Illinois, the University of California at Berkeley, the University of Florida, and the Wharton School at the University of Pennsylvania.

Professor Levy has written more than one hundred articles and books. His articles have appeared in leading journals, including the *American Economic Review, Management Science, Econometrica,* the *Review of Economics and Statistics,* the *Journal of Finance,* and the *Journal of Financial and Quantitative Analysis.* He is also the editor of the journal *Research in Finance* and an associate editor of the *Journal of Finance.*

He has served as a consultant to numerous firms in the United States and Israel. His consulting work has focused mainly on the application of statistical tools and quantitative methods to economic and business problems faced by firms and governments.

BUSINESS STATISTICS

FUNDAMENTALS AND APPLICATIONS

MOSHE BEN-HORIM
HAIM LEVY

RANDOM HOUSE   **BUSINESS DIVISION**   **NEW YORK**

TO OUR FAMILIES

First Edition
987654321
Copyright © 1983 by Random House, Inc.

Library of Congress Cataloging in Publication Data

Ben-Horim, Moshe.
    Business statistics.

    Includes index.
    1. Commercial statistics. 2. Statistics.
I. Levy, Haim. II. Title.
HF1017.B35 1983   519.5'024658   82-15060
ISBN 0-394-33022-6

Manufactured in the United States of America
Cover art and design by Dana Kasarsky Design.

PREFACE

The use of statistical analysis in business and government has become increasingly common in recent years. In more and more areas, decision-making is based on statistical analysis rather than on rules of thumb and subjective evaluations. This trend is likely to continue, enhanced by rapid technological progress in the area of mini and personal computers, whose sales are growing at a very rapid rate. These relatively inexpensive computers can handle analyses that required large, expensive computers only a few years ago. Thus, it is possible for business and government to use a wide variety of statistical analyses to evaluate sales, cash management, inventory, marketing, credit and financing, imports and exports, budgets and so forth.

This book attempts to show how statistical data should be handled and analyzed and how the results can be used in decision-making. We have noticed, in teaching statistics in colleges and universities in different parts of the United States and elsewhere, that students often have difficulty grasping relationships between textbook discussions and practical applications. To put it differently, students eagerly look for "a formula that will tell us which formula to use." Not unexpectedly, they soon find out that no such formula exists. What they need, instead, is experience. Only experience can tell us which technique to use for each type of analysis—by exposing us to a variety of situations for which statistical analysis is required.

Our belief is that a textbook should attempt to provide this experience, by using teaching examples drawn from real life. In pursuit of this idea we have tried to provide a wealth of practical examples and applications, both in the text proper and in the study problems at the ends of the chapters. Some of these examples and applications are based on simplified hypothetical data, constructed to demonstrate specific points in clear and uncomplicated fashion. Others use real data concerning events in companies or industries or in economy-wide or worldwide situations, so that students can get the feeling of handling some of the practical problems that come up in various fields of business and economics. These examples and applications have been carefully chosen from fields that the students is likely to encounter in other courses. (In this way we hope to strengthen the link between the statistics course and other courses in the business curriculum.) Particularly important is the exclusion of irrelevant topics such as cards, marbles, urns, and dice; all these have been replaced by relevant business topics, including:

- stock and bond returns
- determination of life-insurance premiums
- use of sales coupons to increase sales
- research and development spending in relation to sales
- the budget requirements of an oil company
- advertising expenditures and consumer awareness
- allocation of indirect costs to various company departments
- the consumption-income relationship
- the relationship between nominal interest rate and inflation
- inflation, unemployment, and stagflation
- the relationship between a stock's average return and its corresponding risk
- sales versus advertising

We have tried to maintain a reasonable depth of analysis without recourse to algebra beyond the high-school level. But since pocket calculators and electronic computers are now widely available, we emphasize (where appropriate) the selection of data and the way computer output should be interpreted.

The book is suitable for a one-semester or a two-semester course. For a two-semester course we recommend using the entire book; the first semester may cover up to Chapter 11 or 12, and the second may cover the rest of the book. For a one-semester course we recommend Chapters 1 through 14, with the possible omission of some sections of these chapters.

Problem-solving is probably the best vehicle for studying the material covered in the text. Thus, a genuine effort has been made to provide problems that are as good as or better than the examples given in the text. Included are problems whose solution requires primarily an intuitive understanding of the material and for which calculations are minimal; problems calling for calculations, for which the student is expected to master the requisite technique; and finally, problems straight out of the business and research world (some of them adapted a bit to fit the level of the book). The problems in the last category may well serve as class projects or take-home examinations, and may provide material for class discussions.

We hope that teachers and students alike will find the book both useful and enjoyable.

Ancillary Materials

For the convenience of teachers and students, the book is supplemented by a number of ancillary materials. The **Solutions Manual** provides very detailed solutions to almost all the end-of-chapter problems. The **Instructional Materials** include chapter-by-chapter notes and discussion of the book's material plus a **test bank** of about 45 problems with full and detailed solutions. There is also a **multiple choice test bank,** which includes more than 500 problems. Finally, a large set of **transparencies** includes chapter summaries and other material.

Acknowledgments

We wish to acknowledge the help of many people who have helped to make this book a reality. Among them are:

Benzion Barlev	James Jones	Stephen Smith
Meir Barnea	Yoram Kroll	Esther Tuval
R. Mark Bisk	Zvi Lerman	Adele Zarmati
Mary S. Broske	Azriel Levy	Zeev Zeimer
Gaitona Calais	Patricia Priola	Dror Zuckerman
John J. DiNome, Jr.	Mary Register	
Susanne Freund	Pushpalatha Shanker	

For their time, care, and pertinent advice, we would especially like to thank our academic reviewers:

Thomas E. Brelsford, Kellogg Community College
Frank P. Buffa, Texas A&M University
Ronald L. Coccari, Cleveland State University
Randolph H. Forsstrom, Bergen Community College
Robert Miller, University of Wisconsin at Madison
Jacqueline Redder, Virginia Polytechnic Institute and State University
Paul H. Rigby, Pennsylvania State University
Paul Sugrue, University of Miami
Charles J. Teplitz, SUNY-Albany
Richard Twark, Pennsylvania State University
Thomas Wedel, California State University, Northridge

We also acknowledge the secretarial assistance we have obtained at the Department of Finance, College of Business Administration, University of Florida; and at the School of Business Administration, The Hebrew University.

Finally, we owe special thanks to Judith Kromm and Paul Donnelly of Random House, who were responsible for editing this book. We would also like to thank Anne Mahoney, David Rothberg, and Della Mancuso of Random House.

CONTENTS

PART II: PROBABILITY AND DISTRIBUTIONS (DEDUCTION) 93

PART III: DECISION-MAKING BASED ON SAMPLES (INDUCTION) 201

P A R T I V: REGRESSION AND CORRELATION (INDUCTION CONTINUED) 333

P A R T V: OTHER SELECTED TOPICS 409

BUSINESS STATISTICS

FUNDAMENTALS AND APPLICATIONS

CHAPTER ONE OUTLINE

1.1 Descriptive Statistics

1.2 Deduction (Probability) and Induction
    (Inference)

1.3 Statistics in Business

1
INTRODUCTION

Statistics is the science that deals with the collection, classification, analysis, and interpretation of numerical facts or data. Values computed from the data are also called statistics. Statistics can be *descriptive* (simply summarizing the data), deductive, or inferential. *Deductive statistics* deals with probability: it is the area of statistics in which properties of specific results are studied on the basis of knowledge of the general population. *Inferential* or *inductive statistics* deals with the opposite relationship: it leads to conclusions or inferences about larger populations of which the data are a sample. All sciences use inferential statistics to add a greater degree of confidence to research results because statistics often allow one to calculate the probability that one's conclusions are in error.

Most of this book is devoted to inferential statistics, but part of it deals with descriptive and deductive statistics. Let us describe in a little more detail the scope of each of the main branches of statistics.

1.1 Descriptive Statistics

Descriptive statistics deals with collection, organization, and presentation of data. The descriptive part of statistics is relatively simple and straightforward; in fact, it is more of an art than science. Nevertheless, it is very often true that the statistician cannot scientifically analyze the data until they have been organized.

The available data in a company and at the local, state, and federal levels of government are frequently overwhelming in quantity. It is easy to read

3

a company's annual report or a government budget in which the data have been organized and properly presented, but we might feel very frustrated if hundreds or thousands of unorganized pieces of data were submitted to us for analysis.

Different types of tables and charts can be used to present the data, and as we shall see, the choice among them depends on the type of data, the purpose of the presentation, and individual preference.

In descriptive statistics we distinguish between data organization and data presentation. Organization involves putting data together in a way that allows convenient access to them. Thus large quantities of information are often stored on computer tapes. Data presentation, however, usually involves compiling tables and drawing charts to visually convey the data or aggregate or average measures of them in a form that will facilitate analysis.

Although most statistics textbooks (including ours) devote more space to statistical inference than to descriptive statistics, we should always remember that we cannot hope to get valid results by analyzing data that are no good to begin with. Examining the data for accuracy also falls within the scope of descriptive statistics.

1.2 Deduction (Probability) and Induction (Inference)

Deduction is used when information is available about a given population (e.g., all the inhabitants of a country or all the electric light bulbs produced by a firm) and we want to draw conclusions about its characteristics. The value of a potential drilling site to an oil company depends on the probability that marketable oil will be found there. This probability can be estimated by studying the results of oil exploration efforts at many other sites that were thought to be potential oil fields. Similarly, the chance of an airline passenger reaching his or her destination on time can be assessed by studying the percentage of flights that have experienced delays in the past. In these examples, our knowledge of the characteristics of the population enables us to deduce the probability of individual events to occur.

Quite often, however, our knowledge about a particular population is limited. For a variety of reasons, such as economic efficiency and the limited time available, we may have to make do with studying only a *sample* of the entire population. By analyzing the sample, we can make inferences about the characteristics of the population as a whole. This process is called *induction*.

Examples of statistical induction are numerous. One is the Labor Department's sample of unemployment. Nationwide unemployment rates in the United States must be frequently and accurately determined for proper implementation of fiscal and monetary policies. It would be extremely costly to engage in a direct examination of the country's entire labor force (the U.S. Census, for example, is carried out only once every ten years). But if it has been properly chosen, even a small sample can provide highly accurate estimates of the population's characteristics. Or take another example, the Consumer Price Index (CPI), a popular measure of the inflation rate that must be frequently and accurately estimated. Here, too, only a small number of products and services are sampled, and statistical induction is used to infer facts concerning the entire population of prices from a highly restricted sample.

Any population—except a totally homogeneous one—can be only approximately represented by a sample. That is to say, when we use a sample to make inferences about the entire population, these inferences will depend in part on which particular members of the population happen to be included in the sample. Consequently, whatever we may infer about the population is subject to some degree of error, whose magnitude must be evaluated. Important as the evaluation of potential errors may be, economists and business managers engaged in decision-making frequently have to go a step further: they must draw clear-cut conclusions and take appropriate action. Drawing such conclusions and deciding upon the actions to be taken is the subject matter of *decision theory*. In real life, we face uncertainty in almost every decision we make. We cannot avoid error in our decisions. The penalty for error can be trivial—e.g., we may lose $1—or serious—e.g., we may choose the wrong location for a $100 million plant—or even fatal—e.g., we may use the wrong drug.

However sophisticated statistical analysis becomes, it cannot eliminate such decision-making errors. But *properly employed* statistical tools ensure that the errors are minimized. For example, think of a business manager who decides to invest $10 million in developing a new product. If he or she invests without prior study, he or she stands to lose money if there is no demand for the product. It is possible, however, to find out whether or not there is a sufficient demand by conducting a small survey. This survey will not enable the manager to eliminate the possibility of loss altogether, but the manager can minimize loss by collecting and carefully analyzing the statistical data and making the best decision, taking all the data into account.

1.3 Statistics in Business

Statistics make a vital contribution to business. Anyone glancing through a local newspaper can verify the fact that a good part of its contents is devoted to statistics, such as daily price changes in stocks and bonds, dividends for various firms, sports scores, temperatures, prices, unemployment rates, and so on.

An entrepreneur starting a new business or planning a new investment can improve the chances for success by collecting and analyzing statistical data. For example, the introduction of a new product calls for a survey of people's income (i.e., their potential purchasing power), and their consumption behavior. Such data may assist the entrepreneur in determining the best location for the firm's factory, the most effective advertising policy, and perhaps the most advantageous pricing policy as well.

When we make business decisions we usually rely upon a combination of census data (i.e., for entire populations) and sample statistics. Take, for example, a bank loan department, which has to fix loan rates for different borrowers. In order to determine proper rates, data must be gathered concerning the borrowers. The data with regard to the borrowers' percentage of default may be taken from the Census of Statistics of all banks, and in addition a sample of the specific bank's borrowers can provide further information. People who have some general characteristic in common (e.g., similar income) may also have similar repayment habits, and analysis of the data enables us to group the borrowers by such relevant characteristics. Thus, the percentage of defaulters will vary from one group to the next and the bank can take appropriate steps: usually, the higher the default percentage for a given group of borrowers, the higher the interest the bank demands.

PART 1

DESCRIPTIVE STATISTICS

The area of descriptive statistics deals with alternative methods of presenting statistical data effectively. The presentation of data may be a final goal; or it may be a step toward further analysis.

Chapter 2 discusses the use of tables and charts in the presentation of data. Chapter 3 discusses the concept of population, and shows how population data are organized in a frequency distribution. In Chapter 4, statistical measures of various population characteristics are developed; these measures are used in statistical analyses of populations.

The applications, examples, and problems in Part 1 deal with stock and bond returns, the analysis of cumulative frequency distributions for insurance purposes, the use of charts in financial reporting, and a number of other topics.

CHAPTER TWO OUTLINE

2.1 Data Collection

2.2 Data Organization
Editing Data
Data Classification
Tabulation

2.3 Data Presentation: Tables
Parts of the Table
Common Types of Tables

2.4 Data Presentation: Charts and Graphs
Line Charts
Component-Part Line Charts
Bar Charts
Component Bar Charts
Grouped Bar Charts
Pie Charts

2.5 A Word of Caution

Key Terms

statistical data
primary source
secondary source
sampling
editing
mutually exclusive
    classes
collectively exhaustive
    classes
tabulation
analytical tables
reference tables
table number
table title
table headnote
table boxhead
table stub
table stub head
table stub body
table cell
table field

table footnotes
table source note
one-way classification table
cross-classified table
percentage table
chart title
chart headnote
chart scales
graphs
chart footnotes
chart source note
time-series line chart
component-part line
    chart
horizontal bar chart
vertical bar chart
duo-directional
    horizontal bar chart
component bar chart
grouped bar chart
pie chart

2
HANDLING STATISTICAL DATA: COLLECTION, ORGANIZATION, AND PRESENTATION

Statistical data form the foundation of virtually all applied statistical analyses. Since an analysis can be only as good as the data it uses, great care must be taken to ensure that the data are appropriately handled: that the relevant data are collected and screened for errors, and that they are well organized and effectively presented.

A distinction is usually made between *statistical data* and data that are not statistical. **Statistical data** are *facts that bear some sort of measurable relationship to each other* (such as prices, costs, weights, etc.), whereas data that are not statistical are basically arbitrary (such as item numbers in a sales catalogue).

Numerous decisions in business and in government are based on statistical data. As time goes on, the complexity of these decisions is increasing, causing an ever-growing need for statistical data in a variety of areas—finance, marketing, management, accounting, insurance, production, fiscal and monetary economics, quality control, and many more.

We may distinguish the following steps in any statistical study:

1. The collection of relevant data
2. Organization
3. Presentation
4. Analysis
5. Interpretation

This chapter deals with data handling, a subject that includes the first three of these steps: collection, organization, and presentation.

2.1 Data Collection

After a problem has been clearly identified and a decision made to perform a statistical analysis in order to find a solution, relevant data should be collected in preparation for the analysis. The quality of statistical analysis can be no better than the data on which it is based, so the data must be as reliable and as accurate as possible. Therefore we always prefer *published* to *unpublished* data. To the extent that reliable published data are available, it is much more economical to use them than to conduct one's own survey. If possible, it is always better to collect published data from their **primary source**—that is, from *the source in which the data were originally published*—than from a **secondary source**, which merely *reproduces the data*.

In addition to published data, there are numerous data files (that is, collections of data) available to the public on computer tapes, particularly in the fields of finance, accounting, and economics. Again, a good search for such tapes should be conducted before one sets out to collect one's own data.

Despite the abundance of published business and economics data, there are many situations in which a special survey must be conducted to obtain the data for a study if it is to be meaningful. One highly efficient method consists of **sampling**, that is, *observing only part of all the existing data*. We discuss sampling in detail in Chapter 9, so further discussion of it is not necessary here. Instead, we will proceed to the next important topic, the organization of statistical data.

2.2 Data Organization

After the data have been collected, they need to be organized. Three procedures must be carried out before data can be presented and analyzed: editing, classification, and tabulation. We shall consider each one of these in turn.

EDITING DATA

Editing involves *a close examination of the statistical data that have been collected.* The purpose of such an examination is to determine whether the data contain errors, to correct any errors that are detected, and to check on any data that appear to be unreliable. Ample editing time should be allowed if the data have been collected via a questionnaire; and editing is of great importance when the data consist of direct observations (the number of defective items found when the quality of a given product is checked, the frequency of busy signals received when a specified number of telephone calls are made, and so on).

When the data consist of records of direct observations, the editing process usually involves looking over the data to determine whether the figures "make sense" and are within a "reasonable range." It is also desirable to find a pattern of internal consistency in the data, if such a pattern exists.

DATA CLASSIFICATION

The best way to classify data depends primarily on the purpose of the study being undertaken, and on the nature of the variables involved in it. Given the goal and the variables that will be studied, one can determine the most workable classification method.

Obviously, data may be classified according to a host of variables; the multitude of possibilities makes generalizations about variables very difficult. Nevertheless, for the purpose of classification, we can distinguish among four major types of variables. There are variables related to *time,* such as days, years, age, and so on; variables related to *place,* such as states, cities, and towns; *quantitative* variables such as units sold, costs of production, and payroll deductions; and finally there are *qualitative* variables, such as color, style, and sex.

It is essential that the data be classified into mutually exclusive and collectively exhaustive classes. **Mutually exclusive classes** are *classes that are chosen so that each answer or figure will belong to only one class.* **Collectively exhaustive classes** are *classes that include all the possible answers or the entire range of possible values of the variable of interest.* Suppose, for example, that we wish to classify firms according to number of employees. The classes 0–20 employees, 21–50 employees, 51–100 employees, and 101 or more employees are both mutually exclusive and collectively exhaustive. They are mutually exclusive because at a given point in time, each firm's number of employees can fit into *only one* of these classes. They are also collectively exhaustive because *no firm could have a number of employees that did not fit into any* of the classes.[1]

TABULATION

After the data are obtained and the classification system is determined, there remains the task of tabulation as a preparation for the final presentation and analysis. **Tabulation**—*the enumeration and recording of the data in each class for each variable*—can be done either manually or by computer. Manual tabulation is suitable for smaller studies when the data are not too voluminous to handle. When a large volume of data is involved, tabulation by computer may be more efficient. Computer tabulation has the advantages of higher speed and credibility, but these advantages must be contrasted with the relatively high setup and overhead cost of the computer approach.

2.3 Data Presentation: Tables

Statistical data can be presented verbally or by means of tables and charts. Verbal description alone may be sufficient if the data contain very little information. When a large amount of information is to be conveyed, however, verbal description is usually less efficient than a table or a chart.

The construction of a statistical table is determined by the type of data to be presented and the purpose of the presentation. Generally speaking, we distinguish between analytical tables and reference tables. **Analytical tables** are *tables that lead to analytical study.* They must be relatively short so that the reader does not lose sight of the forest among the trees. **Reference tables** *hold more information than analytical tables;* they can be used to verify details. Analytical and reference tables are constructed in basically the same way. The form of a typical cross-classified table is given in Figure 2.1.

[1] The following is an example of classification that is neither mutually exclusive nor collectively exhaustive: 0–20, 21–50, 50–100. It is not mutually exclusive because if the number of employees is 50, this item of data could be recorded in two classes (21–50 and 50–100) rather than just one. It is also not collectively exhaustive because none of the classes applies to firms with more than 100 employees.

Figure 2.1

The functional parts of a cross-classified statistical table

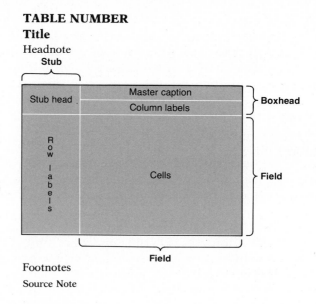

TABLE NUMBER

Title

Headnote

PARTS OF THE TABLE

The **table number** is important for easy reference and citation. The numbers can be assigned consecutively throughout a report, an article, or a book, or they can contain a chapter number followed by a consecutive subnumber within the chapter. "Table 4.7," for example, refers to Table 7 of Chapter 4.

The **title** provides the reader with a short statement describing the table's contents. It should not be too long, or so short as to omit essential information about the table's contents. The title should answer the following questions: (1) *What* are the data in the table? (2) *Where* do the data apply? (3) *When* do the data apply? (4) *How* are the data classified?

A **headnote** contains additional information about the table, not included in the title. Usually the headnote names the units in which the table's figures are measured: percentages, dollars, or any other unit of measurement.

The **boxhead** contains the master caption, which states the way the columns are classified, and the column labels. If the master caption merely repeats information already given in the title, it is usually omitted.

The **stub** contains the **stub head,** stating the way in which the rows are classified, and the **body** of the stub, in which the individual rows are briefly labeled.

A **cell** is the intersection of a column and a row; this is where the individual item of data is inserted.

The **field** of the table is the total collection of cells.

Footnotes may be added to present additional information that is not stated in the title, headnote, captions, or stub. Footnotes are commonly placed below the stub.

The **source note** provides an accurate description of the origin of the data presented in the table. It provides the reader with a way to verify the data or to obtain additional information, and gives the original data collector due recognition. The source note may be omitted only if the source of the data is quite trivial. For example, when data are collected in a survey and then presented in a report, it is not necessary to have a separate source note for each table stating that the source consists of the data collected in the study.

COMMON TYPES OF TABLES

Table 2.1 is probably the simplest of all tables, a **one-way classification table** that presents some data in an organized manner. The rows present specific years, and the field consists of a single column showing the consumer installment loans outstanding in the United States at the ends of the years 1976 through 1980.

Table 2.2 contains more information on these loans than Table 2.1 does. It is a **cross-classified table,** which groups the loans in the years 1976 through 1980 according to type of loan.

In some tables, such as 2.2, a total of the content of the rows is relevant, and in others a total of the content of the columns may be of interest; sometimes, as in Table 2.3, totals are given for both the columns and the rows.

TABLE 2.1

**Consumer Installment Loans
Outstanding at End of Year, 1976–80**
(millions of dollars)

December 31 of year	Loans outstanding
1976	194.0
1977	230.8
1978	275.6
1979	311.1
1980	306.8

Source: Moody's *Bank & Finance Manual*, 1981, vol. 2, p. a13.

TABLE 2.2

Consumer Installment Loans Outstanding at End of Year, by Type of Loan, 1976–80
(millions of dollars)

December 31 of year	Type of Loan				Total
	Automobile	Mobile home	Revolving credit	Other	
1976	$ 67.7	$14.6	$17.2	$ 94.5	$194.0
1977	82.9	15.1	39.3	93.5	230.8
1978	102.5	16.0	47.1	110.0	275.6
1979	115.0	17.4	55.5	123.2	311.1
1980	115.6	17.4	53.9	119.9	306.8

Source: Moody's *Bank & Finance Manual*, 1981, vol. 2, p. a13.

TABLE 2.3

Employees of American Telephone & Telegraph Company, by Sex and Occupation, December 31, 1980

Occupation	Male	Female	Total
Officials and managers	131,551	73,255	204,806
Professionals	48,485	17,790	66,275
Technicians	67,152	17,824	84,976
Office and clerical workers	57,123	329,539	386,662
Others	232,249	85,370	317,619
Total	536,560	523,778	1,060,338

Source: American Telephone & Telegraph Company, *Annual Report, 1980*, p. 12.

The table represents the employees of the American Telephone and Telegraph Company at the end of 1980, by sex and occupation. The sum on the bottom row shows the total of male employees (536,560), the total of female employees (523,778), and the total number of employees in the company (1,060,338). The column headed "Total" shows the total number of employees in the various occupations. At the end of 1980, there were 204,806 officials and managers, 66,275 professional employees, and so on. The total number of employees, in all the various occupations, is 1,060,338—obviously the same as the total number of male and female employees.

Percentage Tables. **Percentage tables** *show percentages of totals*, either alone or together with the absolute figures. It is essential that the totals of which the percentages are part be clearly identified. Consider, for example, Table 2.3. This table can be presented in two separate percentage tables. One, shown in Table 2.4, presents the percentage of men and women in each occupation. Note that the inclusion of the right-hand "Total" column, which demonstrates that the total percentage of men and women in each occupation equals 100 percent, removes all doubt about what the percentages represent: they represent the percentages of men and women employed by the company in any given occupation. The percentages on the bottom row (that is, 50.6 percent for men and 49.4 percent for women) *are not the total, nor are they the simple average of their respective columns.* They are in fact the weighted averages of the columns (a concept we shall discuss in Chapter 4); stated more simply, they represent the percentages of men and women in the total number of people employed by the company. Men comprised 50.6 percent and women comprised 49.4 percent of the people employed by AT&T on December 31, 1980.

Another percentage table—one that reveals different information—can be derived from the data of Table 2.3. If we look at all the male employees first and determine their occupational distribution in percentages, and then do the same thing for the female employees and for all employees, we get Table 2.5. Notice how different the figures in Table 2.5 are from those in Table 2.4. For example, in the first cell on the left on the top line, we have 64.2 percent in Table 2.4 and 24.5 percent in Table 2.5. These two figures represent completely different things. Among all the employees in the occupational category "officials and managers," 64.2 percent were men (and 35.8 percent were women), but among all the men employed by the company, only 24.5 percent were in this particular category, and 75.5 percent (100.0 − 24.5 = 75.5) were in other occupations.

TABLE 2.4

Employees of American Telephone & Telegraph Company, by Sex and Occupation, December 31, 1980
(percentage)

Occupation	Male	Female	Total
Officials and managers	64.2%	35.8%	100.0%
Professionals	73.2	26.8	100.0
Technicians	79.0	21.0	100.0
Office and clerical workers	14.8	85.2	100.0
Others	73.1	26.9	100.0
Total	50.6%	49.4%	100.0%

Source: American Telephone & Telegraph Company, *Annual Report, 1980*, p. 12. See also Table 2.3.

TABLE 2.5

Employees of American Telephone & Telegraph Company, by Sex and Occupation, December 31, 1980
(percentage)

Occupation	Male	Female	Total
Officials and managers	24.5%	14.0%	19.3%
Professionals	9.0	3.4	6.3
Technicians	12.5	3.4	8.0
Office and clerical workers	10.6	62.9	36.4
Others	43.4	16.3	30.0
Total	100.0%	100.0%	100.0%

Source: American Telephone & Telegraph Company, *Annual Report, 1980*, p. 12. See also Table 2.3.

2.4 Data Presentation: Charts and Graphs

Charts and graphs are also frequently used to present data. A pictorial expression of the data often has a greater impact on the reader than a table would have, and may help the reader gain a better understanding of the data involved. Like tables, charts and graphs may be used to express the results of analyses. The principal parts of a chart are the **title, headnote, scales, graphs, footnotes,** and **source note**. The title and headnote are presented above the chart, the footnotes and source note below it; all serve the same purpose as those of a table. The horizontal scale, known as the *X*-axis or *X*-scale, is often (though not always) used to classify the data; the vertical scale, known as the *Y*-axis or *Y*-scale, is used to measure the magnitude of the relevant data. Figure 2.2 illustrates the principal parts of a typical chart.

It is difficult—perhaps impossible—to determine the "most appropriate" type of chart or graph for any type of data. It is often very useful to use two or more complementary charts or graphs for illustration, each bringing to the fore a separate fact or set of facts.

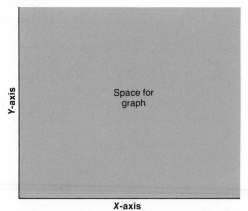

Figure No.
Title
Headnote

Y-axis

Space for graph

X-axis

Footnotes

Source Note

Figure 2.2

The functional parts of a chart

LINE CHARTS

As the name indicates, line charts use lines—solid or broken—to represent data. The most common of all line charts is the **time-series line chart,** which simply exhibits the change in the data over a period of time. Figure 2.3 is an example of a time-series line chart. Here two graphs are exhibited, representing two series on one chart. One of the graphs represents the value of new

Figure 2.3

New issues of corporate bonds and common stocks, 1970–80

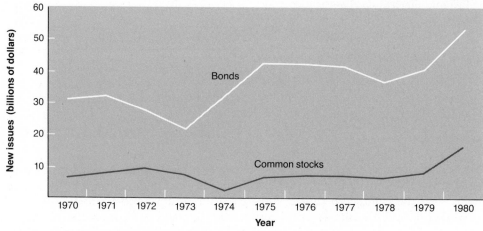

Source: *Moody's Industrial Manual,* 1981, vol. 1, p. a31.

issues of common stocks in the United States during 1970–80, and the other represents the value of new issues of bonds. To facilitate reading the graph, grid lines are drawn horizontally and vertically. The grid lines must be drawn in a distinct (and often light) color or shade so that the line graph is easily visible against the chart's background.

One draws a time-series line chart by first placing points on the chart to represent the data and then connecting the points by straight line segments. Let us illustrate this procedure through an example.

EXAMPLE 2.1

The aggregate of all imported cars sold in the United States in the period 1975–80 is shown in Table 2.6. Let us draw a time-series line chart for these data.

TABLE 2.6
Imported Cars Sold in the United States, 1975–80
(thousands of cars)

Year	Cars sold
1975	1,564
1976	1,491
1977	2,064
1978	1,993
1979	2,319
1980	2,390

Source: *Ward's Automotive Yearbook,* various issues.

First, we need to determine the scale for both axes. Our choice of scale depends on the relationship between the size of the chart and the number of data points that need to be represented, and their value. In the case at hand, data for six years need to be represented on the chart, so the horizontal width available for the chart should be divided into six equal intervals, each representing one year. The values of the data range from 1,564 to 2,390 (thousands of cars). As the highest value we have is 2,390, we decide to let the vertical axis extend to 2,400. But we must also consider whether or not the scale should start at zero. If it does, about two-thirds of the chart will be unusable, since the lowest value we need to present is 1,564. We therefore use a broken *Y*-axis, and present the range from 1,400 to 2,400 (thousands of cars) in the available space. In this way we focus only on the *relevant* range of values of the variable of interest. Note the jagged marks near the bottom of the vertical scale in Figure 2.4: they tell the reader that the scale does not start at zero.

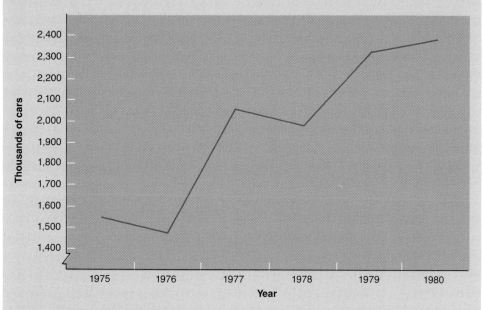

Figure 2.4

Imported cars sold in the United States, 1975–80

Source: *Ward's Automotive Yearbook,* various issues.

Now let us consider another point. Take the first piece of data available: the sales of imported cars in the United States in 1975. Clearly the sales were not made in full at the beginning of the year or at the end of the year, but gradually throughout the year. Consequently, if we plan to use only one point to represent the 1975 sales, that point is best located over the center of the interval representing the year 1975.

This practice of placing the point over the center of the interval is not followed without exception, however. Depending on the nature of the variable measured, the point may be placed over the tick mark representing the beginning or the end of the year.

COMPONENT-PART LINE CHARTS

An interesting extension of the line graph is the **component-part line chart,** applicable when we want to show not only the way a given total has changed over time but also the way its components have changed. Let us simply extend Example 2.1 and look at the components of U.S. car import sales in the period 1975–80.

EXAMPLE 2.2

Table 2.7 shows the sales of imported cars in the United States by make for the period 1975–80. Each make is regarded as a component of the total.

Figure 2.5 displays the data in a component-part line chart, in which the components are represented by four layers distinctively shaded and

TABLE 2.7
Imported Cars Sold in the United States by Make, 1975–80
(thousands of cars)

Year	Make				Total
	Toyota	Datsun	Honda	Others	
1975	278	260	102	924	1,564
1976	347	270	151	723	1,491
1977	493	388	261	922	2,064
1978	442	399	275	877	1,993
1979	508	472	353	986	2,319
1980	582	517	375	916	2,390

Source: *Ward's Automotive Yearbook,* various issues.

labeled. Note that the amount of each component is represented by the *height of the layer representing that component alone* rather than by the total height from the horizontal axis.

Figure 2.5

Imported cars sold in the United States by make, 1975–80

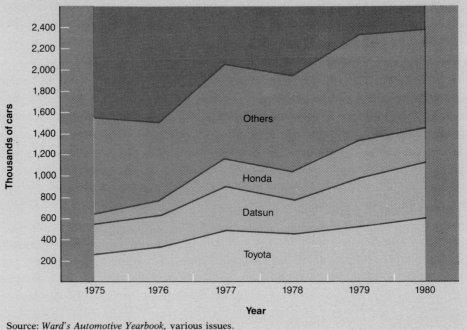

Source: *Ward's Automotive Yearbook,* various issues.

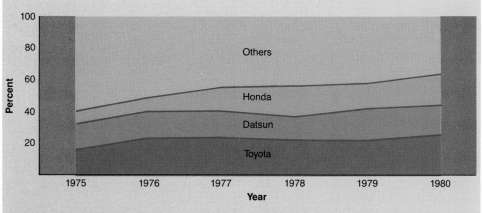

Figure 2.6
Percentage of imported cars sold in the United States, by make, 1975–80

Source: Derived from *Ward's Automotive Yearbook,* various issues.

We may use a similar technique when we wish to show the components as percentages of the total (Figure 2.6). Again, the percentage of a given component is shown by the height of that component's layer only, rather than by the height measured from the horizontal axis. Figure 2.6 illustrates a very convenient method for representing trends in the *relative* magnitudes of the components and their changes over time. (We cannot, of course, use this method to give information about the trend of the total over the years, as we do in Figure 2.5.)

BAR CHARTS

Bar charts may be used to represent data classified by either quantitative or qualitative variables. An example of a **horizontal bar chart** is given in Figure 2.7, which shows sales of the five largest tire and rubber companies in the United States in 1980. The bars are separated for clarity and labeled on the left, in the chart's stub. The bars are organized in order of decreasing

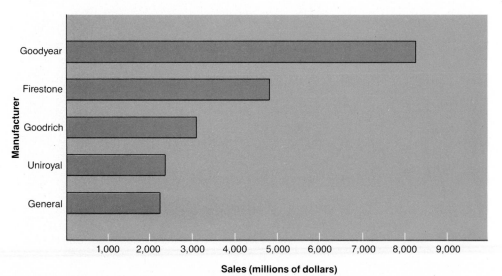

Figure 2.7
Sales of five largest tire and rubber companies, United States, 1980

Source: *Value Line Investment Survey,* April 3, 1981. © 1981 by Arnold Bernhard & Co., Inc., publishers.

length from top to bottom; that is, the longest bar is shown at the top of the chart and the shortest at the bottom. This order may be reversed, of course, so that the shortest bar is at the top and the longest at the bottom.

All the bars are of the same width and each company's sales are indicated by the bar's length. The lengths of the bars are measured against grid lines that run from top to bottom and are marked either at the top or at the bottom of the chart, or both.

The scale showing the value of the variable of interest (sales, in this case) should start at zero to eliminate the possibility of misinterpretation. After all, charts are designed to bring out similarities and differences in data, so great care must be taken to avoid ambiguity.

Sometimes, however, when we start the scale at zero, differences between the bars may be less visible than we would like. In such cases, we may use other devices to clarify the differences. For example, suppose we want to use a bar chart to represent new housing units started for the years 1975–80. These data are represented in Figure 2.8, a **vertical bar chart.** It is identical to a horizontal bar chart in all respects except that the vertical and horizontal

Figure 2.8

New private housing units started, 1975–80

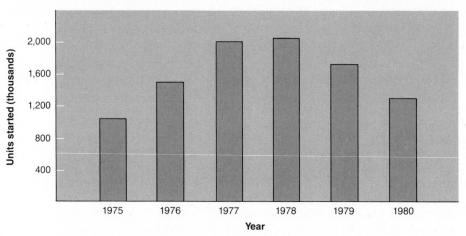

Source: Council of Economic Advisers, *Economic Indicators,* July 1981, p. 19.

Figure 2.9

Annual percentage change in new private housing units started, 1976–80

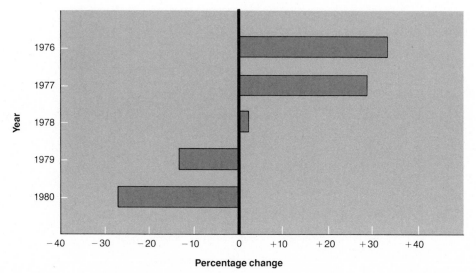

Source: Council of Economic Advisers, *Economic Indicators,* July 1981, p. 19.

scales have been reversed. If we wish to display the variations in the data more clearly, we may simply focus directly on the *changes* in the variable, measured either in absolute amounts or in terms of percentage change. Figure 2.9 is a **duo-directional horizontal bar chart,** where the percentage changes in new housing units started in the years 1975–80 are plotted. The arrangement of the bars should follow the sequence of time, if relevant, from top to bottom or vice versa. If such order is not relevant and no other dominant criterion for ordering is applicable, the bars should be arranged by length in either decreasing or increasing order.

COMPONENT BAR CHARTS

Like line charts, bar charts may be adapted to show components of totals. In Figure 2.10, a **component bar chart,** vertical bars are used to show the number of unemployed people in four industry groups by duration of unemployment in June 1981. Each component is marked by a distinct shade. When no other particular order is called for, the bars are organized in order of total length.

The limitation of a component bar chart is that the *relative* magnitudes of the components are hard to discern because the totals are unequal. When

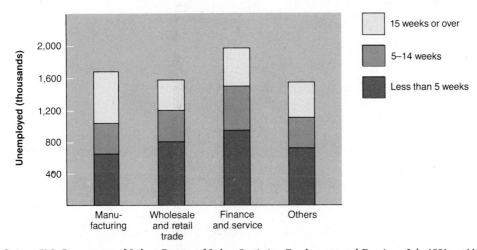

Figure 2.10

Unemployed persons, by industry and duration of unemployment, June 1981

Source: U.S. Department of Labor, Bureau of Labor Statistics, *Employment and Earnings*, July 1981, p. 44.

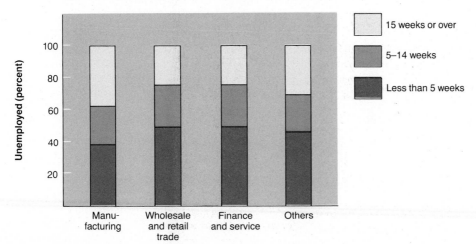

Figure 2.11

Percentage distribution of unemployed persons, by industry and duration of unemployment, June 1981

Source: U.S. Department of Labor, Bureau of Labor Statistics, *Employment and Earnings*, July 1981, p. 44.

the relative magnitude is of interest, a percentage component bar chart may be drawn, like that in Figure 2.11. Here all three bars have the same length, so that their components may be more readily compared.

GROUPED BAR CHARTS

The **grouped bar chart,** illustrated in Figure 2.12, provides essentially the same information as the component bar chart, but the components are separated (and grouped). They are measured from the horizontal axis. The

Figure 2.12

Unemployed persons, by industry and duration of unemployment, June 1981

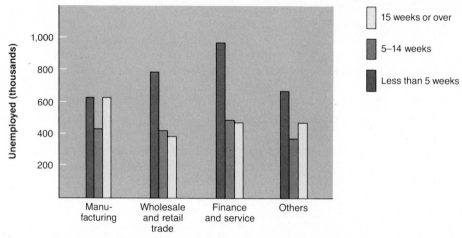

Source: U.S. Department of Labor, Bureau of Labor Statistics, *Employment and Earnings*, July 1981, p. 44.

advantage of Figure 2.12 over 2.10 is that it permits the magnitude of each component to be more easily seen. The totals, however, may be more readily compared when the data are presented in a component bar chart (Figure 2.10).

PIE CHARTS

The **pie chart** is another pictorial device that may be used to show the proportions of a total represented by its component parts. The components are

Figure 2.13

U.S. federal budget, 1980: receipts and outlays
(millions of dollars)

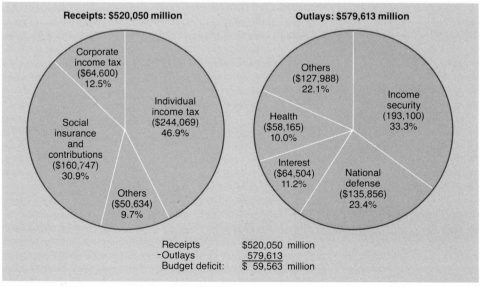

Source: *Federal Reserve Bulletin*, August 1981, p. A29.

expressed as percentages of the whole and are shown as portions of a circle, like the wedges of a pie. Sometimes the absolute values of the components are also presented. With 360 degrees in the circle, the angle of each component is determined thus:

Angle of component (in degrees) =

$$\text{number of units in component} \cdot \frac{360}{\text{total number of units}}$$

The pie chart should not be subdivided into too many components. Each component should be clearly labeled, with the label preferably positioned horizontally. This type of chart is particularly useful when two related pies are presented side by side. Such is the case in Figure 2.13, showing the U.S. budget for 1980 broken down into receipts and outlays.

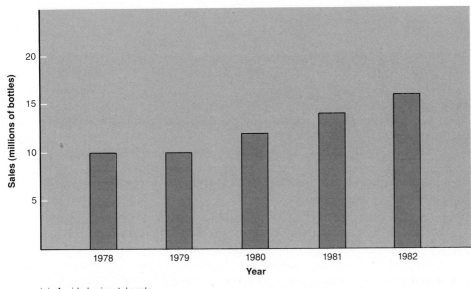

(a) A wide horizontal scale

Figure 2.14

The impact of the horizontal scale choice on a bar chart

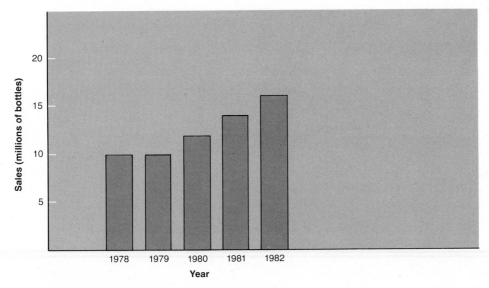

(b) A narrow horizontal scale

2.5 A Word of Caution

Graphic presentation can help present data in an attractive, easy-to-understand way that allows the reader to grasp the facts at a glance.

One should, however, be aware that it is quite easy to use graphs and charts to give false impressions of what the data represent, thus misleading the reader. Mispresentation of this kind is common; here we will give one example to illustrate this point.

Suppose a soft-drink manufacturer wants to present the sales growth of his company over the last few years. The data, let us assume, are:

Year	Number of bottles sold (in millions)
1978	10
1979	10
1980	12
1981	14
1982	16

The data can be presented in a vertical bar chart. And the bar chart can be set up in several different ways—two of which we show in Figures 2.14*a* and 2.14*b*. These two charts give quite different impressions of the company's sales growth, even though they in fact represent the very same data. The difference between the charts is in the horizontal scale: the chart in Figure 2.14*b* gives the impression of a much faster growth than that in Figure 2.14*a*. A similar impact on a chart could be caused by differences in the vertical scale. As an example, consider Figure 2.15. In panel (*a*) the scale is much more condensed than that of panel (*b*). As a result, the line in panel (*b*) looks a lot steeper than that of panel (*a*).

It is not possible to say, in either of these instances, which of the panels exhibits a more "correct" presentation; how the data are presented is largely a matter of choice. Nevertheless, it is clear that each presentation makes its own distinct impression on the viewer. You should bear this caution in mind if you are developing a presentation for others to view—and if you're on the "receiving end" of a presentation too.

Figure 2.15
The impact of the vertical scale choice on a line chart

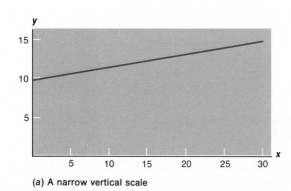

(a) A narrow vertical scale

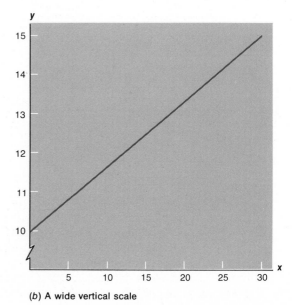

(b) A wide vertical scale

Problems

2.1. Explain briefly each of the following:

1. Headnote.
2. Boxhead.
3. Stub.
4. Field.
5. Two-way classification table.

2.2. Explain the following concepts:

1. Line chart.
2. Component-part line chart.
3. Bar chart.
4. Component bar chart.
5. Duo-directional bar chart.

2.3. What is the advantage of a group bar chart in comparison with a component bar chart? What is the disadvantage?

2.4. Briefly explain the following concepts:

1. Primary source of data.
2. Secondary source of data.
3. Editing.
4. Data classification.

2.5. When data are classified, the classes must be mutually exclusive and collectively exhaustive. Explain these concepts and their importance.

2.6. Can some data be presented on a bar chart that cannot be presented on a line chart? Explain.

2.7. Table P2.7 provides data on end-of-year mortgage debt outstanding in the United States from 1976 through 1980, classified by type of holder, in millions of dollars.

TABLE P2.7

Type of holder	1976	1977	1978	1979	1980
All holders	$889,327	$1,019,688	$1,169,412	$1,326,750	$1,451,840
Major financial institutions	647,650	741,544	848,177	938,567	998,386
Federal and related agencies	66,753	70,175	81,739	97,084	114,300
Individuals and others	174,924	207,969	239,496	291,099	339,154

Source: *Federal Reserve Bulletin,* various issues.

(*a*) Present the data in two component-part time-series line charts, one with the components measured in absolute values and one with them measured in percentages.
(*b*) Present the data in two component bar charts, one with the components measured in absolute values and one with them measured in percentages.

2.8. Table P2.8 provides data on mortgage debt outstanding in the United States at the end of the first quarter of 1981, by type of holder and type of property, in millions of dollars.

TABLE P2.8

All holders		Federal and related agencies	
1 to 4 families	$973,601	1 to 4 families	$ 62,171
Multifamily	139,087	Multifamily	15,117
Commerical	262,140	Commercial	479
Farm	100,115	Farm	38,539
Major financial institutions		Individuals and others	
1 to 4 families	669,355	1 to 4 families	242,075
Multifamily	88,186	Multifamily	35,784
Commercial	228,630	Commercial	33,031
Farm	22,094	Farm	39,482

Source: *Federal Reserve Bulletin,* August 1981, p. A39.

(*a*) Organize the data in a two-way classification table.
(*b*) Present the data in two separate tables, in percentages. Explain the information each table provides.

2.9. Table P2.9 provides data concerning consumer installment credit outstanding at the end of June 1981 by type of credit and by holder, in millions of dollars.

TABLE P2.9

Type of credit	Type of holder	Amount oustanding (millions)
Automobile	Commercial banks	$59,192
	Credit unions	21,847
	Finance companies	38,646
Revolving credit	Commercial banks	29,722
	Retailers	23,384
	Gasoline companies	5,364
Mobile home	Commercial banks	10,179
	Finance companies	3,990
	Credit unions	486
	Savings and loans and others	3,069
Other	Commercial banks	44,217
	Finance companies	40,087
	Credit unions	23,353
	Retailers	4,028
	Savings and loans and others	10,895

Source: *Federal Reserve Bulletin,* August 1981, p. A40.

(*a*) Organize the data in a two-way classification table.
(*b*) Organize the data in two separate tables of percentages. Explain the information each of the tables provides.

2.10. (*a*) Suppose a cross-classified table presents data on health spending in the last five years, classified by type of spending (hospitals, physicians, and so on) and by years. Is it possible to separate the table into two tables, one showing the data by type of spending only and one by years only?
(*b*) Had the data originally been presented in two separate single-classification tables as described in part *a*, would it be possible, on the basis of this information alone, to construct a cross-classified table by type of spending and years? Explain.

2.11. There are many types of banking institutions in which we make deposits and from which we borrow. The deposits in those institutions at the ends of the years 1969–78 are shown in Table P2.11 (in billions of dollars). Present the data in two component line charts, one showing the absolute values of the deposits and one showing their percentages of total annual deposits.

Table P2.11

Year	Savings and loan associations	Commercial banks	Savings banks	Finance companies	Credit unions	Retailers	Other
1969	$134.2	$ 81.8	$ 50.8	$25.0	$10.0	$ 9.1	$ 88.2
1970	142.5	82.9	52.3	25.1	9.7	10.2	99.8
1971	155.8	85.8	50.9	25.0	10.9	8.7	110.1
1972	190.6	126.1	53.4	32.0	9.9	7.6	125.0
1973	208.0	149.1	59.6	33.0	19.1	8.3	129.1
1974	223.4	168.3	69.9	32.9	24.0	20.0	141.0
1975	259.0	176.0	74.7	33.4	19.9	17.4	152.3
1976	298.0	200.0	75.3	34.6	24.5	18.3	174.7
1977	361.1	249.3	79.7	74.3	20.0	35.9	160.6
1978	402.0	303.9	100.3	50.0	52.7	30.7	220.3

Source: *Federal Reserve Bulletin*, various issues.

2.12. A 1976 study of the FORTUNE 500 executives showed that their major fields of study in college had been as shown in Table P2.12 (in percentages). Present the data in a horizontal grouped bar chart (use one group for each of the nine fields of study presented).

TABLE P2.12

Field of study	Undergraduate	Graduate
Humanities	11.0%	0.0%
Business	29.5	46.5
Economics	20.0	7.3
Engineering	21.0	11.0
Physical sciences	6.3	4.4
Social sciences	6.6	0.7
Education	0.9	1.6
Law	3.2	26.5
Other	1.5	2.0
	100.0%	100.0%

Source: Based on material that originally appeared in Charles G. Burck, "Group Profile of the FORTUNE 500 Chief Executive," *Fortune*, May 1976.

2.13. Table P2.13 shows the market share of refrigerator sales, by make, in the United States in the years 1974–79.

TABLE P2.13

Year	GE/Hotpoint	Whirlpool	White Consolidated	Frigidaire	Others	Total
1974	26.0	22.8	11.0	12.1	28.1	100.0
1975	27.5	22.2	14.2	10.0	26.1	100.0
1976	29.2	23.0	15.0	10.0	22.8	100.0
1977	30.0	25.0	15.0	9.2	20.8	100.0
1978	30.0	26.5	15.9	9.0	18.6	100.0
1979	30.0	27.0	25.0	0.0*	18.0	100.0

* Frigidaire was acquired by White Consolidated in 1978.

Data from *Business Week*, May 7, 1979, p. 96.

(*a*) Present the data in a grouped bar chart.
(*b*) Present the data in a component bar chart.
(*c*) Present the data in a component-part line chart.

2.14. In 1979 G. A. Barnett of the Bank of Bermuda published a paper that focuses on the average rate of return (that is, the average percentage of profit) obtained by investment in world bond markets in the years 1970–79 and on the risk of those investments (measured by a statistical measure called "standard deviation," which we shall introduce later, in Chapter 4). The average rate of return and the risk of investment in bonds of nine countries are given in Table P2.14.

TABLE P2.14

Country	Average rate of return	Risk
Australia	5.2%	3.4%
United States	7.3	1.5
United Kingdom	7.6	3.8
Canada	7.6	2.1
France	10.0	3.1
Japan	13.3	2.4
Netherlands	13.9	3.1
Switzerland	15.9	3.0
West Germany	16.5	2.9

Source: G. A. Barnett, "The Best Portfolios Are International," *Euromoney*, April 1979.

(*a*) Present the average rates of return on a horizontal bar chart, with the shortest bar at the top and the longest at the bottom.
(*b*) Present the risks on a horizontal bar chart, with the shortest bar again at the top and the longest bar at the bottom.
(*c*) In view of the fact that a high average rate of return and a low risk level are appreciated by investors, was France's performance better than that of the United Kingdom? Why?

2.15. Tables P2.15*a* and P2.15*b* show 1978 sources of revenue and distribution of revenues of the Aluminum Company of America (ALCOA). Present the data in two separate pie charts.

TABLE P2.15*a*
Sources of Revenues
(millions)

	Revenues
Bauxite, alumina, and chemical products	$ 369.2
Primary aluminum products	229.9
Flat-rolled products	1,871.9
Extruded, rolled, and drawn products	527.5
Other aluminum processing sales	410.2
Finished products and other	643.1
Other income	20.7
Equity earnings	57.2
Total	$4,129.7

TABLE P2.15*b*
Distribution of Revenues
(millions)

	Distribution
Cost of materials, services, etc.	$2,124.9
Wages, salaries, and employee benefits	1,215.3
Depreciation and depletion	227.5
Taxes	249.3
Earnings reinvested in the business	243.9
Dividends	68.8
Total	$4,129.7

Source: Aluminum Company of America, *Annual Report, 1978.*

CHAPTER THREE OUTLINE

3.1 The Population and Related Concepts
Setting Up the Frame: Some Points to Remember

3.2 The Frequency Distribution
Setting Up the Frequency Distribution

3.3 Graphic Display of Frequency Distributions
The Histogram
The Frequency Polygon

3.4 Cumulative Frequency Distribution: "Less Than" and "More Than" Distributions
How Cumulative Frequency Graphs Are Constructed

3.5 Relative Frequency and Cumulative Relative Frequency Distributions

3.6 Determining the Width of Classes in a Frequency Distribution

3.7 Application: Analysis of Cumulative Frequency Distributions for Insurance Purposes

Key Terms

frame
observation
population
sample
frequency distribution
classes
tally table
histogram

frequency polygon
"less than" cumulative frequency distribution
"more than" cumulative frequency distribution
ogive

3
THE POPULATION AND ITS FREQUENCY DISTRIBUTION

The segment of statistics dealing with the collection, organization, and presentation of data, discussed in Chapter 2, is known as **descriptive statistics.** We shall continue that discussion here; at this point, however, we turn to topics that lead directly to statistical analysis.

3.1 The Population and Related Concepts

When we study a given body of data, the need for statistical analysis emerges almost inevitably. But the type of analysis required and the extent and accuracy of the statistical study depend on the characteristics of the data—and, of course, on the goals of the study and the resources available.

Clearly, data vary greatly in type from study to study. Some studies involve a limited amount of data and some involve enormous amounts. Some involve data of nominal measurement, such as income, weight, distance, production, cost, profit; others involve ordinal data (rankings). Still other data are strictly descriptive and qualitative, as in those concerning sex. Yet there are certain similarities in the way we approach all types of data. There is a general terminology, for example, that is used with most types of data; we shall discuss this terminology here.

The first concept we need to discuss is that of the **frame.** Any collection of data relates to some elementary units, and the frame is *the totality of all elementary units.* We use the term **observation** to refer to *an item of information of interest concerning the elementary units in the frame,* and we use the term **population** to refer to *the totality of all the individual observations of interest.* To illustrate, consider the annual income of all adult residents of Montana.

The list of all of the adult residents of Montana is the frame; the annual income of a Montana resident is an observation; and the total body of data concerning the annual incomes of all adult residents of Montana is the population.

Clearly, more than one population can be defined within the same frame. In the preceding example, within the frame of Montana residents, we focused on the population of annual incomes of all adult residents. But using the same frame, we can also define the population according to homeownership status: each element (that is, each adult resident) in the frame will be identified as either a homeowner or not. The term "population" thus has a broader meaning in statistical contexts than in everyday usage. Ordinarily we use "population" to refer to a group of human beings or a collection of physical objects; in statistical usage, however, the term refers to the total body of items of any sort that we wish to subject to statistical analysis. It can relate to "yes" or "no" answers, quantitative measurements of variables, qualitative descriptions of the units in the frame ("defective" and "not defective," or "poor," "average," "good"), and so on.

SETTING UP THE FRAME: SOME POINTS TO REMEMBER

The goal of the analysis must be unambiguously and explicitly defined before any further steps can be taken. Once the goal is stated, a suitable and workable frame must be identified and relevant data collected. Clearly, the choice of the frame depends on the goal of the study. Does our goal require us to include all of the adult residents of Montana, or only those who are in the labor force? The frame should be carefully defined to suit the goal in mind.

We often find it difficult, sometimes impossible, to actually enlist the frame or the population desired. For example, we may wish to encompass within a frame all the credit applicants at a large personal finance company over the last 10 years. Clearly, our ability to construct this frame depends on the records the company maintains, but we should try to obtain as complete a frame as possible before continuing with the analysis. If records are kept for only a few years and then discarded, our frame will be incomplete. As we shall see later, in many cases only a portion of the population is actually observed. This portion is called a **sample.** After the data for the population or the sample are collected, they must be analyzed and summarized, and appropriate conclusions must be drawn. Figure 3.1 shows the steps of a statistical analysis.

Figure 3.1
The steps of
a statistical
analysis

Goal → Frame → Population → Sample → Analysis and summary of sample results → Interpretation and conclusion

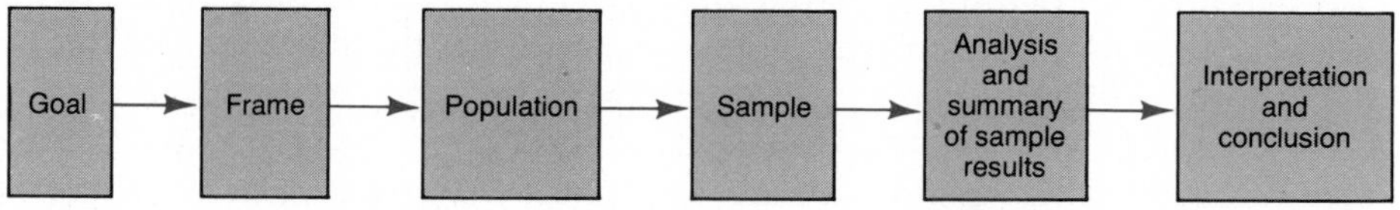

3.2 The Frequency Distribution

The population, which is basically a body of data, is the target of the statistical analysis. At this point in our discussion, we shall assume that the data for the *entire population* is available. (Samples and their role in statistical analysis are discussed in later chapters.) We note at the outset that before any conclusions can be drawn and before any generalizations can be made about the population, it is essential to provide an efficient summary of the data by means of tables, charts, and, when applicable, some quantitative measures reflecting the important characteristics of the data. In the rest of this chapter, therefore, we shall discuss one of the most conventional ways to summarize quantitative data, the frequency distribution. Additional ways will be discussed in Chapter 4.

The idea behind the frequency distribution is to provide a convenient summary of the data by listing all the possible values of the variable and then noting the *frequency* of observations for each value in the population. For example, suppose the population data consist of the end-of-period value of amounts of $100 invested in 60 different common stocks. The data are given in Table 3.1, and they should be interpreted in the following way: If we had invested $100 in stock number 1 and waited until the end of the period (say one year), the value of our investment would have been $105. If we had invested $100 in stock number 2, the value at the end of the period would have been $97. Thus our investment would have appreciated in value by $5 in the first case (stock 1) and depreciated by $3 in the second case (stock 2). The data concerning other stocks should be interpreted in a similar way.

While the various observations in the population have been assigned sequential numbers for reference (stock 1, stock 2, and so on), one can easily see that the end-of-period values do not appear in any particular order. If we wish to see the data presented in order, we can set up the **frequency distribution**—*a summary of the data made by organizing the variable of interest in intervals of increasing value. The intervals in the frequency distribution* are often referred to as **classes.**

SETTING UP THE FREQUENCY DISTRIBUTION

To set up the frequency distribution, we first develop a **tally table,** which serves as a *worksheet for preparation of the frequency distribution.* In our tally table, Table 3.2, we show the end-of-period values of the stocks in our

TABLE 3.1
End-of-Period Value of $100 Investment in 60 Common Stocks

Stock	Value	Stock	Value	Stock	Value	Stock	Value
1	$105.0	16	$106.4	31	$109.5	46	$ 81.4
2	97.0	17	111.5	32	112.7	47	111.3
3	121.5	18	115.6	33	99.0	48	116.6
4	110.9	19	100.3	34	128.6	49	99.0
5	108.8	20	90.4	35	87.6	50	104.3
6	96.3	21	109.3	36	107.4	51	127.0
7	85.1	22	100.5	37	86.7	52	113.3
8	88.0	23	112.5	38	112.3	53	108.6
9	110.4	24	133.3	39	91.4	54	111.3
10	100.7	25	148.4	40	113.6	55	119.9
11	103.6	26	107.9	41	104.4	56	106.9
12	109.6	27	114.3	42	119.8	57	124.2
13	131.0	28	103.7	43	140.8	58	110.8
14	116.2	29	113.9	44	97.0	59	118.0
15	118.2	30	91.3	45	98.2	60	157.1

TABLE 3.2
Tally Table for Stock-Value Data

Class	Tally
$ 80.0– 84.9	|
85.0– 89.9	||||
90.0– 94.9	|||
95.0– 99.9	卌 |
100.0–104.9	卌 ||
105.0–109.9	卌 卌
110.0–114.9	卌 卌 ||||
115.0–119.9	卌 ||
120.0–124.9	||
125.0–129.9	||
130.0–134.9	|
135.0–139.9	
140.0–144.9	|
145.0–149.9	|
150.0–154.9	
155.0–159.9	|

example organized in classes with a width of $5 each. Note that the first class (the class $80.0–84.9) includes all values from $80.0 up to $85.0 but not including $85.0. It is only for convenience that we list it as $80.0–84.9. The same, obviously, applies to the other classes as well.

To derive the tally table (Table 3.2) from Table 3.1, we simply go down the various observations of Table 3.1 and place a tally mark beside the class to which each observation belongs. We then place a check mark beside the observation in the original list of data. In our example, the first observation is $105.0, so the tally mark is placed in class $105.0–109.9 (Table 3.2) and a check mark (not shown) is placed beside the first observation in Table 3.1. When the tally table (Table 3.2) is completed, it shows a list of the frequencies (that is, the numbers) of observations alongside the classes into which the observations are grouped. And this gives us our frequency distribution: since the number of observations in each class corresponds to the number of tally marks in the class, the frequency distribution may be easily constructed from the tally table by a simple count of the marks in each class. Table 3.3 is the frequency distribution for our example.

How does the frequency distribution compare with Table 3.1? Tables 3.1 and 3.3 are two different descriptions of the same data, but Table 3.1 provides more details than Table 3.3. For example, from Table 3.3 we know that one of the stocks had a value of between $80.0 and $84.9. The table does not tell us, however, that stock's specific value. If we want to know this extra detail, we can turn to Table 3.1: the stock's value was $81.40 (see stock 46).

If the raw-data table (Table 3.1) provides more information than the frequency distribution (Table 3.3), why are we interested in the frequency distribution table? The answer is, of course, that although the raw data present the most information, it is hard to see any systematic pattern in those data before they are organized. This is particularly true when the number of observations is large. The frequency distribution, although it conceals some details, is appealing to the analyst, since it reveals the general pattern of the population. We easily see, for example, that the most frequent classes in Table 3.3 are the $110.0–114.9 and $105.0–109.9 classes, and that the frequency tends to decline as we move out to outer classes (away from the center) on either side. We also note that the declining frequency with movement away from the center is not without exceptions.

TABLE 3.3
Frequency Distribution of
End-of-Period Stock Values

Class	Frequency
$ 80.0– 84.9	1
85.0– 89.9	4
90.0– 94.9	3
95.0– 99.9	6
100.0–104.9	7
105.0–109.9	10
110.0–114.9	14
115.0–119.9	7
120.0–124.9	2
125.0–129.9	2
130.0–159.9	4
Total	60

3.3 Graphic Display of Frequency Distributions

THE HISTOGRAM

Graphic display of data is very useful as a means of description, as we have seen in Chapter 2. A large variety of diagrams can be used to present data visually, but the most common type used to represent a frequency distribution is the **histogram,** *a series of bars whose widths indicate class intervals and whose areas indicate corresponding frequencies.* Figure 3.2 is a histogram showing the end-of-period stock-value data of Table 3.3. The frequency in each class is represented by a bar whose area is proportional

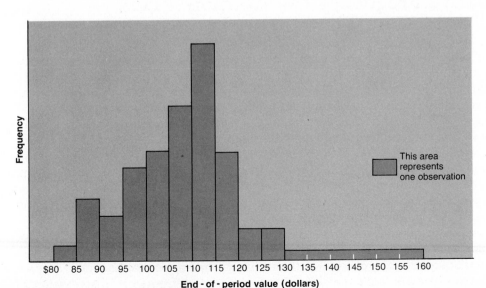

Figure 3.2

Frequency distribution of end-of-period stock values: histogram to the frequency in the class. The horizontal axis measures the range of classes of the variable.

Let's look more closely at our statement that the *area* of the bars is proportional to the frequency they represent. Note that the last class in Table 3.3 is six times as wide as any other class. Since the total frequency in this range is 4, we can determine the height to be 4/6 = 0.667; that is, in the range $130.0–159.9, the *average frequency* is 0.667 per standard width of $5.0. Again, consider the first interval on the left. Its width is $5.0 and its height is 1 (because the frequency in that interval is equal to 1). The total area of the bar, then, is equal to 5·1 = 5. The width of the last interval on the right is $30 and the height is 0.667, so that the total area is 30·0.667 = 20. In other words, if an area of 5 units represents 1 observation, a 20-unit area proportionally represents 4 observations.

THE FREQUENCY POLYGON

The **frequency polygon** is an alternative way of displaying the frequency distribution. In Figure 3.3*a* we reproduce the end-of-period value histogram, but this time we add a bold dot to all the midpoints of the top portions of the histogram's bars. Two additional bold dots are placed on the horizontal axis one-half standard width to the left of the lowest class and one-half standard width to the right of the highest class. From this we construct the frequency polygon, shown in Figure 3.3*b*, where the histogram bars have been removed and the bold dots of Figure 3.3*a* are connected with line segments. The frequency polygon of 3.3*b* basically represents the same data as the histogram of 3.3*a*, and the choice between them is largely a matter of personal taste. The histogram may look more appealing to some, and the polygon may better suit the tastes of others.

If we wish to give additional emphasis to the general structure of the data, even at the cost of giving up some details, we approximate the frequency polygon by a smooth curve such as the one in Figure 3.3*c*.

3.4 Cumulative Frequency Distribution: "Less Than" and "More Than" Distributions

In Table 3.4 we present two more types of frequency distributions for the data of Table 3.3: the "less than" and the "more than" cumulative distributions. For each starting and ending value of the classes of Table 3.3, the **"less than" cumulative frequency distribution** provides the *total of all the frequencies of values that are less than or equal to that value,* and the **"more than" cumulative frequency distribution** provides the *total of all the frequencies of values that are greater than that value.* The *graph of a cumulative frequency distribution* is called an **ogive.** The "less than" and "more than" ogives for the data in Table 3.4 are shown in Figure 3.4. As we can see, the ogives are obtained by connecting the cumulative values of the starting and ending values of the classes by straight lines.

A potential investor in stocks who is interested in end-of-period values may wish to know how many of the stocks' end-of-period values were less than or equal to the original $100 invested. The answer to this question may be read directly from the "less than" cumulative frequency distribution in

Figure 3.3

Frequency
distribution of
end-of-period
stock values:
three graphic
representations

This area
represents
one observation

(a) Histogram

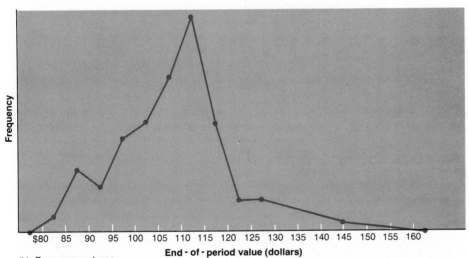

(b) Frequency polygon

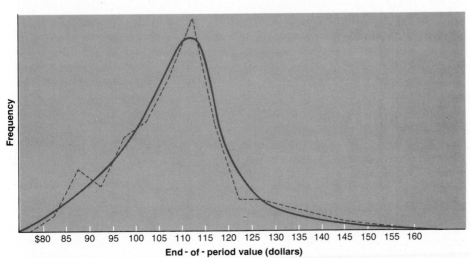

(c) Smoothed polygon

TABLE 3.4
"Less Than" and "More Than" Cumulative
Frequency Distributions of End-of-Period
Stock Values

End-of-period value	Cumulative frequency distribution	
	"Less than"	"More than"
$ 80.0	0	60
85.0	1	59
90.0	5	55
95.0	8	52
100.0	14	46
105.0	21	39
110.0	31	29
115.0	45	15
120.0	52	8
125.0	54	6
130.0	56	4
160.0	60	0

either the table or the graph. Out of 60 stocks, 14 turned out to have end-of-period values less than or equal to $100; the number of stocks whose end-of-period values were less than or equal to $120 was 52. If on the other hand we are interested in the frequency of stocks whose values were more than, say, $110 at the end of the period, we look up the value of the "more than" cumulative frequency distribution corresponding to $110 and discover that 29 stocks had end-of-period values greater than $110.

HOW CUMULATIVE FREQUENCY GRAPHS ARE CONSTRUCTED

A word of explanation is in order concerning the way the graphs in Figure 3.4 are constructed and read. Let us start with the "less than" ogive of Figure 3.4a. Above each of the end-of-period values given in Table 3.4 we place a dot at the level indicated by the "less than" column. The dots are then connected by line segments. The line is also extended horizontally to the right at the ogive's highest point. To read off the number of stocks with end-of-period values less than, say, $105, we raise a vertical line from that value toward the ogive, and read off the required cumulative frequency from the vertical axis. Twenty-one stocks had end-of-period values less than or equal to $105. This figure corresponds to the frequency indicated in Table 3.4. The figure also shows that the "less than" cumulative frequency at the value $108, for example, is 27. This number assumes an even accumulation of frequency in each class interval. Since the interval $105.0–109.9 has a frequency of 10 stocks, we assume an accumulation of two stocks per $1 change in the end-of-period value. It follows that if 21 stocks are accumulating up to the value $105, 23 stocks are accumulating up to the value $106, 25 stocks are accumulating up to $107 and 27 stocks up to $108.

Figure 3.4b is constructed and read in a similar manner. Each value in the "more than" column of Table 3.4 is represented by a dot. The dots are then connected by line segments, and the proper "more than" frequency is read off the graph. For an end-of-period value of $108, for example, we get 33 stocks. Note that there are 27 stocks with end-of-period values less than or

Figure 3.4
"Less than" and
"more than"
ogives of
end-of-period
stock values

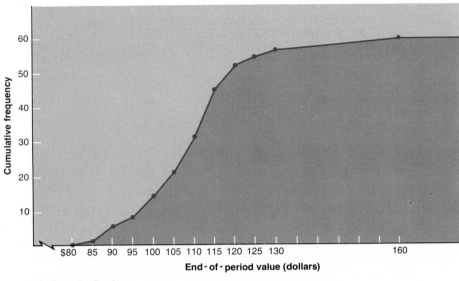

(a) "Less than" ogive

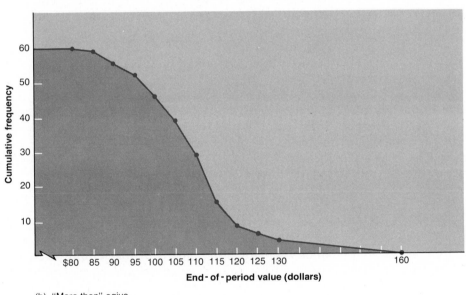

(b) "More than" ogive

equal to $108, and indeed 27 + 33 = 60, which is the total number of stocks considered.

3.5 Relative Frequency and Cumulative Relative Frequency Distributions

The frequency distribution shows the *number* of observations in each of the variable's classes. The sum of the frequencies in all the classes equals the total number of observations in the population. If we divide the frequency in each class by the total number of observations, we obtain the *ratio* of the number of observations in the class to the total population size. Let us take an example. Table 3.5 provides the frequency and relative frequency distri-

TABLE 3.5

Frequency and Relative Frequency Distributions of Auto Collision Damages in City *A*, 1982

(1) Damages (dollars)	(2) Frequency	(3) = (2) ÷ 380 Relative frequency
Up to $1,000	56	0.15
$ 1,001–2,000	128	0.34
2,001–3,000	115	0.30
3,001–5,000	81	0.21
Total	380	1.00

butions of auto collision damages in city *A* during 1982 according to the amount of damages.[1] We obtain the "Relative frequency" column by dividing each of the frequencies in column 2 by the total population size, 380. So the first number in the "Relative frequency" column is 56/380 = 0.15, the second number is 128/380 = 0.34, and so on, meaning that 15 percent of the collision damages in city *A* in 1982 were for amounts of up to $1,000, 34 percent were for amounts of $1,001 to $2,000, and so on. The total of the "Relative frequency" column is 1.00 (that is, 100 percent), meaning that all of the observations were tabulated.

The relative frequency distribution is particularly useful when we compare the way two or more populations are distributed. In Table 3.6, 1982 collision damage data are provided for cities *A* and *B*. Even though both of the frequency distributions are listed side by side over the same categories, it is hard to make a meaningful comparison directly from Table 3.6, because the totals of the populations differ. When the relative frequencies of both cities are contrasted, however, the differences and similarities become evident. The relative frequency distributions are shown in Table 3.7. A brief glance at the table is enough to show that in both cities a substantial portion of the damages are for amounts of $2,000 or less, and that the proportion of damages declines as amounts rise. A more careful look at Table 3.7 reveals that the major difference between the cities is that the relative frequency of the lower classes is greater in city *B*, while the relative frequency of the higher classes is greater in city *A*. This difference is further emphasized by the "less than" cumulative relative frequency distribution presented in

TABLE 3.6

Frequency Distributions of Auto Collision Damages in City *A* and City *B*, 1982

Damages (dollars)	City A	City B
Up to $1,000	56	304
$ 1,001–2,000	128	591
2,001–3,000	115	431
3,001–5,000	81	272
Total	380	1,598

[1] The classes of Table 3.5 should be understood to be as follows: up to $1,000.00, $1,000.01 to $2,000.00, $2,000.01 to $3,000.00, and so on. This is a little different from the classes of Table 3.3. We present the two classification types because both are commonly used.

TABLE 3.7
Relative Frequency Distributions of Auto Collision Damages in City *A* and City *B*, 1982
(percent)

Damages (dollars)	City A	City B
Up to $1,000	0.15	0.19
$ 1,001–2,000	0.34	0.37
2,001–3,000	0.30	0.27
3,001–5,000	0.21	0.17
Total	1.00	1.00

Table 3.8. The "less than" cumulative relative frequency distribution shows the *total* of the relative frequencies up to and including a specified value of the variable of interest, and is obtained directly from the relative frequency distribution by continuous accumulation.

The cumulative relative frequency distribution is a very useful descriptive measure of the population. It shows, for instance, that in city *A*, 49 percent of all damages in 1982 were for amounts less than $2,000, 79 percent of the damages were of amounts less than $3,000, and so on. A comparison of two or more cumulative relative frequency distributions sometimes proves useful. In our example, for each value on the stub of Table 3.8, the proportion in city *B* is greater than that in city *A*, meaning that for each such value the proportion of damages in city *B* of amounts *less than* that value is greater. For example, 56 percent of the damages were of amounts less than $2,000 in city *B* as compared to only 49 percent in city *A*; 83 percent of the damages were of amounts less than $3,000 in city *B* compared to only 79 percent in city *A*; and so on. We thus see a definite pattern: claims tend to be of lower amounts in city *B* than in city *A*.

Both the relative frequency and the cumulative relative frequency distributions are often presented diagrammatically by a histogram and an ogive, similar to the presentation of their respective nonrelative counterparts. Though the diagrams contain precisely the same information as the tables, they often give extra emphasis to facts that may otherwise be overlooked. Figure 3.5 presents a histogram of the relative frequency distributions of damages in city *A* and city *B* in 1982. Figure 3.6 shows the respective cumulative relative frequency distributions. Let us compare Figure 3.5 with the data of Table 3.7. In particular, notice the difference between the relative frequencies in category $2,001–3,000 and category $3,001–5,000. The relative

TABLE 3.8
Cumulative Relative Frequency Distributions ("Less Than" Type) of Auto Collision Damages in City *A* and City *B*, 1982
(percent)

Damages (dollars)	City A	City B
Up to $1,000	0.15	0.19
Up to $2,000	0.49	0.56
Up to $3,000	0.79	0.83
Up to $5,000	1.00	1.00

frequency in category $2,001–3,000 for city A is 0.30, in category $3,001–5,000, 0.21—over half as much. In Figure 3.5, however, the height of the bar in the latter category is *substantially less* than half the height of the bar over class $2,001–3,000. The reason is trivial: class $3,001–5,000 is twice as wide as class $2,001–3,000. Thus the average of the relative frequency per $1,000 interval in class $3,001–5,000 is 0.21/2 = 0.105. Thus, although Table 3.7 and Figure 3.5 reflect the same information, the figure complements the table: details that are hard to read directly from Table 3.7 are easily revealed by Figure 3.5.

Figure 3.5

Histograms of relative frequency distributions of auto collision damages in city *A* and city *B*, 1982

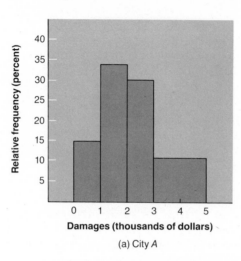

(a) City *A*

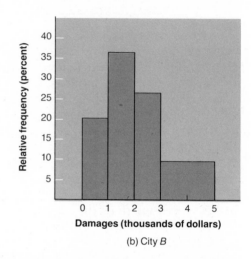

(b) City *B*

Figure 3.6

Cumulative frequency distributions ("less than" type) of auto collision damages in city *A* and city *B*, 1982

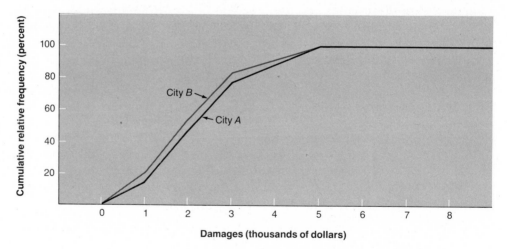

3.6 Determining the Width of Classes in a Frequency Distribution

Up to this point we have assumed that the class limits are given, and simply focused on deriving and interpreting the various types of frequency distributions. Now it is time for us to discuss the way the class width is determined.

There is no formula or rigid rule concerning the determination of class width, but we can offer some valuable guidelines. The key to this problem is the trade-off between the amount of detail revealed by a large number of classes, on the one hand, and the general population pattern revealed by a smaller number of classes, on the other hand. As we reduce the number of classes in a frequency distribution, we eliminate details, and at the same time the general population pattern becomes clearer. At a given point, however, a further reduction of the number of classes will not only conceal more details but make it more and more difficult to learn about the overall shape of the population distribution. At exactly what level of aggregation this negative trade-off occurs we cannot say, as the level differs from one case to another. Experience and judgment must be used to determine an optimal breakdown into class intervals. Let us consider an example that will illustrate the trade-off.

EXAMPLE 3.1

The weekly orders of a large department store are regularly teletyped to a regional warehouse, where they are processed and the merchandise is shipped to the store. Although the store management maintains a carefully planned inventory policy, stockouts are unavoidable because the store carries virtually thousands of items for which the demand varies from day to day. When stockouts occur, or when the store's inventory of a given item reaches a low level, a priority order (that is, an emergency order) is then teletyped. Such orders are usually sent in the late afternoon or evening and the merchandise is supposed to reach the store during the following night, to be displayed on the shelf or stored in the storage room the next morning. In Table 3.9, we show the

TABLE 3.9
Elapsed Time between Sending of Priority Orders and Arrival of Merchandise on 40 Working Days
(hours)

Day of priority order	Delivery time	Day of priority order	Delivery time
1	5.2	21	8.4
2	4.3	22	9.8
3	7.9	23	10.0
4	9.5	24	9.5
5	7.7	25	8.9
6	22.0	26	9.5
7	10.0	27	10.5
8	8.3	28	12.0
9	10.8	29	12.5
10	9.0	30	7.6
11	11.3	31	10.0
12	8.2	32	20.0
13	11.2	33	10.0
14	10.5	34	9.2
15	12.0	35	12.4
16	11.5	36	11.0
17	17.0	37	13.0
18	8.5	38	8.0
19	13.0	39	10.5
20	9.6	40	13.4

number of hours that elapse between the typing of the priority orders and the arrival of the merchandise in the store on 40 specific working days.

Considering a frequency distribution with half-hour intervals first, we construct a tally table (Table 3.10). From the tally table we derive the frequency distribution (Table 3.11).

TABLE 3.10
Tally Table for Frequency Distribution of Data in Table 3.9
(half-hour intervals)

Class	Tally
4.1–4.5	|
4.6–5.0	
5.1–5.5	|
5.6–6.0	
6.1–6.5	
6.6–7.0	
7.1–7.5	
7.6–8.0	||||
8.1–8.5	||||
8.6–9.0	||
9.1–9.5	||||
9.6–10.0	|||| |
10.1–10.5	|||
10.6–11.0	||
11.1–11.5	|||
11.6–12.0	||
12.1–12.5	||
12.6–13.0	||
13.1–22.0	||||

TABLE 3.11
Frequency Distribution Derived from Table 3.10

Class	Frequency
4.1–4.5	1
4.6–5.0	0
5.1–5.5	1
5.6–6.0	0
6.1–6.5	0
6.6–7.0	0
7.1–7.5	0
7.6–8.0	4
8.1–8.5	4
8.6–9.0	2
9.1–9.5	4
9.6–10.0	6
10.1–10.5	3
10.6–11.0	2
11.1–11.5	3
11.6–12.0	2
12.1–12.5	2
12.6–13.0	2
13.1–22.0	4
Total	40

Consider now Table 3.12, a more condensed frequency distribution, in which one-hour intervals are used.

TABLE 3.12
Frequency Distribution Based on One-Hour Intervals

Class	Frequency
4.1–5.0	1
5.1–6.0	1
6.1–7.0	0
7.1–8.0	4
8.1–9.0	6
9.1–10.0	10
10.1–11.0	5
11.1–12.0	5
12.1–13.0	4
13.1–22.0	4
Total	40

Finally, suppose we choose only two classes: 4.1–10.0 and 10.1–22.0. In this case our frequency distribution will be as follows:

Class	Frequency
4.1–10.0	22
10.1–22.0	<u>18</u>
Total	40

Figure 3.7 shows the histograms of the frequency distributions obtained

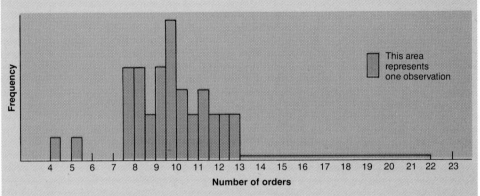

(a) Classification by half–hour intervals

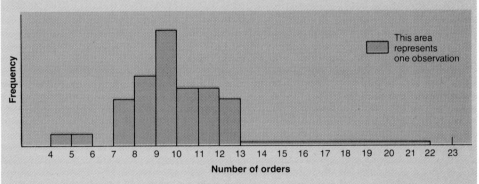

(b) Classification by one–hour intervals

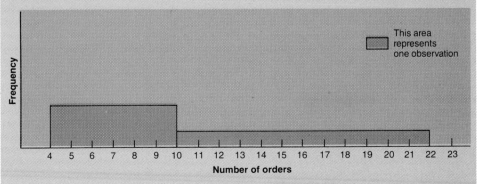

(c) Classification by intervals of 4.1–10.0 and 10.1–22.0 hours

Figure 3.7

Histograms of frequency distributions obtained by use of three alternate classifications

by the use of the three alternate classifications. While all three diagrams present the very same data, the differences between them reflect the trade-off between details and general pattern. While Figure 3.7a shows the most details, Figure 3.7b gives a better idea of the structure of the frequency distribution. Consider, for example, the interval 11.1–12.0. Figure 3.7b shows that the frequency in this interval is five observations. Figure 3.7a reveals that three of these observations are in the interval 11.1–11.5, and two of them are in the interval 11.6–12.0. This extra information is "lost" in Figure 3.7b, but in many cases these fine details may not be of much interest to whoever is presented with the data. When the process of aggregation continues, even the general structure of the frequency distribution becomes unclear: in Figure 3.7c, where only two classes are presented, so many details have been eliminated that very little information is revealed. The most extreme case happens, of course, when we present all the population in just one class interval. In this case, the histogram contains only one bar, which shows the range of the population and the total number of observations contained in it. Clearly, such a histogram reveals nothing whatsoever about the internal structure of the population; thus its contribution to the data presentation is extremely limited.

Clearly, judgment must be exercised in deciding which interval width to use in presenting a population frequency distribution. Whatever the width, however, it is generally *desirable* that the widths of all the classes presented be the same, for ease of reading and interpretation. When class intervals are of unequal widths, misinterpretation can result if this fact is overlooked. Nevertheless, although use of unequal class intervals is undesirable, it is occasionally necessary when one is dealing with populations in which the frequency varies greatly from one range to another.

3.7 APPLICATION:

ANALYSIS OF CUMULATIVE FREQUENCY DISTRIBUTIONS FOR INSURANCE PURPOSES

In its simplest form, life insurance consists of a contractual agreement that works in the following way: The insured person pays the insurance company a certain amount of money every year. This amount is called the premium. If the insured dies during the year, the company pays his or her beneficiaries (usually family members) an amount that has been agreed upon in advance. The higher the amount pledged to the beneficiaries in the event of the death of the insured, the higher the premium. If the insured survives, the insurance company keeps the premium and has no further obligations toward the insured.

Clearly, then, the insurance premium is related to the probability of the insured's survival. The greater the chance of survival, the lower the premium charged by the insurance company.

All comparative analyses in this regard are based on the premium per $1,000 insurance benefits. For example, if the insured buys a $50,000 life insurance policy (that is, $50,000 will be paid to the beneficiaries at the death of the insured) and the premium is $250 per year, the premium per $1,000 is $5 $\left(= \dfrac{\$250}{\$50,000} \cdot \$1,000 \right)$. From now on, unless we say otherwise, we shall be considering the premium per $1,000 insurance coverage.

Suppose a U.S.-based insurance company is contemplating a plan to expand its operations and offer life insurance to people in Madagascar. As an exploratory step, the company wants to know if it can offer the insurance in this country at the same premiums it charges in the United States, or whether it should charge higher or lower premiums.

The data for the analysis are given in Table 3.13. It shows the frequency of male deaths by age in the United States and Madagascar.

Male deaths per 100,000 are given for each country, which means that the data show how many males in each group (of 100,000) die on the average in the first year of life (class 0–1), between their first and fifth years (class 1–5), and so on. Figure 3.8 depicts the ogive of deaths per 100,000 males in the United States and Madagascar. As the figure shows, the ogive of the United States is to the right of the ogive of the male population of Madagascar—a clear indication that the male population of the United States enjoys a greater longevity than that of Madagascar. In every age group we find more male deaths at or prior to that age in Madagascar than in the United States.

TABLE 3.13
Death Frequency of Males in U.S. and Madagascar, by Age Group
(per 100,000 males)

Age group[a]	United States		Madagascar	
	Frequency of deaths	Cumulative frequency	Frequency of deaths	Cumulative frequency
0–1	2,060	2,060	17,584	17,584
1–5	352	2,412	5,575	23,159
5–10	229	2,641	1,585	24,744
10–15	246	2,887	858	25,602
15–20	772	3,659	1,371	26,973
20–25	1,061	4,720	1,954	28,927
25–30	955	5,675	1,934	30,861
30–35	1,054	6,729	2,093	32,954
35–40	1,411	8,140	2,487	35,441
40–45	2,111	10,251	3,059	38,500
45–50	3,306	13,557	3,830	42,330
50–55	4,789	18,346	4,884	47,214
55–60	7,085	25,431	6,186	53,400
60–65	9,617	35,048	7,682	61,082
65–70	11,828	46,876	9,262	70,344
70–75	13,836	60,712	10,169	80,513
75–80	14,216	74,928	9,377	89,890
80+	25,072	100,000	10,110	100,000

[a] The class interval 0–1 includes the first birthday; the class interval 1–5 does not include the first birthday but includes the fifth. Other class intervals are defined similarly.

Source (U.S. data only):Samuel H. Preston, Nathan Keyfitz, and Robert Schoen, *Causes of Death: Life Tables for National Populations* (New York: Seminar Press, 1972).

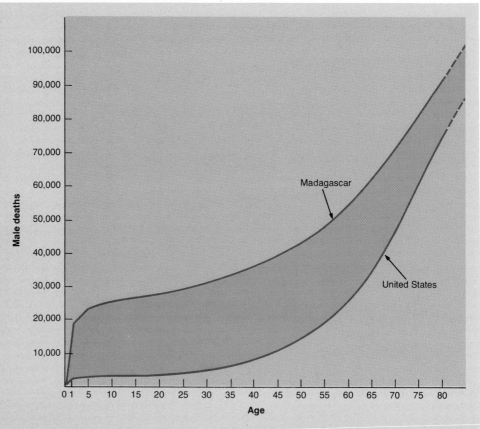

Figure 3.8
Cumulative male death frequencies in the United States and Madagascar (per 100,000 males)

Source: U.S. data from Preston et al., op. cit.

Consider, for example, the 40-year-olds. For every 100,000 males there are 35,441 deaths of males 40 years old or younger in Madagascar, compared to 8,140 in the United States. This relationship (higher cumulative frequency in Madagascar as compared to the United States) holds not only for the age of 40, but for *each and every age we choose to examine*. (Of course, the cumulative frequency in the age interval 80+ is 100,000 in both countries.)

Since the probability of dying at a younger age is greater in Madagascar for any age we choose, the incumbent risk the insurance company takes in Madagascar is greater, and it is obvious that higher premiums must be charged there than in the United States. Using conditional probability (Chapter 5) it is possible to determine how much higher the premiums must be.

Problems

3.1. Explain what a tally table is and what its purpose is.

3.2. Briefly explain each of the following:

    (*a*) Frequency distribution
    (*b*) Cumulative frequency distribution
    (*c*) Relative frequency distribution
    (*d*) Cumulative relative frequency distribution

3.3. Distinguish between the frequency polygon and the ogive.

3.4. What are the advantages and disadvantages of having too many class intervals in a frequency distribution? What are the advantages and disadvantages of having too few?

3.5. The following is a list of the Grade Point Averages (GPA's) of 36 students in a class, rounded off to the nearest tenth:

2.0	3.0	2.8
1.5	3.9	1.9
3.3	4.0	3.0
2.0	3.3	2.3
3.6	2.2	1.5
2.8	2.0	2.5
3.3	2.7	2.8
2.9	2.5	2.7
1.9	3.7	2.4
3.2	2.5	4.0
1.4	2.7	1.9
2.6	2.0	2.4

(*a*) Rearrange the GPA's in ascending order.

(*b*) Construct a frequency distribution, using an interval width of 0.1. How many class intervals do you have?

(*c*) Reconstruct the frequency distribution, this time using an interval width of 0.5. How many class intervals do you have?

(*d*) Reconstruct the frequency distribution one more time, using an interval width of 1.5.

(*e*) Graph the histograms of the frequency distributions for parts *b*, *c*, and *d*, and discuss the differences among them.

3.6. The following are four different class-interval designations for a frequency distribution of employees' hourly pay in a given company.

(*a*)	(*b*)	(*c*)	(*d*)
Up to $4.0	$4.0–7.0	Up to $4.0	Up to $3.99
4.0–6.0	6.0–7.0	5.0–7.0	4.00–5.99
6.0–8.0	8.0–11.0	8.0–10.0	6.00–7.99
8.0–10.0	10.0–13.0	11.0 or more	8.00–9.99
10.0 or more			10.00–11.99
			12.00 or more

Which of the above classifications is appropriate, and what is your specific criticism of each of the other classifications?

3.7. Why is it important to have intervals with equal width in a frequency distribution?

3.8. A food-chain management has gathered information about the maximum length of time (in days) during which blueberry yogurt stays fresh after it has been brought to the store and put in the freezer. The following is the data gathered for 120 blueberry yogurt cups of three different brands (*A*, *B* and *C*) as marked:

Brand A					Brand B				Brand C	
33	33	35	28	32	30	29	28	30	36	35
29	33	32	34	33	29	32	32	32	37	38
34	35	34	33	29	34	26	33	31	34	39
34	31	29	30	33	35	29	30	27	33	35
35	33	34	33	32	30	28	36	29	36	36
32	36	32	31	30	29	35	29	30	36	40
34	34	36	33	32	36	29	31	28	38	37
33	30	32	31	31	28	30	35	31	34	39
35	35	34	32	32	32	28	26	30	37	36
31	33	35	31	32	29	28	26	30	35	38
30	31	36	32	29	25	28	34	27		
				33						

(*a*) Derive the frequency distribution of yogurt freshness for the brands *A*, *B* and *C*, as well as for the entire population combined. When deriving the frequency distribution, do not group the days into intervals. (That is, determine the frequency for 25 days, for 26 days, for 27 days, and so on, without grouping.) Draw histograms of the frequency distributions.

(*b*) Which of the frequency distributions—the separate ones or the one combining the entire population—is more important for the food-chain management? Explain.

(*c*) Rework part *a* once again, but this time derive the *relative* frequency distributions. Are the relative frequency distributions more useful than the frequency distributions when it comes to making comparisons? Explain.

3.9. Table P3.9 presents the percentage change in the average prices of stocks in leading countries around the world for periods of 3, 6, 9 and 12 months, ending on February 28, 1979.

(*a*) Derive the frequency distribution for the 3-month and for the 12-month percentage changes, using intervals of 5 percentage points each, starting at −10 percent. (That is, the first interval is −10.0 to −5.1, the second interval is −5.0 to −0.1, the third interval is 0.0–4.9, and so on.)

(*b*) Draw the "less than" ogives for the 3-month and the 12-month data, on the same diagram. Can you explain the relationship between the two graphs?

TABLE P3.9

Rank		3 months % change	6 months % change	9 months % change	12 months % change	Price/ earnings ratio	Yield
1	Norway	19.1%	8.0%	22.9%	29.9%	17.9	5.1
2	Australia	14.0	8.1	18.9	39.7	10.5	4.5
3	Switzerland	11.9	8.2	8.7	4.0	11.6	2.8
4	Italy	9.5	18.4	20.0	34.1	NA	3.0
5	Canada	8.6	14.7	25.0	39.9	9.0	4.3
6	Belgium	7.2	5.6	7.1	10.5	20.5	11.0
7	United Kingdom	6.4	4.0	8.6	21.8	8.5	5.3
8	Hong Kong	4.8	−25.7	11.6	24.9	11.5	4.7
9	Singapore	4.6	−14.2	11.8	36.1	16.4	2.7
10	Denmark	3.3	−3.7	−0.9	−4.8	5.9	6.0
11	Japan	2.3	5.9	7.4	12.6	18.8	2.1
12	Sweden	1.6	−9.7	−2.0	5.4	11.0	5.2
13	United States	1.5	−6.8	−1.2	10.0	7.9	5.8
14	Netherlands	1.1	−8.3	−1.8	3.1	7.1	7.1
15	Austria	−0.4	−3.2	−1.1	−0.8	NA	3.5
16	Spain	−2.2	−14.1	−13.6	−6.4	NA	11.2
17	Germany	−2.7	−3.8	2.7	0.7	10.5	5.0
18	France	−6.6	−1.4	2.4	37.2	14.8	6.2
	The World Index	2.5	−2.1	3.1	12.6	9.6	5.0

Source: *Institutional Investor*, April 1979.

3.10. The dividends paid out as a percentage of net income by 30 manufacturing firms last year are shown in Table P3.10.

(*a*) Set up a tally table of the data, using intervals of 10.0 percent each and starting at zero.

(*b*) Write down the frequency distribution of the data, and draw the histogram.

(*c*) Draw a polygon of the data.

(*d*) Draw the "less than" and "more than" ogives on one diagram.

(*e*) Over what percentage dividend payout do the ogives intersect? How many firms have a percentage payout lower than the intersection point, and how many firms have a percentage payout greater than that point? Is this a coincidence? Explain.

TABLE P3.10

Firm	Percentage paid out	Firm	Percentage paid out
1	51.1	16	48.8
2	62.7	17	70.5
3	20.1	18	69.3
4	51.2	19	71.4
5	10.2	20	29.3
6	42.8	21	45.2
7	25.9	22	54.4
8	13.9	23	61.7
9	30.8	24	40.0
10	53.4	25	36.5
11	37.6	26	55.6
12	66.0	27	69.4
13	34.1	28	34.9
14	77.9	29	56.8
15	68.2	30	67.6

3.11. One of the questions on an application for a certain job is the age of the applicant. The classes presented are:

Up to 24
25–28
29–32
33–36
37 or older

(*a*) Into what class should an applicant classify himself if he is 28 years and 6 months old?
(*b*) What is the interval width of the classes 25–28, 29–32 and 33–36? What are their midpoints?

3.12. Use the data in Table 3.3 to derive the frequency and relative frequency distributions for the end-of-period stock values using class intervals of $10.

3.13. Use the data of Table 3.8 to derive the "more than" type relative cumulative distribution of auto collision damages in cities *A* and *B*. Explain the implication of their relationship.

3.14. The daily sales of a company in the last 26 working days is given below (in thousands of dollars):

$25	$29
27	37
18	42
34	32
19	37
20	43
41	19
37	28
18	36
22	31
22	30
32	29

(*a*) Construct a frequency distribution of the sales, using six class intervals.
(*b*) Draw the frequency polygon of the distribution.

CHAPTER FOUR OUTLINE

Key Terms

population parameters
location parameters
dispersion parameters
measures of degree of asymmetry
average
mean
arithmetic mean
weighted mean
median
percentiles
first quartile
third quartile

fractile
mode
bimodal distribution
skewness
variance
standard deviation
coefficient of variation
interfractile range
interquartile range
range
coefficient of skewness
proportion
geometric mean

4
POPULATION PARAMETERS

While the frequency distribution described in Chapter 3 is an effective tool with which to illustrate a population both quantitatively and graphically, additional devices are required for describing the distribution of a population easily and clearly. There are two major reasons that such additional devices are needed. First, despite the relative compactness of a frequency distribution, we always welcome further compactness if it can be achieved without significant loss of information. Maximum compactness is particularly appreciated when the problem at hand involves more than one distribution. Second, it is very advantageous from an analytical point of view to be able to separate out and measure *individual characteristics* of the population. The frequency distribution does not separate out individual characteristics.

Thus, we take the additional step of determining certain **population parameters**—*quantitative measures of specific characteristics of a population*, such as:

1. Location (or central tendency).
2. Dispersion.
3. Degree of asymmetry.

The advantage of using these population parameters is that with a separate measure for each characteristic of a population, we can much more conveniently measure and describe differences between populations as well as variations that occur in one population over time. These variations can then be used to formulate answers to policy questions: the amount by which wages should be allowed to increase in contract negotiations, the degree of additional risk an investment institution should be allowed to undertake, and the like. (The frequency distribution does not directly answer these questions.)

In this chapter, we shall discuss the major **location parameters:** the *arithmetic mean,* the *weighted mean,* the *median,* the *mode,* and *percentiles.* The *geometric mean*—also a location parameter—is introduced and contrasted with the arithmetic mean in Appendix 4B. We shall also discuss the major **dispersion parameters:** the *variance,* the *standard deviation,* and the *interfractile range,* and touch upon a measure of relative dispersion, the *coefficient of variation.* We shall look at a third category of population parameters, **measures of degree of asymmetry,** and discuss two major measures, *skewness* and *coefficient of skewness.* Finally, we consider *proportions.*

4.1 The Arithmetic Mean

The number of years an American young person spends in school before leaving to get a job varies from one person to another and from one population group to another. Consequently, no single number can encompass *all* the information about this variable. This is the case, of course, with virtually all populations as long as the population elements are not identical to one another. If we say, however, that the *average* number of years Americans spend in school is 12, this figure in itself contains a great deal of information about the subject of interest. In fact, many will argue that among all the *single* valued parameters of a population, the *average* contains the most information.[1] *The population's* **average** *is also known as the population's* **mean.** *It measures the central tendency of the population—or, in other words, the population's location.* Although there are several types of means, we shall concentrate upon two: the *arithmetic mean,* which is by far the most important, and the *geometric mean* (discussed in Appendix 4B).

To obtain the **arithmetic mean** of a population, we first *add up the values of all the observations, then divide the sum by the number of observations in the population.* Let us demonstrate this calculation with an example.

EXAMPLE 4.1

Ten marketing courses are given currently at a certain Louisiana college. The number of students in these classes are 50, 45, 36, 23, 67, 18, 33, 31, 29, and 42. The mean number of students in a marketing class at that college at the present time is:

$$\frac{50 + 45 + 36 + 23 + 67 + 18 + 33 + 31 + 29 + 42}{10} = \frac{374}{10}$$

$$= 37.4 \text{ students}$$

Following common practice, let us use the letter X to denote the set of data that is given, and a subscript to identify each piece of data individually. Thus, the ten observations of Example 4.1 are respectively denoted $X_1, X_2, X_3, \ldots, X_{10}$, where $X_1 = 50, X_2 = 45, X_3 = 36$, and so on. The symbols $X_1, X_2,$

---

[1] Of course, this statement is judgmental and is conceivably open to argument.

X_3, and so on are referred to as "*X* sub one," "*X* sub two," "*X* sub three," or for short, "*X* one," "*X* two," "*X* three."

The subscripts of *X* constitute a series of consecutive integers known as an *index*, which is traditionally denoted by the letter i.[2] With this notation established, we can more generally say that the ith element of the data is denoted by X_i, where i takes on the values 1, 2, 3, and so on.

Denoting the number of population observations by *N* and the arithmetic mean by the Greek letter μ (pronounced *mu*), we may write:

$$\mu = \frac{X_1 + X_2 + X_3 + \cdots + X_N}{N} \tag{4.1}$$

To provide for a more compact expression than Equation 4.1, we introduce an additional mathematical notation: the *summation sign*. To denote summation we use the Greek letter Σ (capital sigma). Specifically, the expression

$$\sum_{i=1}^{N} X_i \tag{4.2}$$

should be read as follows: "sum of X_i, i going from 1 to N." That is to say, the expression stands for the sum of all the X_i's, where i runs consecutively from the value $i = 1$ through the value $i = N$. Equation 4.3 should further clarify the notation:

$$\sum_{i=1}^{N} X_i = X_1 + X_2 + X_3 + \cdots + X_N \tag{4.3}$$

When confusion is not likely to arise, we may drop the index and use the shorter expression

$$\Sigma X \tag{4.4}$$

which actually means precisely the same as Expression 4.2. Using 4.1 and 4.4, we may now simply write the equation for obtaining the arithmetic mean:

THE ARITHMETIC MEAN

$$\mu = \frac{X_1 + X_2 + X_3 + \cdots + X_N}{N} = \frac{\Sigma X}{N} \tag{4.5}$$

Appendix 4A summarizes the rules of operating with the summation sign, Σ. You would do well to become familiar with it.

CALCULATING THE ARITHMETIC MEAN FROM GROUPED DATA

Sometimes we need to compute the arithmetic mean without having full information about the values X_i. With only partial information available, the statistician must make the best of the data at hand. A common occurrence

---

[2] When more than one index is required, the letters j and k are also frequently used.

is the need to compute the mean from a frequency distribution of the data. In this situation the specific X_i values are unknown, but information is available about the frequencies in the class intervals and about the class interval *midpoints*. Denoting the frequencies by $f_1, f_2, f_3, \ldots$, the class midpoints by X_1, X_2, X_3, and the number of class intervals used by M, one calculates the mean from the grouped data in the following way:

THE MEAN CALCULATED FROM GROUPED DATA

$$\mu = \frac{f_1 X_1 + f_2 X_2 + f_3 X_3 + \cdots + f_M X_M}{f_1 + f_2 + f_3 + \cdots + f_M} = \frac{\Sigma f X}{\Sigma f} = \frac{\Sigma f X}{N} \qquad (4.6)$$

To illustrate the use of Equation 4.6, let us consider an example.

EXAMPLE 4.2

Table 4.1 is a frequency distribution of the "age" of accounts receivable of MONEY, Inc. (the "age" is determined by the number of days the account is past due). What is the average age of the past-due accounts of MONEY, Inc.?

Consider the first interval of 1–20 days past due. There are 80 past-due accounts in this age interval, but we do not have a breakdown of the specific age of any of those accounts. Should we assume that the average age within this interval is 5 days? Should we assume 15 days? Some other value? When no additional information is available, the most appealing assumption is that the average *within* each interval is equal to the *interval's midpoint*. The midpoint of the first interval is 10 days, that of the second interval is 30 days, and so on.

TABLE 4.1
Frequency Distribution of Age of Accounts Receivable of MONEY, Inc.
(days past due)

Age interval	Number of accounts
1–20	80
21–40	40
41–60	30
61–80	30
81–180	20
Total	200

Now that we have chosen the interval's midpoint to represent the average in each class, we proceed to calculate the average age using Equation 4.6:

$$\mu = \frac{(80 \cdot 10) + (40 \cdot 30) + (30 \cdot 50) + (30 \cdot 70) + (20 \cdot 130)}{200}$$

$$= \frac{8{,}200}{200}$$

$$= 41 \text{ days}$$

It is often convenient to develop a work sheet to calculate the mean from grouped data, as illustrated in Table 4.2.

TABLE 4.2
Calculating the Mean from Grouped Data

(1) Age interval	(2) Interval midpoint (X)	(3) Frequency (f)	(4) fX
1–20	10	80	800
21–40	30	40	1,200
41–60	50	30	1,500
61–80	70	30	2,100
81–180	130	20	2,600
Total		$N = \Sigma f = 200$	$\Sigma fX = 8,200$

$$\text{Mean} = \mu = \frac{\Sigma fX}{N} = \frac{8,200}{200} = 41$$

Note that the arithmetic average calculated from grouped data is an *approximation*, since the use of the midpoint for each class interval is only an estimate of the true class-interval average. When the number of classes is large and each class interval is relatively narrow, the approximation is good. If the class intervals are wide and the number of intervals is small, however, the approximation may be quite wide of the mark. In Example 4.2, for instance, the interval 81–180 days is quite wide. For a midpoint we used 130 days, because other information was lacking. If additional information could be obtained to improve the estimate, the average age could be calculated more accurately. Obtaining additional information is essential when we have an open-end class interval. Suppose, for example, the last interval were the open-end interval "81 or more." Such an interval has no midpoint, so the best available information must be used before the average age of past-due accounts is calculated by means of Equation 4.6.

4.2 The Weighted Mean

The arithmetic mean computed from grouped data is in essence a **weighted mean**: it is *an average of the interval midpoints, where each interval midpoint is weighted proportionally to the frequency of its interval.* To illustrate, notice that Equation 4.6 can be rewritten as follows:

$$\mu = \frac{f_1 X_1 + f_2 X_2 + f_3 X_3 + \cdots + f_M X_M}{\Sigma f}$$

$$= \frac{f_1}{\Sigma f} X_1 + \frac{f_2}{\Sigma f} X_2 + \frac{f_3}{\Sigma f} X_3 + \cdots + \frac{f_M}{\Sigma f} X_M$$

(4.7)

On the right-hand side of Equation 4.7 we have a weighted average, where each weight is simply equal to the relative frequency of the interval. When we use Equation 4.7, we find that the mean age of the accounts receivable of MONEY, Inc. (Example 4.2) is

$$\mu = \left(\frac{80}{200} \cdot 10\right) + \left(\frac{40}{200} \cdot 30\right) + \left(\frac{30}{200} \cdot 50\right)$$
$$+ \left(\frac{30}{200} \cdot 70\right) + \left(\frac{20}{200} \cdot 130\right) = 41 \text{ days}$$

The weighted average is applicable not only to data grouped in intervals but to discrete data as well. Consider a class of 28 school children, all of whom are 7 years old, and two 41-year-old teachers. A simple average of 7 and 41 is $(7 + 41)/2 = 24$, but even with a very modest degree of common sense, we would reject this number out of hand as the average age of the children and the teachers combined. The number 24 was obtained without any regard for the fact that there are many more 7-year-old children in the group than 41-year-old teachers. We can rectify this logical error by using the weighted average that gives each of the numbers 7 and 41 a weight corresponding to its frequency in the group. The total number of people in the group (that is, children and teachers) is 30 (= 28 + 2). The fraction of 7-year-old children in the group is $\frac{28}{30}$ and the fraction of 41-year-old teachers is $\frac{2}{30}$. Thus, the weighted average is $\left(\frac{28}{30} \cdot 7\right) + \left(\frac{2}{30} \cdot 41\right) = 9.3$ years. This average is obtained also when we explicitly average the ages of all the children and teachers:

$$\mu = (7 + 7 + 7 + \cdots + 7 + 41 + 41)/30 = 9.3$$

In general, a weighted average of the values $X_1, X_2, X_3, \ldots, X_N$ is μ_w and is given by the equation:

FORMULA FOR A WEIGHTED AVERAGE

$$\mu_w = w_1 X_1 + w_2 X_2 + w_3 X_3 + \cdots + w_N X_N = \Sigma w_i X_i \qquad (4.8)$$

---

where the w_i's are the corresponding weights: the relative frequency of each X_i in the population. By definition, then, the sum of the weights is equal to 1.0:

$$w_1 + w_2 + w_3 + \ldots + w_N = \Sigma w = 1.0$$

Before going on, let us give one more example of the weighted average.

EXAMPLE 4.3

A firm owns $10 million worth of assets, which were supplied from two sources. The owners of the firm (that is, its shareholders) supplied $6 million, and $4 million was supplied by banks in the form of loans. The $6 million is the *equity capital* of the firm and the $4 million is its *debt capital*. Both the owners and the banks expect some return on their funds. Suppose that the owners require $600,000 a year in return on their money, which amounts to 10 percent on the equity capital, and the banks require $200,000 in return, which amounts to 5 percent on the debt capital. From the point of view of the firm, it is said that the cost of equity capital is 10 percent; likewise, the cost of debt capital is 5 percent. Financial theory shows us that the relevant cost of capital for investment decisions is neither the 10 percent cost of equity capital nor the 5 percent of debt capital. Rather, the firm's analysis should focus on the weighted average cost of capital. What is the weighted average cost of capital in this example? Since the weight of equity capital in relation to total capital is 6,000,000/10,000,000 = 0.60, and similarly the weight of debt capital is 4,000,000/10,000,000 = 0.40, the weighted average cost of capital is:

$$\mu = (0.6 \cdot 10\%) + (0.4 \cdot 5\%) = 6\% + 2\% = 8\%$$

4.3 The Median

Although the arithmetic average is a very useful parameter for measuring the location of the population at hand, we need not go through lengthy demonstrations to show that in itself, the average conveys only a very partial description of the population. To use just one example, consider the annual income of six individuals, *A, B, C, D, E,* and *F,* as shown in Table 4.3. The average annual income is $222,000/6 = $37,000, which hardly reflects the annual income of the majority of this group of individuals. The number $37,000, taken alone, provides no clue to the fact that five of the six individuals involved have incomes below $7,500.

TABLE 4.3
Annual Income of Six Individuals

Individual	Income
A	$ 6,000
B	7,000
C	5,000
D	6,600
E	7,400
F	190,000
Total	$222,000

It is clear that additional population parameters must be developed to supplement the information provided by the average. The next parameter we shall discuss is the median.

DEFINITION: THE MEDIAN

The **median** is defined as *the value above and below which lie an equal number of population elements.*

When we rearrange the above annual incomes in ascending order we get Table 4.4. The total number of observations in Table 4.4 is even (6) and we can easily identify the two central values as $6,600 and $7,000. In fact, any number in the range between $6,600 and $7,000 qualifies as the median by

TABLE 4.4
Median Value of Incomes in Table 4.3

	Individual	Annual income
Three population elements below the median	C A D	$ 5,000 6,000 6,600
		← median income is $6,800
Three population elements above the median	B E F	7,000 7,400 190,000

our definition, but the midpoint between them is customarily considered to be the median. Thus we identify the median as

$$\text{Median} = (6{,}600 + 7{,}000)/2 = \$6{,}800$$

If the number of population elements is odd (rather than even), the median is equal to the central value. For example, if only individuals *A, B, C, D*, and *E* were in the population, the median would have been $6,600, since this would be the only value complying with our definition of median:

Individual	Annual income	
C	$ 5,000	
A	6,000	
D	6,600	← median
B	7,000	
E	7,400	

The unique merit of the median—a merit not shared by the mean—is that it is not affected by extreme values. Changing the annual income of *F* from $190,000 all the way down to $8,000 will leave the median unchanged, whereas the arithmetic average will be greatly affected. For this reason the average and the median should be considered as complementing each other, not as substituting for each other: given the average, the median often provides *additional* information about the population's location, and vice versa.

CALCULATING THE MEDIAN FROM GROUPED DATA

Let us begin our explanation of median calculation from grouped data with an example.

EXAMPLE 4.4

The loans outstanding at the Bay Area Bank are classified by size, as shown in Table 4.5. The total number of loans outstanding (N) is 600. The median, therefore, is that amount above which 300 ($= N/2$) of the loans are in smaller amounts and 300 loans are in greater amounts. Looking at the *cumulative* number of loans, we realize that 290 loans are in amounts less than or equal to $10,000, and 390 loans in amounts less than or equal to $25,000. Thus the median is evidently between $10,000 and $25,000. In other words, the class interval $10,001–25,000 is the one

TABLE 4.5
Frequency Distribution of Bank Loans, by Amount

Amount of loan	Frequency of loans	Cumulative frequency of loans
Up to $5,000	180	180
$ 5,001–10,000	110	290
10,001–25,000	100	390
25,001–50,000	90	480
50,001–100,000	70	550
100,001 and over	50	600
Total	600	

containing the median. Once we have located the class containing the median, our job is reduced to simple interpolation. Since there are 100 loans outstanding within an interval of $15,000 width ($10,001–25,000), the average density in the interval containing the median is one loan per $150 interval: $15,000/100 = $150. So having no other information, we assume that there are 290 loans in amounts up to $10,000; 291 loans in amounts up to $10,150; 292 loans in amounts up to $10,300; 293 loans in amounts up to $10,450; and so on. Starting off again at $10,000, we need to add 10 (300 − 290 = 10) more intervals of $150 each to get to the median: $10,000 + $\left(\dfrac{\$15,000}{100} \cdot 10\right)$ = $11,500. Thus the median is $11,500.

In Figure 4.1, we look at the median from another viewpoint. Here we display a "less than" ogive. Lacking any data concerning the specific distribution within the various classes, we assume an even and smooth distribution, reflected in the figure by the straight-line segments of the ogive. To find the median, we need to identify the distance AE, where the

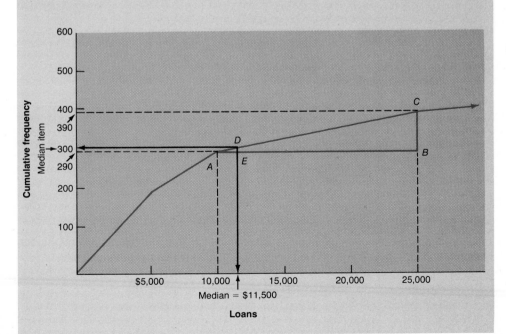

Figure 4.1
Median value
of loans in
Table 4.5

point E is directly underneath D, which in turn is at the same level as the median item: $N/2$ (300 in our example). The following triangles are similar: ABC and AED. From here we get:

$$\frac{DE}{CB} = \frac{AE}{AB}$$

or

$$\frac{10}{100} = \frac{AE}{\$15,000}$$

so that

$$AE = \frac{\$15,000}{100} \cdot 10 = \$1,500$$

Adding the $1,500 to the $10,000, which is the lower class limit containing the median, we obtain once again:

$$\text{Median} = \$10,000 + \$1,500 = \$11,500$$

Finally we should note that the median appears diagrammatically at a point exactly underneath the intersection of the "less than" and "more than" ogives, as Figure 4.2 exhibits. The intersection between the graphs occurs at the loan amount above which 300 (=$N/2$) of the loans are in smaller amounts and 300 are in greater amounts. As explained earlier, this is exactly the median.

Figure 4.2

The location of the median shown at the intersection of the "less than" and "more than" ogives

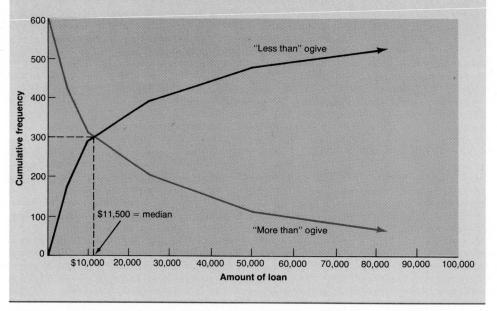

Now that we have seen the calculation of the median in an example, let us treat it more generally. To do so we need the following notations:

F Frequency accumulated up to the lower limit of the class containing the median.

W The width of the class interval containing the median.

f_M The frequency of the class containing the median.

L_M The lower limit of the class containing the median.

N The total number of observations in the population.

Using these notations, we find the median by the following equation:

$$\text{Median} = L_M + \frac{N/2 - F}{f_M} \cdot W \tag{4.9}$$

For Example 4.4, $L_M = \$10,000$, $N/2 = 300$, $F = 290$, $f_M = 100$, and $W = \$15,000$, so that again we get:

$$\text{Median} = \$10,000 + \frac{300 - 290}{100} \cdot \$15,000 = \$11,500$$

4.4 Percentiles and Fractiles

Just as we would often like to measure the median in order to obtain the value below which 50 percent of the population elements are located, so might we also wish to find the values below which some other percentages of the population elements are located. Those values are called **percentiles.** For example, the 10th percentile is that value below which 10 percent of the population elements are located. The 25th percentile is that value below which 25 percent of the population elements are located. The 25th percentile is also known as the **first quartile.** The 75th percentile—that value below which 75 percent of the population elements are located—is also known as the **third quartile.** The median, by this definition, is the 50th percentile.

Instead of referring to a value by the percentage of population elements below the value, we may choose to refer to it by the *fraction* of the population elements below the value, and term the value a **fractile** rather than a percentile. Thus the 10th percentile is the 0.10th fractile, the 75th percentile is the 0.75th fractile, and so on (see Figure 4.3).

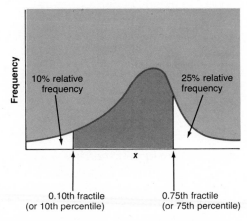

Figure 4.3

Illustration of percentiles and fractiles

Locating a percentile involves a method similar to that used in locating the median. It is illustrated in Example 4.5.

EXAMPLE 4.5

Suppose the average annual net profits of 280 firms in the past 10 years are given by the frequency distribution in Table 4.6.

TABLE 4.6
Frequency Distribution of Average Annual Net Profits of 280 Firms

Average annual profit	Frequency	Cumulative frequency
Up to $25,000	14	14
$ 25,001–50,000	30	44
50,001–75,000	50	94
75,001–100,000	70	164
100,001–150,000	57	221
150,001–200,000	24	245
200,001–300,000	12	257
300,001–400,000	13	270
400,001–500,000	3	273
500,001–1,000,000	3	276
1,000,001 and over	4	280
Total	280	

To find the first quartile, we look at the cumulative frequency to locate the class that contains it. The first quartile in our example is that annual profit below which 70 of the values are located (70 is 25 percent of the 280 firms in the population). Since 44 firms had profits below $50,000 and 94 firms had profits below $75,000, it is clear that the first quartile is somewhere in the range of $50,000–75,000. Within this range the average rate of accumulation of frequency is $25,000/50 = $500 ($25,000 is the width of the interval containing the first quartile and 50 is the frequency of firms in that range). The $500 average rate of accumulation implies that on the average one additional firm is accumulated in this range per $500 advance on the annual profit scale. So if 44 firms had annual profits below $50,000, we can assume that 45 firms had profits below $50,500; 46 firms had profits below $51,000; 47 firms had profits below $51,500; and so on. What is the profit below which we estimate a frequency of 70 firms? With 70 − 44 = 26 more firms needed to be accumulated in the range $50,000–75,000, and with one firm accumulated per $500 advance on the annual profit scale, we need to advance $500 · 26 = $13,000 to get to the first quartile. Thus the first quartile is equal to $50,000 + $13,000 = $63,000.

In a similar fashion we can locate the 50th percentile (that is, the median), the third quartile, and other percentiles as well. We urge you to verify that the median is $91,428.57 and the third quartile is $140,350.88.

A percentile is a useful measure in determining the location of a population. By using percentiles we can conveniently locate the values of specified percentages of the population. As we shall soon see, when two or more percentiles are considered simultaneously, they can also serve as useful measures of population dispersion.

4.5 The Mode

Another parameter measuring the population's location, or central tendency, is the **mode.**

DEFINITION: THE MODE

The **mode** is defined as the *most frequently occurring value in the population.*

When the data are grouped, the *midpoint of the interval with the greatest frequency is considered to be the mode.* For example, the most frequent class of Example 4.5 is $75,001–100,000, since its frequency is 70, which is greater than any of the other class frequencies. The mode, then, is the midpoint of that interval, $87,500.

A notable advantage of the mode reveals itself when we deal with categorical data. Here the average and the median are simply meaningless. Consider the following data on annual sales of cars of various models by a car dealer:

Model	Number of cars sold
Sedan	700
Hatchback	701
Wagon	449
Total	1,850

The mode here is the hatchback, since it was sold more frequently than any other model. With "model" being a nonquantitative variable, the average and the median become completely irrelevant.

Another case in which the mode is a useful measure of central tendency occurs when the data are quantitative but involve a limited number of discrete values, such as the number of rooms in a house, the number of children in a household, the number of times per year firms pay dividends to their shareholders, and so on. To see the usefulness of the mode for this kind of data, consider Example 4.6.

EXAMPLE 4.6

A population of 1,000 households is examined for the number of cars owned by each household, and the following frequency distribution is established:

Number of cars owned by households	Frequency
0	40
1	750
2	180
3	30
	1,000

The average number of cars for this population is 1.2 cars (verify!), but 1.2 cars per household is somewhat difficult to interpret because *none* of the households owns 1.2 cars. They own either 0, 1, 2, or 3 cars. The mode of this population is 1, which simply means that one car is the most frequent number of cars per household. Compared to the mean, this figure is somewhat easier to interpret.

Though the mode has certain advantages, it also has some significant shortcomings. First, it says very little about the population apart from the most frequent value. Consider again the car-dealer example. Although the hatchback is the most frequently sold model, the sedan is sold almost as frequently, yet the mode gives no indication of this fact. Wagons are also sold in great frequency (though materially less than the other two models). This fact, too, is not reflected in the mode. Second, when grouped data are involved, the model can be artificially changed by rearrangement of the class intervals. Reexamine Example 4.5 and combine the classes $25,001–50,000 and $50,001–75,000 into one class: $25,001–75,000. The frequency in this class will be the sum of the frequencies in the two original classes (30 and 50, respectively). Thus the frequency in the new class interval is 80 (30 + 50 = 80). While the mode was previously considered to be the midpoint of class $75,001–100,000 (that is, $87,500), it will now be redefined as the midpoint of the interval $25,001–75,000 (that is, $50,000). A change in the grouping of the data will then significantly affect the mode, though the data remain precisely the same. Only our presentation of the data has changed. Such oversensitivity to grouping is highly undesirable in a parameter.

4.6 Bimodal Distributions

A population with more than one mode is referred to as a **bimodal distribution.** Figure 4.4 shows two bimodal distributions. The one in Figure 4.4*a* has two equally frequent modes, while the one in 4.4*b* has two values that occur with significantly greater frequency than neighboring values yet do not occur at the same frequency. Both distributions are considered bimodal. A bimodal

Figure 4.4

Examples of bimodal distributions

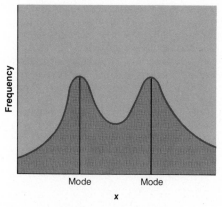

(a) Two modes of equal frequency

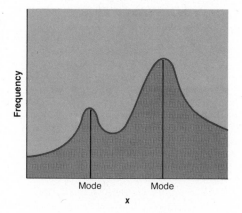

(b) Two modes of unequal frequency

distribution is usually heterogeneous in some sense, indicating that two groups in the population differ from each other in some way. Example 4.7 should clarify this point.

EXAMPLE 4.7

The frequency distribution of the rates of return on 900 stocks and bonds traded on stock exchanges in a given year in the United States is given in Table 4.7.[3] A glance at the table and at Figure 4.5, which graphically displays the distribution, suffices to show that the distribution is bimodal.

TABLE 4.7
Frequency Distribution of Rates of Return on 900 Securities

Rate of return	Number of securities
(−89.9)–(−50.0)	8
(−49.9)–(−25.0)	16
(−24.9)–0.0	27
0.1–4.0	81
4.1–8.0	220
8.1–12.0	116
12.1–16.0	75
16.1–20.0	100
20.1–24.0	80
24.1–28.0	60
28.1–32.0	45
32.1–36.0	38
36.1–200.0	34
Total	900

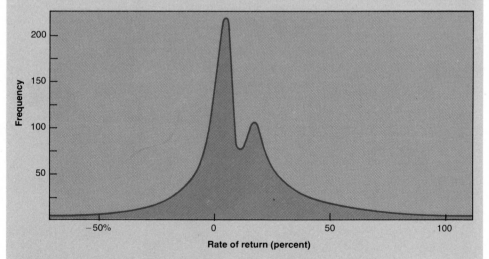

Figure 4.5

Bimodal frequency distribution of rates of return on 900 securities

[3] This is not the place to go into detail on the way rates of return are calculated. Briefly, however, if we invest $100 at the beginning of the year and the *value* of our investment (i.e., the price of the security plus any interest or dividends obtained during the year) rises to $115, the rate of return is 115/100 − 1 = 0.15, or 15 percent. Similarly, if it drops to $88, our rate of return is 88/100 − 1 = −0.12, or a loss of 12 percent on our investment, and so on. Thus, the rate of return is the rate of profit or loss we make on our investment.

One mode occurs at the midpoint of the interval 4.1–8.0 percent, and another occurs at the midpoint of the interval 16.1–20.0 percent. Furthermore, the distribution is such that there is a gradual clustering of

TABLE 4.8
Frequency Distribution of Rate of Return on Stocks and Bonds

Rate of return (percent)	Number of securities		
	Stocks	*Bonds*	*Total*
(−89.9)–(−50.0)	8	0	8
(−49.9)–(−25.0)	16	–	16
(−24.9)–0.0	22	5	27
0.1–4.0	31	50	81
4.1–8.0	40	180	220
8.1–12.0	51	65	116
12.1–16.0	75	–	75
16.1–20.0	100	–	100
20.1–24.0	80	–	80
24.1–28.0	60	–	60
28.1–32.0	45	–	45
32.1–36.0	38	–	38
36.1–200.0	34	–	34
Total	600	300	900

Figure 4.6
Frequency distributions of rates of return on 600 stocks and on 300 bonds (bimodal distribution of Figure 4.5 broken down into two unimodal distributions)

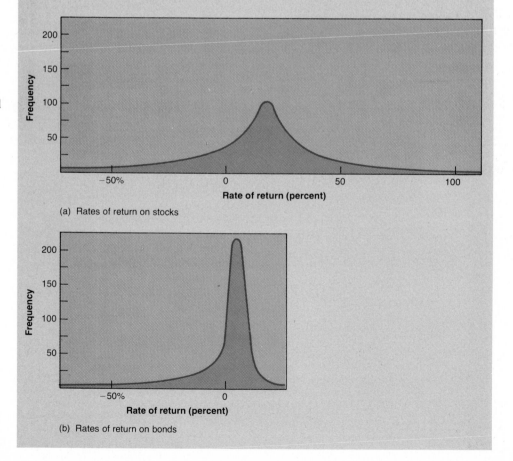

(a) Rates of return on stocks

(b) Rates of return on bonds

the population around these two central points as we approach them from the right or the left. This pattern suggests that there is some *economic reason* for the appearance of this distribution. The reason is that there are two *types* of securities making up the group of 900 securities: stocks and bonds. Table 4.8 shows the frequency distribution of the separate rates of return on stocks and bonds.

The separate frequency distributions for stocks and bonds are also shown in Figure 4.6. When the two types of securities are considered separately, each of the distributions emerges as a unimodal distribution.

When two or more distinct population groups can be identified, it is generally a good idea to study the groups separately. In that way the characteristics of the population emerge with special clarity.

4.7 Contrasting the Mean, the Median, and the Mode

As we have seen, the analysis of a frequency distribution is significantly more meaningful when the population of interest has a reasonable degree of homogeneity. Assuming a reasonably homogeneous population, let us consider three common shapes of frequency distributions and their respective relationships to the mean, the median, and the mode.

In Figure 4.7 we see symmetrical, positively skewed, and negatively skewed distributions. (The word **skewed** means *more developed on one side or in one direction than the other.*) The positively skewed distribution has a long right tail and the negatively skewed distribution has a long left tail. The mean, median, and mode of the symmetrical distribution coincide at one value. In the skewed distributions, however, the means, medians, and modes occupy separate locations. The mode of the positively skewed distribution (Figure 4.7b) is to the left of the median, and the median appears to the left of the mean. The mean appears farthest to the right because observations with extreme values on the right (not countered by extreme values on the left) are incorporated in the calculation of the mean and thus pull the mean value rightward. (Hence, the long tail on the right.) In contrast, these extreme values have no impact on the mode or the median. The order of the

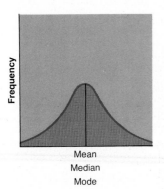

(a) Symmetrical distribution

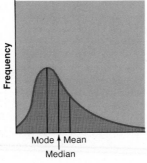

(b) Positively skewed distribution

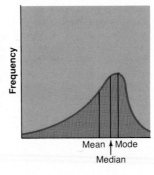

(c) Negatively skewed distribution

Figure 4.7

Symmetrical, positively skewed, and negatively skewed frequency distributions parameters of the negatively skewed distribution is exactly the opposite: the mode appears to the right of the median, which in turn is to the right of the mean.

We saw earlier that the three location parameters, mean, median, and mode, complement one another: each imparts information not given by the other two parameters. Although in most situations we can use all three measures to enlarge our understanding of population, sometimes one or two of them may be irrelevant. Consider first a garment manufacturer who wishes to specialize in the production of pants of only *one size*. The net profit per pair of pants sold is the same for all pairs, regardless of size. A survey of the demand for pants shows the following figures:

Size	Demand (pairs per month)
26	100
28	200
30	300
32	200
34	100

In which size should the manufacturer specialize? The answer must be determined by the mode, that is, size 30. By specializing in this size, the manufacturer will face the greatest demand and therefore will make the greatest profit. The median and arithmetic mean have no significance in this problem.

Now consider a different situation, one in which the median, not the average or the mode, is relevant. Consider a government plan to subsidize housing for large families in neglected urban areas. Suppose the frequencies of the sizes of families considered for the program are as shown in Table 4.9.

Our concern here is to limit the number of families according to the limiting criterion: family size. The mean (5.35) and the mode (5.0) are not really pertinent. The median (or any other percentile) is much more germane, since if we know the median we can easily determine the number of families that will be entitled to subsidized housing. This figure in turn will have a direct bearing on the size of the program's budget.

Finally, suppose we wish to compare the performance of two institutional investors who invest in various stocks. The average rate of return, rather than the median or the mode, is the relevant measure here, since total performance is affected by the specific rate of return on each stock included

TABLE 4.9
Frequencies of Family Sizes

Size of family	Frequency
2	20,000
3	25,000
4	60,000
5	80,000
6	50,000
7	40,000
8	20,000
9	10,000
10	10,000

in the investment portfolio. Only the mean takes into account the specific values of the rates of return.

We see, then, that although the mean, the median, and the mode are all location parameters, sometimes one or two of them may be irrelevant.

4.8 Measures of Population Variability: Variance and Standard Deviation

So far we have developed measures known as location parameters, or parameters that measure a population's central tendency. As important as these measures are, when they stand alone they give only limited information about the population. Additional parameters are needed to measure additional population characteristics. In particular, measures of the variability of the population are needed for a better representation of that population. There are several popular variability measures, the most common of which are the variance and the standard deviation.

The **variance** is defined as *the average squared deviation of the population elements from their arithmetic mean,* and is commonly denoted by the square of the Greek letter sigma: σ^2. Thus σ is read "sigma" and σ^2 is read "sigma squared." The formula for the variance is:

FORMULA FOR THE VARIANCE

$$\sigma^2 = \frac{\Sigma(X - \mu)^2}{N} \qquad\qquad (4.10)$$

The **standard deviation** is defined as *the (positive) square root of the variance* and is indicated by σ. The formula for the standard deviation is:

FORMULA FOR THE STANDARD DEVIATION

$$\sigma = \sqrt{\frac{\Sigma(X - \mu)^2}{N}} \qquad\qquad (4.11)$$

Let's look at an example that shows how these measures of variability are used.

EXAMPLE 4.8

The following is a list of withdrawals from a savings account during the past year:

Withdrawal number (i)	Withdrawal amount (X)
1	$310
2	460
3	350
4	1,060
5	20
6	200

The population here is composed of the six withdrawal amounts, and the variance and standard deviation of the withdrawal amounts are calculated in Table 4.10.

TABLE 4.10
Calculation of the Variance and the Standard Deviation

(1) Withdrawal amount (X)	(2) Deviation from the mean $(X - \mu)$	(3) Squared deviation from the mean $(X - \mu)^2$
\$ 310	\$ −90	\$ 8,100
460	60	3,600
350	−50	2,500
1,060	660	435,600
20	−380	144,400
200	−200	40,000
$\Sigma X = \$2,400$	$\Sigma(X - \mu) = \$0$	$\Sigma(X - \mu)^2 = \$634,200$

$$\text{Mean} = \mu = \frac{\Sigma X}{N} = \frac{\$2,400}{6} = \$400$$

$$\text{Variance} = \sigma^2 = \frac{\Sigma(X - \mu)^2}{N} = \frac{634,200}{6} = 105,700$$

$$\text{Standard deviation} = \sigma = \sqrt{\frac{\Sigma(X - \mu)^2}{N}} = \sqrt{105,700} = \$325.12$$

Since the variance is the average squared deviation from the mean, our first step is to determine the population mean, μ. The mean savings withdrawal as shown in Table 4.10 is \$400. Once the mean has been calculated, we continue to develop columns 2 and 3. We obtain column 2 by recording the deviation of each X value from the mean, μ, and column 3 by squaring each term in column 2 individually. We obtain the variance (105,700) by adding the figures in column 3 and dividing the sum (634,200) by the number of observations, 6. We should take note of the fact that columns 1 and 2 are measured in dollars, while column 3 and the variance are measured in dollars squared. Why do we square the $X - \mu$ terms just to obtain magnitudes measured in such unappealing units as dollars squared? The answer lies in the fact that when we measure deviation, we are not interested in the *direction* of a deviation from the mean but rather in its *magnitude*. In more technical terms, this is the problem: the sum of the deviations $X - \mu$ is always equal to zero (see Table 4.10). By squaring the deviations, we find one good way to eliminate the problem: the sum of the squared deviations is never zero, if dispersion exists. To solve the problem of working with dollars squared, we take the square root of the variance and obtain the standard deviation (\$325.12), which is measured in dollars.

By simple algebraic manipulation, we can show that the variance as given in Equation 4.10 can also be written in the following form, known as the "shortcut" formula:[4]

SHORTCUT FORMULA FOR THE VARIANCE

$$\sigma^2 = \frac{1}{N} \Sigma X^2 - \mu^2 \qquad (4.12)$$

The shortcut standard deviation formula is:

SHORTCUT FORMULA FOR THE STANDARD DEVIATION

$$\sigma = \sqrt{\frac{1}{N} \Sigma X^2 - \mu^2} \qquad (4.13)$$

The calculation of the variance and the standard deviation by use of Equations 4.12 and 4.13 is presented in Table 4.11.

TABLE 4.11
Variance and Standard Deviation Calculated by the Shortcut Method

(1) Withdrawal amount (X)	(2) X²
$ 310	$ 96,100
460	211,600
350	122,500
1,060	1,123,600
20	400
200	40,000
ΣX = $2,400	ΣX² = $1,594,200

$$\text{Mean} = \mu = \frac{\Sigma X}{N} = \frac{\$2,400}{6} = \$400$$

$$\text{Variance} = \sigma^2 = \frac{1}{N}\Sigma X^2 - \mu^2 = \frac{1}{6}\$1,594,200 - \$400^2 = 105,700$$

$$\text{Standard deviation} = \sigma = \sqrt{105,700} = \$325.12$$

---

[4] Since $(X - \mu)^2 = X^2 + \mu^2 - 2X\mu$, we also know that $\Sigma(X - \mu)^2 = \Sigma(X^2 + \mu^2 - 2X\mu) = \Sigma X^2 + \Sigma\mu^2 - 2\mu\Sigma X$. Also, since μ is a constant, the expression $\Sigma\mu^2$ is equal to $N\mu^2$. From the definition $\mu = \Sigma X/N$ we know that $\Sigma X = N\mu$, and we therefore get: $\Sigma(X - \mu)^2 = \Sigma X^2 + N\mu^2 - 2\mu N\mu = \Sigma X^2 + N\mu^2 - 2N\mu^2$, or $\Sigma(X - \mu)^2 = \Sigma X^2 - N\mu^2$. Equation 4.10 follows directly, since

$$\sigma^2 = \frac{\Sigma(X - \mu)^2}{N} = \frac{\Sigma X^2 - N\mu^2}{N} = \frac{1}{N}\Sigma X^2 - \mu^2$$

CALCULATING VARIANCE AND STANDARD DEVIATION FROM GROUPED DATA

In Section 4.1 we explained how the mean can be calculated from grouped data. The basic approach was to assume that all the observations in a class interval have the value of the class midpoint. We can use the same approach to calculate the variance and standard deviation of grouped data. The variance formula applicable to grouped data is as follows:

$$\sigma^2 = \frac{\Sigma f(X - \mu)^2}{N} \tag{4.14}$$

The standard deviation is obtained by taking the (positive) square root of the variance. Thus:

$$\sigma = \sqrt{\frac{\Sigma f(X - \mu)^2}{N}} \tag{4.15}$$

The equivalent shortcut formulas are then:

$$\sigma^2 = \frac{\Sigma fX^2 - N\mu^2}{N} = \frac{1}{N}\Sigma fX^2 - \mu^2 \tag{4.16}$$

$$\sigma = \sqrt{\frac{1}{N}\Sigma fX^2 - \mu^2} \tag{4.17}$$

Example 4.9 shows how to use Equations 4.16 and 4.17.

EXAMPLE 4.9

Suppose we want to calculate the standard deviation of the rate of return on the 300 bonds presented in Example 4.7. Here again are the rates of return for the bonds:

Rate of return (percent)	Frequency
(−24.9)–0.0	5
0.1–4.0	50
4.1–8.0	180
8.1–12.0	65
	300

First we determine the midpoint of each class interval and denote it by X. They are shown in Table 4.12, which is in fact a work sheet for the calculation of the variance. Given the totals of columns 3, 4, and 6, we can easily compute the mean, variance, and standard deviation, as shown at the bottom of the table. For the bond population we find that $\sigma^2 = 11.85$ (percent squared) and $\sigma = 3.44$ percent. Perform similar calculations for the rate of return on the stocks of Example 4.7, and verify that for the stocks the parameters are: $\mu = 20.33$ percent, $\sigma^2 = 846.53$, and $\sigma = 29.10$ percent.

TABLE 4.12
Calculating Variance and Standard Deviation from Grouped Data

(1) Rate of return (percent)	(2) Class midpoint (X)	(3) Frequency of bonds (f)	(4) fX	(5) X²	(6) fX²
(−24.9)–0.0	−12.5	5	−62.50	156.25	781.25
0.1–4.0	2.0	50	100.00	4.00	200.00
4.1–8.0	6.0	180	1,080.00	36.00	6,480.00
8.1–12.0	10.0	65	650.00	100.00	6,500.00
Total		$\Sigma f = N = 300$	$\Sigma fX = 1,767.50$		$\Sigma fX^2 = 13,961.25$

$$\text{Mean} = \mu = \frac{\Sigma fX}{N} = \frac{1,767.50}{300} = 5.89\%$$

$$\text{Variance} = \sigma^2 = \frac{1}{N}\Sigma fX^2 - \mu^2 = \left(\frac{1}{300}\ 13,961.25\right) - 5.89^2 = 11.85$$

$$\text{Standard deviation} = \sigma = \sqrt{11.85} = 3.44\%$$

By comparing the means and standard deviations of the stock and bond populations of our example, we can gain insight into the usefulness of these parameters. Looking at the mean only, we realize that the rate of return on the stocks is materially higher than the rate of return on the bonds (20.33 percent versus 5.89 percent). Given this difference, one might wonder why anyone would invest in bonds when the average rate of return on stocks is so much higher. The standard deviation reveals information that provides the clue to the answer. Although the mean rate of return on the stocks is higher, the variability of the rate of return on the stocks is also substantially higher. Thus investors who are attracted by the high average rate of return on stocks might at the same time be deterred by the variability of that rate. By taking account of both the average and the variance of the rates of return, the investor gets a fuller description of the distributions of the rates of return, and thus of the relative attractiveness of those investments.

4.9 The Coefficient of Variation

The **coefficient of variation,** *C,* is defined as *the ratio of the standard deviation to the average of the population and is expressed by the following formula:*

THE COEFFICIENT OF VARIATION

$$C = \frac{\sigma}{\mu} \tag{4.18}$$

The coefficient measures the *relative variability* and is sometimes used to evaluate changes that have occurred in a population over time, or to evaluate differences between populations. Suppose that the average earnings per share (EPS) of U.S. industrial firms in 1980 was $3.00 and the standard deviation was $0.50. Suppose also that the equivalent figures for 1982 were $4.00 and $0.70, respectively. Clearly, both the average EPS and its standard deviation rose during the period 1980–82, but the coefficient of variation reveals that the variability in EPS across the industrial firms rose faster than the average: the coefficient for 1980 is $C_{80} = 0.5/3 = 0.1667$ and for 1982 it is $C_{82} = 0.7/4 = 0.1750$.

The coefficient of variation is also useful when we compare the variability of two or more distinctive populations. For example, in a comparison of Japanese workers' income (measured in Japanese yen) with American workers' income, the coefficient of variation can be helpful.

Similarly, when we compare the variability of two or more populations of widely varying magnitudes, the coefficient of variation can be most appropriate. Let's say we want to compare uniformity of size among people and among ants. Obviously absolute variance of size is greater among people than the average: the coefficient for 1980 is $C_{80} = 0.5/3 = 0.1667$ and for variance of size relative to their own average size. It is quite possible that ants will show greater *relative* variation.

Although the coefficient of variation is appealing as a measure of relative variability, it must be used with care. In many situations, the standard deviation is a more relevant variability measure. Thus differences between populations should first be analyzed in terms of the mean and the standard deviation separately; then the coefficient of variation may be used as a complementary measure.

4.10 Interfractile Range as a Measure of Dispersion

Although the variance and standard deviation are by far the most popular variability measures, they are by no means the only ones. Let us briefly discuss one more variability measure: the **interfractile range.** The interfractile range is *the range of values between two specified fractiles.* The most common is the **interquartile range,** which is *the range between the first quartile and the third quartile.* Thus the interquartile range contains the middle 50 percent of the population. Figure 4.8 shows the interquartile range. *The interfractile*

Figure 4.8

The inter-quartile range

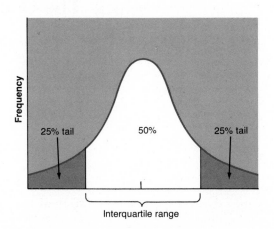

range between the 0th and the 1.0th fractile, within which all the population is concentrated, is known as the **range.** The range is a common measure of population variability.

4.11 Measures of Asymmetry

Some population characteristics will not be revealed by the variability measures we have been discussing. The degree of asymmetry is an obvious example. We mentioned earlier that some populations have symmetrical distributions while others can be negatively or positively skewed. The degree of skewness is an important population characteristic for which we need to develop a quantitative measure. One such measure is μ_3, known as the population's third moment, and is defined as follows:

THIRD MOMENT OF A POPULATION

$$\mu_3 = \frac{\Sigma(X - \mu)^3}{N}$$

(4.19)

When the distribution is symmetrical, μ_3 is equal to zero. When the distribution is skewed to the left, μ_3 is negative (and thus it will be said to be negatively skewed); when the distribution is skewed to the right, μ_3 is positive (and it will be said to be positively skewed). For grouped data, the formula is as follows:

THIRD MOMENT OF A POPULATION (GROUPED DATA)

$$\mu_3 = \frac{\Sigma f(X - \mu)^3}{N}$$

(4.20)

The value μ_3 is measured in the third power of the units of X. If X is measured in dollars, then μ_3 is measured in cubic dollars. To revert to the original units of X, and to obtain a measure that is somewhat more easily interpreted, we often *scale* the third moment by the magnitude σ^3. As a result, we get:

THE COEFFICIENT OF SKEWNESS

Coefficient of Skewness $= \dfrac{\mu_3}{\sigma^3}$

(4.21)

μ_3/σ^3 is expressed as a number without units.

EXAMPLE 4.10

The ages of the 2,000 students of the business school of a certain university occur in the frequencies shown in Table 4.13. To calculate the distribution's skewness, we develop Table 4.14.

TABLE 4.13
Frequency Distribution of Ages of Students

Age interval	Frequency
17–18	19
19–20	708
21–22	542
23–24	356
25–29	260
30–34	25
35–39	20
40–44	33
45–49	17
50–54	8
55–59	12
	2,000

TABLE 4.14
Calculation of Parameters: Grouped Data

(1) Age interval	(2) Midpoint (X)	(3) Frequency (f)	(4) fX	(5) X − μ	(6) (X − μ)²	(7) (X − μ)³	(8) f(X − μ)²	(9) f(X − μ)³
17–18	18.0	19	342.0	−5.5	30.25	−166.375	574.75	−3,161.12
19–20	20.0	708	14,160.0	−3.5	12.25	−42.875	8,673.00	−30,355.50
21–22	22.0	542	11,924.0	−1.5	2.25	−3.375	1,219.50	−1,829.25
23–24	24.0	356	8,544.0	0.5	0.25	0.125	89.00	44.50
25–29	27.5	260	7,150.0	4.0	16.00	64.000	4,160.00	16,640.00
30–34	32.5	25	812.5	9.0	81.00	729.000	2,025.00	18,225.00
35–39	37.5	20	750.0	14.0	196.00	2,744.000	3,920.00	54,880.00
40–44	42.5	33	1,402.5	19.0	361.00	6,859.000	11,913.00	226,347.00
45–49	47.5	17	807.5	24.0	576.00	13,824.000	9,792.00	235,008.00
50–54	52.5	8	420.0	29.0	841.00	24,389.000	6,728.00	195,112.00
55–59	57.5	12	690.0	34.0	1,156.00	39,304.000	13,872.00	471,648.00
Total		$\Sigma f = N =$ 2,000	$\Sigma fX =$ 47,002.5				$\Sigma f(X - \mu)^2 =$ 62,966.25	$\Sigma f(X - \mu)^3 =$ 1,182,558.63

$$\text{Mean} = \mu = \frac{\Sigma fX}{N} = \frac{47,002.5}{2,000} = 23.50$$

$$\text{Variance} = \sigma^2 = \frac{\Sigma f(X - \mu)^2}{N} = \frac{62,966.25}{2,000} = 31.48$$

$$\text{Standard deviation} = \sigma = \sqrt{31.48} = 5.61$$

$$\text{Skewness} = \mu_3 = \frac{\Sigma f(X - \mu)^3}{N} = \frac{1,182,558.63}{2,000} = 591.28$$

$$\text{Coefficient of skewness} = \frac{\mu_3}{\sigma^3} = \frac{591.28}{176.56} = 3.35$$

The positive value of the skewness (either μ_3 or μ_3/σ^3) reflects the fact that some students are substantially older than the bulk of the school's students, whereas no students are substantially younger. The conclusion drawn is that the distribution is not symmetrical, but rather positively skewed (that is, skewed to the right), as the skewness and coefficient of skewness indicate. Note that since σ^3 is always positive, the skewness, μ_3, and the coefficient of skewness, μ_3/σ^3, always have the same sign.

4.12 Proportions

Sometimes our interest is focused on the proportion of population members that have some common characteristic, defined in either quantitative or qualitative terms. The **proportion,** *p,* is given as *the ratio of the population members with the specified characteristic* to the total population. This may be expressed by the formula:

FORMULA FOR PROPORTION

$$p = \frac{\text{Number of population members in category}}{\text{Population size}}$$

We might, for instance, be interested in the proportion of students in Example 4.10 over the age of 30. We obtain the proportion, *p,* by dividing the number of students over 30 (115) by the total number of students in the population (2,000):

$$p = \frac{115}{2,000} = 0.0575 \quad \text{or} \quad 5.75\%$$

Proportion is particularly useful in dealing with qualitative data. For example, when factory products are classified as "nondefective" and "defective," our interest is often focused on the proportion of defective products produced by the factory.

APPENDIX 4A:
The Summation Operation

Several algebraic rules apply to the summation operation. We shall list here those that are relevant to the material covered in this book.

RULES FOR SUMMATION OPERATIONS

Rule 1: Let X and Y be two variables. Then:

$$\sum_{i=1}^{N} (X_i + Y_i) = \sum_{i=1}^{N} X_i + \sum_{i=1}^{N} Y_i$$

Rule 2: Let a be a constant. Then:

$$\sum_{i=1}^{N} a = \underbrace{a + a + \cdots + a}_{N \text{ times}} = Na$$

Rule 3:

$$\sum_{i=1}^{N} (a + X_i) = \sum_{i=1}^{N} a + \sum_{i=1}^{N} X_i = Na + \sum_{i=1}^{N} X_i$$

Rule 4:

$$\sum_{i=1}^{M} aX_i = a \sum_{i=1}^{N} X_i$$

APPENDIX 4B:
The Geometric Mean

Donna Rogers paid $1,000 in personal income tax two years ago and an additional $3,000 a year ago. The total for the two years was $4,000 and the arithmetic annual average was $4,000/2 = $2,000. Had Miss Rogers paid $2,000 in the first year and an additional $2,000 in the second, her total tax payment for the two years would have been exactly the same: $4,000.

Consider now an investment made by Miss Rogers two years ago. She invested $100 and the value of that investment declined 20 percent in the first year. The value, then, was $80 at the end of the first year. During the second year the value of the investment rose 50 percent—from $80 to $120. The change over the two-year period was an increase of 20 percent in the value of the investment (from $100 to $120). The arithmetic average of the annual percentage changes was $(-20\% + 50\%)/2 = 15\%$. Had Miss Rogers earned 15 percent on her investment in the first year and 15 percent again in the second year, *the final value of the investment at the end of the second year would not have been $120.* It would have been $132.25, calculated as follows: the original $100 would have increased in value to $115 at the end of the first year (an increase of 15 percent: $100 · 1.15 = $115) and to $132.25 at the end of the second year (an additional increase of 15 percent: $115 · 1.15 = $132.25).

The second example clearly shows that when percentage changes of a variable's value over time are involved, the arithmetic mean of the percentage changes can be misleading. It should be replaced by the **geometric mean.**

Let the value of the variable at the *beginning* of the first period be denoted by V_0. Let the value at the *end* of the first period be V_1, at the *end* of the second period be V_2, and so on. The rate of change in the variable's value in period i, denoted by R_i, is given as follows:

Period (i)	Rate of change
1	$R_1 = \dfrac{V_1}{V_0} - 1$
2	$R_2 = \dfrac{V_2}{V_1} - 1$
3	$R_3 = \dfrac{V_3}{V_2} - 1$
.	. .
.	. .
.	. .
n	$R_n = \dfrac{V_n}{V_{n-1}} - 1$

The geometric mean of the n periods is given by μ_g:

$$\mu_g = \sqrt[n]{(1 + R_1)(1 + R_2) \cdots (1 + R_n)} - 1$$

$$= \sqrt[n]{\frac{V_1}{V_0}\frac{V_2}{V_1} \cdots \frac{V_{n-1}}{V_{n-2}}\frac{V_n}{V_{n-1}}} - 1 = \sqrt[n]{\frac{V_n}{V_0}} - 1$$

Reconsidering Miss Rogers' investment, we get

Period	Rate of change
1	$R_1 = \dfrac{80}{100} - 1 = -0.2$
2	$R_2 = \dfrac{120}{80} - 1 = 0.5$

so that the geometric mean is

$$\mu_g = \sqrt[2]{(1 - 0.2)(1 + 0.5)} - 1 = \sqrt[2]{1.2} - 1 = 0.095$$

If Miss Rogers' investment's value had appreciated 9.5 percent in each of the two years, it would have had a value of $109.50 at the end of the first year ($109.50 = $100 · 1.095) and $119.90 at the end of the second year ($119.90 = $109.5 · 1.095). The deviation from $120 is due to rounding.

If you still have any lingering doubts about the applicability of the geometric mean, you will certainly be convinced after considering another of Miss Rogers' investments. She invested $100 and the value of the investment declined 50 percent (down to $50) in the first year. During the second year the value of the investment rose 100 percent, to $100. All in all, over the two-year period she gained nothing. She started with $100 and still had only $100 after two years. The arithmetic average of the annual percentage change in the value of the investment was $(-50\% + 100\%)/2 = 25\%$. This figure gives the erroneous impression that the value of her investment increased over the two years. The geometric average is as follows:

Period	Rate of change
1	$R_1 = \dfrac{50}{100} - 1 = -0.5$
2	$R_2 = \dfrac{100}{50} - 1 = 1.0$

so that

$$\mu_g = \sqrt[2]{(1 - 0.5)(1 + 1.0)} - 1 = \sqrt[2]{1} - 1 = 0$$

Indeed, where the rate of change of a variable over time is concerned, the geometric mean provides a correct average, whereas the arithmetic mean does not.

Problems

4.1. The arithmetic average, the median, and the mode are three location parameters. They do not measure precisely the same thing. What are the differences among them?

4.2. Demonstrate, using your own numerical example, that the median is not affected by extreme values in the population.

4.3. Calculate the median of the frequency distribution of Example 4.5. Show that the median is located exactly underneath the intersection of the "less than" and "more than" ogives.

4.4. Calculate the 30th, 40th, 65th, and 85th percentiles of the frequency distribution of Example 4.5.

4.5. What are the unique characteristics of the mode, not shared by the other location parameters?

4.6. Multimodal distributions are usually not good subjects for statistical analysis. Why?

4.7. When the mean appears to the left of the median, what is implied with respect to the distribution's skewness?

4.8. If a variable X is measured in days, what are the units by which its variance is measured? What are the units by which its standard deviation is measured? Explain.

4.9. If a variable X is measured in dollars, what are the units by which the third moment, μ_3, is measured? What are the units by which the coefficient of skewness, $\dfrac{\mu_3}{\sigma^3}$, is measured? Explain.

4.10. Figure P4.10 depicts electricity production in the United States by type of fuel and by region in 1978, in percentages.

  (*a*) Calculate the mean and variance of the percentage of each of the five energy sources across the 10 regions. Which source of energy exhibits the highest variability?

  (*b*) Compare the means you obtained in part *a* with those labeled "Total for nation" in Figure P4.10. Are they the same? If not, has the *Newsweek* statistician erred? Explain.

4.11. The following table presents a summary of the lifetimes of 6,000 tires of a certain brand (measured in miles):

Lifetime (thousands of miles)	Number of tires
15.0–19.9	600
20.0–24.9	1,100
25.0–29.9	1,800
30.0–34.9	1,200
35.0–39.9	700
40.0–49.9	600
	6,000

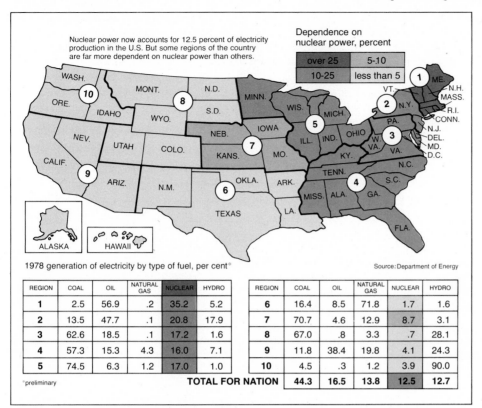

Figure P4.10

U.S. electricity production by type of fuel and region, 1978 (percent)

Nuclear power now accounts for 12.5 percent of electricity production in the U.S. But some regions of the country are far more dependent on nuclear power than others.

Dependence on nuclear power, percent

| over 25 | 5-10 |
| 10-25 | less than 5 |

1978 generation of electricity by type of fuel, per cent*

Source: Department of Energy

REGION	COAL	OIL	NATURAL GAS	NUCLEAR	HYDRO
1	2.5	56.9	.2	35.2	5.2
2	13.5	47.7	.1	20.8	17.9
3	62.6	18.5	.1	17.2	1.6
4	57.3	15.3	4.3	16.0	7.1
5	74.5	6.3	1.2	17.0	1.0

REGION	COAL	OIL	NATURAL GAS	NUCLEAR	HYDRO
6	16.4	8.5	71.8	1.7	1.6
7	70.7	4.6	12.9	8.7	3.1
8	67.0	.8	3.3	.7	28.1
9	11.8	38.4	19.8	4.1	24.3
10	4.5	.3	1.2	3.9	90.0
TOTAL FOR NATION	**44.3**	**16.5**	**13.8**	**12.5**	**12.7**

*preliminary

Source: *Newsweek*, April 16, 1979. Copyright 1979 by Newsweek, Inc. All Rights Reserved. Reprinted by Permission.

(*a*) Find the median lifetime of this tire population and briefly explain its meaning.

(*b*) Calculate the average lifetime of the tires. Is the average different from the median? What is implied about the population's skewness?

(*c*) Calculate the skewness and coefficient of skewness of the distribution.

4.12. The Beta Company sells three kinds of products: 1, 2, and 3. The net profit per item and total number of items sold of each kind are as follows:

Kind of product	Net profit per item	Number of items sold
1	$ 20	10,000
2	720	5,000
3	1,200	900
		15,900

What is the average net profit per item?

4.13. The production workers of a given firm are ranked *A*, *B*, *C*, and *D*. The following is the annual salary per worker by rank and the percentage distribution of ranks:

Rank	Annual salary per worker	Workers in rank (percent)
A	$13,000	6.2%
B	12,000	21.8
C	10,500	56.4
D	9,000	15.6
		100.0%

What is the weighted average annual salary of the workers?

4.14. Good and Fast Service Bank, Inc., is providing free checking account service. Since costs have surged in recent months, however, management has decided that a monthly charge is inevitable and has initiated a study of checking account balances to determine a fair monthly charge. The study shows the following frequency distribution by average monthly balance:

Average monthly balance (dollars)	Frequency
$ 0.00– 99.99	2,000
100.00– 199.99	2,500
200.00– 499.99	10,000
500.00– 799.99	2,500
800.00– 999.99	1,100
1,000.00–5,000.00	1,900
	20,000

Calculate the following:

(a) The average monthly balance for the 20,000 accounts.
(b) The median.
(c) The variance of the monthly balance.
(d) The mode.

4.15. The following is a frequency distribution of 200 companies over their last year's earnings (in millions of dollars):

Net earnings	Frequency
$(−4.00)–(−0.01)	20
0.00– 3.99	50
4.00– 7.99	60
8.00–11.99	30
12.00–15.99	40
	200

Calculate the following:

(a) The mean net earnings.
(b) The variance and standard deviations.
(c) The coefficient of variation.
(d) The 15th percentile.
(e) The interquartile range.
(f) The mode.

4.16. Twenty employees are paid the following monthly salaries:

$397	$444
480	567
830	457
610	390
642	610
560	480
437	478
425	667
550	565
330	602

(a) Compute the range.
(b) Compute the first and third quartiles and determine the interquartile range.
(c) Compute the variance and standard deviation.

4.17. Four hundred businesses in a given city had the following profit distribution in the last year:

Profit (thousands of dollars)	Frequency
(-100)–(-51)	5
(-50)–(-1)	35
0–49	260
50–99	47
100–149	34
150–199	19
	400

(*a*) Find the population's average profit.

(*b*) Find the median profit, and show on a chart that the location of the median can be found by the intersection of the "less than" and "more than" ogives.

(*c*) What is the variance of the profit of the above population?

(*d*) Determine the population coefficient of variation.

(*e*) Is the profit distribution skewed? Can you support your answer by calculating the skewness?

4.18. Table P4.18 is a frequency distribution of people who dined in a given restaurant in a period of 90 days:

TABLE P4.18

Number of people	Frequency of days
0–24	2
25–49	1
50–74	10
75–99	10
100–124	8
125–149	16
150–174	13
175–199	15
200–224	5
225–249	4
250–274	3
275–299	3
	90

For the population in Table P4.18, calculate:

(*a*) The mean.

(*b*) The median.

(*c*) The standard deviation.

(*d*) The interquartile range.

(*e*) The range between the 10th and 90th percentiles.

4.19. The monthly rates of return on investment in stocks of two industries are given below for the last five years:

Rate of return (%)	Food company stocks	Oil company stocks
(-5)–(-1)	5	5
0–4	5	5
5–9	40	20
10–14	5	20
15–19	5	10
	60	60

(*a*) Calculate the mean of the rate of return on the stocks in each of the two industries for the last five years.

(*b*) Calculate the variance of rates of return on each of the two distributions.

(*c*) Can you think of an economic reason that the means and variances of the rates of return on these two industries' stocks might be related to one another, as in this problem?

4.20. A population is examined for its quartiles and median. Suppose the first quartile, Q_1, is 60 and the third quartile, Q_3, is 120. Suppose also the median is 90.

(*a*) Is it necessarily true that the population is symmetrical? If your answer is positive, explain your reason. If it is negative, give a numerical example to demonstrate your answer.

(*b*) If $Q_1 = 60$ and $Q_3 = 120$ but the median is 80, is it possible that the skewness

$$\mu_3 = \frac{1}{N} \Sigma(X - \mu)^3$$ is equal to zero? Explain and demonstrate numerically.

4.21. Community Savings is a savings bank that makes mortgage loans to community residents. The bank loan portfolio includes loans made at a variety of interest rates, as follows:

Amount of loans (thousands of dollars)	Interest rates
$500	$6\frac{1}{2}\%$
800	7
50	$8\frac{3}{4}$
1,070	$9\frac{3}{4}$

What interest rate does the bank earn on its loan portfolio?

4.22. Marketable securities are securities that can be sold by the original purchaser. The time to maturity is the length of time that must elapse before a security matures. The amount (in millions of dollars) of marketable government bonds with various times to maturity in the years 1977, 1978, and 1979 was as follows:

Year	Up to 1 year	1–4 years	5–9 years	10–19 years	20–30 years
1977	161,329	113,319	33,067	8,428	10,531
1978	163,819	132,993	33,500	11,383	14,805
1979	181,883	127,574	32,279	18,489	20,304

Source: *U.S. Treasury Bulletin,* April 1980.

For each of three years (1977, 1978, and 1979), calculate:

(*a*) The average time to maturity of the above bonds.

(*b*) The standard deviation of the time to maturity.

(*c*) The skewness and coefficient of skewness of the time to maturity.

4.23. The following table shows ownership of outstanding U.S. public debt (notes and bonds issued by the federal government and held by the public):

Held by	Amount held in March 1980 (billions of dollars)
Commercial banks	$75.5
Insurance companies	0.6
Mutual savings banks	0.8
Corporations	7.5

Source: *U.S. Treasury Bulletin,* April 1980.

What location parameter is appropriate to use in describing the data? Why?

4.24. The yearly compensation (salary and other benefits) of financial analysts in a large company is shown in Table P4.24. For this population, calculate:

(a) The mean.
(b) The standard deviation.
(c) The coefficient of variation.
(d) The skewness.
(e) The 15th percentile.
(f) The median.

TABLE P4.24

Compensation (thousands of dollars)	Frequency
$ 0.0– 9.9	2
10.0–19.9	13
20.0–29.9	15
30.0–39.9	5
40.0–49.9	5
50.0–59.9	6
60.0–69.9	1
70.0–79.9	1
	48

4.25. Of 11,194 residents of a city, 5,677 do not subscribe to any daily newspaper; 5,049 residents are subscribers to one daily newspaper and 468 subscribe to two daily papers. What is the average number of daily papers per resident to which the city residents subscribe? What is the variance? In what units is the variance measured?

4.26. Suppose the frequency distribution of the net profits of some 300 firms is constructed for 1981 and 1982. Suppose the two frequency distributions are exactly the same. Is it possible to determine the mean *change* of net profit of these firms between 1981 and 1982 by comparing the two distributions? Is it possible to determine the variance of the change? The skewness of the change? Explain by means of a numerical example.

4.27. The average household electric bill in a given population is $75 per month. The standard deviation is $15. The average number of household members in the same population is 3.2 and the standard deviation is 0.6. The households exhibit greater variability with respect to which of the two variables? Explain your answer.

4.28. Commercial banks hold some of their deposit liabilities in cash. Consider the following table, which contains information about the percentage of funds held in cash by commercial banks of various sizes.

Total deposits (millions of dollars)	Percent of all banks	Percent of deposits held in cash
$0–49	20%	10.5%
50–99	25	9.9
100–149	18	9.0
150–199	13	9.1
200–249	10	8.0
250–299	9	8.2
300–349	5	8.0
	100%	

What percentage of all the deposits in the above banks is held in cash?

4.29. Consider the following prices for Midwest Corporation's stock at the end of ten trading days:

Day	Price
1	42
2	$42\frac{1}{2}$
3	$41\frac{3}{4}$
4	$41\frac{1}{4}$
5	$40\frac{3}{4}$
6	42
7	42
8	$42\frac{3}{4}$
9	$43\frac{1}{2}$
10	$44\frac{1}{4}$

(a) Calculate the mean, variance, standard deviation, and skewness of the price.

(b) For each day (starting at day 2) calculate the price change from the previous day. Calculate the mean, variance, standard deviation, and skewness of the price changes.

(c) For each day (starting at day 2), calculate the percentage change from the previous day by computing the price change as a percentage of the previous day's price. For example, the price change in day 2 is $42\frac{1}{2} - 42 = \frac{1}{2}$. The percentage change is $\frac{\frac{1}{2}}{42} = 0.0119$, or 1.19 percent. Calculate the mean, variance, standard deviation, and skewness of the percentage change.

4.30. In 1980 the chairmen of U.S. oil companies were compensated (in salary and other benefits) as shown in Table P4.30.

(a) Compute the average compensation of the chairmen.

(b) What is the median compensation?

(c) What is the standard deviation of the compensation of the chairmen?

TABLE P4.30

Company	Chairman	Compensation (thousands of dollars)
Atlantic Richfield Co.	Robert O. Anderson	$754
Conoco, Inc.	Ralph E. Bailey	614
Exxon Corp.	C. C. Garvin, Jr.	640
Gulf Oil Corp.	Jerry McAfee	781
Mobil Corp.	Rawleigh Warner, Jr.	498
Phillips Petroleum Co.	W. F. Martin	350
Shell Oil Co.	John F. Bookout	740
Standard Oil Co. of California	H. J. Haynes	450
Standard Oil Co. (Indiana)	J. E. Swearingen	380
Standard Oil Co. (Ohio)	A. W. Whitehouse, Jr.	486
Texaco, Inc.	Maurice F. Granville	453
Union Oil Co. of California	Fred L. Hartley	440

Data from *Business Week*, May 11, 1981, pp. 58–78.

4.31. The use of coupons by consumers was studied by a marketing team of a large firm that was reevaluating its policy on distribution of coupons. The distribution of coupon redemption by 650 residents in a one-month period was as shown in Table P4.31:

(*a*) Determine the average number of coupons redeemed per resident.

(*b*) What is the total value of the coupons redeemed and what is their average value?

TABLE P4.31

Number of coupons redeemed in one month	Coupon value				
	5¢	10¢	15¢	20¢	Total
0	129	406	473	553	1,561
1	83	18	16	10	127
2	212	118	98	40	468
3	162	96	51	27	336
4	64	12	12	20	108
Total	650	650	650	650	2,600

4.32. Tom invested $12,000 in the stock market for one year. Six thousand dollars was invested in bonds, on which his rate of return was 6 percent; $3,000 was invested in stock on which his rate of return was negative: −3 percent; and $3,000 was invested in stock on which the rate of return was 15 percent. What was Tom's average rate of return on his investment? Did you use the arithmetic average, the weighted average, or the geometric average? Explain your choice.

4.33. Tony invested $12,000 in bonds for one year. At the end of the year, his investment was worth 6 percent more: $12,720. He reinvested the money in stock for another year, and his return this time was −3 percent, so his investment at the end of the second year was worth $12,338.40. He then reinvested for another year and his return was 15 percent, so at the end of the third year the investment was worth $14,189.16. What was Tony's average rate of return per year?

4.34. Would your answer to Problem 4.33 be any different if the return in the second year were −30 percent (rather than −3 percent) and in the third year 59.357 percent (rather than 15 percent)? Explain.

4.35. The gross national product (GNP) of Fluctuania has fluctuated over the last few years. The percentage change of GNP over the previous year's level for six recent years is given below. Find the arithmetic mean and the geometric mean of the percentage change. Which is more appropriate?

Year	Percentage change in GNP
1977	−50%
1978	+50
1979	−25
1980	+25
1981	−40
1982	+22

4.36. Verify your answer to Problem 4.35 in the following way: assume that Fluctuania's GNP in 1976 was $10 billion, and derive the BNP level for *each* of the years 1977 through 1982, once by applying the actual annual percentage changes as given in Problem 4.35, once by applying to each year a percentage change equal to the arithmetic mean, and once by applying to each year a percentage change equal to the geometric mean. Of the last two methods, which results in a 1982 GNP level identical to that obtained by applying the actual GNP percentage changes?

4.37. Table P4.37 presents a frequency distribution of the percentage change of average stock prices of 32 industries for the week of October 12, 1981.

TABLE P4.37

Percentage change in average stock prices	Frequency of industries
(−8.00)–(−7.01)	1
(−7.00)–(−6.01)	0
(−6.00)–(−5.01)	3
(−5.00)–(−4.01)	2
(−4.00)–(−3.01)	2
(−3.00)–(−2.01)	6
(−2.00)–(−1.01)	4
(−1.00)–(−0.01)	6
0.00 – 0.99	4
1.00 – 1.99	1
2.00 – 2.99	1
3.00 – 3.99	0
4.00 – 4.99	1
5.00 – 5.99	1
	32

Source: *Barron's*, October 19, 1981, p. 126. Reprinted by permission of *Barron's*, © Dow Jones & Company, Inc. (Oct. 19, 1981). All Rights Reserved.

(*a*) Calculate the average percentage change in average stock prices.

(*b*) Calculate the variance and standard deviation of the percentage change.

(*c*) Determine the median and mode.

4.38. Thirty-one stores located in various areas around the country were surveyed for the price of a given product. Table P4.38*a* shows the price frequency distribution, and Table P4.38*b* shows the same data with the prices organized in intervals of $2 each.

TABLE P4.38*a*

Price	Frequency
$25.00	1
26.00	3
27.89	4
28.59	2
29.00	3
29.99	3
30.00	3
30.99	7
31.00	2
32.90	1
33.00	1
34.00	1
	31

TABLE P4.38*b*

Price interval	Frequency
$25.00–26.99	4
27.00–28.99	6
29.00–30.99	16
31.00–32.99	3
33.00–34.99	2
	31

(a) Denote the price of the product by X and calculate its mean and variance, once using Table P4.38a and once using Table P4.38b.

(b) Explain the difference in the results obtained by using the two tables. Which gives more accurate results? Which is easier to calculate?

(c) Use Table P4.38b to calculate the median and the first and third quartiles.

(d) Compute the skewness (third moment) and the coefficient of skewness.

4.39. The number of employees of three firms is given below, by age group:

	Firm A	Firm B	Firm C
29 and under	550	50	12,222
30–39	799	27	10,479
40–49	437	32	8,334
50 and over	242	10	6,860

(a) Determine the proportion of employees 40 years or older in each of the three firms.

(b) What is the proportion of employees 40 years or older in the three firms combined?

(c) Of all of the employees in the three firms, what proportion is younger than 30 and employed by firm A?

4.40. The arithmetic mean and the geometric mean are two location parameters. The geometric mean is not always applicable. When is it applicable?

PART 2

PROBABILITY AND DISTRIBUTIONS (DEDUCTION)

We use deductive statistics when we draw conclusions about an individual observation, given complete information about the population. In Part 2 of this book, we develop the methods by which such conclusions can be drawn.

Chapter 5 introduces and explains basic probability concepts. Chapter 6 deals with the concept and analysis of a random variable. Chapter 7 presents some important discrete distributions: Chapter 8 is devoted to the normal distribution—a distribution that plays a central role in statistical analysis.

The applications, examples, and problems presented in Part 2 concern the determination of insurance premiums by means of the conditional probability concept; budgeting research and development; the need for capital of a company drilling for oil; the "safety first" investment principle; and many other topics.

CHAPTER FIVE OUTLINE

5.1 The Random Experiment and the Sample Space
Possible Outcomes of the Random Experiment
Basic Outcomes and the Sample Space

5.2 Events
Venn Diagrams
Parts of the Venn Diagram

5.3 The Meaning of Probability

5.4 The Addition Rule

5.5 Conditional Probability and the Multiplication Rule
The Joint Probability Table

5.6 Independent Events
Graphic Representation of Independence and Dependence of Events

5.7 More on Conditional Probability: Bayes' Theorem
What Is Conditional Probability Good For?

5.8 Symmetrical Sample Spaces and Counting Techniques
The Addition Rule and the Multiplication Rule for Counting
Permutations
Combinations

5.9 Application: Determining Insurance Premiums by the Use of Conditional Probabilities

Key Terms

random experiment
basic outcome
sample point
sample space
univariate sample space
bivariate sample space
multivariate sample space
event
simple event
composite event
Venn diagram
intersection
empty event
union
complement

partition
objective probability
subjective probability
addition rule
conditional probability
marginal probabilities
joint probabilities
multiplication rule
Bayes' theorem
independent event
total probability equation
symmetrical sample spaces
permutations
combinations

5
BASIC PROBABILITY CONCEPTS

It is difficult to imagine a world of certainty, in which the outcomes of future processes and phenomena are known in advance. Such a world, if it existed, would doubtless be excruciatingly dull. We are apparently lucky, then, to live in the more exciting world of uncertainty. Paradoxically, however, we are constantly seeking to minimize the uncertainty facing us and to foresee the outcomes of processes subject to chance. We seem to like uncertainty, but would prefer a little less of it than we usually have.

While the total elimination of uncertainty is usually impossible, prudent assessment of the chances involved in uncertain events is important if we are to make "good" decisions. While we may have to face uncertainty regarding the *outcomes* of some phenomena, we are often able to eliminate uncertainty with respect to the *chances* of such outcomes. Thus if we inquire about the chances or *probabilities* of the outcome of a given process, and find that outcome A has a 10 percent chance of occurring and outcome B has a 90 percent chance, we are better able to assess the process's probable results than we would be if these chances were themselves unknown.

Probability assessment is easy in some simple situations, but requires careful examination in the complex problems that are typical of the business world. This chapter is devoted to the elementary concepts and rules of probability; these will provide the basis for a better understanding of uncertain processes.

5.1 The Random Experiment and the Sample Space

First, let us consider the concept of the **random experiment**—*any process that has an uncertain outcome or outcomes.* The outcomes of such an experiment will generally differ from one run of the experiment to the next.

When we deposit a coin in a coffee machine, we are conducting an experiment. We may get coffee in return, our money back, or neither. Similarly, tomorrow's weather may be viewed as a random experiment with a variety of possible outcomes. And predicting next year's level of profit of a company is an interesting business example of a random experiment.

POSSIBLE OUTCOMES OF THE RANDOM EXPERIMENT

While our primary objective in this chapter is to determine how *probable* a given outcome or set of outcomes is, we shall first be concerned with all the outcomes that are *possible* as a result of the experiment.

The possible outcomes of some experiments are trivially determined. When we flip a coin, for example, the possible outcomes are simply "heads" and "tails."

In many situations, however, the possible outcomes of an experiment are not trivial. They are a function of what we have decided to define as an "outcome." Consider, for example, the change in price of Stock *A* on the next trading day on the New York Stock Exchange. The number and scope of the possible outcomes of this experiment depend on our choice of definitions. We may choose to differentiate only among the following outcomes: "the price goes up," "the price is unchanged," and "the price goes down." If we wish, however, we may select different classifications for the possible outcomes of the same experiment distinguishing, for example, between "the stock's price goes up by 3 percent or more," "the stock's price goes up by less than 3 percent," etc.

In any random experiment, our particular choice of classification should depend on the nature and goal of the study in which we are engaged. For that reason we often talk about the *outcomes of interest* of a random experiment.

BASIC OUTCOMES AND THE SAMPLE SPACE

Each of the possible outcomes of interest in a random experiment is called a **basic outcome.** We note that *one and only one* basic outcome will occur as a result of the experiment. A *basic outcome* is frequently referred to as a **sample point,** and *the set of all possible sample points* of an experiment is known as the **sample space.** We distinguish among *univariate, bivariate,* and *multivariate* sample spaces. A **univariate sample space** is one in which *the sample points are defined by classification of the outcome of one variable,* such as the price change of one stock on a given trading day. A **bivariate sample space** is one in which *the sample points are defined by cross-classification of two variables.* The combination of price changes of two stocks during a given trading day defines a bivariate sample space; so does the price change of a single stock on two particular trading days. A **multivariate sample space** is one in which *the sample points are defined by cross-classification of more than two variables.*

As a simple example of a univariate sample space, consider the classification of people by the brand of toothpaste they most often use. The sample space includes the following sample points:

Colgate
Crest
Close-Up
Ultra Brite
Other

Let us now consider a bivariate sample space. Suppose we have a two-way classification of college students by seniority (freshman, sophomore, junior, senior) and by cumulative grade-point average (GPA). The cross-classification is illustrated in Table 5.1.

Twelve basic outcomes are shown in Table 5.1. For easy reference, we denote each combination of seniority and GPA by the notation $O_1, O_2, \ldots, O_{12}$. So, for example, O_6 is the notation for a sophomore with a GPA in the range 2.00–2.99.

The twelve basic outcomes shown in Table 5.1 form a sample space, since every student must fit into one (and only one) of the twelve outcomes. Note that some basic outcomes may be redundant, in the sense that no student will have their characteristics. The outcome O_4, for instance, may be redundant in that we expect no senior student to have a cumulative GPA below 2.0.

Finally, let us consider a multivariate sample space—one defined by a cross-classification of more than two variables. Here is a simple example: we may classify U.S. citizens into groups defined by sex, age, occupation, and location of residence.

As we have said, our main interest in this chapter is in the probability of the various outcomes. We need to look at combination of outcomes (that is, events), however, before we can consider their probability.

TABLE 5.1
Example of a Bivariate Sample Space: College Students by Seniority and Grade-Point Average (GPA)

Seniority GPA	Freshman	Sophomore	Junior	Senior
0.00–1.99	O_1	O_2	O_3	O_4
2.00–2.99	O_5	O_6	O_7	O_8
3.00–4.00	O_9	O_{10}	O_{11}	O_{12}

5.2 Events

In Section 5.1 we explained how we classify the random experiment's results into what we termed basic outcomes. Quite often we will be interested in questions involving the possibility or the probability of occurrence of results consisting of more than just one basic outcome. To facilitate the handling of problems of this sort, let us consider the concept of event. An **event** is *the result of a random experiment consisting of one or more basic outcomes.* If an event consists of only *one basic outcome*, it is a **simple event;** if it consists of *more than one basic outcome*, it is a **composite event.**

EXAMPLE 5.1

A newsstand operator is selling *Easy Life* magazine, and we wish to observe two days of operations. Suppose we define a bivariate sample space in the following way:

Number of magazines sold on first day

		0–5	6–10	11 or more
Number of magazines sold on second day	0–5	O_1	O_2	O_3
	6–10	O_4	O_5	O_6
	11 or more	O_7	O_8	O_9

Many events involving one or more basic outcomes may be defined over this sample space. Some examples are given in Table 5.2.

TABLE 5.2
Description and Basic Outcomes of Four Events

Event	Description	Basic outcomes of event
A	The number of magazines sold on the first day is less than or equal to 10.	$O_1, O_2, O_4, O_5, O_7, O_8$
B	The number of magazines sold on the second day is greater than or equal to 11.	O_7, O_8, O_9
C	The number of magazines sold on each of the two days does not exceed 5.	O_1. Event C consists of only one basic outcome and thus it is a *simple event*.
D	The number of magazines sold on the first day is greater than or equal to 6 and the number sold on the second day is between 6 and 10 (inclusive).	O_5, O_6

VENN DIAGRAMS

It is often useful to describe events in the sample space on a **Venn diagram** such as the one presented in Figure 5.1, showing the events of Table 5.2. The rectangle itself represents the *sample space*, while the points in it stand for the *basic outcomes*. The sample space is denoted by U, which stands for the "universal set," and the basic outcomes of the two-day sale of magazines are presented by nine points within U.

We shall frequently use a more schematic Venn diagram, in which the basic outcomes are omitted and only relevant events are plotted in the sample space. Events are described by areas enclosed in circles or other geometrical forms. Figure 5.2, for example, shows events A and D of Table 5.2 along with all the basic outcomes in the sample space. As we can see, events A and D overlap, since they have a common basic outcome, O_5.

In general, we distinguish among the following three possible relationships between two events:

U

Figure 5.1

Venn diagram
showing nine
different sample
points

1. The two events have no basic outcomes in common. If this is the case, the events are said to be *mutually exclusive*. Consider, for example, the events "a person is less than 30 years old" and "that person is 30 or older." Since no one person can be both younger than 30 years and (at the same time) 30 years or older, these two events are mutually exclusive. Another example is "driving a Buck" and "driving a Chevy" (same person, same time). Figure 5.3 presents two mutually exclusive events, *A* and *B*. The basic outcomes are not presented by dots in this diagram and will not be presented from here on. The two events in Figure 5.3 do not overlap because, being mutually exclusive, they cannot be observed together.

2. One event entirely overlaps the other. In Figure 5.4 event *A* overlaps the entire event *B*, so that all of the basic outcomes in *B* are in *A* as well; not all the basic outcomes in *A*, however, are in *B*. Notationally we write $B \subset A$, meaning that all of *B* is included in *A*. As an example, suppose that *B* represents all companies in the steel industry that have up to \$100 million in net assets and *A* represents all companies in the steel industry with up to \$200 million in net assets. Obviously all the companies that belong to *B* also belong to *A*, while companies that belong to *A* are not necessarily included in *B*.

3. The two events partially overlap one another. Figure 5.5 shows this kind of relationship between events. Here some basic outcomes are common to *A* and *B*, but some basic outcomes in *A* are not in *B*, and others are in *B* but not in *A*. As an example, suppose the experiment

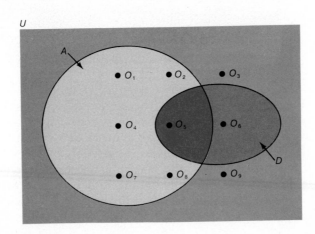

U

Figure 5.2

Events *A* and *D*
and their basic
outcomes

Figure 5.3

Two mutually
exclusive events

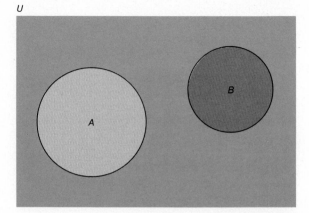

Figure 5.4

One event
entirely
overlapping
another

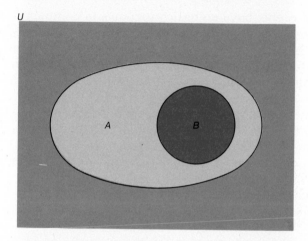

Figure 5.5

Two partially
overlapping
events

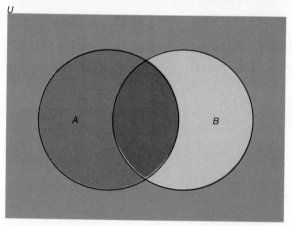

involves the random selection of people in Cincinnati for a marketing survey. Suppose *A* is the event "the person selected is a woman" and *B* is the event "the person selected is the parent of at least two children." In this case, *A* and *B* relate to one another as depicted in Figure 5.5: some of the people are women who are parents of at least two children (and thus belong to both *A* and *B*); some are women who do not have at least two children (and thus belong to *A* but not to *B*); and finally, some are parents of at least two children, but are men (and so they belong to *B* but not to *A*).

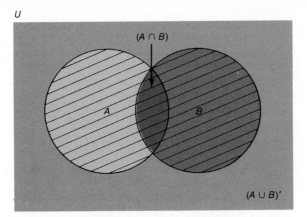

Figure 5.6

Intersection, union, and complement of two events

(A ∪ B) is represented by the hatched area.

Case 3 may be regarded as the more general case, with cases 1 and 2 being considered special cases of 3.

PARTS OF THE VENN DIAGRAM

The area in which two (or more) events overlap is called the **intersection** of the events: $A \cap B$ (read "A intersection B," or "A and B"). In Figure 5.6 the intersection is represented by the darkest region.

An *event that does not include any basic outcome* is an impossible event and is referred to as the **empty event.**[1] The symbol we use for the empty event is $\varnothing$. If events A and B are mutually exclusive, they have no basic outcomes in common, and it follows that in that case we have $A \cap B = \varnothing$. Suppose Xerox employees' wages range between $6,000 and $100,000 annually. Select an employee and define the following event: "the employee's wage is $150,000 annually." The event is empty, since none of the employees earns more than $100,000 per year.

The *area that encompasses the overlapping as well as the nonoverlapping sections of A and B combined* is called the **union** of events: $A \cup B$ (read "A union B," or "A and/or B"). The union of A and B is the entire hatched area in Figure 5.6.

The **complement** of a given event (denoted by a prime next to the event's symbol) is the set of *all basic outcomes that are not included in the event*. If A is the event "the train arrived on time," then A' is the event "the train did not arrive on time." The complement of $A \cup B$ in Figure 5.6, $(A \cup B)'$, is represented by the unhatched section of the sample space.

Consider Figure 5.7. Here the sample space, U, is *partitioned* by events A_1, A_2, A_3, A_4, and A_5. The events form a **partition** because they are *mutually exclusive* and also *collectively exhaustive* in the sense that their union constitutes the entire sample space. For example, the annual salary of an individual selected at random out of a given population will be in one of the following categories: $0–$10,000 (event A_1), $10,001–$20,000 (event A_2), $20,001–$30,000 (event A_3), $30,001–$40,000 (event A_4), $40,001 and over

---

[1] The reason an empty event is impossible is that basic outcomes were defined so that one of them must occur as a result of the experiment. Note that the empty event is not really an "event" by our definition: we defined event to be a result of a random experiment consisting of one or more basic outcomes. The term "event" is therefore used as an exception to the general definition.

Figure 5.7

Sample space partitioned by five mutually exclusive events

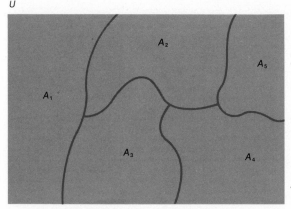

(event A_5). Obviously events A_1, A_2, A_3, A_4, and A_5 are mutually exclusive and collectively exhaustive. (Note, as a trivial example, that an event and its complement are mutually exclusive and collectively exhaustive.)

Before proceeding, we shall discuss another example and demonstrate the use and the usefulness of the concepts developed so far.

EXAMPLE 5.2

For the purpose of examining the proposed expansion of a major southeastern airport, data on passenger arrivals and departures have been amassed. The daily passenger arrivals and departures are grouped into four classes of interest: 0–1,000, 1,001–5,000, 5,001–10,000, and 10,001 or more. This classification defines a bivariate sample space, as shown in Table 5.3.

TABLE 5.3
Air Traffic, by Arrivals and Departures

Arrivals / Departures	0–1,000	1,001–5,000	5,001–10,000	10,001 or more
0–1,000	O_1	O_2	O_3	O_4
1,001–5,000	O_5	O_6	O_7	O_8
5,001–10,000	O_9	O_{10}	O_{11}	O_{12}
10,001 or more	O_{13}	O_{14}	O_{15}	O_{16}

Let us define several events in the sample space referring to the "results" of a daily operation.

- A There were no more than 1,000 arrivals.
- B There were no more than 10,000 arrivals.
- C There were 1,001–5,000 arrivals.
- D There were 1,001–5,000 departures.
- E There were 10,001 or more departures.
- F There were no more than 1,000 departures.

Table 5.4 identifies the basic outcomes, descriptions, and diagrammatic presentations of some relationships among these events.

TABLE 5.4
Events, Descriptions, and Venn Diagrams of Relationships Among Events

Event	Basic outcomes included	Description	Venn diagram
B'	O_4, O_8, O_{12}, O_{16}	10,001 or more arrivals, regardless of number of departures.	
$B \cap F$	O_1, O_2, O_3	0–1,000 departures and up to 10,000 arrivals.	
$A \cup F$	$O_1, O_2, O_3, O_4, O_5, O_9, O_{13}$	0–1,000 arrivals and/or 0–1,000 departures.	
$(A \cup F)'$	$O_6, O_7, O_8, O_{10}, O_{11}, O_{12}, O_{14}, O_{15}, O_{16}$	1,001 or more arrivals and also 1,001 or more departures.	
$A \cap B$	O_1, O_5, O_9, O_{13}	Up to 1,000 arrivals.	
$D \cap F$	None	This is an empty event (ϕ). D and F cannot be observed on the same day: they are mutually exclusive.	
$A \cap E'$	O_1, O_5, O_9	0–1,000 arrivals and up to 10,000 departures.	
$(C \cap D)'$	All basic outcomes except O_6	All arrival and departure combinations except 1,001–5,000 of both.	

5.3 The Meaning of Probability

We distinguish between two types of random experiment: one that can be repeated many times under the same circumstances (such as flipping a coin) and one that cannot be repeated many times under the same conditions (such as the process that generates McDonald's profit next year). The outcomes of these two types of experiments lead us to define *two types of probability*. When the experiment can be repeated numerous times under identical conditions, we talk about *objective* probability; when conditions change each time an experiment is repeated, we speak of *subjective* probability. The **objective probability** of an event is its *frequency of occurrence over many repeated experiments*. In many cases, objective probability cannot be assigned to events. Many events in business and economics can be assigned only **subjective probability.** It is the probability we assign to the outcome of experiments based on the available information. The growth of GNP next year is an experiment that can be observed only once, and thus cannot be assigned an objective probability. But with information concerning many economic variables, subjective probability may be assigned. Since individual and group (such as corporate management) behavior is a function of subjective probabilities, however, subjective probability is as important as objective probability. For both objective and subjective probabilities, the following properties and rules hold:

PROPERTY 1

$$0 \le P(A) \le 1$$

The probability of event *A* is greater than or equal to zero and less than or equal to one.

PROPERTY 2

$$P(U) = 1$$

The probability of the entire sample space, *U*, is equal to 1. This means that if the experiment takes place, one of the basic outcomes must occur.

PROPERTY 3

When $A \cap B = \varnothing$, $\qquad P(A \cup B) = P(A) + P(B)$

The probability of the union of two *mutually exclusive* events, *A* and *B*, is equal to the sum of the probability of the individual events.

$$P(A \cup A') = P(A) + P(A') = P(U) = 1$$

By combining Properties 2 and 3 we find that every event is mutually exclusive of its complement and their union exhausts the entire sample space.

5.4 The Addition Rule

In many situations it becomes necessary to calculate the probability of a combination of events. Venn diagrams help to explain the basic probability rules. First, since the rectangle *U* represents the entire sample space, its area is equal to 1 (of any desired unit). Second, the areas of all the events in the sample space are proportional to their probability. All events with a 25 percent probability of occurrence take 25 percent of the rectangle's area, and so on. Thus area represents probability in a Venn diagram.

Consider Figure 5.8, in which the union of events A and B (that is, $A \cup B$) is divided into three mutually exclusive subareas:

Area 1: The part of A that is outside of B (i.e., $A \cap B'$).
Area 2: The part of B that is outside of A (i.e., $A' \cap B$).
Area 3: The intersection of A and B (i.e., $A \cap B$).

The following two equations follow directly from our definitions:

$$P(A) = \text{Area 1} + \text{Area 3}$$
$$P(B) = \text{Area 2} + \text{Area 3}$$

The probability of the union of A and B [that is, $P(A \cup B)$] is represented by the sum of Areas 1, 2, and 3. We may obtain this probability by adding the probabilities of the two events A and B and subtracting the probability of the intersection. If we add up the probabilities of A and B without subtracting the probability of the intersection, we will be double-counting Area 3, since $P(A) + P(B) = (\text{Area 1} + \text{Area 3}) + (\text{Area 2} + \text{Area 3})$. Thus to get the probability of the union, we have to subtract the probability of the intersection (Area 3), which is expressed by the following:

ADDITION RULE FOR TWO EVENTS

$$P(A \cup B) = P(A) + P(B) - P(A \cap B) \qquad (5.1)$$

EXAMPLE 5.3

Suppose a country music singer is nominated as both Female Vocalist of the Year and Best Entertainer of the Year. Let A be the event "the singer is voted Female Vocalist of the Year," and let B be the event "the singer is voted Best Entertainer of the Year." Suppose also that the following probabilities are assumed: $P(A) = 0.10$; $P(B) = 0.20$; $P(A \cap B) = 0.03$.

What is the probability that the singer will win *at least one* of the two awards—that is, that she will win the first award *and/or* the second? We need to find the probability of the union of the events A and B: $P(A \cup B) = ?$ The answer is given by Equation 5.1:

$$P(A \cup B) = P(A) + P(B) - P(A \cap B) = 0.10 + 0.20 - 0.03 = 0.27$$

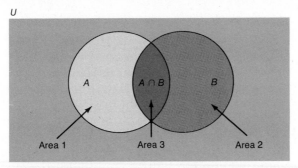

U

Figure 5.8

A graphic representation of the addition rule

Event A consists of the sum of Area 1 + Area 3.
Event B consists of the sum of Area 2 + Area 3.
It follows that $P(A \cup B) = P(A) + P(B) - P(A \cap B)$.

When the events A and B are mutually exclusive, so that $P(A \cap B) = 0$, the addition rule reduces to

$$P(A \cup B) = P(A) + P(B) \qquad \text{when } A \cap B = \emptyset \tag{5.2}$$

In this case, the generalization of the addition rule to n events follows immediately:

$$P(A_1 \cup A_2 \cup A_3 \cup \cdots \cup A_n)$$
$$= P(A_1) + P(A_2) + P(A_3) + \cdots + P(A_n) \tag{5.3}$$

Equation 5.3 applies, of course, only when all n events have no intersection.

5.5 Conditional Probability and the Multiplication Rule

Frequently we are interested in **conditional probability**—*the probability that an event will occur on condition that some other event has occurred.* Take, for example, the probability that a new novel will become a best-seller. Given no information about the novel, we would estimate its chances of becoming a best-seller as rather slim. Given additional information regarding the novel, however, we might well reassess our (subjective) estimate of the probability that the novel will become a best-seller. For example, if we learn that the author of the novel has already written two best-sellers in the past, we undoubtedly would raise our estimate of the probability that the new novel will be a best-seller.

Consider another example. Suppose 2 percent of all corporate bonds outstanding today will default in the next year and there exist some financial services that rank the quality of corporate bonds. To simplify matters, suppose a given financial service is using three ranks: A for high-quality bonds, B for average-quality bonds, and C for poor-quality bonds. Let us denote the event "the company will default on its bonds" by DF. We might write:

$$P(DF) = 0.02$$

If we know the quality of the bond, however, the probability of default may change. Past records may show the following probabilities:

$$P(DF \mid A) = 0.005$$
$$P(DF \mid B) = 0.018$$
$$P(DF \mid C) = 0.045$$

where the probability of the event written to the left of the vertical line is *conditional upon* the occurrence of the event written on its right. For example, $P(DF \mid A)$ (read "the probability of DF conditional upon A," or "the probability of DF given A") means the probability of default of bonds rated A. If the bond is an A bond, the default probability is one-half of 1 percent; if it is a B bond, the probability is 1.8 percent; and if it is a C bond, the probability is 4.5 percent. (Note that these conditional probabilities do not add up to $P(DF) = 0.02$. They average out to 0.02 when properly weighted as we shall see when discussing Bayes' theorem.)

THE JOINT PROBABILITY TABLE

A *joint probability* distribution table will give us further insight into the meaning of conditional probability and help us to develop the relevant equations for calculating it. Consider the breakdown of a company's accounts receivable by balance and age (that is, the length of time the account is outstanding) presented in Table 5.5.

Let us denote four events as follows:

Y   The account is young (i.e., up to 30 days).
O   The account is old (i.e., 31 days or more).
S   The account has a small balance (i.e., up to $500).
L   The account has a large balance (i.e., $501 or more).

If one selects an account at random, the probabilities of the occurrence of the above events may be simply determined from the *margin* of Table 5.5. For example, to determine the probability of selecting a young account at random, we divide the number of young accounts (680) by the total number of accounts (800); to determine the probability of selecting a small account at random, we divide the number of small accounts (600) by the total number of accounts (800); and so on. The resulting probabilities are:

$$\text{Classification by age} \begin{cases} P(Y) = \dfrac{680}{800} = 0.8500 \\ \\ P(O) = \dfrac{120}{800} = 0.1500 \end{cases} \text{add up to 1.0}$$

$$\text{Classification by balance} \begin{cases} P(S) = \dfrac{600}{800} = 0.7500 \\ \\ P(L) = \dfrac{200}{800} = 0.2500 \end{cases} \text{add up to 1.0}$$

These are called **marginal probabilities,** since the events (Y, O, S and L) are classified *by one variable only*. For example, the event L is defined by the account's balance but not by its age, and so its probability is determined from the *margin* of the table. Such events as Y ∩ S and O ∩ L, on the other hand, are defined by *two-way classifications*—balance and age—and their probabilities are known as **joint probabilities.** To determine the probability of the event Y ∩ S, for example, we find the ratio of the accounts in Y ∩ S to

TABLE 5.5
Accounts Receivable, by Age and Balance

Balance \ Age	Y Up to 30 days	O 31 days or more	Total
S Up to $500	490	110	600
L $501 or more	190	10	200
Total	680	120	800

the total number of accounts in the sample space. The joint probabilities in our example are as follows:

$$
\left\{
\begin{array}{l}
P(Y \cap S) = \dfrac{490}{800} = 0.6125 \\[2em]
P(Y \cap L) = \dfrac{190}{800} = 0.2375 \\[2em]
P(O \cap S) = \dfrac{110}{800} = 0.1375 \\[2em]
P(O \cap L) = \dfrac{10}{800} = 0.0125
\end{array}
\right\} \text{add up to 1.0}
$$

Turning now to **conditional probability,** we may ask, for example, what is the probability that the account is "large" *given* that it is "young"? If the account is known to be young, our focus must shift from the entire sample space with its 800 accounts to the more restricted group of 680 young accounts (see Figure 5.9). The conditional probabilities of our example are as follows:

$$
\left\{
\begin{array}{l}
P(L \mid Y) = \dfrac{190}{680} = 0.2794 \\[2em]
P(S \mid Y) = \dfrac{490}{680} = 0.7206
\end{array}
\right\} \text{add up to 1.0}
$$

$$
\left\{
\begin{array}{l}
P(L \mid O) = \dfrac{10}{120} = 0.0833 \\[2em]
P(S \mid O) = \dfrac{110}{120} = 0.9167
\end{array}
\right\} \text{add up to 1.0}
$$

$$
\left\{
\begin{array}{l}
P(Y \mid L) = \dfrac{190}{200} = 0.9500 \\[2em]
P(O \mid L) = \dfrac{10}{200} = 0.0500
\end{array}
\right\} \text{add up to 1.0}
$$

Figure 5.9

Venn diagram showing conditional probability

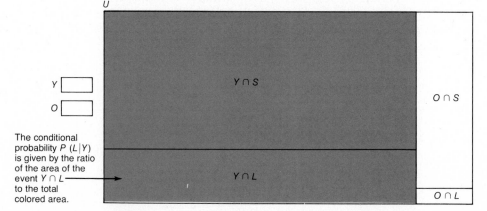

The conditional probability $P(L \mid Y)$ is given by the ratio of the area of the event $Y \cap L$ to the total colored area.

$$\left\{ \begin{array}{l} P(Y\,|\,S) = \dfrac{490}{600} = 0.8167 \\[4mm] P(O\,|\,S) = \dfrac{110}{600} = 0.1833 \end{array} \right\} \text{ add up to 1.0}$$

Consider another example involving conditional probability.

EXAMPLE 5.4

A given car is produced in two models: Model 1 and Model 2. Each model is produced in two designs: hatchback and wagon. Thirty percent of all the cars produced are Model 1s, and 20 percent of all the cars produced are wagons. But 50 percent of Model 1s are wagons. Suppose a car is selected at random.

Let event *A* be "the car selected is a Model 1" and let *B* be the event "the car selected is a wagon." The probabilities of *A* and *B* are:

$$P(A) = 0.30$$

$$P(B) = 0.20$$

These probabilities are reflected in Figure 5.10 by the relative area of the events in *U*. They are marginal probabilities. In addition we know that $P(B\,|\,A) = 0.50$. This is a conditional probability. Given the condition *A*, that the car selected is a Model 1, the probability of *B* rises to 0.50. This probability is presented in Figure 5.10: the intersection of *A* and *B* takes 50 percent of the total area of *A*. Thus if our interest is limited to Area *A*, there is a 50 percent chance of randomly selecting a wagon.

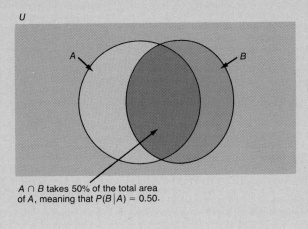

A ∩ B takes 50% of the total area
of A, meaning that P(B|A) = 0.50.

Figure 5.10

Venn diagram showing conditional probability

The last example leads us to the first important rule of conditional probability:

$$P(B\,|\,A) = \frac{P(A \cap B)}{P(A)} \qquad (5.4)$$

where $P(A) \neq 0$.

A similar argument will lead us to the related rule:

$$P(A \mid B) = \frac{P(A \cap B)}{P(B)} \qquad (5.5)$$

where $P(B) \neq 0$.

Let us multiply Equation 5.4 by $P(A)$ and Equation 5.5 by $P(B)$. That yields the following:

MULTIPLICATION RULES FOR CONDITIONAL PROBABILITY

$$P(A \cap B) = P(B \mid A)P(A) \qquad (5.6)$$

$$P(A \cap B) = P(A \mid B)P(B) \qquad (5.7)$$

Since the left-hand sides of Equations 5.6 and 5.7 are identical, the right-hand sides of these formulas must be equal to each other as well, so that we get

$$P(A \mid B)P(B) = P(B \mid A)P(A) \qquad (5.8)$$

Dividing both sides of Equation 5.8 by $P(B)$, we derive the following useful formula for conditional probability problems:

BAYES' THEOREM

$$P(A \mid B) = \frac{P(B \mid A)P(A)}{P(B)} \qquad (5.9)$$

Bayes' theorem is named for the statistician Thomas Bayes. A more detailed version of it is presented later in this chapter.

In Example 5.4, given the probabilities $P(A) = 0.30$, $P(B) = 0.20$, and $P(B \mid A) = 0.50$, we may use Equation 5.9 to find

$$P(A \mid B) = \frac{P(B \mid A)P(A)}{P(B)} = \frac{0.50 \cdot 0.30}{0.20} = 0.75$$

so that the probability that a randomly selected car is a Model 1, given that it is a wagon, is 0.75. We urge you to verify Equations 5.6 through 5.9 by using probabilities derived from Table 5.5, above.

Let us illustrate the usefulness of Equation 5.9 by another example.

EXAMPLE 5.5

An oil company is planning to drill for oil in two locations, field A and field B. Experts estimate the probability of finding oil of acceptable quality to be 20 percent in field A and 25 percent in field B. They also state that if oil of acceptable quality is found in field A, the probability of finding the same in field B is 80 percent (that is, $P(B \mid A) = 0.80$). If oil of acceptable quality is found in field B, what is the probability that such oil will be found in field A?

Let *A* stand for "oil of acceptable quality is found in field A." Let *B* stand for "oil of acceptable quality is found in field B."

We may write: $P(A) = 0.20$, $P(B) = 0.25$, $P(B \mid A) = 0.80$. The problem is $P(A \mid B) = ?$ Using Equation 5.9, we get

$$P(A \mid B) = \frac{P(B \mid A)P(A)}{P(B)} = \frac{0.80 \cdot 0.20}{0.25} = \frac{0.16}{0.25} = 0.64$$

And here is another way to approach the problem. While *A* and *B* take 20 and 25 percent respectively of the entire sample space (see Figure 5.11), they interrelate in such a way that 80 percent of *A* is overlapped by event *B*, representing the probability $P(B \mid A) = 0.80$. The area $A \cap B$ is then equal to 0.16 (that is, 80 percent of 0.20). To get $P(A \mid B)$ we find the relative area of $A \cap B$ within the area *B*, and thus we get $0.16 / 0.25 = 0.64$.

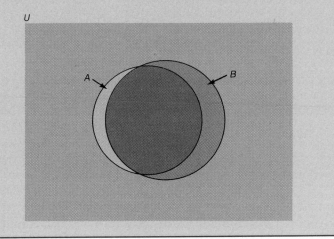

Figure 5.11
Venn diagram:
$P(B \mid A) = 0.80$

5.6 Independent Events

Two events, *A* and *B*, are said to be **independent** of one another if and only if the following relationship holds true:

$$P(A \mid B) = P(A) \tag{5.10}$$

In plain English, we say that *A* and *B* are independent if and only if the fact that *B* has occurred does not affect the chance that *A* will occur. The recovery of patient *A*, who has had a heart attack in San Francisco, may be regarded as independent of the recovery of patient *B* in Toronto at the same time; *B*'s recovery or failure to recover has no effect on *A*'s recovery. If, on the other hand, an experimental marketing technique for a new brand of cigarettes is being tried in New York, the results certainly have implications for the probability of success in a different region of the United States. Thus, if *SNY* stands for the event "success in New York" and *SAG* stands for "success in Atlanta, Georgia," then $P(SAG \mid SNY) \neq P(SAG)$, and the two events are not independent.

From Equation 5.7 we know that $P(A \cap B) = P(A \mid B)P(B)$. When events A and B are independent, we have $P(A \mid B) = P(A)$. Substituting this result into Equation 5.7 yields $P(A \cap B) = P(A)P(B)$, and thus we summarize the **multiplication rule** as follows:

In general

$$P(A \cap B) = P(A \mid B)P(B)$$

If A and B are independent, the general formula applies but is reduced to

$$P(A \cap B) = P(A)P(B) \tag{5.11}$$

The multiplication rule can be easily extended to more than two events. The probability of the intersection of n independent events is equal to the product of the probabilities of these events.

It is interesting to note the relationship between independence and mutual exclusiveness. Specifically, are two mutually exclusive events also independent? (At this point you should be well equipped to answer the question on your own!) The answer is that two mutually exclusive events are necessarily *dependent* on each other. Logically, if the events are mutually exclusive, the occurrence of one of them must *exclude* the occurrence of the other; if the events were independent, the occurrence of one of them *would have no effect* on the probability that the other would occur. If you are in Texas, you cannot be in Illinois at the same time, and thus the two events are mutually exclusive. It is therefore clear that whether you are in Illinois *depends* on whether or not you are in Texas. By contrast, the results of two flips of a coin are independent, since the result of the first flip can in no way affect the result of the second flip. Diagrammatically, two mutually exclusive events have *no intersection*, while two independent events have an intersection with an area given by Equation 5.11. Note, that if two events A and B are independent, then the following pairs are also independent: A and B', A' and B, A' and B'.

GRAPHIC REPRESENTATION OF INDEPENDENCE AND DEPENDENCE OF EVENTS

Suppose the probabilities of events B_1 and B_2 in Figure 5.12 are 0.60 and 0.40, respectively:

$$P(B_1) = 0.60$$

$$P(B_2) = 0.40$$

Suppose also that event A, which intersects both B_1 and B_2, has a probability of 0.50:

$$P(A) = 0.50$$

As we can see, the bulk of A is concentrated in event B_1. If we have the information that B_1 will occur, then the probability that A will also occur is

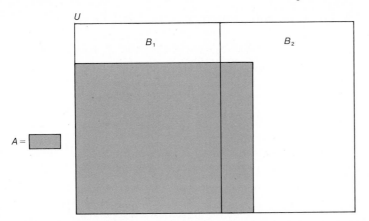

Figure 5.12

Venn diagram:
events A and B_1
are dependent;
events A and B_2
are dependent

greater than 0.50: $P(A \mid B_1) > 0.50$. Diagrammatically, the probability $P(A \mid B_1)$ is given by the ratio of the portion of A that overlaps B_1 to the total area of B_1. Similarly, if we have the information that B_2 will occur, then the probability that A will occur is less than 0.50: $P(A \mid B_2) < 0.50$. We can see this in Figure 5.12 by the fact that the colored area in B_2 is less than 50 percent of the total area of B_2. Events A and B_1 are dependent, and so are events A and B_2.

Figure 5.13, by contrast, depicts a situation of statistical independence between A and B_1 and between A and B_2. The proportion of A in B_1 is equal to the proportion of A in B_2 (and consequently both are equal to the proportion of A in the entire sample space). Any information we may have about the occurrence or nonoccurrence of B_1 (or B_2) does not alter the probability of A.

5.7 More on Conditional Probability: Bayes' Theorem

Let us proceed now to discuss a more generalized form of Bayes' theorem. Consider Figure 5.14, in which the sample space is *partitioned* by the events B_1, B_2, B_3, and B_4—in other words, events B (that is, B_1, B_2, B_3, and B_4) are mutually exclusive and collectively exhaustive. Event A intersects the B events. Because events B_1, B_2, B_3, and B_4 are mutually exclusive and thus have no intersection among themselves, event A may be expressed as the sum of its intersections with the B events:

$$P(A) = P(A \cap B_1) + P(A \cap B_2) + P(A \cap B_3) + P(A \cap B_4) \qquad \textbf{(5.12)}$$

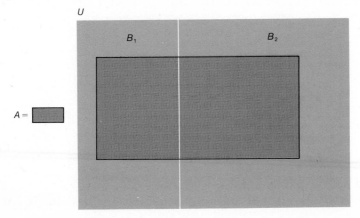

Figure 5.13

Venn diagram:
A is statistically
independent of
B_1 and B_2

By the multiplication rule (Equations 5.6 and 5.7), we may write:

$$P(A \cap B_1) = P(A \mid B_1)P(B_1)$$
$$P(A \cap B_2) = P(A \mid B_2)P(B_2)$$
$$P(A \cap B_3) = P(A \mid B_3)P(B_3) \qquad \textbf{(5.13)}$$
$$P(A \cap B_4) = P(A \mid B_4)P(B_4)$$

and by substituting Equation 5.13 in 5.12, we get the following well-known equation:

TOTAL PROBABILITY EQUATION

$$P(A) =$$
$$P(A \mid B_1)P(B_1) + P(A \mid B_2)P(B_2) + P(A \mid B_3)P(B_3) + P(A \mid B_4)P(B_4) \quad \textbf{(5.14)}$$

Sometimes the probabilities on the right-hand side of Equation 5.14 are available, but we need to calculate the probability of one of the B events conditional upon A. Suppose, for example, we want to find $P(B_2 \mid A)$. By Equation 5.4 we get

$$P(B_2 \mid A) = \frac{P(A \cap B_2)}{P(A)} \qquad \textbf{(5.15)}$$

Given the total probability equation (5.14), and remembering that by the multiplication rule we get $P(A \cap B_2) = P(A \mid B_2)P(B_2)$, we may provide a formula for $P(B_2 \mid A)$ as follows:

$$P(B_2 \mid A) = \frac{P(A \mid B_2)P(B_2)}{P(A \mid B_1)P(B_1) + P(A \mid B_2)P(B_2) + P(A \mid B_3)P(B_3) + P(A \mid B_4)P(B_4)}$$

$$\textbf{(5.16)}$$

Equation 5.16 is a generalized form of **Bayes' theorem.**

Figure 5.14
Venn diagram
demonstrating
Bayes' theorem

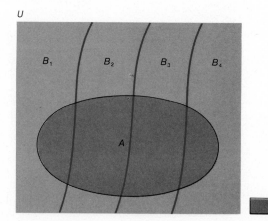

EXAMPLE 5.6

Delicious Coffee, Inc., produces four types of coffee, each of which is produced in two grades, regular and deluxe. All the coffee is packed in 10-ounce jars. The percentages of each type of coffee produced are as follows:

Type	Percentage
1	20
2	20
3	10
4	50
Total	100%

The percentage distribution of each type of coffee in regular and deluxe grades is as follows:

Type	Regular	Deluxe	Total
1	80%	20%	100%
2	60	40	100
3	70	30	100
4	95	5	100

One coffee jar is selected at random from the entire production for inspection. Determine the probability that the jar selected contains deluxe coffee. Also, if it is known that the jar contains deluxe coffee, what is the probability that it contains Type 4 coffee?

To obtain the required probabilities let us introduce the following notation:

T_1 The selected jar contains Type 1 coffee.
T_2 The selected jar contains Type 2 coffee.
T_3 The selected jar contains Type 3 coffee.
T_4 The selected jar contains Type 4 coffee.
D The selected jar contains deluxe coffee.

Our first step is to calculate the probability that the coffee selected is deluxe. Notationally, we are looking for $P(D)$. Using the *total probability formula*, we write:

$$P(D) = P(D \mid T_1)P(T_1) + P(D \mid T_2)P(T_2) + P(D \mid T_3)P(T_3) + P(D \mid T_4)P(T_4)$$

Each of the probabilities on the right-hand side is provided in the question. For example, $P(D \mid T_1)$ is the probability that the coffee is deluxe *given* that the selection is made from Type 1 coffee. Since 80 percent of Type 1 is produced in regular grade and 20 percent in deluxe grade, it is obvious that $P(D \mid T_1) = 0.20$. Similarly we determine the rest of the conditional probabilities to be $P(D \mid T_2) = 0.40$, $P(D \mid T_3) = 0.30$, and $P(D \mid T_4) = 0.05$. The distribution of the production by coffee types provides us with the following probabilities: $P(T_1) = 0.20$, $P(T_2) = 0.20$, $P(T_3) = 0.10$, and $P(T_4) = 0.50$. When all these probabilities are substituted in the total probability equation we get

$$P(D) = (0.20 \cdot 0.20) + (0.40 \cdot 0.20) + (0.30 \cdot 0.10) + (0.05 \cdot 0.50)$$

$$= 0.040 + 0.080 + 0.030 + 0.025 = 0.175$$

There is a 17.5 percent chance of obtaining a jar containing deluxe coffee when the selection is made at random. To obtain $P(T_4 \mid D)$ we use the generalized form of Bayes' theorem:

$$P(T_4 \mid D) = \frac{P(D \mid T_4)P(T_4)}{P(D \mid T_1)P(T_1) + P(D \mid T_2)P(T_2) + P(D \mid T_3)P(T_3) + P(D \mid T_4)P(T_4)}$$

The denominator is, by the total probability formula, equal to $P(D)$, which was calculated to equal 0.175. The numerator is the product of $P(D \mid T_4) = 0.05$, and $P(T_4) = 0.50$. We therefore get

$$P(T_4 \mid D) = \frac{P(D \mid T_4)P(T_4)}{P(D)} = \frac{0.05 \cdot 0.50}{0.175} = \frac{0.025}{0.175} = 0.143$$

Thus the probability of selecting Type 4 coffee given that the selection is made out of all the deluxe production is 14.3 percent. Figure 5.15 illustrates the situation. The total shaded area in the rectangle is 17.5 percent of the total, and of the total shaded area, T_4 accounts for 14.3 percent.

Figure 5.15

Venn diagram showing the conditional probability $P(T_4 \mid D)$

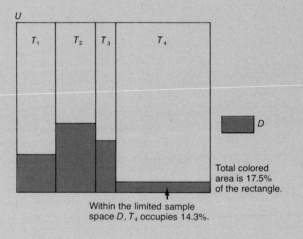

EXAMPLE 5.7

Two types of electric hair blowers are manufactured by Handy Electric Accessories, Inc.: the Handy Blower and the Super Handy Blower. Forty percent of the blowers produced by the company are of the Super model. Both models carry a warranty for one year. The company's records show that 3 percent of the Handy Blowers and 1 percent of the Super Handy Blowers are being returned within the warranty period (the first year after purchase) for repair or replacement. When a blower is brought in for repair or replacement, what is the probability that it will be a Super Handy Blower?

Denote the following events:

> *HB* The blower is a Handy Blower.
> *SHB* The blower is a Super Handy Blower.
> *R* The blower is being returned within the warranty period.
>
> The information given above may now be easily summarized as follows:
>
> $$P(HB) = 0.60 \qquad P(SHB) = 0.40$$
>
> $$P(R \mid HB) = 0.03 \qquad P(R \mid SHB) = 0.01$$
>
> and the question is $P(SHB \mid R) = ?$ Using the generalized form of Bayes' theorem, we get
>
> $$P(SHB \mid R) = \frac{P(R \mid SHB)P(SHB)}{P(R \mid HB)P(HB) + P(R \mid SHB)P(SHB)}$$
>
> so that
>
> $$P(SHB \mid R) = \frac{0.01 \cdot 0.40}{0.03 \cdot 0.60 + 0.01 \cdot 0.40} = \frac{0.004}{0.022} = 0.1818$$
>
> The required probability is thus 18.18 percent.

WHAT IS CONDITIONAL PROBABILITY GOOD FOR?

The primary reason for dealing with conditional probabilities is that they are helpful in decision-making. With respect to bonds, for example, the conditional probability of default given the bond rating of financial services is valuable information for current and potential bondholders. A few major financial service institutions (Moody's, Standard and Poor's, and others) compile an enormous amount of data in order to keep up with the quality of hundreds of bond issues in the United States so that they can sell this information to institutions and individuals interested in investing in the bond market. Collecting and computing the data costs many millions of dollars every year. This information is bought by investors so that they can improve their bond-trading decisions. Another example of the use of conditional probabilities in decision-making is in drilling for oil. Clearly, the drilling can take place without a prior seismic test. However, seismic tests are often conducted for improving the decision to drill or not to drill. What is relevant for the drilling decision is the conditional probability of striking oil given that the seismic test indicates that there is oil in the well (otherwise, why bother with the test?) Yet another example where extensive use is made with conditional probability is the determination of insurance premiums. For example, the life insurance premium of a 50-year-old man is determined on the basis of the probability of his survival through age 51 *given* that he has already survived to the age of 50.

5.8 Symmetrical Sample Spaces and Counting Techniques

As we saw earlier, in Section 5.3, the third property of probability is that if two events are mutually exclusive, the probability of their union is equal to the sum of their probabilities. Since sample points are mutually exclusive by definition, and since any event in U is the union of all the sample points included in that event, it follows that the probability of any event is equal to the sum of the probabilities of the sample points included in it. The problem is, however, that the number of basic outcomes in the event is sometimes very large and the probabilities of the basic outcomes themselves are often unknown.

A **symmetrical sample space** is a *set of basic outcomes, all of which have the same probability.* The probability of each sample point in a symmetrical sample space may be fairly easily obtained; therefore, adding up the probabilities of the basic outcomes in order to derive the events' probabilities is a very reasonable thing to do even when the number of basic outcomes involved is very large.

Symmetrical sample spaces are common in gambling, and outside it as well. Each of the six possible results (sample points) of rolling a balanced die has the same probability. Each ticket of a lottery is normally given the same chances of being chosen for the big prize as any other ticket. Likewise, quality-control problems involve the random selection of items, so that all items have equal chances of being inspected.

When all the sample points in the sample space have equal probability, the probability of each sample point is equal to $\frac{1}{n(U)}$, where $n(U)$ is the number of sample points in the sample space, U. For example, when we roll a fair die, $n(U)$ is equal to 6 and the probability of any sample point is $\frac{1}{n(U)} = \frac{1}{6}$. The general formula and rule for finding probability for sample points in a sample space is:

$$P(A) = \frac{n(A)}{n(U)}$$

If an event, A, in a symmetrical sample space contains $n(A)$ sample points $[0 \leq n(A) \leq n(U)]$, it follows from our analysis that the probability A is equal to $\frac{n(A)}{n(U)}$.

To illustrate, suppose a box contains 30 calculators, 3 of which are defective. If a calculator is selected at random (for inspection, say), the probability that the calculator chosen will be defective (call this event D) is 0.10, since in this case $n(D) = 3$ and $n(U) = 30$, so the probability of D is

$$P(D) = \frac{n(D)}{n(U)} = \frac{3}{30} = 0.10$$

THE ADDITION RULE AND THE MULTIPLICATION RULE FOR COUNTING

In some experiments, the number of sample points in the events is more difficult to determine. In fact, some simple experiments involve symmetrical sample spaces with many millions of sample points, and probability calculations by counting can be carried out only if an efficient technique of counting sample points (in the sample space and in particular events) is available. We turn now to a discussion of some counting rules that, when considered together, will provide such a technique.

RULE 1 FOR COUNTING (ADDITION RULE)

If two operations are *mutually exclusive,* and if the first operation can be performed in N_1 ways and the second operation can be performed in N_2 ways, then the total operation (that is, either the first *or* the second operation) can be performed in $N_1 + N_2$ ways.

The addition rule applies only in the case of *mutually exclusive* operations, that is, when only one of the two operations takes place. As an illustration, suppose one item has to be selected for inspection out of eight items that have been produced today and six items that were produced yesterday. Since only *one* item is to be selected, the two operations (that is, selection from today's production and selection from yesterday's production) are mutually exclusive: if the item chosen is from today's production, no item will be chosen from yesterday's production, and vice versa. Following the first rule, we ascertain that the total number of possible selections is $6 + 8 = 14$. The addition rule can easily be extended to include K different mutually exclusive operations, with $N_1, N_2 \cdots N_K$ ways in which each can be performed. The total number of ways in which the total operation can be performed is $N_1 + N_2 + \cdots + N_K$.

RULE 2 FOR COUNTING (MULTIPLICATION RULE)

If one operation can be performed in N_1 ways, and if thereafter another contingent operation can be performed in N_2 ways, then the two operations can be performed in $N_1 \cdot N_2$ ways.

Unlike the addition rule, the multiplication rule applies to a situation in which *both* operations will take place (thus the operations are *not* mutually exclusive). Consider, for example, a group made up of five economists and four psychologists. Suppose that one economist and one psychologist have to be selected to work on a given project. The number of possible selections of an economist out of the five available is certainly five, and similarly there are four possible selections of a psychologist. The total number of ways to select the team is $20 = 5 \cdot 4$. Denoting the first economist by E_1, the second E_2, and so on, and similarly the first psychologist by P_1, the second by P_2, and so on, we list all the 20 possible selections:

$$
\begin{array}{lllll}
E_1, P_1 & E_2, P_1 & E_3, P_1 & E_4, P_1 & E_5, P_1 \\
E_1, P_2 & E_2, P_2 & E_3, P_2 & E_4, P_2 & E_5, P_2 \\
E_1, P_3 & E_2, P_3 & E_3, P_3 & E_4, P_3 & E_5, P_3 \\
E_1, P_4 & E_2, P_4 & E_3, P_4 & E_4; P_4 & E_5, P_4
\end{array}
$$

The multiplication rule of counting can easily be generalized to refer to any number of operations that are not mutually exclusive.

PERMUTATIONS

When the number of ways to perform each operation of interest is known, we apply Rule 1 if the operations are mutually exclusive and Rule 2 if they are not. Now how do we determine the number of ways we can perform a given operation? Obviously, it depends on the type of operation we are considering. Let us look at some rules that may be helpful in certain situations.

RULE 3 FOR COUNTING (PERMUTATIONS OF *n* OBJECTS)

There are *n*! (read "*n* factorial") different ways to arrange *n* distinct objects, where $n! = n(n-1)(n-2) \cdots (3)(2)(1)$ and where $0! = 1$ by definition. *Each arrangement of the objects is called a* **permutation**, and we say that *n*! is the total number of permutations of *n* objects.

Take, for example, the letters *A*, *B*, *C*, and *D*. There are four letters ("objects") in all. To determine the number of possible arrangements, note that we have four ways to select the first letter. Once the first letter has been selected, there are three ways to perform the second operation, selecting the second letter from the remaining three letters. Similarly there are two ways to select the third letter, and only one way to select the last letter. Using the multiplication rule (Rule 2), we find that the total number of ways to make the selection (that is, to arrange the letters) is

$$4! = 4 \cdot 3 \cdot 2 \cdot 1 = 24$$

Figure 5.16 is a "tree diagram" describing the 24 possible arrangements of 4 letters where each arrangement is seen as a branch of the "tree." Note that the number *n*! increases very rapidly with *n*. Try to calculate, for example, the number 20! to see how large a number that is.

Figure 5.16

Tree diagram: permutations of the letters *A*, *B*, *C*, and *D*

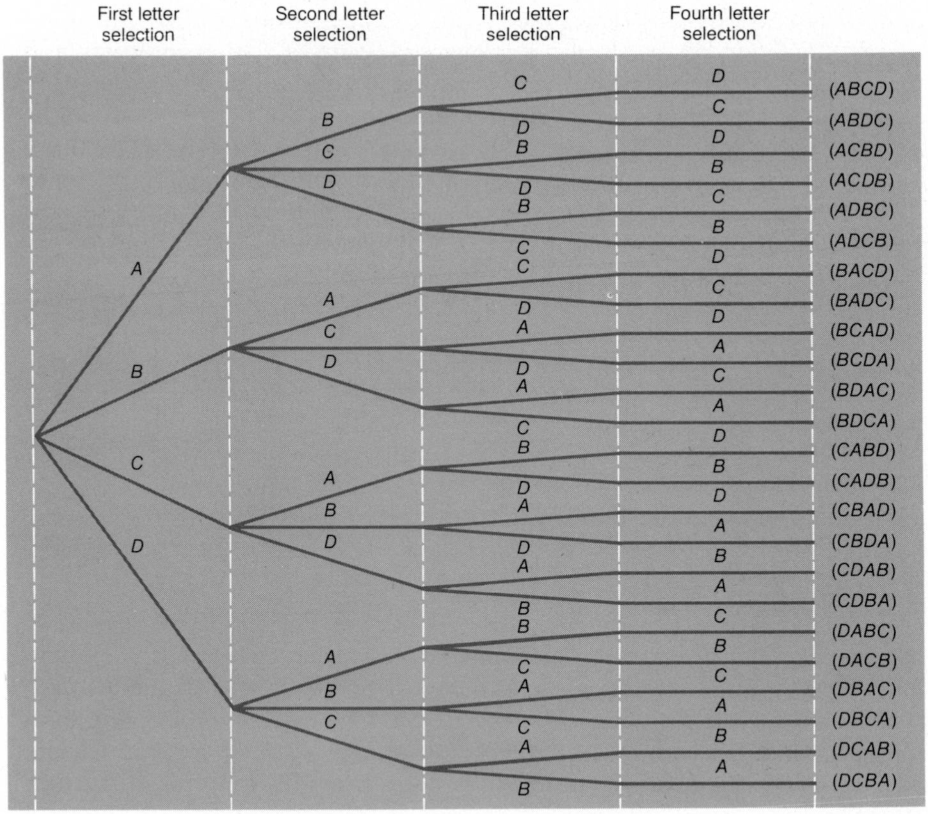

RULE 4 FOR COUNTING (PERMUTATIONS OF *r* OBJECTS TAKEN OUT OF *n* OBJECTS)

The number of ways in which we can select and arrange *r* objects when the selection is made from a group of *n* objects ($0 \leq r \leq n$) is denoted by P_r^n and is given by the formula:

$$P_r^n = \frac{n!}{(n-r)!}$$

Consider again the four letters *A*, *B*, *C*, and *D*. This time suppose we want to select two letters at a time and arrange them ($n = 4, r = 2$). According to the rule we should have $P_2^4 = \dfrac{4!}{(4-2)!} = \dfrac{4 \cdot 3 \cdot 2 \cdot 1}{2 \cdot 1} = 12$ different arrangements are as follows:

Available objects	*A, B, C, D*					
Combinations of distinct objects (six combinations)	(*A, B*)	(*A, C*)	(*A, D*)	(*B, C*)	(*B, D*)	(*C, D*)
Possible permutations (two permutations per each combination)	*A, B* *B, A*	*A, C* *C, A*	*A, D* *D, A*	*B, C* *C, B*	*B, D* *D, B*	*C, D* *D, C*

Let us illustrate the use of Rule 4 by an additional example.

EXAMPLE 5.8

Consider eight candidates eligible to serve on a student government committee at a given university. Assume that three nominees will be selected out of the eight candidates, and assume also that their order of selection matters (perhaps because one will serve as committee chairperson and the others as "number 2" and "number 3" people). In how many different ways can this selection be made?

$$P_3^8 = \frac{8!}{(8-3)!} = \frac{8 \cdot 7 \cdot 6 \cdot 5 \cdot 4 \cdot 3 \cdot 2 \cdot 1}{5 \cdot 4 \cdot 3 \cdot 2 \cdot 1} = 8 \cdot 7 \cdot 6 = 336$$

There are 336 different ways of making this selection.

COMBINATIONS

The fifth and last rule of counting that we shall consider concerns the number of combinations (rather than permutations) of *r* objects taken out of *n* objects.

RULE 5 FOR COUNTING (COMBINATIONS OF *r* OBJECTS TAKEN OUT OF *n* OBJECTS)

The number of ways in which we can select different combinations of *r* objects out of a group of *n* distinct objects ($r \leq n$) is denoted by C_r^n or (more commonly, perhaps) by $\binom{n}{r}$, and is given by the formula:

$$\binom{n}{r} = \frac{n!}{(n-r)!\, r!}$$

Rule 5 deals with **combinations** of r distinct objects taken out of n distinct objects. Each combination of r objects can be rearranged in $r!$ permutations, but they will all be counted as just one combination. We can best explain this concept by an example. Below we have listed all possible permutations of three letters taken out of the five letters A, B, C, D, and E. We have arranged the permutations in such a way that each column provides all possible arrangements of the *same* letter combination. In the first column, for instance, we have listed all possible arrangements of the letters A, B, and C, in the second column all the possible arrangements of the letters A, B, and D, and so on.

ABC	ABD	ABE	ACD	ACE	ADE	BCD	BCE	BDE	CDE
ACB	ADB	AEB	ADC	AEC	AED	BDC	BEC	BED	CED
BAC	BAD	BAE	CAD	CAE	DAE	CBD	CBE	DBE	DCE
BCA	BDA	BEA	CDA	CEA	DEA	CDB	CEB	DEB	DEC
CAB	DAB	EAB	DAC	EAC	EAD	DBC	EBC	EBD	ECD
CBA	DBA	EBA	DCA	ECA	EDA	DCB	ECB	EDB	EDC

All of these permutations count as only *one* combination since they consist of the same letters.

While in each column there are six different arrangements of letters, they differ only in the *order* in which the letters are listed, and thus should count as only one combination. This being the case, the number of combinations of r objects taken out of n objects must equal P_r^n divided by $r!$, where $r!$ is, of course, the number of possible ways we can arrange r objects. In our example, where $n = 5$ and $r = 3$, we get

$$P_3^5 = \frac{5!}{(5-3)!} = \frac{5!}{2!} = 5 \cdot 4 \cdot 3 = 60$$

which is the total number of arrangements listed. For $r!$ we get

$$3! = 3 \cdot 2 \cdot 1 = 6$$

which is the number of possible arrangements we obtain for each group of three letters by merely changing their order. Therefore, the number of *combinations* is

$$C_3^5 \equiv \binom{5}{3} = \frac{P_3^5}{3!} = \frac{60}{6} = 10.$$

where the symbol $\equiv$ indicates identity by definition. In the general case we find

$$\binom{n}{r} = \frac{P_r^n}{r!} = \frac{\dfrac{n!}{(n-r)!}}{r!} = \frac{n!}{(n-r)!\,r!} \tag{5.17}$$

which is the formula provided by Rule 5. It is time to turn to examples.

EXAMPLE 5.9

The First Nebraska Bank is giving out gifts to depositors. Eligible depositors may choose 2 out of 15 gifts. How many possible selections can a depositor make? Obviously, the nature of the problem is such that it requires calculation of the number of possible *combinations* of 2 gifts out of the available 15. This is so because the order of selecting the gifts makes no difference here, and therefore we do not wish to count the selection of "gift 1 and gift 2," for example, as a separate selection from "gift 2 and gift 1." Once we recognize the fact that we ought merely to concern ourselves with the number of combinations, we can use Rule 5 to get:

$$\binom{15}{2} = \frac{15!}{(15-2)!\,2!} = \frac{15!}{13!\,2!} = \frac{15 \cdot 14}{2} = 105$$

Thus there are 105 possible combinations.

EXAMPLE 5.10

Out of four students, a freshman, a sophomore, a junior, and a senior, we are to choose two to do a certain task. What is the probability that the freshman and the senior will be chosen (call this event A) if the selection is made at random?

First we should find out how many possible selections of two people out of the four are possible:

$$\binom{4}{2} = \frac{4!}{2!\,2!} = \frac{4 \cdot 3 \cdot 2 \cdot 1}{(2 \cdot 1) \cdot (2 \cdot 1)} = 6$$

Since the selection of the freshman and the senior is as *equally probable* a selection as any other, we get $n(A) = 1$ and $n(U) = 6$, and because the sample space is symmetrical, the probability of A is $P(A) = \dfrac{n(A)}{n(U)} = \dfrac{1}{6}$.

5.9 APPLICATION:
DETERMINING INSURANCE PREMIUMS BY THE USE OF CONDITIONAL PROBABILITIES

Insurance companies make extensive use of probability and conditional probability theory in determining the price (that is, the premium) they charge their policyholders. The basis for computing these probabilities for life insurance are the "life tables" that display the frequency of death in each age group per 100,000. Customarily, separate tables are compiled for males and females, since a female's chance for survival in each age group is greater than a male's. Table 5.6 is a life table for males in the United States. The table was constructed on the basis of the age distribution of males who had died during a period of several years. It may be interpreted also as anticipated death distribution by age of those who were born in a given year, although there is a continuous trend toward longer life.

TABLE 5.6

Death Frequencies of Males in the United States, by Age, and Determination of Life-Insurance Premiums

(1) Age group[a]	(2) Frequency of death (per 100,000 males)	(3) "More than" cumulative frequency[b] (per 100,000 males)	(4) = (2) ÷ (3) Conditional probability of death	(5) = (4) · $10,000 Required premium for 5 years[c]
0–1	2,060	100,000	0.0206	$ 206
1–5	352	97,940	0.0036	36
5–10	229	97,588	0.0023	23
10–15	246	97,359	0.0025	25
15–20	772	97,113	0.0079	79
20–25	1,061	96,341	0.0110	110
25–30	955	95,280	0.0100	100
30–35	1,054	94,325	0.0112	112
35–40	1,411	93,271	0.0151	151
40–45	2,111	91,860	0.0230	230
45–50	3,306	89,749	0.0368	368
50–55	4,789	86,443	0.0554	554
55–60	7,085	81,654	0.0868	868
60–65	9,617	74,569	0.1290	1,290
65–70	11,828	64,952	0.1821	1,821
70–75	13,836	53,124	0.2604	2,604
75–80	14,216	39,288	0.3618	3,618
80+	25,072	25,072	1.0000	10,000

[a] The class interval 0–1 includes the first birthday. The class interval 1–5 does not include the first birthday, but includes the fifth. Other class intervals are defined similarly.
[b] The numbers in column 3 show the frequency of death at ages greater than or equal to the age group.
[c] Except for age groups 0–1 and 80+.

Source: Samuel H. Preston, Nathan Keyfitz, and Robert Schoen, *Causes of Death: Life Tables for Natural Populations* (New York: Seminar Press, 1972).

How does the insurance company use the table to determine the required premium for a term life-insurance policy? Three major factors affect the desirability to the company of insuring the potential client, and hence the determination of the premium. First, the company may require the potential client to have a physical examination, for if the client is in poor health, his conditional probability of death is higher than average, and the company might therefore refuse to insure him. Indeed, most life-insurance plans do, in fact, require a physical examination before the issuance of an insurance policy. Once the good health of the potential client has been confirmed by an examination, the premium is determined as a function of the applicant's sex and age. Different premiums are charged for males and females on the basis of their different death frequencies. Suppose now a healthy 50-year-old man is interested in a one-year term life-insurance policy in the amount of $10,000; that is to say, a policy under which the insurance company will have to pay $10,000 to his beneficiaries if he dies during the next year.

To determine the premium that the company will charge, we first define the following events:

A   The client will die sometime between the ages of 50 and 55.
B   The client will die sometime after his fiftieth birthday.

Since the client is 50 years old, it is certain that event *B* will occur. Hence the relevant probability for the insurance company is that of death in the age interval 50–55 (event A) *given* that event B will occur. By Equation 5.5 we know that

$$P(A \mid B) = \frac{P(A \cap B)}{P(B)}$$

The event $A \cap B$ is the intersection of death in the age interval 50–55 and in the (open) interval 50+. The intersection simply means death in the age interval 50–55. The probability of this event is

$$P(A \cap B) = \frac{4{,}789}{100{,}000} = 0.04789$$

(see Table 5.6, column 2). The probability of event B is

$$P(B) = \frac{86{,}443}{100{,}000} = 0.86443$$

(see Table 5.6, column 3). Hence, the conditional probability that will determine the premium is

$$P(A \mid B) = \frac{P(A \cap B)}{P(B)} = \frac{0.04789}{0.86443} = 0.0554$$

(see Table 5.6, column 4). The required premium should be

$$\text{Premium} = 10{,}000 \cdot 0.0554 = \$554$$

A few comments are called for at this juncture. First, the premiums presented in column 5 of Table 5.6 are for a five-year period (except for the intervals 0–1 and 80+). The annual premium is one-fifth of that amount. For example, the \$554 for men in the age group 50–55 should be sufficient to cover the policyholder for a five-year period. The annual premium is 554/5 = \$110.80. Second, the premiums in the tables are for \$10,000 coverage. Other coverages require proportional adjustment. An annual premium of \$110.80 for \$10,000 coverage will be raised to \$1,108.00 (= \$110.80 · 10) if the coverage is raised from \$10,000 to \$100,000. Third, insurance companies have administrative costs and other expenses, so that the actual premium is higher than that shown. Finally, although premiums in the table are calculated for all age groups, insurance companies don't usually insure children or people over 65 years of age.

Problems

5.1. Four people will be asked whether they have purchased a new house in the past year. All answers are expected to be either yes or no. The sample space consists of the people's answers. Identify all the sample points in the sample space.

5.2. A newsstand operator has bought 16 magazines. Let A be the event "The number of magazines sold is 1, 4, 8, or 16," and let B be the event "The number of magazines sold is 0, 2, 4, 6, 8, 10, 12, or 14." Identify the events:

        (a) $A \cap B$
        (b) $A \cup B$
        (c) $(A \cup B)'$
        (d) A'

5.3. A car dealer's stock consists of two cars, *a* and *b*. We define the following events:

> A   Car *a* is sold.
> B   Car *b* is sold.

Express events *C* through *G* below, using events *A* and *B* and their complement.

>   C   Both cars are sold.
>   D   Neither of the cars is sold.
>   E   At least one car is sold.
>   F   Car *a* is sold, but not car *b*.
>   G   One (and only one) car is sold, but which one is not specified.

5.4. *A* and *B* are two mutually exclusive events and $P(A) > \dfrac{1}{2}$. Show that $P(B) < \dfrac{1}{2}$.

5.5. The probability that a buyer will enter a given furniture store within the next fifteen minutes is 0.90. If in the last fourteen minutes no buyer has entered the store, is the chance that a buyer will enter the store in the next minute especially high? Explain.

5.6. *A* and *B* are two mutually exclusive events.
What is the probability *P(A* and *B* are two independent events)?
Explain.

5.7. Show that for each pair of events, *A* and *B*, in a sample space, $P(A \cap B) \le P(A)$ and also $P(A \cap B) \le P(B)$.

5.8. Mr. Stanford holds corporate bonds of two corporations, *A* and *B*. Consider the following events:

> AD   The price of bond *A* drops.
> BD   The price of bond *B* drops.

(*a*) Describe the events listed below in terms of the bond prices.

> AD'
> $AD \cap BD$
> $AD' \cap BD$
> $AD \cup BD$
> $AD \cup BD'$

(*b*) Identify each of the events in part *a* on a Venn diagram.
(*c*) Suppose $P(AD) = 0.40$, $P(BD) = 0.50$, and $P(AD \cap BD) = 0.20$. Find $P(AD \cup BD)$ and $P[(AD \cap BD)']$.

5.9. *C* and *D* are two events and $P(C) = \dfrac{7}{12}$ and $P(D) = \dfrac{3}{4}$. Is it possible that *C* and *D* are mutually exclusive?

5.10. Let *A* and *B* be events in a sample space.

(*a*) Is it possible that $P(A) = \dfrac{1}{4}, P(B) = \dfrac{1}{3}$, and $P(A \cup B) = \dfrac{7}{12}$?

(*b*) Is it possible that $P(A) = \dfrac{1}{5}, P(B) = \dfrac{2}{7}$, and $P(A \cup B) = \dfrac{1}{2}$?

(*c*) Is it possible that $P(A) = \dfrac{1}{10}, P(B) = \dfrac{1}{5}$, and $P(A \cap B) = \dfrac{4}{10}$?

5.11. Let *A* be the event "The sales of store *a* have been above \$1 million in 1979" and let *B* be the event "The sales of store *b* have been above \$1 million in 1979." It is given that $P(A) = \frac{1}{2}$, $P(B) = \frac{1}{2}$, and $P(A \mid B) = \frac{3}{4}$.

(*a*) Explain the meaning of the following: A', $A \cap B'$.

(*b*) Find $P(B \mid A)$.
(*c*) Find $P(A \cup B)$.

5.12. The following are probabilities of several events in a given sample space: $P(A) = 0.20$, $P(B) = 0.40$, $P(C) = 0.10$, $P(D) = 0.50$, $P(A \cap B) = 0.08$, $P(A \cap D) = 0.00$, $P(C \cup D) = 0.60$.

(*a*) Are A and B mutually exclusive? Are they independent?
(*b*) How would you characterize the relationship between A and D?
(*c*) How would you characterize the relationship between C and D?
(*d*) Give an example of two independent events and of two mutually exclusive events from the business world.

5.13. A large pharmaceutical company employs several teams in an effort to find a way to make the production of a certain type of drug economical. Management estimates the probability that a team will be successful within a year at 8 percent. How many teams should management employ if it wishes to make the probability of discovering at least one economical process equal to 80 percent? 95 percent? 100 percent? Assume that each team's success is independent of the success of the others.

5.14. A market researcher interviews ten families in order to determine whether they regularly purchase a certain product. The families' purchasing habits are independent. Assuming that the probability of getting a positive answer is 0.40, determine the probability that all ten families will give positive answers, and the probability that all ten families will give negative answers.

5.15. (*a*) Suppose $P(A) = 0.40$, $P(B) = 0.30$, $P(A \cap B) = 0.10$. Find $P(A \mid B)$ and $P(B \mid A)$.
(*b*) Suppose $P(A) = 0.60$, $P(B) = 0.80$, $P(A \mid B) = 0.70$. Find $P(B \mid A)$.
(*c*) Suppose $P(A) = 0.20$, $P(B) = 0.60$, $P(A \cup B) = 0.70$. Find $P(B \mid A)$.

5.16. Show that if A and B are independent, then A' and B' are also independent.

5.17. A and B are two mutually exclusive events. It is given that $P(A) = 0.20$ and $P(B) = 0.40$. Find:

(*a*) $P(A \mid B)$
(*b*) $P(B \mid A)$
(*c*) $P(A \cap B)$
(*d*) $P(A \cup B)$

5.18. Two candidates are running for office in a given city. The probability that the first will be elected is 0.70; the probability that the second will be elected is 0.30. If the first is elected, the probability that a new air terminal will be built is 0.60; if the second is elected, the probability is 0.40. What is the probability that a new air terminal will be built?

5.19. Two machines, $m1$ and $m2$, produce 60 percent and 40 percent respectively of the total production of Gamma Company. $m1$ produces 3 percent defects and $m2$ produces 4 percent defects. An item is picked up at random from the production of the machines.

(*a*) What is the probability that the item chosen is defective?
(*b*) If the item chosen is known to be good (not defective), what is the probability that the item selected is a product of $m1$?

5.20. Box a holds two 60-watt bulbs, three 75-watt bulbs, and two 100-watt bulbs. Box b holds five 60-watt bulbs, one 75-watt bulb, and three 120-watt bulbs. One bulb is picked at random out of one of the two boxes. Given the selection procedure, the probability that the bulb selected is is from box a is $\frac{1}{3}$, and the probability that the bulb selected is from box b is $\frac{2}{3}$.

(*a*) What is the probability that the bulb selected is a 60-watt bulb?
(*b*) If it is known that the bulb that was picked is a 60-watt bulb, what is the probability that it was from box a?

5.21. Of a given group of people, 40 percent are Republicans, 45 percent are Democrats, and 15 percent are independent. If 50 percent of the Republicans, 20 percent of the Democrats, and 30 percent of the independents are against a certain proposed tax bill, what is the probability that a voter selected at random from the above group is against the bill?

5.22. The probability that a firm will develop new inventions naturally increases as the R&D budget increases. In a survey of the pharmaceutical industry it was found that firms that spend less than $2 million annually on R&D have a 20 percent chance of obtaining more than three patents during the next year, whereas firms that spend $2 million or more have a 40 percent chance of obtaining more than three patents during the next year. Eighty percent of the firms in the industry spend less than $2 million on R&D.

Suppose you randomly select a firm in the pharmaceutical industry, and you find out that it obtained more than three patents during the past year. What is the probability that the firm's annual R&D budget is over $2 million?

5.23. Eighty percent of the students applying for jobs through the placement office of Good College are graduate students, while 20 percent are undergraduates. A graduate student applying through the office has a 60 percent chance of getting a job, while an undergraduate has only a 40 percent chance of being placed in a job.

(*a*) What is the probability that a student (randomly selected) who is applying through the office will get a job?
(*b*) If it is known that a student has gotten a job, what is the probability that he or she is a graduate student?

5.24. A market research project involves the study of the nutritional value of food and its effect on the demand for food. Two brands of cake mix are being compared. Sixty-five percent of the people surveyed said they pay no attention to nutritional content when they shop for food; 35 percent said they do pay attention. Of the 65 percent who do not consider nutritional content, 50 percent prefer Brand *A* to *B* and 50 percent prefer *B* to *A*. Of the 35 percent who do pay attention to nutritional content, 80 percent prefer brand *A* to *B* and 20 percent prefer *B* to *A*.

A shopper who intends to buy one of the two brands says she prefers *A* to *B*. What is the probability that she cares for nutritional content?

5.25. Table P5.25 is a cross-classification of a population of 10,000 stockholders by the value of their portfolios and the number of securities in their portfolios.

(*a*) If one of the 10,000 stockholders is selected at random, what is the probability that he owns 6–10 stocks at a total value of $2,000.0–5,999.9?
(*b*) What is the probability that the stockholder selected owns 1–5 stocks?
(*c*) If the stockholder selected is known to own stock at a total value of $11,000.0–20,999.9, what is the probability that he owns 16 or more stocks?
(*d*) If the stockholder selected is known to own 1–5 stocks, what is the probability that the total value of his holdings is $21,000 or more?

TABLE P5.25

		Number of Securities in Portfolio			
		1–5	*6–10*	*11–15*	*16+*
Value of stock	Up to $1,999.9	200	100	50	25
	$2,000.0–5,999.9	100	400	200	35
	$6,000.0–10,999.9	50	1,000	200	70
	$11,000.0–20,999.9	40	1,500	1,100	100
	$21,000 or more	30	3,500	1,000	300

5.26. The Safety Corporation, an international insurance firm with headquarters in New York, operates in many countries. The firm charges insurance premiums to cover its potential loss plus 30 percent to cover administrative expenses. Table P5.26 shows the frequency of death of males and females in Chile per 100,000 people born in a given year.

(a) Calculate the premium the firm will charge males and females in all age groups in Chile (but ignore people younger than 20).

(b) Draw schematic Venn diagrams showing why the premium for females in Chile in the age group 50–55 is higher than in the age group 30–35.

TABLE P5.26

Age group	Males	Females
0–1	11,668	10,313
1–5	2,493	2,557
5–10	576	527
10–15	438	367
15–20	740	568
20–25	1,197	799
25–30	1,502	1,010
30–35	2,108	1,442
35–40	2,661	1,768
40–45	3,461	1,879
45–50	4,169	2,478
50–55	5,460	3,595
55–60	6,523	4,677
60–65	8,403	6,840
65–70	9,673	8,884
70–75	11,667	11,507
75–80	10,567	12,835
80–85	7,897	11,217
85+	8,797	16,737

Source: Samuel H. Preston, Nathan Keyfitz, and Robert Schoen, *Causes of Death: Life Tables for Natural Populations* (New York: Seminar Press, 1972).

5.27. (a) Suppose a 60-year-old American male wants to buy a $100,000 life insurance policy for five years. Using Table 5.6 and assuming that the company does not charge anything above what is needed to cover the risk, what is the insurance premium? Assume that the insurance premium is payable at the beginning of the insurance coverage period.

(b) Rework part a, this time assuming that the man is 65 years old.

(c) Suppose a 60-year-old American man wants to buy a $100,000 life insurance policy for ten years. How much should the insurance company charge? Assume again that the entire premium for the ten years is payable at the beginning of the insurance coverage period.

(d) Is the premium determined in part c equal to the sum of the premiums determined in parts a and b? If not, how can the difference be explained? Explain in detail.

5.28. A committee is to consist of 5 people, to be chosen from a list of 6 Democrats, 4 Republicans, and 2 independents.

(a) How many distinctly constituted committees are possible?

(b) Assuming an equal chance for each of the 12 candidates to be selected, what is the probability that the committee will have no independents?

(c) What is the probability that the committee will consist entirely of Democrats?

5.29. The employees of Electric Apparatus, Inc. wish to elect a committee on which each department will be represented. There are four departments:

Department A   12 employees
Department B   10 employees
Department C   10 employees
Department D   8 employees

      (*a*) How many possibilities are there in selecting the committee membership?

      (*b*) If the employees decide to disregard the principle of departmental representation, how many possibilities exist then?

5.30. A bus company's standby repair crew consists of 3 electricians and 8 mechanics. There are 8 repair jobs to be performed, of which 2 require electrical repair and 6 require mechanical repair.

      (*a*) How many combinations of 2 electricians can be chosen to perform the jobs?

      (*b*) How many combinations of 6 mechanics can be chosen to perform the jobs?

      (*c*) In how many different ways can you assign the electricians and mechanics to their jobs?

5.31. Of 14 tax returns that were filled out by residents of a given state and received by its tax bureau, 2 need to be corrected. If only 4 forms will be randomly selected and carefully studied by the bureau's personnel, how many combinations of 4 is it possible to select among the 14 forms with exactly 1 of them needing to be corrected?

5.32. A truck driver has to deliver merchandise from point *A* to points *B*, *C*, *D*, and *E*. How many different routes can he take? Show your solution on a tree diagram and indicate what counting rule you used to get your answers.

5.33. Thirty-six kinds of ice cream are featured at the Ice Cream Palace. Four friends wish to buy one kind of ice cream each.

      (*a*) In how many different arrangements can they buy their ice cream?

      (*b*) If each decides to buy a *different* kind of ice cream, how many arrangements are possible?

CHAPTER SIX OUTLINE

6.1 Random Variables and Probability Distributions
The Cumulative Distribution Function
Discrete and Continuous Probability Distributions
The Uniform Distribution

6.2 Expected Values of Random Variables

6.3 The Variance and the Standard Deviation of a Random Variable
A Shortcut Formula for the Variance

6.4 Linear Transformation of a Random Variable

6.5 Skewness

6.6 The Relationship between a Population's Frequency Distribution and the Probability Distribution of a Random Variable

Key Terms

random variable
probability distribution
cumulative distribu-
   tion function
discrete random variable
continuous random variable
density function

uniform distribution
expected value (mean)
variance
standard deviation
skewness
coefficient of skewness

6
RANDOM VARIABLES, PROBABILITY DISTRIBUTIONS AND DISTRIBUTION PARAMETERS

6.1 Random Variables and Probability Distributions

Our interest in a random experiment is often diverted from the events that take place in the sample space to a variable (or variables) whose numerical value is determined by the result of the random experiment. The value of such a variable is thus determined by *chance*, and is called a *random variable*. Formally, we define it thus:

DEFINITION: RANDOM VARIABLE

A **random variable** is *a variable whose numerical value is determined by the outcome of a random experiment.*

And we define a related concept, that of the probability distribution, thus:

DEFINITION: PROBABILITY DISTRIBUTION

A **probability distribution** is *a systematic listing of all the possible values a random variable can take on, along with their respective probabilities.*

Let us see an example that will help to clarify these concepts.

EXAMPLE 6.1

Two research teams of a pharmaceutical firm are trying independently to develop totally different types of drugs. The probability of each team's success is 0.40. If the first team succeeds, the anticipated annual sales of the drug it develops are estimated at $400,000; if the second team succeeds, the annual sales of its drug are expected to reach $600,000. For convenience, let us define the two events:

A The first team succeeds in developing its drug.

B The second team succeeds in developing its drug.

There are four possible basic outcomes in the sample space: $A \cap B$, $A' \cap B$, $A \cap B'$, $A' \cap B'$. From the data we know that $P(A) = P(B) = 0.40$, and therefore $P(A') = P(B') = 0.60$. Since events A and B are independent, it follows that A' and B, A and B', and A' and B' are also independent pairs of events (see Chapter 5, Section 5.6), so that the probabilities of their intersections are as follows:

$$P(A \cap B) = P(A)P(B) \quad = 0.4 \cdot 0.4 = 0.16$$

$$P(A' \cap B) = P(A')P(B) \quad = 0.6 \cdot 0.4 = 0.24$$

$$P(A \cap B') = P(A)P(B') \quad = 0.4 \cdot 0.6 = 0.24$$

$$P(A' \cap B') = P(A')P(B') = 0.6 \cdot 0.6 = \underline{0.36}$$
$$1.00$$

Each of the four basic outcomes in the sample space is associated with estimated drug sales, and the value of these sales is the random variable in which we are interested. For example, event $A \cap B$ means that both teams are successful; in this event drug sales are estimated to be $400,000 + $600,000 = $1,000,000. Similarly, event $A' \cap B$ means that the second team is successful while the first team is not. The drug sales associated with this event are $0 + $600,000 = $600,000. Table 6.1 lists the four possible events along with their probabilities and their respective drug sales.

TABLE 6.1
Probabilities of Four Events and Associated Sales

Event	Probability	Drug sales
$A \cap B$	0.16	$1,000,000
$A' \cap B$	0.24	600,000
$A \cap B'$	0.24	400,000
$A' \cap B'$	0.36	0
	1.00	

As we have mentioned, in many instances our principal interest is not directed specifically toward the events that take place in the sample space, but rather toward their implications for other numerical variables whose values are determined by these events. In our example, management will probably be concerned about the probability distribution of the sales that will be generated by the new drugs. We then simply use X to represent the sales of the new drugs. The random variable X has the following (discrete) distribution:

x	P(x)
$ 0	0.36
400,000	0.24
600,000	0.24
1,000,000	0.16
	1.00

Note the distinction between X and x. While X represents the random variable, x is a specific value assumed by X. Since all possible values of the random variable X must be listed in a probability distribution, the sum of all the probabilities must always equal 1.00.

Diagrammatically, the discrete probability distribution of X is presented in Figure 6.1. Here the probability of every possible value of X is represented by a bar with a height equal to the probability of the occurrence of that value.

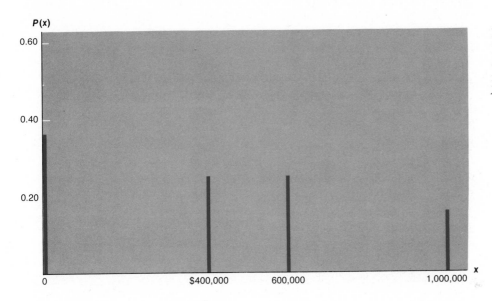

Figure 6.1

Probability distribution of random variable X

THE CUMULATIVE DISTRIBUTION FUNCTION

An interesting function to consider for some types of statistical analysis is the **cumulative distribution function.** Denoting the cumulative distribution function by $F(x)$, we define the function in the following way:

$$F(x) = P(X \leq x) \tag{6.1}$$

For any value of X, such as x, the cumulative function gives us the probability that X will be less than or equal to x. For our two drug teams, the probability that X will be less than or equal to \$200,000, say, is 0.36, the probability that X will be less than or equal to \$600,000 is 0.84, and so on. Systematically, we may write the cumulative distribution function as follows:

$$F(x) = \begin{cases} 0.00 & x < 0 \\ 0.36 & \$0 \le x < \$400,000 \\ 0.60 & \$400,000 \le x < \$600,000 \\ 0.84 & \$600,000 \le x < \$1,000,000 \\ 1.00 & \$1,000,000 \le x \end{cases}$$

where the range of x is listed on the right and the respective value of $F(x)$ is listed on the left. Note that the function $F(x)$ must be defined over the entire range $(-\infty, \infty)$, and so the intervals listed on the right should cover all of this range in a systematic order ranging from the lowest to the highest values of X. Diagrammatically the function as depicted in Figure 6.2 looks fairly simple. The function starts at zero on the left and rises to 1 on the right.

The cumulative function $F(x)$ provides us with an answer to the question "What is the probability that X will be *less than or equal to* some value x?" Because the probability distribution of our example is discrete, the cumulative function is a "step function." At each such step, the cumulative probability is read off the top of the vertical line. In Figure 6.2, the probability

Figure 6.2

Cumulative distribution function of X, $F(x)$

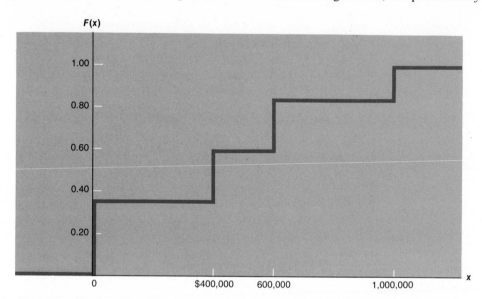

$F(\$400,000)$, for example, is 0.60, not 0.36. Another interesting cumulative function provides the answer to a second question: "What is the probability that X will be *greater than* some value x?" Denoting this function by $G(x)$, we define it as follows:

$$G(x) = P(X > x) \tag{6.2}$$

Comparing Equations 6.2 and 6.1, we realize that the following relationship holds:

$$G(x) = 1 - F(x) \tag{6.3}$$

Thus the probability under $G(x)$ is equal to 1 minus the corresponding probability under $F(x)$. To illustrate, let's calculate the probability that X will be greater than $\$500,000$ in our drug example. In this discrete example, if X is greater than $500,000 it can be equal either to $600,000 (probability of 24 percent) or to $1,000,000 (probability of 16 percent); thus the combined probability is $0.24 + 0.16 = 0.40$. We obtain the same probability if we use Equation 6.3. Since $F(\$500,000) = 0.60$, we get

$$G(\$500,000) = 1 - F(\$500,000) = 1.00 - 0.60 = 0.40$$

The function $G(x)$ for the drug example is presented graphically in Figure 6.3. On the vertical lines, the probability is read off the lowest point. For example, $G(\$400,000) = 0.40$, not 0.64.

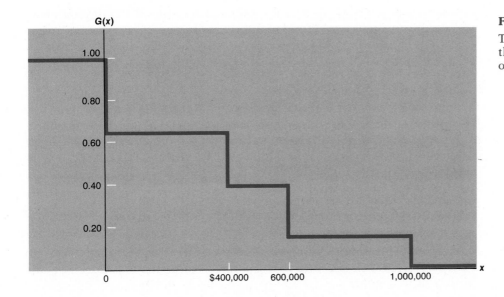

Figure 6.3

The "greater than" function of X, $G(x)$

DISCRETE AND CONTINUOUS PROBABILITY DISTRIBUTIONS

Our discussion so far has dealt only with discrete probability functions. A random variable, however, may be either *discrete* or *continuous*. **A discrete random variable** may obtain its value out of some finite number, or out of a countable infinite number of possible outcomes. **A continuous random variable,** on the other hand, may assume any numerical value on a continuous scale within a given interval and thus have an infinite number of possible outcomes. The interval within which the variable may obtain its values is called the *range*. It may be narrow or as wide as the entire range of real numbers: $-\infty$ to ∞. Such variables as weight, speed, volume, distance, and time are continuous. Regardless of the width of the interval, a continuous random variable assumes only one of the infinite number of possible values, and so the probability that such a variable will obtain a particular value is practically zero. For example, the probability that the weight of the coffee in any "10 oz." coffee jar is precisely equal to 10 ounces is practically zero, as we can be quite sure that there is some deviation between the actual coffee weight and the weight stated on the label. The deviation could be as tiny as one-thousandth of an ounce or one-millionth of an ounce, but some deviation is sure to exist.

Consider the amount of fluid poured into soda bottles by a bottling machine. The machine is adjusted to fill each bottle with 64 ounces of soda, but in fact there are slight deviations from one bottle to another. Past records show that the amount of soda per bottle varies within the range of 63 to 65 ounces, with the bulk of bottles containing $63\frac{1}{2}$ to $64\frac{1}{2}$ ounces. The amount of soda in the next bottle filled by the machine is a continuous random variable whose **density function** (the term used for the *probability distribution of a continuous variable*) is depicted in Figure 6.4. It is important to note that here, unlike the case of a discrete random variable, the height of the function does *not* indicate the probability. The probability is read off the graph in a different way, as we shall soon see.

A property of all density functions is that the area under the function over the entire range of the variable is equal to 1 (measured in any desired units). This is analogous to the equivalent property of discrete random variables, whose probabilities always total 1. Thus the area under the bell-shaped function in Figure 6.4 is equal to 1. The probability that the random variable will obtain values within any interval is represented diagrammatically by the area under the density function over that interval. For example, the probability that the amount of soda in any given bottle will be somewhere between $63\frac{1}{2}$ and $64\frac{1}{2}$ ounces is shown in Figure 6.4 by the darkest area. If the darkest area is equal to 0.75, then there is a 75 percent chance that the next bottle filled will contain between $63\frac{1}{2}$ and $64\frac{1}{2}$ ounces of soda. Twelve and a half percent of the bottles are filled with less than $63\frac{1}{2}$ ounces of soda, while the remaining $12\frac{1}{2}$ percent are filled with more than $64\frac{1}{2}$ ounces; thus the two white tails of the density function are equal to 0.125 each. We noted earlier that the probability of any single value of a continuous random variable is equal to zero. Diagrammatically, over a single value (say, $6\frac{3}{4}$ ounces) one can draw only a vertical line, but the *area* of a line is zero, and therefore so is the probability of that value.

Figure 6.4

Density function of the continuous variable *X*

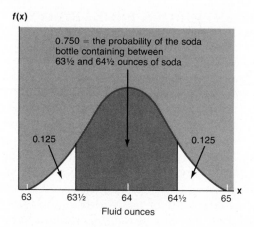

THE UNIFORM DISTRIBUTION

To clarify further the concept of a density function, let us now focus on a simple but common one: the **uniform distribution.**

Assume that the variable *X* can take on any value in the range from 10 to 20, and that the probability that the variable will assume a value within any interval in this range is the same as the probability that it will assume a

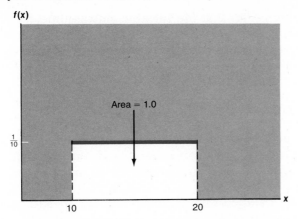

Figure 6.5

Density function
of a uniformly
distributed
random variable

value in any other interval of the same width in that range. For example, the
probability that *X* will assume a value in the range 11 to 13 is the same as
the probability that it will assume a value in the range 15 to 17 (since both
intervals have the same width). If the above conditions hold, then *X* is
uniformly distributed and the area under the density function is a rectangle
of the type shown in Figure 6.5. The rectangle's area is equal to 1, meaning
that *X* is sure to take on a value in the range 10 to 20. Formally we have
$P(10 \leq X \leq 20) = 1$.

Since the width of the range is 10 units $(20 - 10 = 10)$, the height of the
density function must be $\frac{1}{10}$ so as to make the rectangle's area equal to 1
$(10 \cdot \frac{1}{10} = 1)$. With this in mind, one can easily determine probabilities under
uniform probability distribution. To illustrate, let us find the probability
that *X* will assume a value between 14 and 18. This probability is graphically
shown by the darkest area in Figure 6.6, which is equal to the width, 4,
$(= 18 - 14)$, times the height, $\frac{1}{10}$, or $4 \cdot \frac{1}{10} = 0.40$. Thus we conclude that the
probability is 0.40. Note that 0.40 is the ratio of the darkest area in Figure
6.6 to the total area in the range (10–20), since the latter is equal to 1.

It seems in order to note at this point that we occasionally treat some
discrete random variables as continuous and vice versa. For example,
variables measured in dollars—profit, net return on investment, a firm's
operating costs, and the like—are theoretically discrete because in any finite
range the possible outcomes are determined by the number of cents within
the interval (cent being the smallest monetary unit). Because of the large

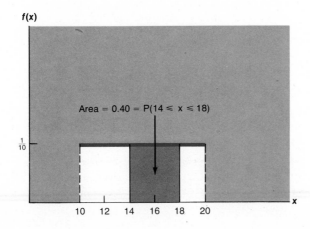

Figure 6.6

Uniformly
distributed
random variable
X, showing
probability that
X will assume a
value between
14 and 18

number of possible values that the variable may assume, however, it is often regarded as continuous. On the other hand, a variable such as age, which is basically continuous, is often broken down into age intervals and treated like a discrete variable.

6.2 Expected Values of Random Variables

Now that we have become familiar with the concept of a probability distribution for a random variable, we can go on to consider its expected value. In many situations we may be interested in the value of the variable "on the average."

Let us consider a game in which the prizes are $0, $10, and $20, and the probabilities are 0.50, 0.25, and 0.25, respectively. This information is summarized in the following probability distribution for the prize X:

Prize (x)	P(x)
$ 0	0.50
10	0.25
20	0.25

To find the "expected prize," we simply multiply each possible prize by its corresponding probability and total the products. This procedure gives us a weighted average value of the prize, in which the probabilities are used as weights. Denoting the *expected value* of X (or the *mean* of X) by $E(X)$, we may simply write

$$E(X) = \$0(0.50) + \$10(0.25) + \$20(0.25) = \$0.00 + \$2.50 + \$5.00 = \$7.50$$

Seven dollars and fifty cents is what one can expect to win per game, on the average, if the game is repeated many times.

Another convenient and conventional notation for the mean of a random variable is μ, which we used earlier for a population mean. So turning to the formal definition, we can state:

THE EXPECTED VALUE OF A DISCRETE RANDOM VARIABLE

The **expected value** or **mean (μ)** of a discrete random variable is the *sum of the products of all possible values of the random variable and their corresponding probabilities.* The expected value of a discrete random variable can be expressed as follows:

$$\mu \equiv E(X) = \Sigma x P(x) \tag{6.4}$$

Again, x refers to each value that X assumes.

Let us further illustrate the calculation of the mean of a random variable with examples.

EXAMPLE 6.2

One of the people questioned in a survey on consumers' anticipation of next year's rate of inflation (X) has given the following probability distribution:

x (percent)	P(x)
4	0.05
5	0.06
6	0.20
7	0.30
8	0.25
9	0.10
10	0.04
	1.00

This consumer, then, believes that there is a 5 percent chance that next year's inflation rate will be 4 percent, a 6 percent chance that next year's inflation rate will be 5 percent, a 20 percent chance that the rate will be 6 percent, and so on. The mean rate anticipated by this individual is 7.1 percent, calculated as follows:

$$E(X) = \Sigma x P(x)$$

$$\mu \equiv 4(0.05) + 5(0.06) + 6(0.20) + 7(0.30) + 8(0.25) + 9(0.10) + 10(0.04)$$

$$= 0.20 + 0.30 + 1.20 + 2.10 + 2.00 + 0.90 + 0.40 = 7.10$$

Thus the one rate of inflation that would best represent this individual's anticipation of next year's inflation rate is 7.1 percent.

EXAMPLE 6.3

A stockholder who has $10,000 worth of stock in a given company is unsure about the amount of cash dividends the firm will decide to distribute. His expectations are probabilistic: he assumes a 25 percent chance that his dividends will amount to $500, a 50 percent chance that they will amount to $700, and a 25 percent chance that they will amount to $1,000. The expected value of the dividends he anticipates is equal to the sum of the products of the dividends and their respective probabilities, so that

$$E(X) = \$500(0.25) + \$700(0.50) + \$1,000(0.25)$$

$$= \$125 + \$350 + \$250 = \$725$$

6.3 The Variance and the Standard Deviation of a Random Variable

While the mean of a probability distribution is a very useful and informative piece of data, we are often interested in other measures of a distribution in addition. The *variance* and the *standard deviation* are measures of the dispersion of a random variable around its mean. While it is interesting to know that the expected number of children of a randomly selected couple in Massachusetts after ten years of marriage is 1.93, this figure does not reflect the degree to which the couples differ from one another with respect to the number of children they have. The variance and standard deviation reflect this degree of diversity.

Technically, the **variance** is the *expected value of the squared deviation of a random variable from its mean.* A general formula for the variance, which applies to both discrete and continuous variables, is

$$\sigma^2 \equiv \text{var}(X) = E(X - \mu)^2 \qquad (6.5)$$

The symbol σ^2 (sigma squared) is the common notation for the variance.

In the case of a discrete random variable, the formula can be modified somewhat, as follows:

$$\sigma^2 = \Sigma(x - \mu)^2 P(x) \qquad (6.6)$$

To illustrate, consider the example that follows.

EXAMPLE 6.4

To determine how frequently a cigarette machine should be restocked, the average as well as the dispersion of the number of cigarette packs bought (X) has to be determined. Suppose the probability distribution of X is as follows:

x	$P(x)$
10	0.10
11	0.15
12	0.25
13	0.25
14	0.20
15	0.05
	1.00

We first calculate the mean number of cigarette packs bought:

$$\mu = 10(0.10) + 11(0.15) + 12(0.25) + 13(0.25) + 14(0.20) + 15(0.05)$$
$$= 1.00 + 1.65 + 3.00 + 3.25 + 2.80 + 0.75 = 12.45$$

The variance may now be easily calculated by means of Equation 6.6:

$$\text{var}(X) = \Sigma(x - \mu)^2 P(x)$$

$$\sigma^2 \equiv (10.00 - 12.45)^2 \cdot 0.10 + (11.00 - 12.45)^2 \cdot 0.15$$
$$+ (12.00 - 12.45)^2 \cdot 0.25 + (13.00 - 12.45)^2 \cdot 0.25$$
$$+ (14.00 - 12.45)^2 \cdot 0.20 + (15.00 - 12.45)^2 \cdot 0.05$$
$$= 0.600 + 0.315 + 0.051 + 0.076 + 0.481 + 0.325 = 1.848$$

The **standard deviation,** which we denote by σ (sigma), is also a measure of the dispersion of a probability distribution and is simply equal to *the positive square root of the variance.*[1] In our last example we get $\sigma = \sqrt{1.848} = 1.359$. While the mean number of cigarette packs bought is 12.45, there is a standard deviation of 1.359 packs around that mean.

A SHORTCUT FORMULA FOR THE VARIANCE

Equation 6.5 may be simplified in the following way:

$$\sigma^2 = E(X - \mu)^2 = E(X^2 + \mu^2 - 2X\mu) = E(X^2) + E(\mu^2) - E(2X\mu)$$

Since μ is a constant, we get[2]

$$\sigma^2 = E(X^2) + \mu^2 - 2\mu E(X) = E(X^2) + \mu^2 - 2\mu^2 = E(X^2) - \mu^2$$

We thus conclude that

$$\sigma^2 = E(X^2) - \mu^2 \tag{6.7}$$

where $E(X^2)$ is simply the expected value of the variable X^2. We should note that Equation 6.7 applies to both discrete and continuous random variables. To find the variance in Example 6.4 by means of Equation 6.7, we first calculate $E(X^2)$ as follows:

$$E(X^2) = 10^2(0.10) + 11^2(0.15) + 12^2(0.25) + 13^2(0.25)$$
$$+ 14^2(0.20) + 15^2(0.05) = 100(0.10) + 121(0.15)$$
$$+ 144(0.25) + 169(0.25) + 196(0.20) + 225(0.05)$$
$$= 10.00 + 18.15 + 36.00 + 42.25 + 39.20 + 11.25 = 156.85$$

so that by Equation 6.7 we get

$$\sigma^2 = 156.85 - (12.45)^2 = 156.85 - 155.00 = 1.850$$

and $\sigma = \sqrt{1.850} = 1.360$: we obtained the same results earlier by means of Equation 6.6. (The deviation here is due to rounding.)

---

[1] Just as in the case of a population variance, we note here that the standard deviation is expressed in more easily interpreted units than those of the variance. To illustrate, consider the following distribution of X:

x	$P(x)$
$10	$\frac{1}{2}$
20	$\frac{1}{2}$
$\mu = \$15$	

$$\sigma^2 = (10 - 15)^2 \cdot 0.50 + (20 - 15)^2 \cdot 0.50 = 25 \cdot 0.50 + 25 \cdot 0.50 = 25$$

The variance is equal to 25 "squared dollars." When we take the standard deviation we get $\sigma = \sqrt{25} = \$5$. Thus the standard deviation is easier to interpret intuitively, since it is expressed in the same units as the data.

[2] See the discussion in Section 6.4.

6.4 Linear Transformation of a Random Variable

It often becomes necessary to find the expected value and the variance of a transformed random variable. Assume, for example, that the net return on a $100 investment (after one year, say) is the random variable X, whose probability distribution (rather simplified, of course) is given below:[3]

x	$P(x)$
−$10	1/8
5	1/8
12	1/2
15	1/8
22	1/8

The expected net return on the $100 investment is thus

$$E(X) = (-\$10) \cdot \frac{1}{8} + \$5 \cdot \frac{1}{8} + \$12 \cdot \frac{1}{2} + \$15 \cdot \frac{1}{8} + \$22 \cdot \frac{1}{8} = \frac{\$80}{8} = \$10$$

Given that $10 represents the expected return on the $100 investment, one might be interested in finding the expected return on an $850 investment. It can easily be shown that the expected value of the return on the investment is $85, which equals 8.5 times the return on $100. This example may be generalized. If $E(X) = \mu$, and b is a constant,[4] then:

The expected value of the multiple of a random variable is equal to the multiple of the expected value.

$$E(bX) = b\mu \tag{6.8}$$

It can also be shown that for a constant a, the following rule applies:[5]

$$E(a + X) = a + \mu \tag{6.9}$$

Combining the results of Equations 6.8 and 6.9 we get

$$E(a + bX) = a + b\mu \tag{6.10}$$

If, for instance, there is a $3 fixed transaction cost involved in an investment (so that this cost is not a function of the amount invested) and $850 is

---

[3] Net return of −$10 means that only $90 was returned on the $100 investment, thus there was a net loss of $10. Similarly, net return of $22 means that $122 was returned on the $100 investment.

[4] In the case of a discrete random variable we can easily show that Equation 6.8 follows from the simple rules of summation: $E(bX) = \Sigma(bX)P(X) = b\Sigma XP(X) = bE(X)$.

[5] In the case of a discrete random variable, Equation 6.9 follows directly from the simple rules of summation, as follows:

$$E(a + X) = \Sigma(a + X)P(X) = \Sigma aP(X) + \Sigma XP(X) = a\Sigma P(X) + \Sigma XP(X)$$

Equation 6.9 follows directly, since $\Sigma P(X) = 1$ and $\Sigma XP(X) = \mu$.

invested in the stock, the expected net return after transaction cost may be calculated in the following way (by Equation 6.10):

$$E(-3 + 8.5X) = -3 + 8.5E(X) = -3 + (8.5 \cdot 10) = \$82$$

While the expected return before the transaction cost is \$85, because of the \$3 transaction cost (independent of volume) the expected net return is \$85 − \$3 = \$82, as we found when we used Equation 6.10.

While adding a constant (either positive or negative) to a random variable has an effect on the expected value of a variable, as shown by Equations 6.9 and 6.10, there is no such effect on the variance. Consider a group of students standing barefoot on the floor of a gym. Let their heights be the variable under consideration. The average height and variance of height for the group may be easily calculated. Now suppose each of the students is given identical pairs of sneakers to wear. The height of each student measured from the floor to the top of his or her head is now greater by, say, one inch. Since all the sneakers are identical by assumption, each student is now one inch taller. What happens to the average height? It increases one inch. What happens to the variance? Nothing! It remains unchanged since all heights relate to *one another* and to the *average height* in the same way as they did originally. We thus conclude:[6]

$$\text{var}(a + X) = \text{var}(X) = \sigma^2 \qquad \textbf{(6.11)}$$

When we multiply a variable by a constant, the variance of the product is[7]

$$\text{var}(bX) = b^2\sigma^2 \qquad \textbf{(6.12)}$$

Combining Equations 6.11 and 6.12, we obtain:

$$\text{var}(a + bX) = b^2\sigma^2 \qquad \textbf{(6.13)}$$

and the standard deviation of $(a + bX)$ is:

$$SD(a + bX) = \sqrt{b^2\sigma^2} = |b|\sigma \qquad \textbf{(6.14)}$$

where $|b|$ is the absolute value of b.

---

[6] The proof is rather simple. From Equation 6.5 we know that $\sigma^2 = E(x - \mu)^2$. The mean of the variable $a + X$ is equal to $a + \mu$, as indicated by Equation 6.9. The variance of $a + X$ is

$$\sigma^2_{a+X} = E[(a + X) - (a + \mu)]^2 = E[a + X - a - \mu]^2 = E(X - \mu)^2 = \sigma^2_X$$

[7] To derive Equation 6.12, we again go back to Equation 6.5:

$$\sigma^2_{bx} = E[(bx) - E(bX)]^2 = E\{b[X - E(X)]\}^2$$
$$= E\{b^2[X - E(X)]^2\} = b^2E(X - \mu)^2 = b^2\sigma^2$$

6.5 Skewness

The mean and the variance of a probability distribution are important parameters reflecting location and dispersion. While they give us considerable information about the distribution, often there are additional characteristics of interest to us which are exhibited neither in the mean nor in the variance. The degree of asymmetry, called **skewness,** is not reflected in these parameters. Skewness (μ_3) is defined as follows:

FORMULA: SKEWNESS

$$\mu_3 = E(X - \mu)^3 \tag{6.15}$$

To see the importance of skewness and its meaning, take a look at Figure 6.7. Here three continuous probability distributions are drawn with the same mean and variance but different skewnesses. The distribution drawn in part (a) is positively skewed ($\mu_3 > 0$), the one in part (b) is symmetrical and so its skewness is equal to zero ($\mu_3 = 0$), and the one in part (c) is negatively skewed ($\mu_3 < 0$).

Figure 6.7

Continuous probability distributions with same mean and variance but different skewnesses

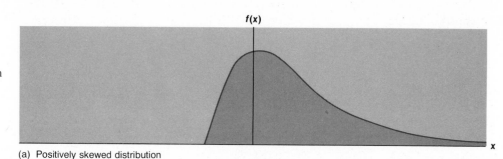

(a) Positively skewed distribution

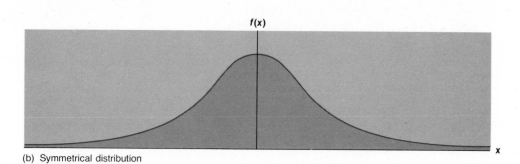

(b) Symmetrical distribution

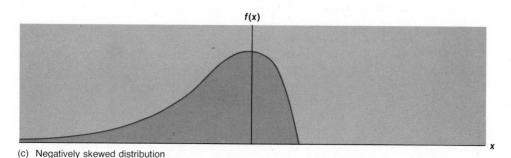

(c) Negatively skewed distribution

It is impossible to discriminate between the three distributions in terms of their means and variances: their means and variances are equal. The distributions are not the same, however, and so the information augmented by μ_3 is often relevant and sometimes vital for the analysis of the random variables' behavior. Take Mr. Young Fearful, for example, a young college graduate in business administration who has $30,000 available for investment in a new business. Mr. Fearful considers three potential investment alternatives and soon realizes that profits in the "real world" can never be known in advance. His collection of information indicates, though, that the profit probability distributions of his three alternatives follow the three patterns presented in Figure 6.7. To avoid the risk of severe loss, on the one hand, and to maintain at least some chance of extremely high profits, on the other, he chooses the investment whose profit probability distribution is positively skewed. Note that information dealing with the mean and the variance only would not suffice to inform Mr. Fearful as to which of the probability distributions leaves him more vulnerable to severe losses.

Most people, whether fearful or not, will probably agree with Mr. Fearful's choice. This, however, does not imply that in general, positive skewness is always a desirable feature for a probability distribution to have. If, for example, the curves in Figure 6.7 are thought of as alternative cost distributions, most people would undoubtedly prefer the negatively skewed distribution, so as to avoid the chance of incurring very high costs. Thus, the attractiveness of positive or negative skewness depends on the type of variable under consideration; but in all cases, as we can see, the skewness reveals information about the distribution not revealed by the mean or the variance.

As was the case with the population's frequency distribution, skewness as defined in Equation 6.15 is difficult to interpret numerically. A useful index that is often easier to handle is **the coefficient of skewness,** which is expressed thus:

FORMULA: THE COEFFICIENT OF SKEWNESS

$$\text{Coefficient of skewness} = \frac{E(X - \mu)^3}{\sigma^3} = \frac{\mu_3}{\sigma^3} \tag{6.16}$$

Let us turn now to an example that will illustrate the calculation of skewness and the coefficient of skewness. Our example involves a distribution famous for its high positive skewness: the distribution of lottery prizes.

EXAMPLE 6.5

One hundred thousand tickets are issued weekly under a state lottery program. The tickets are offered to the general public for $3 each. The following is a list of prizes given in the lottery:

Number of prizes	Amount of each prize
1	$10,000
2	5,000
500	100
10,000	3
89,497	0

Find the expected value as well as the standard deviation and skewness of the prize distribution. Also, find the expected loss involved for a single lottery ticket.

The solution is straightforward. We start by calculating the expected prize $E(X)$:[8]

$$E(X) = 0.00001(10,000) + 0.00002(5,000)$$
$$+ 0.005(100) + 0.1(3) + 0.89497(0)$$
$$= \$1$$

The expected value of the prize is $1; therefore, the expected loss is $3.00 − $1.00 = $2.00, since the tickets are sold for $3. The variance is calculated as follows:

$$\sigma^2 \equiv \text{Var}(X) = 0.00001(10,000 - 1)^2 + 0.00002(5,000 - 1)^2$$
$$+ 0.005(100 - 1)^2 + 0.1(3 - 1)^2 + 0.89497(0 - 1)^2$$
$$= 1,549.90 \text{ (squared dollars)}$$

and the standard deviation is

$$\sigma = \sqrt{1,549.90} = \$39.37$$

Let us now proceed with the calculation of the skewness:

$$\mu_3 = 0.00001(10,000 - 1)^3 + 0.00002(5,000 - 1)^3 + 0.005(100 - 1)^3$$
$$+ 0.1(3 - 1)^3 + 0.89497(0 - 1)^3$$
$$= 12,500,352 \text{ (cubic dollars)}$$

To get the coefficient of skewness we divide by σ^3:

$$\text{Coefficient of skewness} = \frac{\mu_3}{\sigma^3} = \frac{12,500,352}{(39.37)^3} = 204.85$$

Both μ_3 and the coefficient of skewness indicate a strong positive skewness of the prize distribution.

[8] Since there are 100,000 tickets outstanding weekly, the probability of getting the first prize is 1/100,000 = 0.00001, the probability of getting the second prize is 2/100,000 = 0.00002, and so on.

6.6 The Relationship between a Population's Frequency Distribution and the Probability Distribution of a Random Variable

In Chapters 3 and 4 we presented the concept of a population, its frequency distribution, and its parameters. In this chapter we have discussed the notion of a random variable, its probability distribution, and its parameters. There is a delicate distinction between these two formulations (populations and random variables), and it is to the clarification of this distinction that this section is devoted.

Suppose a small town of 100 adult residents has the following annual income frequency distribution:

Income (X)	Frequency (f)	Relative frequency f / 100
$ 5,000	10	0.10
12,000	60	0.60
15,000	30	0.30
$\Sigma f = N = 100$		$\Sigma \dfrac{f}{N} = 1.00$

Figure 6.8 shows the relative frequency distribution of the town's income. The above data show that 10 percent of the town's population earn $5,000 annually, 60 percent earn $12,000, and 30 percent earn $15,000. *These are descriptive data.* Now suppose we would like to select *one person at random* out of the town's population. The income of that person, before the selection, is a random variable. There is a 10 percent chance that his income is $5,000, a 60 percent chance that it is $12,000, and a 30 percent chance that it is $15,000. The probability distribution of the random variable does not differ in any way from the relative frequency distribution shown in Figure 6.8. We only need to replace the words "Relative frequency" by "Probability" to obtain a probability distribution. Thus, the sole difference between the frequency distribution and the probability distribution is in the differing interpretations given to them. When dealing with the population, we can state with *full certainty* that 10 percent of the population have incomes of $5,000, 60 percent have incomes of $12,000, and so on. There is no such certainty with respect to the random variable, since the relative frequencies are interpreted here as the probability of occurrence of *uncertain* outcomes. We should note that the mean income computed from the frequency distribution ($12,200, as you can verify) is equal to the expected income of the random variable. This is similarly true with respect to other parameters, such as the variance and the skewness, but the same distinction between a population's frequency and a random variable's probability is carried over to them as well.

Consider a newly formed firm whose management is trying to forecast its sales for the following year. Past data are not available, and sales cannot be known with certainty in advance. Unlike the situation in the previous example, it is clear that next year's sales is a random variable, and no parallel population exists whose relative frequency is equal to the probability distribution of the random variable. The probability distribution of this random variable is subjective and has nothing to do with any population's frequency distribution.

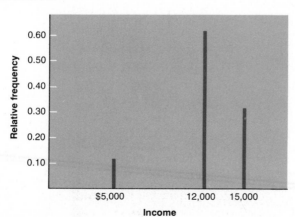

Figure 6.8

Relative frequency distribution of town income

Problems

6.1. In a football game between Team A and Team B, the probability concerning the game's outcome is as follows:

Event	Probability
Team A will win	0.60
Team A will not win	0.40
	1.00

   (a) Have we defined a random variable above?
   (b) If not, define one related to the game, and write down its probability distribution.

6.2. Is the cumulative probability distribution of a discrete random variable discrete or continuous? Why?

6.3. Over each of the sample spaces of the following random experiments, define two random variables:

   (a) A coin is flipped once, and a head or a tail is observed.
   (b) Two coins are flipped and heads and tails are observed.
   (c) Next Tuesday's weather.

6.4. Classify the following random variables as discrete or continuous:

   (a) The number of children present in a classroom at a given time.
   (b) The daily volume of gasoline sold at a given gas station.
   (c) Tomorrow's temperature.
   (d) The daily number of customers coming into a furniture store.

6.5. Meat orders are placed periodically by a large meat store. The delivery period (that is, the time that elapses between the placement of the order and the delivery of the meat) is a random variable, X (measured in days), with the following probability distribution:

$$P(X = x) = \begin{cases} \dfrac{1}{10} \cdot (7 - x) & x = 3, 4, 5, 6 \\ 0 & \text{otherwise} \end{cases}$$

   (a) Is the probability distribution discrete or continuous?
   (b) Graph the probability distribution.
   (c) Write down and graph the cumulative distribution function.

6.6. Consider the following distributions of $X_1, X_2,$ and X_3:

x_1	$P(x_1)$	x_2	$P(x_2)$	x_3	$P(x_3)$
0	$\frac{1}{4}$	0	$\frac{3}{8}$	0	$\frac{1}{2}$
10	$\frac{1}{4}$	10	$\frac{1}{8}$	30	$\frac{1}{2}$
20	$\frac{1}{4}$	20	$\frac{1}{8}$		
30	$\frac{1}{4}$	30	$\frac{3}{8}$		

   (a) Find the means of $X_1, X_2,$ and X_3.
   (b) Determine *without any calculations* which of the distributions has the greatest variance. Explain how you reached your decision.
   (c) Verify your answer to part b by calculating the variances.

6.7. Calculate the mean, variance, and skewness of the following distributions:

x_1	$P(x_1)$	x_2	$P(x_2)$
10	$\frac{1}{4}$	0	$\frac{1}{8}$
20	$\frac{1}{4}$	20	$\frac{1}{4}$
30	$\frac{1}{4}$	30	$\frac{1}{4}$
40	$\frac{1}{4}$	50	$\frac{3}{8}$

6.8. Hetty and Taily play the following game: Hetty starts off with one cent and Taily with two cents. Hetty and Taily flip, *simultaneously,* one coin each. If both coins fall on the same side, Taily gives Hetty one cent. If the coins fall on different sides, Hetty gives Taily one cent. The game continues until either Hetty or Taily has all three coins.

(*a*) Write down the probability distribution of X, where X is the number of flips required to end the game.

(*b*) Write down the cumulative probability distribution of X.

6.9. Compute the mean, variance, standard deviation, and skewness of the variables $X_1, X_2, X_3, X_4,$ and X_5, whose probability distributions are given below:

x_1	$P(x_1)$	x_2	$P(x_2)$	x_3	$P(x_3)$
10	$\frac{1}{2}$	10	$\frac{3}{4}$	10	$\frac{9}{10}$
20	$\frac{1}{2}$	20	$\frac{1}{4}$	20	$\frac{1}{10}$

x_4	$P(x_4)$	x_5	$P(x_5)$
10	$\frac{1}{4}$	10	$\frac{1}{10}$
20	$\frac{3}{4}$	20	$\frac{9}{10}$

6.10. Compute the mean, variance, standard deviation, and skewness of the variables $X_1, X_2,$ and X_3, whose probability distributions are given below:

x_1	$P(x_1)$	x_2	$P(x_2)$	x_3	$P(x_3)$
100	$\frac{1}{2}$	90	$\frac{1}{2}$	70	$\frac{1}{2}$
120	$\frac{1}{2}$	130	$\frac{1}{2}$	150	$\frac{1}{2}$

6.11. A painting service is bidding for a contract on which it hopes to make a profit of $40,000. The cost of preparing the bid and submitting it is $3,000, and the probability that the company will get the contract is 0.50. What is the expected profit, its standard deviation, and its skewness?

6.12. Illinois Chemicals, Inc., produces a certain quantity of a given chemical every day. The quantity produced is a random variable (X) having a uniform distribution in the range 16–19 pounds.

(*a*) Write down and draw the density function of X.

(*b*) Write down and draw the cumulative probability distribution.

(*c*) Calculate the probability that the production on a certain day will be between 17 and 17.5 pounds.

6.13. The German state lottery is often advertised in American magazines. An ad appearing in the January 8, 1979 issue of *Business Week* is an example. According to the ad, a ticket will cost you 618 German marks (DM). While the actual lottery is a bit more involved, we will assume that there are 106,643 prizes, as shown in Table P6.13. Find the expected value of the prize, assuming that you buy one ticket, and compare it to the ticket's price. Also calculate the standard deviation and the coefficient of skewness of the prize. Note that 300,000 tickets are sold altogether, each having an equal chance of winning.

TABLE P6.13

Number of prizes	Amount (DM)
193,357	0
6,000	200
6,600	300
7,800	400
8,730	500
73,260	600
390	700
420	800
2,850	1,000
150	2,000
120	3,000
135	5,000
96	10,000
24	25,000
12	50,000
35	100,000
21	1,000,000
300,000	

6.14. Referring back to Problem 6.13, what is your expected profit (or loss) if you decide to join with a friend, buy two tickets, and share the cost and all prospective prizes equally?

6.15. Write down the probability distribution of a discrete random variable of your choice. Make sure the function has the following parameters:

$$\mu = 70, \quad \sigma = 25, \quad \mu_3 = 0$$

6.16. The following is a company's anticipated sales distribution:

x	$P(x)$
$20,000	0.20
25,000	0.20
30,000	0.20
35,000	0.20
40,000	0.20
	1.00

(a) Find the expected value of the sales and the variance of sales.

(b) Find the expected value and variance of sales when the latter is expressed in thousands of dollars.

(c) If there is a fixed cost of $15,000 and variable cost of 50 cents per unit, what is the expected value and variance of the profit? Assume that the company sells each unit of production for $1.

Hint: If the price of the product is denoted as P, the variable cost per unit as C, the number of units sold as X, fixed cost as F, and profit as π, π is given by:

$$\pi = (P - C)X - F.$$

6.17. The Alpha Company employs salespersons to market its products. Each of the salespersons is paid $120 a week plus 15 percent commission on the sales he or she makes. The number of items a salesperson sells in one week is a random variable whose probability distribution is the same for all salespersons and is as follows:

x	P(x)
10	0.15
11	0.15
12	0.40
13	0.15
14	0.15
	1.00

The price of each item sold is $100.

(a) Find the expected weekly income of one salesperson.

(b) Find the standard deviation of the weekly income of one salesperson.

(c) Answer parts a and b again, this time assuming that the weekly fixed income has risen from $120 to $150.

(d) What are the answers to a and b if the weekly income is $200 plus 20 percent commission?

6.18. A medical insurance plan for individuals to cover the expenses of visits to their doctors is based on an insurance premium of $120 per year. Suppose each visit of the insured to a doctor costs $25, of which 20 percent is paid by the insured and 80 percent by the insurance company. Also assume that each policy contains a $50 deductible clause, so that the first two visits to a doctor are not covered by the plan. Suppose the probability distribution describing the chances that an individual will visit a doctor a given number of times per year is as shown in Table P6.18. Find the company's expected income (i.e., $120 minus the expected payments to cover the cost of doctor visits) from one policy and its variance.

TABLE P6.18

Number of doctor visits	Probability
0	0.10
1	0.30
2	0.20
3	0.08
4	0.07
5	0.06
6	0.05
7	0.04
8	0.04
9	0.04
10	0.02
	1.00

6.19. The ABC Company supplies a given item to the DEF Company. According to the contract between the companies, ABC will supply exactly 10,000 items at a price of $100 each. The production cost of the items consists of $200,000 in fixed costs plus additional variable costs in the amount of $C per item, where C is a random variable having the following probability distribution:

c	P(c)
$6	0.30
7	0.40
8	0.20
9	0.10
	1.00

(*a*) Calculate the skewness of the variable cost per unit, C.

(*b*) Calculate the skewness of total variable costs (i.e., variable cost per unit times the number of units produced).

(*c*) ABC's net profit is equal to the revenues less variable costs and less fixed costs. Find the skewness and the coefficient of skewness of the net profit. Ignore taxes. What is the relationship between the skewness of C and the skewness of the net profit?

Hint: Avoid unnecessary calculations and use the fact that the net profit is a linear transformation of the variable cost per unit, C.

CHAPTER SEVEN OUTLINE

Key Terms

Bernoulli process
binomial probability
   distribution
geometric probability
   distribution
Poisson probability
   distribution

7

IMPORTANT DISCRETE DISTRIBUTIONS

Some probability distributions (discrete and continuous) are more common than others, and hence deserve our special attention. In this chapter we focus on three important discrete probability distributions: the binomial, the geometric, and the Poisson probability distributions. All of these distributions have both theoretical and practical applications. We shall begin by describing the Bernoulli process, and in doing so lay the foundation for the introduction of the binomial and geometric distributions.

7.1 The Bernoulli Process

Many business-related processes consist of a number (great or small) of repetitive independent trials: making a series of telephone calls, responding to signals requiring immediate attention (such as firefighter response to fire alarms), periodically examining the quality of items produced by a given machine, and so on.

Let us examine an isolated trial in one of these processes. The first process, for example, is described as "making a series of telephone calls." One trial in this process is "making a telephone call." The result of this trial may be described in a variety of ways such as "reached a wrong number" versus "reached the correct number," "operator's assistance was required" versus "operator's assistance was not required," and so on. Suppose we are interested in just one aspect of the result: whether or not the call was long-distance. In this case we may describe the trial as a dichotomy (that is, as having two and only two possible outcomes), since the outcome must be that the call was either long-distance or not. No other outcomes are possible according to this classification. Thus the two outcomes are mutually exclusive and collectively exhaustive, and one and only one of them will occur in any given trial. The trials of the other processes may be classified according to similar complementary outcomes. For example, firefighters' responsiveness to a fire alarm could be thought of as a process in which each response is considered to be one trial with two possible outcomes. The outcomes we consider depend on our interest. We may wish to consider the following pair of outcomes: "at most one fire engine responded" versus "more than one fire engine responded." Alternatively, we might consider "the alarm was false" versus "the alarm was not false."

Let us leave examples for a time and return to more general discussion. First, to facilitate reference to the trial's outcomes we name them "successes" and "failures," although the outcomes may involve no accomplishments or disappointments of any kind. So if "the call was long-distance" is termed a "success," then "the call was not long-distance" is called a "failure," and vice versa.

Now let us recapitulate. We are dealing with a process involving a number of repetitive independent trials, the outcome of each being termed either a "success" or a "failure." When we say that these trials are independent and repetitive, we indicate that they are of the same nature—and hence if the probability of "success" in one trial is p, so is the probability of "success" in all the other trials. *The process generating this series of independent trials is called a* **Bernoulli process;** it is named after Daniel Bernoulli, an eighteenth-century Swiss mathematician.

The letter p is customarily used to denote the probability of success for any given trial ($0 \leq p \leq 1$). The probability of "failure" is denoted by q ($0 \leq q \leq 1$) and is computed as follows:

$$q = 1 - p \tag{7.1}$$

From a statistical point of view, there are two interesting random variables related to the Bernoulli process. The first is the number of successes that occur when the process consists of a predetermined number of trials. The probability distribution of this random variable is called the **binomial probability distribution.** The second is the number of trials that it takes to obtain the first success. The probability distribution here is known as the **geometric probability distribution.** We begin with a discussion of the binomial probability distribution.

7.2 The Binomial Probability Distribution

Suppose n trials of a Bernoulli process are observed. The trials are therefore independent of one another, each having an equal probability of success, p, and each resulting in either success or failure. It is sometimes of interest to look at the number of successes as a random variable for which we wish to develop a probability distribution. The probability distribution for this random variable is known as a *binomial probability distribution.* Denoting the random variable by X, we write

$$X \sim B(n, p)$$

where the symbol $\sim$ stands for "is distributed" and B stands for "binomially."

Thus, $X \sim B(n, p)$ is read: "the random variable X is distributed binomially with the parameters n and p."

If x denotes a specific value of the variable X, our immediate interest is in determining the probability that X will take on that value—that is, in determining $P(X = x)$.

Let us derive the probability distribution by means of an example.

EXAMPLE 7.1

Twenty percent of all clothing items ordered from a mail-order firm are returned within a week of receipt. Suppose six orders are received and we want to determine the probability distribution of the number of returns.

The number of returns is a binomial random variable since all of the following hold true:

1. The experiment consists of six *independent* trials.
2. There are *only two possible outcomes* of interest for each trial: the order will either be returned or not.
3. The probability that a given order will be returned is *the same* as for any other order.

In this example, $n = 6$ and $p = 0.2$. It follows that $q = 1 - p = 1 - 0.2 = 0.8$. Using compact statistical notation, we write $X \sim B(6, 0.2)$. The possible values that X can take on are 0, 1, 2, 3, 4, 5, and 6. Let us proceed to determine the probability of each of these outcomes. First, consider the probability that there will be no returns—that is, $P(X = 0)$. The probability that the *first order* will not be returned is equal to 0.80. The probability that the *second order* will not be returned is also 0.80. Since we assume that the returns of the first and the second orders are independent, it follows that the probability of the intersection of the two events (that is, that *neither* order will be returned) is equal to the product of the probability of the two events: $0.80 \cdot 0.80 = 0.64$. By the same token, the probability that none of the six orders will be returned is equal to

$$0.80 \cdot 0.80 \cdot 0.80 \cdot 0.80 \cdot 0.80 \cdot 0.80 = 0.80^6 = 0.26214$$

or

$$P(X = 0) = q^n = 0.80^6 = 0.26214$$

Now let us generalize the results of our example and then proceed logically to derive the desired binomial probability distribution. Suppose we denote a success in a given trial by S and a failure by F. The event that all six trials will result in failures may be described as

$$\{F, F, F, F, F, F\}$$

This is the only event that will make X equal to zero, and as we have seen, $P(X = 0) = q^n = 0.26214$. There are six distinguishable events that will make X equal to 1:

$$\{S, F, F, F, F, F\}$$
$$\{F, S, F, F, F, F\}$$
$$\{F, F, S, F, F, F\}$$
$$\{F, F, F, S, F, F\}$$
$$\{F, F, F, F, S, F\}$$
$$\{F, F, F, F, F, S\}$$

The location of the letter S in each pair of braces signals the sequential trial in which the success has occurred. For example, the first series describes a success in the first trial followed by five failures; the second series describes a failure in the first trial, success in the second, and failure in all the four succeeding trials; and so on.

The number of events that make $X = 1$ is $\binom{6}{1}$, since there are $\binom{6}{1}$ possible combinations of one S and five F's. Recalling that $\binom{n}{r} = \dfrac{n!}{(n - r)!\, r!}$, we get $\binom{6}{1} = \dfrac{6!}{5!\, 1!} = 6$. Each of the six events has a probability of $(0.80)^5(0.20) = 0.065536$, since each involves one success with a probability of 0.20 and five failures with a probability of 0.80 each. The probability of $X = 1$ is thus 6 times 0.065536, or

$$P(X = 1) = 6 \cdot 0.065536 = 0.39322$$

To find the probability that X equals 2, we first find the number of possible combinations of two successes in six trials. There are $\binom{6}{2} = \dfrac{6!}{4!\, 2!} = \dfrac{6 \cdot 5}{2} = 15$ such combinations:

$$\{S, S, F, F, F, F\} \quad \{F, S, F, F, F, S\}$$
$$\{S, F, S, F, F, F\} \quad \{F, F, S, S, F, F\}$$
$$\{S, F, F, S, F, F\} \quad \{F, F, S, F, S, F\}$$
$$\{S, F, F, F, S, F\} \quad \{F, F, S, F, F, S\}$$
$$\{S, F, F, F, F, S\} \quad \{F, F, F, S, S, F\}$$
$$\{F, S, S, F, F, F\} \quad \{F, F, F, S, F, S\}$$
$$\{F, S, F, S, F, F\} \quad \{F, F, F, F, S, S\}$$
$$\{F, S, F, F, S, F\}$$

Each of the above combinations has a probability of $(0.20)^2(0.80)^4 = 0.016384$ of occurrence, so that the probability for X to equal 2 is

$$P(X = 2) = \binom{6}{2}(0.20)^2(0.80)^4 = 15 \cdot 0.016384 = 0.24576$$

We now generalize as follows:

THE PROBABILITY DISTRIBUTION OF A BINOMIAL RANDOM VARIABLE

$$P(X = x) = \binom{n}{x} p^x q^{(n-x)} \qquad (7.2)$$

Using Equation 7.2, we get the following probability distribution and cumulative distribution of X:

x	$P(x)$	$\Sigma P(x)$
0	0.26214	0.26214
1	0.39322	0.65536
2	0.24576	0.90112
3	0.08192	0.98304
4	0.01536	0.99840
5	0.00154	0.99994
6	0.00006	1.00000
	1.00000	

Both functions are shown in Figure 7.1.

Let us see another example involving the binomial distribution.

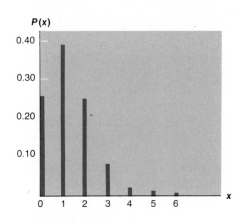
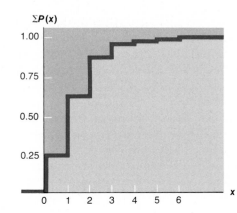

Figure 7.1

Probability distribution and cumulative probability distribution of a binomial random variable with $p = 0.2$ and $n = 6$

EXAMPLE 7.2

A test consisting of 20 true–false questions is taken by a student who is *totally unprepared* and therefore decides to make purely random choices between the true and false answers. The number of correct answers the student will give is a binomial random variable with $n = 20$ as the number of trials and $p = 0.5$ as the probability of success:

$$X \sim B(20, 0.5)$$

Suppose we want to calculate the probability that the student will give exactly five correct answers. We shall follow Equation 7.2 and get

$$P(X = 5) = \binom{20}{5}\left(\frac{1}{2}\right)^5\left(\frac{1}{2}\right)^{15} = 15{,}504 \cdot 0.03125 \cdot 0.00003052 = 0.0148$$

Suppose now that at least eight correct answers are required to pass the test. The probability that the student will pass is found as follows:

$$P(X \geq 8) = 1 - P(X \leq 7)$$
$$= 1 - [P(X = 0) + P(X = 1) + \cdots + P(X = 7)]$$

where each probability within the brackets may be separately calculated by means of Equation 7.2. Calculations show that

$$P(X = 0) + P(X = 1) + \cdots + P(X = 7) = 0.1316$$

so that

$$P(X \geq 8) = 1 - 0.1316 = 0.8684$$

Do not rejoice over this result; most professors will pass their students only if they answer correctly at least 12 of 20 questions.

Although calculating binomial probabilities by means of Equation 7.2 is not overly complicated, particularly with the increasing availability of scientific hand calculators, it does become a tedious task at times, especially when cumulative probabilities are required. To ease the task, binomial probabilities for various combinations of n and p values have been calculated and tabulated. Some of these tables provide the probability distribution, others the cumulative probabilities. Table A.1 in Appendix A gives the cumulative probability for a variety of n and p values. For convenience, a portion of the table is reproduced in Figure 7.2.

To find a cumulative probability, we locate the n and p values in the left-hand column and the top row, respectively. On the second column from the left the values of x are given. Figure 7.2 shows, for example, that the cumulative value for $p = 0.25$, $n = 6$, and $x = 3$ is 0.9624. Figure 7.2 warrants a few additional comments. First, only selected values of n and p are presented. For other values not appearing in the table we need either to perform calculations or to use more detailed tables. If approximation suffices, interpolation may be used. Second, the values of p presented in the binomial table are less than or equal to 0.50. For values greater than 0.50 we use $1 - p$ for the probability of "success" and find the probability $1 - P(X \leq n - x - 1)$ instead. For example, suppose we want to find the probability $P(X \leq 2)$, assuming that $n = 4$ and $p = 0.70$. Since $p > 0.50$, we first use the table to find $P(X \leq 4 - 2 - 1) = P(X \leq 1)$, using $p = 1 - 0.70 = 0.30$. This probability equals 0.6517, and we then compute $P(X \leq 2)$ for $n = 4$ and $p = 0.70$ to equal $1 - 0.6517 = 0.3483$. Third, for each value of n, the values of x listed are $0, 1, \ldots, (n - 1)$. The value $x = n$ is not listed, since it is clear that the

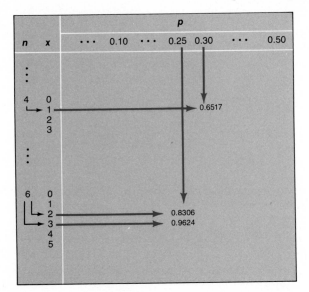

Figure 7.2

A schematic description of the binomial cumulative probability distribution table

cumulative probability for $x = n$ is 1.00. Finally, the (noncumulative) probability for any listed value of n, p, and x is obtained by subtracting the cumulative probability of the value $x - 1$ from that of x. For example, to find the probability $P(X = 3)$ assuming that $n = 6$ and $p = 0.25$, we subtract $P(X \leq 2) = 0.8306$ from $P(X \leq 3) = 0.9624$, where the last two numbers are read off the table, and we get

$$P(X = 3) = P(X \leq 3) - P(X \leq 2) = 0.9624 - 0.8306 = 0.1318$$

7.3 Properties of the Binomial Distribution

In this section we shall discuss the mean, the variance, and the skewness of a binomial distribution. If a binomial experiment consists of 100 trials, each of which has 0.15 probability of success, the expected number of successes is $0.15 \cdot 100 = 15$. In general, if the probability of success is p and the number of trials is n, the expected value of the binomial variable is np. In short:

THE EXPECTED VALUE OF A BINOMIAL VARIABLE

If $X \sim B(n, p)$, then

$$E(X) = np \tag{7.3}$$

To see that the equation does indeed work, consider a binomial random variable with $n = 3$ and $p = 0.3$:

$$X \sim B(3, 0.3)$$

The probability distribution of X is given by Equation 7.2 and is calculated as follows:

x	$P(x)$
0	$\binom{3}{0}(0.3^0)(0.7^3) = 0.343$
1	$\binom{3}{1}(0.3^1)(0.7^2) = 0.441$
2	$\binom{3}{2}(0.3^2)(0.7^1) = 0.189$
3	$\binom{3}{3}(0.3^3)(0.7^0) = 0.027$
Total	1.000

Computing the expected value of X from the above probability distribution, we obtain

$$E(X) = 0 \cdot 0.343 + 1 \cdot 0.441 + 2 \cdot 0.189 + 3 \cdot 0.027 = 0.900$$

which is the same value we obtain when we use Equation 7.3, since $np = 3 \cdot 0.3 = 0.900$.

The variance of a binomial random variable is given by npq:

THE VARIANCE OF A BINOMIAL RANDOM VARIABLE

$$\text{var}(X) = npq \tag{7.4}$$

and as always, the standard deviation is obtained by taking the square root of the variance:

THE STANDARD DEVIATION OF A BINOMIAL RANDOM VARIABLE

$$SD(X) = \sqrt{npq} \tag{7.5}$$

Our next step is to examine Equation 7.5 to see if it "makes sense." First we realize that when $p = 0$ or $q = 0$, the variance is also equal to zero, since if $p = 0$, we know in advance that *all* trials, without exception, will result in "failure," and therefore there is no room for any variability. Similarly, if $q = 0$, *all* trials must result in "success," and hence again there is no variance in outcome. For any other values of p, the variance, as given in Equation 7.4,

Figure 7.3

The relationship between the value of p and the standard deviation of X when $X \sim B(n, p)$

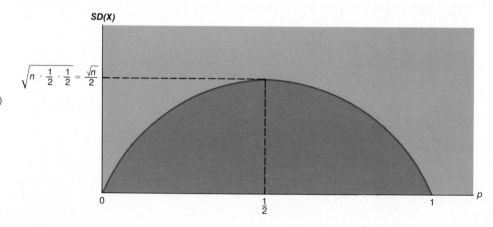

is not equal to zero, and as a result the outcome of one trial can differ from that of the next. In other words, the outcomes have a certain degree of variability. It can be shown that the variance reaches its maximum when $p = q = \frac{1}{2}$.

Graphically, the relationship between the value of p and the standard deviation of X (for a given value of n) is shown in Figure 7.3.

Obviously, we can compute the variance of a binomial random variable directly by Equation 6.6 or 6.7. Using Equation 6.7 for the variance of $X \sim B(3, 0.3)$, we proceed as follows:

x	x^2	$P(x)$
0	0	0.343
1	1	0.441
2	4	0.189
3	9	0.027
		1.000

$$E(X^2) = \Sigma x^2 P(x) = 0 \cdot 0.343 + 1 \cdot 0.441 + 4 \cdot 0.189 + 9 \cdot 0.027$$

$$= 0.000 + 0.441 + 0.756 + 0.243 = 1.440$$

and var $(X) = E(X^2) - \mu^2 = 1.44 - 0.90^2 = 1.44 - 0.81 = 0.63$. Indeed, by Equation 7.4 we obtain var $(X) = npq = 3 \cdot 0.3 \cdot 0.7 = 0.63$.

To summarize

If $X \sim B(n, p)$ then

$$E(X) = np$$

$$\text{var}(X) = npq$$

$$SD(X) = \sqrt{npq}$$

It can be shown that a binomial distribution is symmetrical when $p = q = \frac{1}{2}$, and asymmetrical in all other cases. For any given value of p not equal to $\frac{1}{2}$, however, the greater the value of n, the less the relative skewness of the distribution. This important property of the binomial distribution is shown in Figure 7.4 for three values of p. Note that for all combinations of n and p such that the two conditions $np \geq 5$ and $nq \geq 5$ are met, the distribution is not only close to symmetrical but also bell-shaped, and it resembles the shape of a normal distribution (see Chapter 8). This fact will aid us substantially in approximating binomial probabilities once we have concluded our discussion of the normal distribution.

7.4 The Number of Successes in a Binomial Experiment Expressed in Terms of Proportions

A binomial variable measures the number of successes in a Bernoulli process spanning n independent trials. Often, however, it is convenient to express results not in terms of the number of successes but rather in terms of the

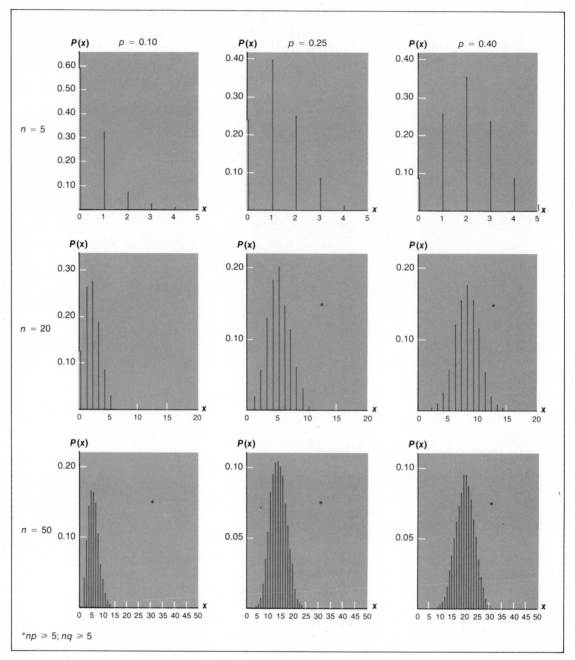

Figure 7.4

The shape of the binomial distributions for various combinations of *n* and *p*

proportion of trials resulting in success. We can easily do so by dividing the number of successes in the experiment (X) by the number of trials (n). Thus, the variable $\dfrac{X}{n}$ measures the proportion of trials resulting in success. For example, if 20 trials resulted in 8 successes, the experiment's result may be expressed as "8 successes" or as "40 percent successes" ($\frac{8}{20} = 0.40$, or 40 percent).

In any given binomial experiment, there is a one-to-one relationship between the value X and the value $\frac{X}{n}$. Consequently, the probability distribution of $\frac{X}{n}$ is identical to that of X and is given by substituting the value $\frac{X}{n}$ in Equation 7.2. Thus

$$P\left(\frac{X}{n} = \frac{x}{n}\right) = \binom{n}{x} p^x q^{n-x} \tag{7.6}$$

As an example, consider $X \sim B(10, 0.6)$. The probability that $X = 4$ is equal to the probability that $\frac{X}{n} = \frac{4}{10} = 0.4$, since both result from the same event (that is, 4 successes out of the 10 trials). Therefore we get

$$P(X = 4) = P\left(\frac{X}{n} = 0.4\right) = \binom{10}{4}(0.6^4)(0.4^6) = 0.1115$$

We can calculate the mean and variance of the variable $\frac{X}{n}$ by implementing previously derived equations. Using the equations applicable to transformed variables (see Chapter 6)—that is, $E(bX) = b\mu$ and $\text{var}(bX) = b^2 \text{var}(X)$—we stipulate $b = \frac{1}{n}$ and find that

$$E\left(\frac{X}{n}\right) = E\left(\frac{1}{n} \cdot X\right) = \frac{1}{n} \cdot np = p$$

$$\text{var}\left(\frac{X}{n}\right) = \text{var}\left(\frac{1}{n} \cdot X\right) = \frac{1}{n^2} \cdot \text{var}(X) = \frac{1}{n^2} \cdot npq = \frac{pq}{n}$$

We may summarize as follows:

If $X \sim B(n,p)$ then

$$E\left(\frac{X}{n}\right) = p$$

$$\text{var}\left(\frac{X}{n}\right) = \frac{pq}{n} \tag{7.7}$$

$$SD\left(\frac{X}{n}\right) = \sqrt{\frac{pq}{n}}$$

Here again, we notice that the variance of $\frac{X}{n}$ equals zero when either $p = 0$ or $q = 0$, and reaches its maximum when $p = q = \frac{1}{2}$.

It is interesting to note that even when p is not equal to 0 or 1, the variance of the proportion approaches zero as the number of trials, n, increases.

7.5 The Geometric Distribution

Suppose we observe a Bernoulli process and instead of counting the number of successes for a predetermined number of trials, as we do in the case of a binomial distribution, we focus on the number of trials necessary to achieve the first success.

The number of trials is the random variable and it can take on the integer values 1, 2, 3, 4, 5, . . . , and so on, and is known as the **geometric distribution.** This probability distribution can be calculated without difficulty. Let us denote the random variable by X. For X to equal 1 there must be a success in the first trial. The probability for that is obviously equal to p by definition. For X to equal 2 one must obtain a failure in the first trial and a success in the second. The probability of obtaining a failure in the first trial is q by definition and the probability of obtaining a success in the second is once again p. Because the two events are independent, the combination of a failure followed by a success has the probability of qp. Similarly, for X to equal 3 one must obtain two failures followed by a success, and the probability for that is $qqp = q^2p$.

Generally for X to equal x, on must obtain $x - 1$ failures followed by one success, and the probability for that is

---

THE GEOMETRIC PROBABILITY DISTRIBUTION

$$P(X = x) = q^{x-1}p \qquad\qquad x = 1, 2, 3, \ldots \qquad (7.8)$$

This geometric probability distribution is summarized in Table 7.1.

TABLE 7.1

A Description of the Geometric Probability Distribution

Value of geometric random variable X	Description of experiments' results	Probability
1	S	p
2	F, S	qp
3	F, F, S	q^2p
4	F, F, F, S	q^3p
.	.	.
.	.	.
.	.	.
x	F, F, F, F, . . ., F, S $(x-1)$ times	$q^{(x-1)}p$
.	.	.
.	.	.

EXAMPLE 7.3

Consider the responsiveness of firefighters to emergencies. One way of looking at this example is to consider the emergency calls as a Bernoulli process in which every minute, say, is the time frame of a trial. That is, we assume that two calls cannot come in during the same minute. Within any minute, an emergency call may come in ("success") or not ("failure"). Suppose the success probability is 0.01. What is the probability that the first emergency will occur during the fifth minute?

For the first emergency call to be received during the fifth minute, we need to observe four "failures" followed by a "success." The probability of this occurrence is given by the geometric distribution:

$$P(X = 5) = q^4 p = 0.99^4(0.01) = 0.0096$$

7.6 The Poisson Probability Distribution

The **Poisson probability distribution** is another important discrete distribution, usually applied in the case of *discrete* independent "successes" on a *continuous* scale. In particular, it is employed when "successes" occur within specified units of time, space, or volume—those dimensions being measured, of course, on a continuous scale. The Poisson probability distribution is applicable to examples of car arrivals at toll booths, calls in a telephone exchange, the number of flaws in a piece of fabric or in an item of glassware, and so on.

Consider, for example, the number of car arrivals at a given toll booth. If there are 180 arrivals per hour ("on the average"), the probability that a car will arrive at the booth at a given second is $180/3,600 = 0.05$, since there are 3,600 seconds in an hour. If we now concern ourselves with a period of, say, one minute ($= 60$ seconds) and ignore the possibility of two arrivals in the same second, we may describe our experiment as a binomial experiment in which the number of trials is 60 (seconds) and the probability of "success" (arrival) in any given trial is 0.05. This probability is given by

$$P(X = x) = \binom{60}{x}(0.05^x)(0.95^{60-x})$$

If, for example, we wanted to find $P(X = 5)$, we would obtain

$$P(X = 5) = \binom{60}{5}(0.05^5)(0.95^{55}) = 0.1016$$

The Poisson probability distribution provides an alternative to the binomial distribution, as we shall now demonstrate with the toll booth example. Its formula is given by the following:

POISSON PROBABILITY DISTRIBUTION

$$P(X = x) = \frac{\lambda^x e^{-\lambda}}{x!} \qquad x = 0, 1, 2, 3, \ldots \tag{7.9}$$

where λ is the only parameter of the Poisson probability distribution and is equal to the average number of "successes" in the relevant unit of time, space, or volume, and e, the base of natural logarithms, is a constant approximately equal to 2.71828.

In our example we find that $\lambda = np = 60 \cdot 0.05 = 3$, where np, as we recall, is the mean of a binomial variable. There are therefore on the average three arrivals per minute, so that

$$P(X = 5) = \frac{3^5 \cdot 2.71828^{-3}}{5!} = \frac{243 \cdot 0.049787}{120} = 0.1008$$

Similarly, if we wish instead to find the probability that the number of arrivals in a given minute would equal exactly 10, we use Equation 7.9:

$$P(X = 10) = \frac{3^{10} \cdot 2.71828^{-3}}{10!} = \frac{59,049 \cdot 0.049787}{3,628,800} = 0.00081$$

For your convenience, a table for e^x and e^{-x} is furnished in Table A.3 in Appendix A at the back of this book. Indeed, most scientific hand calculators nowadays can generate these numbers by a push of a button. But for further convenience, a cumulative Poisson distribution table is provided in Appendix A, Table A.2. A schematic portion of the cumulative Poisson table is presented in Figure 7.5. It shows that the cumulative probability $P(X \leq 6)$ for a Poisson distribution when $\lambda = 5.0$ is equal to 0.76218. Other values may be easily found in Table A.2 in a similar fashion. To find the (noncumulative) probability, we use the relationship

$$P(X = x) = P(X \leq x) - P(X \leq x - 1)$$

as we did for the binomial distribution. For example, assuming again that $\lambda = 5$ and examining $P(X = 6)$, we calculate as follows:

$$P(X = 6) = P(X \leq 6) - P(X \leq 5) = 0.76218 - 0.61596 = 0.14622$$

The following is another illustration of the use of the Poisson distribution.

EXAMPLE 7.4

Electric bills in a given residential area are based on actual readings of electric meters by Southern Electric Company employees. In 1 in every 100 cases, the bill is incorrect. If 800 bills are being processed, what is the probability that 6 bills are in error? Assume that errors in different bills are independent of one another. The problem here involves a binomial distribution with 800 trials and 0.01 "success" probability, so that the expected number of errors is $np = 800 \cdot 0.01 = 8$. The probability may be obtained using the Poisson distribution. Substituting $\lambda = np = 8$ in Equation 7.9, we get

$$P(X = 6) = \frac{8^6 \cdot e^{-8}}{6!} = \frac{262,144 \cdot 0.0003355}{720} = 0.122$$

We earlier saw that the binomial and the Poisson distributions are in a way alternatives to each other. But which is more appropriate to use? The answer depends on the particular situation at hand. In some cases the binomial distribution provides the correct probability, in others the Poisson distribution is appropriate. Let us return to our toll booth example. We assumed 180 arrivals per hour ("on the average"), which is equivalent to $180/3,600 = 0.05$ arrivals per second. The binomial trial in this example is described as observing the toll booth for one second. A success is defined as

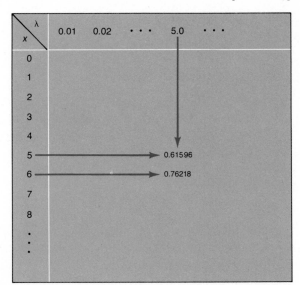

Figures 7.5

A schematic description of the Poisson cumulative distribution table

having an arrival during the second, while no arrival is defined as a failure. *We ignore the possibility of two arrivals in the same second, but that possibility does in fact exist.* It is clear that the possibility of two successes in one trial *contradicts the nature of the binomial process.* We did not, however, run into any contradiction with the Poisson distribution when we estimated 180 arrivals per minute, since the Poisson distribution is defined as successes occurring on a continuous scale. If the process is such that only one success per trial can occur (as when we flip a coin and success is defined as "heads facing up"), the binomial distribution is appropriate. If, however, the trial is defined as a discrete unit on a continuous scale such as time, space, or volume, then two or more "successes" can occur in the same trial, and a switch to the Poisson distribution is called for. Technically, however, the binomial and Poisson distributions are good approximations of one another when the binomial p is small and its n is large.

To see how well the Poisson distribution approximates the binomial, consider a binomial distribution with $n = 20$ and $p = 0.6$. The probabilities of some of the possible outcomes of this variable are listed in Table 7.2, along with the probability as calculated from the Poisson probability distribution (assuming $\lambda = np = 20 \cdot 0.06 = 1.2$).

TABLE 7.2

An Example of the Proximity of Binomial and Poisson Distributions
($n = 20$; $p = 0.06$)

X	Probability by binomial distribution	Probability by Poisson distribution
0	0.2901	0.3012
1	0.3703	0.3614
2	0.2246	0.2169
3	0.0860	0.0867
4	0.0233	0.0260
5	0.0048	0.0062
6	0.0008	0.0012
.	.	.
.	.	.
.	.	.

7.7 APPLICATION 1:

BUDGETING RESEARCH AND DEVELOPMENT

Almost every large company invests significant amounts of money in the development of new products and production processes. For example, General Motors spent $2.22 billion on research and development (R&D) in 1980, and in the same year the Ford Motor Company spent $1.68 billion, IBM spent $1.52 billion, and United Technologies spent $660 million. Some companies spent more than 10 percent of sales on R&D. A comprehensive list of R&D expenditures by U.S. companies is published in *Business Week* once a year, around June or July. Governments also allocate sizable budgets for this purpose when new products are considered important for reasons of public health, safety, or well-being.

In the late 1950s, a strategy known as the "parallel-path strategy" was evolved by the RAND Corporation, one of America's think tanks. Richard Nelson should be credited with providing a framework in which the costs and benefits of the parallel-path strategy can be studied analytically.[1] The problem dealt with is the development of a project at minimum cost.

Suppose n teams are working independently of one another in attempts to develop new products. Suppose further that each team needs a $5 million annual budget and has a 10 percent chance of succeeding. Clearly, the more teams employed, the greater the chance that at least one of them will succeed in developing a new product. It is also clear, however, that the larger the number of teams, the bigger the budget. The correct determination of the number of teams that should work on a project depends on a clear presentation of the appropriate data. Let us start by posing the following question: How much should the firm spend on R&D if it wants to achieve a 95 percent chance of discovering at least one new product?

Since the teams are assumed to be working independently, the binomial distribution can be used. The probability of at least one discovery is

$$P(X \geq 1) = 1 - P(X = 0) = 1 - \binom{n}{0} p^0 q^{(n-0)} = 1 - q^n$$

where X is the number of discoveries and n is the number of teams working on the discovery of new products.

Recalling that the probability of success (p) is equal to 0.10 and that the probability of failure (q) is equal to $1 - 0.10 = 0.90$, we get

$$P(X \geq 1) = 1 - q^n = 1 - 0.90^n$$

It is obvious that if the firm wishes to ensure at least one discovery, it must employ an infinite number of teams, since as n approaches infinity, the probability $P(X \geq 1)$ approaches 1.0. An infinite number of teams calls for an infinite budget. In other words, no matter how large the budget, success is never certain. (Indeed, despite the huge budgets allocated to cancer research, the cure continues to elude researchers.) A realistic budget, implying a finite number of teams, always entails the risk that no discoveries will be made. If the firm will be satisfied with only a 95 percent chance of at least one

[1] Richard Nelson, "Uncertainty, Learning, and the Economics of Parallel Research and Development Effort," *Review of Economics and Statistics*, November 1961.

discovery, however, the number of teams required can be financed within a budget that is wholly feasible. In this case we have

$$P(X \geq 1) = 1 - 0.90^n \geq 0.95$$

or

$$0.90^n \leq 0.05$$

It can be shown that the minimum value of n that solves the above inequality is 29.[2] Since each team needs a \$5 million budget, the total budget must be at least $5 \cdot 29 =$ \$145 million in order to achieve a 95 percent chance of getting at least one discovery.

Table 7.3 shows the number of teams and corresponding budgets required for R&D, assuming different specified probabilities of at least one discovery. Figure 7.6 shows the same data diagrammatically. Given Table 7.3 and Figure 7.6, policy-makers can clearly see how the probability of at least one discovery is directly related to budget size. They will base their R&D budget on both the importance they assign to the achievement of at least one discovery and the availability of funds.

TABLE 7.3
Number of R&D Teams and Total Budget Required for Alternative Specified Probabilities of at Least One Discovery when $p = 0.10$

Required probability for at least one discovery	Number of teams	Budget per team (millions of dollars)	Total R&D budget (millions of dollars)
0.50	7	\$5	\$ 35
0.60	9	5	45
0.70	12	5	60
0.80	16	5	80
0.90	22	5	110
0.95	29	5	145
0.99	44	5	220
1.00	∞	5	∞

Huge budgets are allocated annually for cancer research. The history of cancer research is very similar to the case we are dealing with above, but here, unfortunately, the success probability is particularly small. As a result, the budget needed rises sharply with the slightest increase in probability for at least one discovery. As an illustration, suppose that the cancer research case is characterized by a probability $p = 0.0001$ and

---

[2] One way to solve for n is to take the logarithm on both sides of the inequality

$$0.90^n \leq 0.05$$

so that

$$n \log 0.90 \leq \log 0.05$$

It follows that

$$n \geq \frac{\log 0.05}{\log 0.90} = \frac{-1.30103}{-0.04576} = 28.43$$

or $n = 29$. Note that we change the inequality direction because log 0.90 is negative. Another way of solving for n is by trial and error.

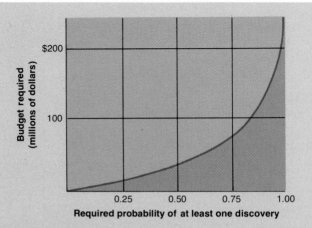

Figure 7.6
Required budget as a function of the probability of at least one discovery when $p = 0.10$

each team's budget is $2 million. To get a rather moderate 50 percent chance for at least one breakthrough in cancer research, we need n teams such that

$$P(X \geq 1) = 1 - 0.9999^n \geq 0.50$$

or

$$0.9999^n \leq 0.50$$

which means that n has to be equal to 6,932 and the budget must be $2 \cdot 6{,}932 = 13{,}864$, or $13.864 billion.[3] This is of course a staggering sum of money that can be budgeted only over a span of several years, a fact that helps to explain the present situation in cancer research. Since the success probability of each team is so small ($p = 0.0001$) and the number of trials (that is, teams) so large, the Poisson probability distribution approximates the binomial very well.

We saw that 6,932 teams require a huge budget. Suppose the real budget allows for the employment of only 1,000 teams. Stipulating $\lambda = np = 1{,}000 \cdot 0.0001 = 0.10$, we find from Table A.2 in Appendix A that there is a 0.90484 probability of zero breakthroughs under these conditions.

Sadly, the probability of at least one breakthrough would therefore be a scant $1 - 0.90484 = 0.09516$.

[3] To get 6,932 we solve

$$0.9999^n \leq 0.50$$

$$n \log 0.9999 \leq \log 0.50$$

$$n \geq \frac{\log 0.50}{\log 0.9999} = \frac{-0.30102999}{-0.00004343} = 6{,}931.4$$

and n should be at least 6,932.

7.8 APPLICATION 2:

OIL DRILLING AND THE NEED FOR CAPITAL

Drilling an oil field is a very risky venture from a financial standpoint. It is true that there are some areas of the Middle East where you can't build a sand castle without striking black gold, but in most regions of the world oil remains a scarce and precious commodity. The financial risk entailed in drilling for oil stems from the ever-present possibility of striking a dry hole.

Drilling is expensive, and a series of consecutive dry holes could lead to bankruptcy. Consequently, a responsible firm must take account of the probability of striking a series of dry holes in order to take the appropriate preparatory steps.

Suppose that drilling a well costs \$10 million. If a well is found to be dry, the company loses \$10 million. If the firm strikes oil, however, the net profit is \$100 million. Suppose further that the probability of striking oil in each potential oil field is 0.2.

One interesting question that arises is the amount of initial capital a starting company should have in order to avoid bankruptcy by a given probability. The concern here is with the chance of striking a given number of consecutive dry holes, so that the geometric probability distribution is applicable. Denoting a dry hole by D and a wet hole (wet with oil, not water) by W, the event of striking $n - 1$ dry holes followed by 1 wet hole is symbolically expressed in the following way:

$$\underbrace{D, D, D, \ldots, D,}_{n - 1 \text{ times}} W$$

The probability of such a sequence is

$$0.80^{n-1} \cdot 0.20$$

Table 7.4 shows the budget required to finance various numbers of dry holes.

TABLE 7.4

Budget Required to Avoid Bankruptcy in the Face of Various Numbers of Dry Holes when $p = 0.20$

(1) Number of trial at which first wet hole is found (n)	(2) = (1) − 1 Number of dry holes (n − 1)	(3) = $0.80^{(n-1)} \cdot 0.20$ Probability of striking this many dry holes	(4) Cumulative probability	(5) = (1) · \$10 Required budget (millions of dollars)
1	0	0.20000	0.20000	\$ 10
2	1	0.16000	0.36000	20
3	2	0.12800	0.48800	30
4	3	0.10240	0.59040	40
5	4	0.08192	0.67232	50
6	5	0.06554	0.73786	60
7	6	0.05243	0.79029	70
8	7	0.04194	0.83223	80
9	8	0.03355	0.86578	90
10	9	0.02684	0.89262	100
11	10	0.02147	0.91409	110
.	.	.	.	.
.	.	.	.	.
.	.	.	.	.

The cumulative probability of the various numbers of dry holes initially drilled and its relationship to the required budget is also shown in Figure 7.7.

From Table 7.5 and Figure 7.7 we can see that the higher the probability of avoiding bankruptcy, the higher the initial budget requirement. For example, a $70 million budget is required in order to render the firm solvent in the face of six consecutive dry wells and the construction of a seventh. The probability that the number of dry wells the firm encounters will be less than or equal to 6 is 0.79029. Similarly, a $100 million budget gives the firm the opportunity to weather nine dry holes and start the tenth. There is 0.89262 probability that the number of dry holes the firm will strike will be less than or equal to 9.

In Figure 7.7, curves demonstrate the relationship between the budget and the probability of avoiding bankruptcy for $p = 0.10$ and for $p = 0.20$. As expected, the curves show that the required budget for each cumulative probability of avoiding bankruptcy is higher for $p = 0.10$ than it is for $p = 0.20$.

Figure 7.7

Cumulative probability of striking various numbers of dry holes and associated budget required to avoid bankruptcy for alternative values of p

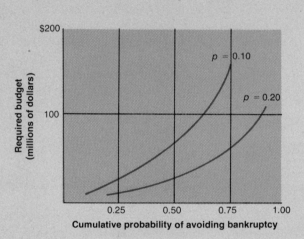

Problems

7.1. Give an example of a Bernoulli process. Explain why it is a Bernoulli process, and define two random variables related to this process, one that has a binomial probability distribution and one that has a geometric probability distribution.

7.2. Let X be a random variable with a binomial probability distribution. Let the parameters of X be $n = 6$ and $p = 0.3$. What are the following probabilities:

$$P(X = 0)$$

$$P(X = 2)$$

$$P(X = 6)$$

$$P(2 \leq X \leq 6)$$

$$P(2 < X < 6)$$

7.3. Let X have a binomial probability distribution with $n = 100$ and $p = 0.02$. Use the Poisson approximation to determine $P(1 < X \leq 3)$, $P(1 \leq X \leq 3)$, $P(1 \leq X < 3)$, and $P(1 < X < 3)$.

7.4. Twenty thousand items contain 1,000 defective items. A sample of 100 is taken without replacement and 4 of them are found to be defective. Using the Poisson probability distribution, determine the *approximate* probability for this sample result, and explain why the probability you have calculated is approximate rather than precise.

7.5. The proportion of defective items produced by a company is 0.04. Batches of 100 items are shipped out. The cost incurred by the company as a result of defective items is as follows:

Number of defective items per batch	Cost incurred
0	$0.00
1	3.00
2	6.00
3	9.00
4 or more	12.00

Use the Poisson approximation in this problem.

(a) What is the probability that 4 or more defective items will be contained in a given batch?
(b) Compute the probability that a given batch will contain exactly 1 defective item.
(c) Compute the expected loss per batch of 1,000 items due to defective items.
(d) Compute the probability that two randomly selected batches out of 100 batches will both be free of defective items.

7.6. A new apartment complex consists of 20 apartments, each rented for one year. There is a 20 percent chance that any one apartment rented for a year will be vacated before the end of the year.

(a) What is the expected number of apartments that will become vacant before the end of the year? What is the variance?
(b) What is the probability that more than 4 apartments will be vacated before the end of the year?

7.7. The probability that a tornado will strike a given area during a period of a year is 0.04. What is the probability that at least one tornado will strike in a period of 30 years? Solve the problem once using the binomial formula and once using the Poisson formula.

7.8. Find the mean and variance of a binomial random variable whose n equals 18 and whose p equals 0.10.

7.9. Find the parameters n and p of a binomial probability distribution for which the mean is 8 and the variance is 1.6.

7.10. A multiple-choice test consists of 6 questions with 4 possible answers to each. Only one of the 4 answers is correct. Suppose a student who is totally unprepared for the test selects his answers at random.

(a) What is the probability that he will score a grade of 50 (out of 100) if each question has an equal weight?
(b) Rework part a, assuming now that the test consists of 10 equally weighted questions.
(c) Suppose that a student must answer at least 50 percent of the questions correctly in order to pass the test. How many questions do we need to include in the test if you want to make the probability of passing the exam by mere chance equal to 5 percent or less? (Assume that the number of multiple-choice questions in the test is even and use the binomial table to find your answer.)

7.11. Five percent of the products made by a certain machine are defective.

(a) If a sample of 8 items is taken, what is the probability that exactly 3 of them are defective?
(b) If a sample of 8 items is taken, what is the probability that the first three items selected will be defective and all the rest will be good?

7.12. Stock market investors have become increasingly interested in the option market in recent years. An option is a security granting its owner the option to buy a specified stock at a specified price during a specified period of time. For example, the financial weekly *Barron's* reported on January 8, 1979, that the owner of a Boeing option would have the option to buy Boeing stock at $80 a share during the period from January through August 1979. The stock's price at the beginning of January 1979 was $74½. Obviously, if the stock's price should rise above $80, the holder of the option would be in a position to make a profit by exercising his option. Suppose there was a probability of ⅛ that the price of Boeing stock would go above $80 in any of the ensuing months, and the price during each month was independent of the price during the other months.

(a) What was the probability that the stock price would remain at $80 or below in all of the 8 months from January through August?

(b) What was the probability that there would be exactly two months when the price would rise above $80?

(c) What was the probability that six months would pass before the price rose above $80?

7.13. A clothing store permits garments to be returned within 10 days of purchase. The store's experience shows a 15 percent chance of return. If 15 garments are sold to 15 (independent) individuals, what is the probability that not more than 20 percent of the garments will be returned? What is the probability that at least 40 percent of the garments will be returned?

7.14. A test consists of 20 true–false questions. If a student decides to answer the questions by a random selection process, what is the student's probability of giving more than 8 correct answers?

7.15. A bank grants loans to firms only if their financial position is thought to be satisfactory. The probability that a firm will be unable to repay its loan after having been found to be solvent is 0.001. Three thousand firms have received loans.

(a) What is the probability that exactly 4 will be unable to repay their loans?

(b) What is the probability that more than 2 will be unable to repay their loans?

7.16. A utility company's records show that the time that elapses between the time customers receive their utility bills and the time they actually pay them is distributed in the form of a Poisson probability distribution with $\lambda = 11.0$. What is the probability that a customer will pay his bill within 10 days of the time he receives it? What is the meaning of λ?

CHAPTER EIGHT OUTLINE

8.1 Introduction

8.2 Calculating Probabilities Involving a Normal Distribution
The Standard Normal Distribution
Using the Normal Distribution Table
Transforming *X* Values into *Z* Values and Vice Versa
Finding Fractiles under the Normal Distribution

8.3 The Normal Approximation to the Binomial Distribution

8.4 Application: The "Safety-First" Principle and Investment Decisions
Baumol's "Expected Gain-Confidence Limit" Criterion

Key Terms

normal distribution
standard normal
  distribution
standardization
fractile
percentile

8
THE NORMAL DISTRIBUTION

8.1 Introduction

One of the most popular probability distributions is the continuous, bell-shaped **normal distribution,** an example of which is shown in Figure 8.1a. The random variable X is measured along the horizontal axis, while the vertical axis specifies corresponding values of the density function. The cumulative distribution is shown in Figure 8.1b.

The normal distribution is symmetrical about the mean (μ), the distribution's mean in this case being identical with its median and mode. The shapes of the normal and cumulative distributions shown in Figure 8.1a and 8.1b indicate that X's value is more likely to fall close to the mean of the distribution than farther out to the left or right, since X is more densely distributed in the immediate vicinity of μ than farther out to either side. The variable can, however, take on any real value in the range $-\infty$ to $+\infty$: although the distribution's tails approach the horizontal axis as X deviates from μ to either side, they never quite touch it.

Many random variables in business and economics have probability distributions that can be fairly well approximated by a normal distribution. Examples are such variables as the rate of return on stocks, the time it takes to perform a given task, and the length of pieces of fabric cut out by a machine. As we shall see later, even discrete random variables often have probability distributions that can be approximated by the normal distribution.

Figure 8.1 depicts one particular normal distribution. There is an infinite number of other normal distributions, differing from each other in mean,

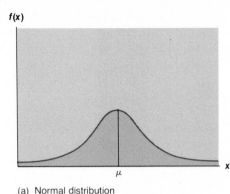

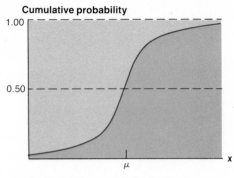

(a) Normal distribution

(b) Cumulative normal distribution

Figure 8.1
The normal distribution

standard deviation, or both. Thus, the normal distribution is specified by its mean and standard deviation, and we often use the expression $X \sim N(\mu, \sigma)$, which is read "X is distributed normally with mean μ and standard deviation σ." Figure 8.2a shows two normal distributions having the same standard deviation but differing means; Figure 8.2b shows two normal distributions with the same mean but differing standard deviations; and Figure 8.2c shows two normal distributions that differ in both mean and standard deviation.

The fact that the normal distribution is specified by its mean and standard deviation means that once we know μ and σ, the entire probability distribution is described.

8.2 Calculating Probabilities Involving a Normal Distribution

To be able to handle problems involving the normal distribution, one must be able to calculate the probability of events concerning the distribution. In this section we shall discuss the method of calculation in detail.

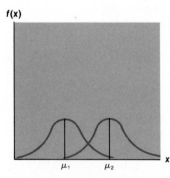

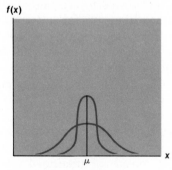

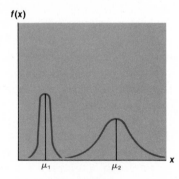

(a) Normal distributions with the same standard deviation but differing means

(b) Normal distributions with same mean but differing standard deviations

(c) Normal distributions with differing means and standard deviations

Figure 8.2
Normal distributions with differing means and standard deviations

Suppose we want to find the probability of a normal random variable, X, whose probability distribution is represented in Figure 8.3 and whose value is in the range from x_1 to x_2. To do this we need to take the integral of the density function over the range from x_1 to x_2. The normal distribution density function, however, is rather cumbersome for integration, so we would like to find a better way of handling the problem.[1]

Imagine that some kind soul graciously calculated the area under a normal curve over all the possible intervals of X and presented his calculations to us in a table. We could then use the table and avoid the need for integration. There is one difficulty, however, with this approach. Since there is an infinite number of normal distributions (differing in their μ and σ combinations), one would actually need an infinite number of tables of normal distributions—one per distribution—for this method to be feasible. Obviously this method is impractical; we need a more efficient way to solve our problem.

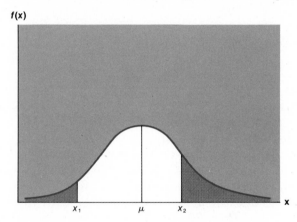

f(x)

x_1 μ x_2

x

Figure 8.3

The probability under a normal distribution over the range from x_1 to x_2

THE STANDARD NORMAL DISTRIBUTION

Normal distributions, although differing from one another, belong to the same family of distributions and thus have common characteristics—a fact we can exploit to our benefit. The calculations needed are rather simple and may be easily illustrated with the help of Figure 8.4. Here two normal distributions are presented, one that we call the **"standard" normal distribution** (its mean is zero, its standard deviation equals 1, and the value of any given point on the distribution is denoted by Z) and another selected arbitrarily (its mean is μ, its standard deviation is σ, and the value of any given point on the distribution is denoted by X). Comparing the two distributions, we find that the areas to the right and left of any X value under the normal distribution of X are equal to the areas to the right and left of the corresponding Z value under the standard normal distribution; "corresponding" here means that the X value and the Z value are situated at the same distance from their respective means *where distance is measured by units of standard deviations*. Thus, the area under the "standard" curve over the interval from 0 to 1 is equal to the area of the other normal curve over the interval from μ to $\mu + \sigma$. Similarly, the area under the standard curve over the interval from 0 to 2 is equal to the area of the other curve over the interval from μ to $\mu + 2\sigma$, and so on. A simple method of evaluating areas

[1] The normal curve is given by the following: $f(x) = \dfrac{1}{\sqrt{2\pi}\sigma} \cdot e^{-\frac{1}{2}[(x-\mu)/\sigma]^2}$.

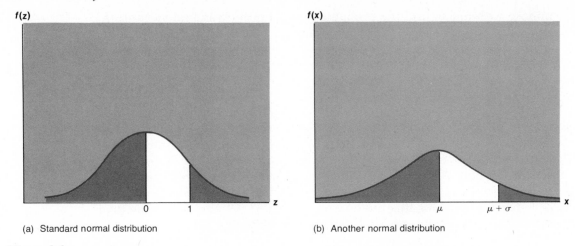

(a) Standard normal distribution   (b) Another normal distribution

Figure 8.4

The "standard" normal and another normal distribution and their relationship

under a normal curve now becomes possible. Suppose we document in a table the areas over alternative intervals for the "standard" distribution. Facing a problem concerning the area (probability) under some other normal distribution, we can now easily find the "corresponding" area under the standard distribution and derive the desired probability from the table.

To illustrate, suppose X is a random variable having a normal distribution with mean 500 and standard deviation 50, or in short, $X \sim N(500, 50)$. And suppose we wish to find the probability that a certain variable will assume a value in the interval from 500 to 575, or, when using symbolic notation, $P(500 \le X \le 575) = ?$ We realize that the value of 500 is the mean of the distribution and thus corresponds to the value 0 of the standard normal distribution. The value 575 is 1.5 standard deviations (of 50 units each) to the right of 500 and therefore corresponds to the value $Z = 1.5$, so that $P(500 \le X \le 575) = P(0 \le Z \le 1.5)$. As probabilities for any normal distribution can be calculated by means of the standard normal distribution, probability tables for Z variables alone are clearly sufficient.

In order to make efficient use of the standard distribution in calculating probabilities under any normal curve, we still need to make two clarifications. First, we need to learn how to use the normal distribution table, so that once a question is formulated in terms of the variable Z, we know how to find the desired probability. Second, we need to find a fast computational method to enable us to switch from any value on the X scale to its corresponding value on the Z scale. Both of these procedures will now be discussed.

USING THE NORMAL DISTRIBUTION TABLE

Here is how we find probabilities concerning the standard normal variable Z. The normal distribution table inside the back cover lists the areas accumulated under the normal curve, from the value zero up to a variety of Z values. For example, to find the area (probability) under the standard normal curve over the interval from 0 to 2, we go down the far left-hand column of the table (the column headed "Z") until we get to the value 2.0. Moving one column to the right, we see the number 0.4772, which is the area under the standard normal curve over the desired range from 0 to 2. We thus conclude that

$$P(0 \leq Z \leq 2) = 0.4772$$

Suppose we want now to find the area over the interval that starts at zero and ends at 1.35. We go down the left-hand column once again, but this time we go down to the value 1.3. To find the second decimal digit, we move to the right until we reach the column headed "0.05." The number we have located states the proper area: 0.4115. Figure 8.5*a* shows the location of the number 0.4115 in the table. Figure 8.5*b* shows how to find the area accumulated under the normal curve over the interval from 0.00 to 0.06.

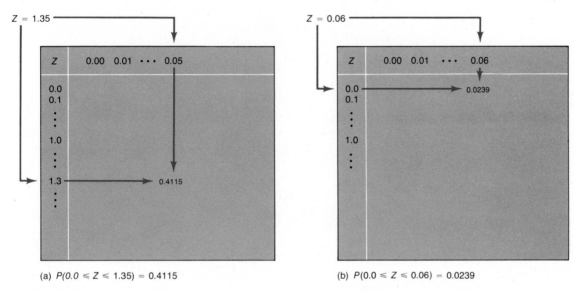

(a) *P(0.0 ≤ Z ≤ 1.35) = 0.4115*   (b) *P(0.0 ≤ Z ≤ 0.06) = 0.0239*

Figure 8.5
Using the normal distribution table

Figure 8.6 serves to demonstrate that a standard normal distribution's Z *score is measured along the horizontal axis and its probability is given by the area under the probability function and over the horizontal domain.*

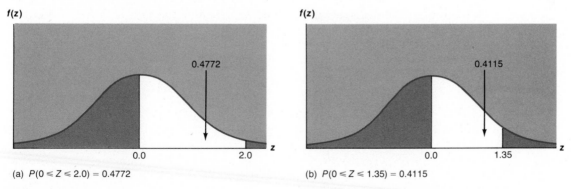

(a) *P(0 ≤ Z ≤ 2.0) = 0.4772*   (b) *P(0 ≤ Z ≤ 1.35) = 0.4115*

Figure 8.6
Graphic presentation of probabilities under the "standard" normal distribution

Once the structure of the normal distribution table becomes clear, one may proceed to derive probabilities over intervals that do not necessarily start at zero and over intervals with negative Z values. We shall illustrate the procedures used with several self-explanatory examples.

EXAMPLE 8.1

$$P(1.00 \leq Z \leq 2.15) = P(0.00 \leq Z \leq 2.15) - P(0.00 \leq Z \leq 1.00)$$

$$= 0.4842 - 0.3413 = 0.1429$$

This procedure becomes clear once we take a look at Figure 8.7. The white area in Figure 8.7 is equal to the difference between the areas over the intervals from 0.00 to 2.15 and from 0.00 to 1.00. The probabilities 0.4842 and 0.3413 are taken directly from the normal distribution table.

Figure 8.7

Showing the probability $P(1.00 \leq Z \leq 2.15)$

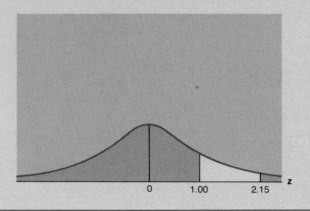

EXAMPLE 8.2

$$P(-1.60 \leq Z \leq 0.00) = P(0.00 \leq Z \leq 1.60) = 0.4452$$

Here we simply take advantage of a well-known property of the standard normal distribution: it is symmetrical around zero. The area from -1.60 to 0.00 in Figure 8.8 is equal to the area over the symmetrical interval in the positive range.

Figure 8.8

Showing the probability $P(-1.60 \leq Z \leq 0.00)$

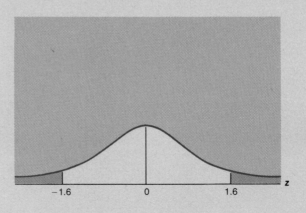

EXAMPLE 8.3

$$P(-0.93 \leq Z \leq -0.04) = P(0.04 \leq Z \leq 0.93)$$

$$= 0.3238 - 0.0160 = 0.3078$$

Once again we take advantage of the symmetry around zero. Then we use the procedure of Example 8.1 again (see Figure 8.9).

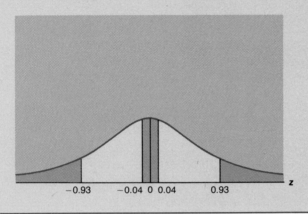

Figure 8.9

Showing the probability $P(-0.93 \leq Z \leq -0.04)$

EXAMPLE 8.4

$$P(Z \geq 0.85) = 0.5000 - P(0 \leq Z \leq 0.85)$$

$$= 0.5000 - 0.3023 = 0.1977$$

The total area to the right of zero in Figure 8.10 is 0.5, and thus the area to the right of $Z = 0.85$ is the complement (to 0.5) of the area over the range from 0.00 to 0.85.

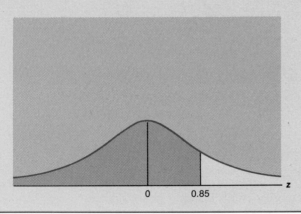

Figure 8.10

Showing the probability $P(Z \geq 0.85)$

EXAMPLE 8.5

$$P(-1.40 \leq Z \leq 2.50) = P(-1.40 \leq Z \leq 0.00) + P(0.00 \leq Z \leq 2.50)$$

$$= P(0.00 \leq Z \leq 1.40) + P(0.00 \leq Z \leq 2.50)$$

$$= 0.4192 + 0.4938 = 0.9130$$

Figure 8.11 makes it quite clear that the area over the range from -1.40 to 2.50 is equal to the sum of the areas over the intervals from -1.40 to 0.00 and from 0.00 to 2.50. The area over the interval from -1.40 to 0.00 is equal to that over the corresponding positive interval (0.00 to 1.40), which is given in the table.

Figure 8.11

Showing the probability $P(-1.40 \leq Z \leq 2.50)$

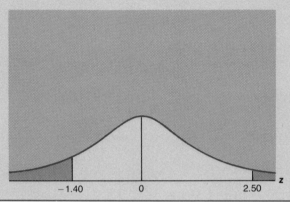

EXAMPLE 8.6

$$P(Z \leq -1.00) = P(Z \geq 1.00) = 0.5000 - P(0.00 \leq Z \leq 1.00)$$

$$= 0.5000 - 0.3413 = 0.1587$$

We first realize that the area to the left of -1.0 in Figure 8.12 is equal to the area to the right of 1.0, and then we obtain the required probabilities from the table.

Figure 8.12

Showing the probability $P(Z \leq -1.00)$

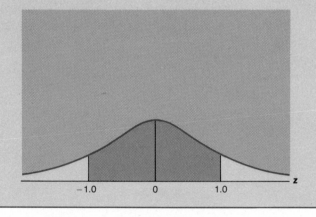

At this point we urge you to use the normal distribution table in the back of the book to verify the following probabilities:

(1) $P(0.00 \leq Z \leq 1.24) = 0.3925$

(2) $P(-2.00 \leq Z \leq -0.55) = 0.2684$

(3) $P(Z \leq -1.44) = 0.0749$

(4) $P(-0.56 \leq Z \leq 0.56) = 0.4246$

(5) $P(Z \leq 1.67) = 0.9525$

(6) $P(Z \le 0) = 0.5000$

(7) $P(2.06 \le Z \le 3.06) = 0.0186$

(8) $P(-1.00 \le Z \le 2.00) = 0.8185$

(9) $P(-1.00 \le Z \le 0.50) = 0.5328$

(10) $P(Z \le -1.50) = 0.0668$

TRANSFORMING X VALUES INTO Z VALUES AND VICE VERSA

We now take another step toward the calculation of probabilities concerning any normal variable. The "trick" to remember here is that when we take a normal variable, X, subtract its mean, and then divide by the standard deviation, we get a standard normal variable.

If

$$X \sim N(\mu, \sigma)$$

then[2]

$$\left(\frac{X - \mu}{\sigma}\right) \sim N(0, 1) \tag{8.1}$$

so we may write

$$Z = \frac{X - \mu}{\sigma} \tag{8.2}$$

By subtracting the mean and dividing by the standard deviation, we get the standard normal variable. This procedure is thus called **standardization.** Note also that from Equation 8.2 it follows that $X - \mu = \sigma Z$, and thus

$$X = \mu + \sigma Z \tag{8.3}$$

As Equation 8.2 can be used to find a Z value given a specific value of X, so can Equation 8.3 be used to find an X value given a specific value of Z. Given a problem concerning the probability of X, such as $P(X \le x)$, where x is any

[2]

$$E\left(\frac{X - \mu}{\sigma}\right) = \frac{1}{\sigma}E(X - \mu) = \frac{1}{\sigma}(\mu - \mu) = 0$$

$$\text{var}\left(\frac{X - \mu}{\sigma}\right) = \frac{1}{\sigma^2}\text{var}(X - \mu) = \frac{1}{\sigma^2}\text{var}(X) = \frac{\sigma^2}{\sigma^2} = 1$$

so that

$$\left(\frac{X - \mu}{\sigma}\right) \sim N(0, 1)$$

real number, we convert X into a Z variable by subtracting μ and then dividing by the standard deviation, σ:

$$P(X \leq x) = P\left(\frac{X - \mu}{\sigma} \leq \frac{x - \mu}{\sigma}\right) = P\left(Z \leq \frac{x - \mu}{\sigma}\right) \qquad (8.4)$$

The last probability on the right may be obtained directly from the normal distribution table (inside back cover).

EXAMPLE 8.7

A machine pours soda into bottles. Suppose experience has shown that the weight of the soda poured is normally distributed and assume that the amount of soda by weight that is poured into each bottle is a normal random variable with a mean of 64 ounces and a standard deviation of 1.5 ounces. What is the probability that the amount of soda that the machine will pour into the next bottle will be more than 65 ounces?

Following Equation 8.4, we write[3]

$$P(X \geq 65.0) = P\left(\frac{X - 64.0}{1.5} \geq \frac{65.0 - 64.0}{1.5}\right) = P(Z > 0.67) = 0.2514$$

To find the probability that the amount of soda will be less than 62.5 ounces, we calculate as follows:

$$P(X \leq 62.5) = P\left(\frac{X - 64.0}{1.5} \leq \frac{62.5 - 64.0}{1.5}\right) = P(Z \leq -1.0) = 0.1587$$

[3] Note that for a continuous distribution the probabilities $P(X > 65)$ and $P(X \geq 65)$ are equal.

EXAMPLE 8.8

Continuing Example 8.7, one may ask, "What is the probability that the amount of soda in the next bottle will differ from the mean of 64 by at least 2 ounces?"

For this to happen, the bottle has to contain either less than 62 ounces or more than 66 ounces. In other words, we are asking $P(X \leq 62) + P(X \geq 66) = ?$ Once again we want to convert the problem into one dealing with the variable Z, and we proceed as follows:

$$P(X \leq 62) = P\left(\frac{X - 64.0}{1.5} \leq \frac{62.0 - 64.0}{1.5}\right) = P(Z \leq -1.33) = 0.0918$$

$$P(X \geq 66) = P\left(\frac{X - 64.0}{1.5} \geq \frac{66.0 - 64.0}{1.5}\right) = P(Z \geq 1.33) = 0.0918$$

so that

$$P(X \leq 62) + P(Z \geq 66) = 0.0918 + 0.0918 = 0.1836$$

FINDING FRACTILES UNDER THE NORMAL DISTRIBUTION

It is sometimes necessary to find fractiles under the normal distribution. For a random variable X, a **fractile** is *a value of X, such that the probability of X's assuming values less than or equal to it, is a specified fraction.* (When the fractile is *expressed as a percentage*, it is called a **percentile.**) As an example, the 0.50 fractile, which is the 50th percentile, of any normal distribution is its mean, μ.

When we have to find a certain fractile for a given normal distribution, we first find the fractile for the standard normal distribution in the standard normal distribution table. For example, the 0.67 fractile is that Z value up to which the accumulated area is equal to 0.67 (accumulation going from left to right). The table shows that $Z = 0.44$ is the 0.67 fractile (see Figure 8.13).

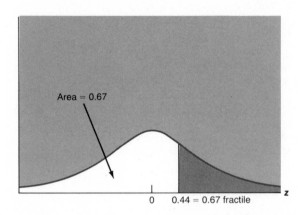

Figure 8.13

The 0.67 fractile on the standard normal distribution

Once the fractile on the standard normal distribution has been determined, we use Equation 8.3 to derive the fractile for the distribution of X. We shall demonstrate this procedure with examples.

EXAMPLE 8.9

The inventory manager of a large firm estimates that the period of time (measured in days) that passes between the ordering of a given item and its actual receipt with regular delivery service (that is, not on an emergency delivery basis) is a random variable having a normal distribution with a mean of six working days and a standard deviation of two working days. Find the 75th percentile of the distribution.

Our first step is to find the 75th percentile on the Z scale. The standard normal distribution table shows that the area over the range from 0.000 to 0.675 is approximately 0.25, but since over the negative range there is an area equal to 0.50, the total area in the range from $-\infty$ to 0.675 is equal to 0.75. It follows that 0.675 is the 75th percentile on the Z scale. Using Equation 8.3, we get

$$X = \mu + Z\sigma = 6.0 + (0.675 \cdot 2) = 6.0 + 1.35 = 7.35 \text{ days}$$

This means that 75 percent of the time, the delivery period is 7.35 days or less.

EXAMPLE 8.10

Find the 20th percentile of the distribution in Example 8.9. We again refer first to the standard distribution. Since it is symmetrical around zero, it follows that the 20th percentile must equal the negative magnitude of the 80th percentile. A glance at the Z table will reveal that the 80th percentile is equal to 0.84, and it follows that the 20th percentile equals -0.84. Equation 8.3 yields

$$X = \mu + Z\sigma = 6 + (-0.84 \cdot 2) = 6 - 1.68 = 4.32 \text{ days}$$

EXAMPLE 8.11

Mrs. Joannie Yost has to be at work at 9 A.M. Commuting time is a random variable having a normal distribution with a mean of 40 minutes and standard deviation of 10 minutes. If she wants to make sure not to be late to work more than 3 percent of the time, at what time does she have to leave home in the morning?

We first have to find the 97th percentile of the Z distribution. From the normal table we find that this value is equal to 1.88. On the relevant (time) distribution, therefore, the 97th percentile is

$$X = \mu + \sigma Z = 40 + (10 \cdot 1.88) = 40 + 18.8 \simeq 59 \text{ minutes}$$

and so she must leave home at 8:01 in the morning.

8.3 The Normal Approximation to the Binomial Distribution

The binomial distribution was discussed in Chapter 7, and one method for its approximation (the Poisson distribution) has been discussed as well. In this section we present yet another approximation procedure useful particularly for approximating the cumulative binomial distribution over a given interval. Here we use the normal probability distribution to approximate the binomial probability distribution.

The idea is quite simple and the procedure for implementing it is straightforward. When we discussed the binomial distribution we noted that in all those cases where $np \geq 5$ and also $nq \geq 5$, the general shape of the binomial distribution resembles the normal distribution shape. Since the mean and the standard deviation of the binomial distribution are np and $\sqrt{npq}$, respectively, we may write

> If $\quad X \sim B(n, p) \quad$ and $\quad np \geq 5 \quad$ and also $\quad nq \geq 5 \quad$ then
>
> $X \simeq N(np, \sqrt{npq})$

where $\simeq$ means "is approximately distributed." Furthermore, we may standardize the variable X to get the following equation:

$$Z = \frac{X - np}{\sqrt{npq}} \simeq N(0, 1) \tag{8.5}$$

Before proceeding, let us illustrate the above argument with the help of a diagram. Let X be a binomial variable with 15 trials and probability of success equal to 0.4, or in short, $X \sim B(15, 0.4)$. We first check on np and nq. In the case at hand, we find that $np = 15 \cdot 0.4 = 6$, which indeed is greater than 5, and $nq = 15 \cdot 0.6 = 9$, which is also greater than 5. In Figure 8.14 we plot the binomial probability distribution as a histogram, along with the normal density function.

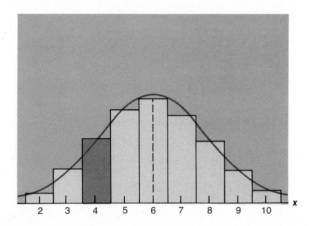

Figure 8.14

Binomial probability distribution, plotted as a histogram, on the background of a normal distribution

The probability of each possible outcome is represented by the *area of the bar centered above it*. The width of each bar is equal to 1, so that the area of the bar is equal to its height. For example, the probability that $X = 4$ is given in the diagram by the area of the bar centered on the value 4 (thus ranging from 3.5 to 4.5). The normal distribution that has been superimposed on the histogram has a mean of $6(= np)$ and a standard deviation of 1.9 $(=\sqrt{npq})$. The area of the darkest box in Figure 8.14, which represents the exact probability that X will equal 4, *is fairly well approximated by the area under the normal curve over the range from 3.5 to 4.5.* Thus we may write

$$P_B(X = 4) \simeq P_N(3.5 \leq X \leq 4.5)$$

where the subscript B means "calculated by the binomial distribution" and the subscript N means "calculated by the normal distribution." In summary, we conclude that when we work under the assumption of a binomial distribution, the probability that X will equal 4 is approximately equal to the probability derived under the assumption of a normal distribution that X will take on values in the interval from 3.5 to 4.5. Since we also know that $X \simeq N(6, 1.9)$, we can proceed as follows:

$$P_B(X = 4) \simeq P_N(3.5 \leq X \leq 4.5)$$

$$= P_N\left(\frac{3.5 - 6.0}{1.9} \leq \frac{X - 6.0}{1.9} \leq \frac{4.5 - 6.0}{1.9}\right)$$

$$= P_N(-1.32 \leq Z \leq -0.79) = 0.1214$$

Assuming a binomial distribution, we find that

$$P(X = 4) = \binom{15}{4}(0.4^4)(0.6^{11}) = 1{,}365 \cdot 0.0256 \cdot 0.003628 = 0.1268$$

Indeed, the approximation is good. To find the probability that X will fall

Figure 8.15

Binomial probability distribution, plotted as a histogram: the area of the six bars is approximated by the area under the normal curve over the interval from 2.5 to 8.5

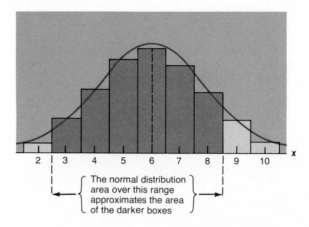

The normal distribution area over this range approximates the area of the darker boxes

within the interval from 3 to 8 by means of the binomial distribution, we would have to go through very extensive calculations, since

$$P(3 \leq X \leq 8) = P(X = 3) + P(X = 4) + \cdots + P(X = 8)$$

and each term to the right of the equals sign has to be calculated by the binomial distribution equation. When we use the normal approximation procedure, however, the calculation is much easier. Figure 8.15 shows that the total area of the histogram bars centered over the values 3 through 8 is fairly well approximated by the normal curve over the range from 2.5 to 8.5, so we may write

$$P_B(3 \leq X \leq 8) \simeq P_N(2.5 \leq X \leq 8.5)$$

$$= P_N\left(\frac{2.5 - 6.0}{1.9} \leq \frac{X - 6.0}{1.9} \leq \frac{8.5 - 6.0}{1.9}\right)$$

$$= P_N(-1.84 \leq Z \leq 1.32) = 0.8737$$

Another example may be helpful.

EXAMPLE 8.12

If 10 percent of all business executives fill out a given marketing survey questionnaire, what is the probability of getting at least 15 questionnaires back out of 200 distributed to executives? What is the probability that between 23 and 33 (inclusive) questionnaires will be filled out?

Assuming that the decision to fill out the questionnaire is made independently by each executive, the number of questionnaires that will be filled out is a binomial random variable: $X \sim B(200, 0.1)$. Our first problem is $P(X \geq 15) = ?$ If we use the normal approximation, the solution comes rather easily. We first calculate the mean and the standard deviation of the variable:

$$E(X) = np = 200 \cdot 0.1 = 20$$
$$SD(X) = \sqrt{npq} = \sqrt{200 \cdot 0.1 \cdot 0.9} = \sqrt{18} = 4.24$$

Now

$$P_B(X \geq 15) \cong P_N(X \geq 14.50)$$

$$= P\left(\frac{X - 20.00}{4.24} \geq \frac{14.50 - 20.00}{4.24}\right)$$

$$= P(Z \geq -1.297) = 0.9032$$

Next, to find the probability that between 23 and 33 questionnaires will be filled out, we compute

$$P_B(23 \leq X \leq 33) \cong P_N(22.5 \leq X \leq 33.5)$$

$$= P\left(\frac{22.50 - 20.00}{4.24} \leq \frac{X - 20.00}{4.24} \leq \frac{33.50 - 20.00}{4.24}\right)$$

$$= P(0.59 \leq Z \leq 3.18) = 0.2769$$

8.4 APPLICATION:

THE "SAFETY-FIRST" PRINCIPLE AND INVESTMENT DECISIONS

Aside from the fact that we haven't the talent for it, most of us would reject Evel Knievel's glamorous way of making a living primarily because of the high probability of what we might term disastrous consequences. A similar attitude is exhibited by firms in their investment decisions. While they search for ways to increase their profits, they tend to reject investment projects that are overly risky. The drive for survival will cause the firms to reject risky projects even at the price of foregoing a potentially high profit. This fact has been recognized by notable economists. In 1952, A. D. Roy wrote:

> Decisions taken in practice are less concerned with whether a little more of this or of that will yield the largest net increase in satisfaction than with avoiding known rocks of uncertain positions or with deploying forces so that, if there is an ambush around the next corner, total disaster is avoided.

He then added:

> For large numbers of people some such idea of a *disaster* exists, and the principle of Safety First asserts that it is reasonable and probable in practice, that an individual will seek to reduce as far as is possible the chance of such a catastrophe occurring.[4]

Consider now three potential investment projects whose returns are denoted by X_1, X_2, and X_3, of which the firm needs to select only one. According to the safety-first rule, the firm will choose the investment that minimizes the probability of receiving income below D, the disaster level. Any income below D means a company catastrophe that may lead to the replacement of current management or to bankruptcy. In short, according to the safety-first rule, the focus is on the probabilities

$$P(X_1 < D)$$

$$P(X_2 < D)$$

[4] A. D. Roy, "Safety First and the Holding of Assets," *Econometrica*, July 1952.

and

$$P(X_3 < D)$$

Note that the value D is determined by the firm's management and varies from one firm to another. In some firms D could equal zero, in others it could be some positive value. For example, if the firm has some debt, D is likely to be positive: unless enough money is generated from the investment, the firm may be unable to pay its debt and may be forced into bankruptcy. At any rate, earnings that do not suffice to cover the interest will cause the stockholders to consider the investment a failure. Suppose the returns X_1, X_2, and X_3 (measured in millions of dollars) are normally distributed as follows:

$$X_1 \sim N(10.0, 10.0)$$

$$X_2 \sim N(8.0, 5.0)$$

$$X_3 \sim N(7.0, 4.2)$$

Of the three investments, we expect the highest profit from X_1 ($10 million), but X_1 is also the riskiest ($10 million standard deviation). The lowest profit is expected from X_3 ($7 million), but its risk as measured by the standard deviation is the lowest ($4.2 million). Investment X_2 is in between, with respect to both expected value and standard deviation. Denoting the respective means and standard deviations by μ_1, μ_2, μ_3, σ_1, σ_2, and σ_3, we have

$$\mu_1 > \mu_2 > \mu_3$$

$$\sigma_1 > \sigma_2 > \sigma_3$$

Suppose that if the income generated from the chosen project is less than $3 million, the firm will face disaster. In that case, $D = 3$, and by the safety-first rule, the investment with the lowest probability $P(X < D)$ should be chosen. Recalling our assumption that X_1, X_2, and X_3 are normally distributed we get

$$P(X_1 < D) = P(X_1 < 3) = P\left(\frac{X_1 - 10}{10} < \frac{3 - 10}{10}\right) = P(Z < -0.70) = 0.2420$$

$$P(X_2 < D) = P(X_2 < 3) = P\left(\frac{X_2 - 8}{5} < \frac{3 - 8}{5}\right) = P(Z < -1.00) = 0.1587$$

$$P(X_3 < D) = P(X_3 < 3) = P\left(\frac{X_3 - 7}{4.2} < \frac{3 - 7}{4.2}\right) = P(Z < -0.95) = 0.1711$$

Thus the second project, which has the smallest probability of disaster, should be selected.

BAUMOL'S "EXPECTED GAIN-CONFIDENCE LIMIT" CRITERION

Although the safety-first criterion is an important rule for investment decisions, we must note that it gives consideration exclusively to the *safety* of investment. William J. Baumol suggested a similar rule, but one that puts more weight on the project's profitability.[5] Baumol asserts that investment in one project whose return is X_1 is better than another whose return is X_2 if the following *two* conditions are met:

$$\mu_1 > \mu_2$$

and

$$L_1 > L_2$$

where

$$L_1 \equiv \mu_1 - k\sigma_1$$

and

$$L_2 \equiv \mu_2 - k\sigma_2$$

where μ_1 and σ_1 are the expected return and standard deviation of X_1, and μ_2 and σ_2 are the corresponding parameters of X_2. The constant k is determined by the firm. The criterion, then, asserts that X_1 is better than X_2 if $\mu_1 > \mu_2$ *and* $L_1 > L_2$. Specifying $k = 2$, for example, and reexamining our three previous investment projects, X_1, X_2, and X_3, we now compute

$$L_1 = \mu_1 - 2\sigma_1 = 10.0 - 2 \cdot 10.0 = -10.0$$
$$L_2 = \mu_2 - 2\sigma_2 = 8.0 - 2 \cdot 5.0 = -2.0$$
$$L_3 = \mu_3 - 2\sigma_3 = 7.0 - 2 \cdot 4.2 = -1.4$$

Comparing X_1 to X_2, we get

$$\mu_1 = 10 > 8 = \mu_2$$

and

$$L_1 = -10 < -2 = L_2$$

so it is not clear which of the two is better. Comparing X_1 to X_3, we find that

$$\mu_1 = 10 > 7 = \mu_3$$
$$L_1 = -10 < -1.4 = L_3$$

Again it is not clear which of the two is a better investment. Finally we compare X_2 and X_3 and see that

$$\mu_2 = 8 > 7 = \mu_3$$

and

$$L_2 = -2.0 < -1.4 = L_3$$

so that none of them can be said to be definitely better than the others.

While the second project is the best of the three according to the safety-first criterion, the expected gain-confidence limit criterion could not, in this particular example, single out any of the three as the best investment project. Selection among the three in this case is left to management or to the financial analyst, who will apply more sophisticated investment analysis.

[5] William J. Baumol, "An Expected Gain-Confidence Limit Criterion for Portfolio Selection," *Management Science*, October 1963.

Problems

8.1. Calculate the following probabilities concerning the standard normal distribution.

(a) $P(Z \geq 0.50)$
(b) $P(Z \leq 2.30)$
(c) $P(Z \geq -1.50)$
(d) $P(0.60 \leq Z \leq 0.65)$

(e) $P(Z \leq -1.10)$
(f) $P(Z \leq -2.50)$
(g) $P(Z \geq -4.00)$

8.2. Calculate the following probabilities for X, where X is normally distributed with a mean of 120 and standard deviation of 20.

(a) $P(X \geq 140)$
(b) $P(X \leq 105)$
(c) $P(X \leq 145)$
(d) $P(107 \leq X \leq 128)$

(e) $P(X \leq 80)$
(f) $P(122 \leq X \leq 130)$
(g) $P(100 \leq X \leq 115)$

8.3. The variable X is normally distributed with a mean of 70 and standard deviation of 8. Find

(a) The 50th percentile.
(b) The 20th percentile.
(c) The 90th percentile.

8.4. The annual income of high-school graduates in their first year at work is normally distributed, with a mean of $6,000 and standard deviation of $800. One youth is picked at random from this population. Find the following probabilities:

(a) That the youth's income is less than $5,000.
(b) That the youth's income exceeds $7,500.
(c) That the youth's income is less than $7,000.
(d) That the youth's income is between $5,500 and $5,900.
(e) That the youth's income is between $5,400 and $8,600.

8.5. If the bottom 5 percent of wage-earners in the population referred to in Problem 8.4 is to receive a pay raise, what is the income level below which a youth will receive a pay raise?

8.6. The random variable X is distributed normally with an unknown mean and standard deviation of 20. Find the mean if $P(X \leq 60) = 0.05$.

8.7. The random variable X is distributed normally with an unknown standard deviation and a mean of 100. Find the standard deviation if $P(X \geq 110) = 0.16$.

8.8. Fly-South Airlines flies several times a week from New York City to Montevideo. All flights are scheduled to leave New York at 10 A.M. The flight's duration is a random variable with normal distribution, a mean of 10 hours, and a standard deviation of a half-hour. Assume that a given plane leaves New York exactly on time.

(a) What is the probability that the plane will arrive at Montevideo *after* 8:30 P.M. New York time?
(b) What is the probability that the arrival time will be between 7:45 P.M. and 8:30 P.M. New York time?
(c) Does the assumption that the flight's duration is normally distributed sound reasonable to you? Explain.

8.9. Mr. Investor has $1,000 that he wishes to invest in common stock. His broker recommends two stocks, A and B. Both have normal distribution of returns, as follows:

$$X_A \sim N(1,100; 40)$$
$$X_B \sim N(1,140; 50)$$

Mr. Investor wants to select only *one* stock of the two. Obviously, stock *B* has a higher expected return ($1,140 versus $1,100 for stock *A*). Stock *B*, however, is more risky in the sense that the standard deviation of *B*'s return is greater than that of *A* ($50 versus $40). Mr. Investor wants to choose that stock which has the smaller probability of returning less than $1,000. Which stock should Mr. Investor choose?

8.10. Mr. Ritchy is not a professional photographer. Ten percent of the pictures he takes are of unsatisfactory quality. If he takes 36 pictures, what is the probability that between 2 and 6 of them (inclusive) are of unsatisfactory quality? Use the normal approximation to the binomial distribution in solving this problem.

8.11. A department store is considering a direct-mail advertising campaign. For simplicity assume that each person who shops at the store brings a profit of $4 before considering the cost of advertising. If printing and mailing the ads cost the store $720 and 10,000 ads are mailed out, what is the probability that the advertising campaign will be profitable to the store? Assume that 2 percent of the people who are mailed the ad go shopping at the store. Use the normal approximation to the binomial distribution to solve this problem.

8.12. If 20 percent of all the clothing purchased at a given department store is returned to the store within seven days of purchase, what is the probability that of 160 articles of clothing sold on a given day, between 25 and 35 (inclusive) will be returned within seven days?

8.13. The XYZ company is extending credit to new customers only if they pass the company's test for reliability. Three percent of new customers who pass the test, however, are found to be unreliable (that is, they pay late or not at all). If 500 new customers are approved for credit after the reliability test, what is the probability that between 10 and 20 (inclusive) customers are unreliable?

8.14. (*a*) Find the probability that $Z \geq 1.96$, where Z is the standard normal random variable.
(*b*) Find the probability that $Z^2 \geq 3.84$. Note that $1.96^2 = 3.84$.

PART 3

DECISION-MAKING BASED ON SAMPLES (INDUCTION)

Up to now, we have discussed the presentation of statistical data, and deduction from population structure to probabilities involving individual observations. Most of the remainder of this book is devoted to induction: using statistical inference to learn about the structure and characteristics of a population by looking at only a part of that population—i.e., a sample. Chapter 9 discusses sampling and estimation. Chapters 10 through 12 deal with various tests of hypotheses.

CHAPTER NINE OUTLINE

9.1 **Introduction**

9.2 **Sample vs. Census**
The Cost of Sampling
Accuracy
The Time Factor
Other Factors

9.3 **Errors Associated with Samples**
Nonsampling Errors
Sampling Errors, Statistics, and Bias

9.4 **Types of Samples**
Convenience Samples
Judgment Samples
Probability Samples

9.5 **The Distribution of the Sample Average $(\overline{X})$**
The Standard Deviation of the Sample
Average $(\overline{X})$
The Sampling Distribution of $\overline{X}$ When the
Population Is Finite
The Sampling Distribution of $\overline{X}$ When X Is
Normally Distributed

9.6 **Interval Estimate vs. Point Estimate**
The Factors Determining e—And What Happens
When They Change
Relationships among the Factors Determining e

9.7 **The Central Limit Theorem**

9.8 **Confidence Intervals for the Population Mean**
The Trade-Off between Confidence Level and
Interval Width

9.9 **Confidence Interval for μ When σ Is Unknown**
The Student t Distribution
Constructing the Confidence Interval

9.10 **Estimation of Proportions**

9.11 **Application 1: The Election Problem**

9.12 **Application 2: Quality Control**

Key Terms

statistical inference
sample
population
census
nonsampling errors
measurement error
sampling errors
statistic
convenience sample
judgment sample
probability samples
simple random sample
strata

stratified random sample
clusters
cluster random samples
standard error
point estimate
interval estimate
tolerable error
risk probability
central limit theorem
confidence interval
confidence level
Student t distribution

9
SAMPLING AND ESTIMATION

9.1 Introduction

Sampling is at the core of wide areas in statistics, and in recognition of that fact we shall devote most of the remainder of this book to **statistical inference,** those methods by which the characteristics of populations and phenomena are studied through the observation of sample data. In this chapter we shall consider the reasons for sampling, describe alternative sampling designs, and discuss point and interval estimation.

A **sample** is a *segment of a population,* and as such is expected to reflect the population, so that by studying the sample we may learn about certain attributes of the total population. A major objective of sampling is to acquire information about a population through an analysis of a sample. More precisely, we take samples in order to obtain estimates of important parameters (such as the mean and the standard deviation) of distributions of random variables under consideration. It is of utmost importance to realize that the population as a whole is only approximately reflected in the sample. Furthermore, results vary from sample to sample, because no two samples contain the same set of observations. If we estimate the average lifetime of a certain brand of AA batteries by measuring the lifetimes of a number of randomly selected batteries of that brand, we can expect to find that the average lifetime of the batteries in the sample will approximate the mean lifetime of *all* the AA batteries of that particular brand. *The sample average, however, will generally differ from the population's mean, and furthermore, had we taken another set of batteries for examination, we probably would have gotten a different average the second time.* All of this indicates that the estimated average (as well as other estimated parameters) is a *random variable.*

9.2 Sample vs. Census

A **population** is defined as *the aggregate of all the items having some specified characteristics*—a definition that encompasses a wide variety of populations, ranging in size from very small to infinite.

It is obviously impossible to study an infinite population in its entirety, and thus studies of such populations must always rely on samples. Studies concerning finite populations can, at least in principle, be carried out by observation of all the members of the population. A *complete enumeration of all the population's members* is called a **census.** But for a variety of reasons, a census is often either impossible or impractical, and so frequently we use samples to study finite populations as well.

THE COST OF SAMPLING

The economic cost of observing a portion of a population is almost invariably less than the cost of observing each and every member of the population. Take, for example, the population of all the parents of high school students in the United States. The planning, data collection, and analysis of a census covering such a population will have to be extremely expensive. Furthermore, it is inevitable that part of the population will inadvertently be excluded: people move, they go on vacation, they refuse to participate. By contrast, a sample of moderate size involves only a fraction of the cost and can provide satisfactory information concerning the population as a whole.

ACCURACY

Because a sample involves fewer population members than a census does, each sample observation is normally of better "quality" than each observation in a census. The number of observations involved in a large census often forces the researcher to devote minimal attention to each one. If the study is to be carried out via a sample of interviews, for example, only a handful of interviewers will be needed, and these can be knowledgeable people. If, on the other hand, the study is to be carried out via a census, more interviewers will have to be employed and some may be less qualified than others. The result with a census is that more observations are provided but the average quality of the observations is likely to be poorer.

THE TIME FACTOR

While the time factor is an important consideration in every kind of research, it is of particular importance in the business world. If we are doing a study designed to find out whether businesses intend to accelerate or decelerate production next month, we cannot afford to spend two months on our investigation. Likewise, when we are surveying consumers to determine the most appealing design for a given product, we cannot spend a great deal of time, because we may find out that a competitor has already come out with a similar product that has proved to be successful. Similar situations are numerous.

A sample has an advantage over a census in that it can be completed more quickly. Also, the processing of the results—the tabulation, typing, coding, punching, and so forth—is a much faster procedure for a sample than it is for a census.

OTHER FACTORS

Among the other reasons for sampling is that in some cases "observing" an item in a population involves the destruction of the item. Worthy of mention here is the well-known anecdote of the joker who struck all his matches in order to test them. To determine the quality or taste of any consumable good it must be used up, and sampling is obviously needed for this purpose.

Also, some populations are not easily accessible. Such is the case when the population encompasses senior corporate management, high government officials, intensive care patients, and the like. Reliance on a sample rather than a census is practically inevitable in those cases.

Having justified reliance on sampling in so many situations, we turn now to the errors associated with sampling and to popular sample designs.

9.3 Errors Associated with Samples

Since a sample is by definition a portion of the population, it is not generally capable of providing a full and flawless reflection of the population as a whole. There will be certain discrepancies between the population and the sample. We distinguish two major sources of such discrepancies: nonsampling errors and sampling errors.

NONSAMPLING ERRORS

Nonsampling errors are *errors that result either from improper selection of sample observations or from erroneous information obtained from the observations.* Let us look at each of these in turn.

Improper Selection

Improper selection of sample observations occurs when a given section of the population has an unduly low or unduly high chance of being selected for the sample. Consider a random sample selected for the purpose of discovering the proportion of people in a developing country who have telephones in their homes. It would surely be absurd to perform the study through telephone interviews. Similarly, we would not go to a church in order to find out the proportion of people who believe in God. Although these are extreme examples, they make an important point. Far too often, samples are selected without careful preparation to ensure full and balanced coverage of the entire population under consideration. To consider a more realistic example, suppose a sample is being selected with the goal of uncovering some common characteristics of investors in common stocks. It will be necessary to get the consent of the stockholders before answers to such questions can be obtained. Yet it is possible that heavy investors may be reluctant to discuss the very matters that are of interest to the researcher. If such is the case, people who own large amounts of common stocks may have an unduly low representation in the sample results, and people with small holdings may have an unduly high representation. How are such difficulties overcome? There are various methods for dealing with them, including the *stratified sample* (which will be discussed later in this chapter).

Erroneous Information

Erroneous information, including misunderstandings, approximations, and incorrect observations, is often referred to as **measurement error.** Consider a sample of TV viewers who are asked how many hours per week they spend watching TV. Approximation is probably the best people can do in response to such a question. Likewise, when a sample requires people's opinions, there is always the risk of some misunderstanding and differences of interpretation.

SAMPLING ERRORS, STATISTICS, AND BIAS

Sampling errors are *errors that result from the chance selection of the sampling units.* They occur when a portion, rather than the entire population, is observed. The larger the sample size, the smaller the magnitude of sampling errors expected; when a census is taken, all sampling errors should disappear.

A **statistic** is a *value computed from the sample data.* Since sample data depend on the specific units that happen to be in the sample, it is clear that a statistic is a random variable. A statistic is usually used for estimation: when we wish to base our estimate of a population parameter on sample data, we compute the parameter's counterpart in the sample (which is a statistic) and use it as an estimate of the parameter. The simplest example is the use of sample data to estimate the population mean. The average lifetime of light bulbs in a sample is the counterpart of the mean lifetime for the entire population of light bulbs out of which the sample has been selected, and thus serves as an estimate of the population's mean. Whereas the population mean is a fixed value, the sample average is a statistic and its value is random (provided the sample observations were selected at random). A "hat" is the common notation used to indicate an estimate. Thus if μ denotes the population means, $\hat{\mu}$ denotes its sample counterpart. Similarly, if σ and p denote the standard deviation and population proportion, respectively, $\hat{\sigma}$ and $\hat{p}$ denote their counterparts (which are statistics) in the sample. Some of the more common parameters, such as μ and σ, have additional notations for their sample counterparts. While μ and σ denote the population mean and standard deviation respectively, $\overline{X}$ and S (in addition to $\hat{\mu}$ and $\hat{\sigma}$) are commonly used for the average and standard deviation of the sample data.

When the expected value of a statistic is equal to its counterpart population parameter, we say that the statistic is *unbiased.* Conversely, when the expected value of the statistic differs from its counterpart population parameter, the statistic is said to be *biased.* In other words, when the sample is not subject to nonsampling errors, the statistic is unbiased, but when nonsampling errors exist, the statistic is biased.

9.4 Types of Samples

Samples are distinguished by the processes used to select them. We speak of convenience samples, judgment samples, and random samples.

CONVENIENCE SAMPLES

A **convenience sample** is one that is selected primarily on the basis of sampling convenience.

Suppose we want to learn about the activities of tourists in the United States (such as dining, lodging, or recreational activities). If we select our sample from those who stay at a nearby hotel, we are using a convenience sample.

While convenience is always an appealing feature, it is clear that a convenience sample is less representative of the whole population than one that is selected in a more diversified manner.

JUDGMENT SAMPLES

A **judgment sample** is one that is selected primarily on the basis of judgment. The primary reason we might want to base our selection of a sample on judgment—rather than on chance—is that we believe such a sample to be more reflective of the population characteristics. We may have good reason for such a belief if the sample size is very small.

For example, if we need to take a sample of public-school students and our resources enable us to sample only two schools, we may be wise to use judgment in the selection of the schools. If we leave the selection of those schools to chance alone, we may select schools that are atypical in some respect. Of course, if the sample size is even moderately large, random selection will prove more advantageous than judgment.

PROBABILITY SAMPLES

Probability samples are selected in such a way that the *probability* of each element in the population being selected is known in advance. The most common sample design of this type is the simple random sample; other well-known designs are the stratified random sample and the cluster random sample.

The Simple Random Sample and the Use of the Random Numbers Table

While the convenience and judgment samples have their advantages, their main deficiency is that they do not enable us to assess sampling errors. Probability samples, on the other hand, allow us to assess the sampling errors involved. Statistical theory is thus more relevant and applicable to these types of samples.

A **simple random sample** is selected in such a way that it *assures equal probability of selection to all samples of the same size.* This definition of simple random selection applies to a finite population only. When the population is infinite, one must interpret the simple random selection as a process that ensures that the observations are independent, and that they are drawn from a population whose composition remains unchanged. A correct procedure for selecting a random sample out of a finite population is to assign a serial number to each of the population elements and to select the sample by drawing a prespecified number of serial numbers at random. A good way of selecting serial numbers at random is to use the random numbers table (see Appendix A, Table A.11). Suppose 150 observations are to be picked from 2,500 population elements, all of which have been assigned serial numbers ranging from 0001 to 2500. Our first step is to enter the table at a page selected blindly, in a totally haphazard way—literally closing the eyes while making the page selection. A number on the page should be selected fortuitously in a similar fashion. The number selected is the starting number. Note that a four-digit number must be read in our example. After the starting number has been selected, we read additional four-digit numbers by moving horizontally or vertically on the page, selecting other pages at random as the need arises. If a number is in the range 0001 through 2500, it is recorded. If it is out of this range, the number is not recorded and we continue reading other numbers. When 150 numbers are recorded, the process is discontinued. The sample is constructed so as to consist of the 150 elements whose serial numbers have been selected.

A distinction must be made here between simple *random* sampling and *haphazard* sampling. In simple random sampling we deliberately make sure that each element in the population is given an equal probability of selection. In haphazard selection, this is not done. Investigating public opinion on the women's liberation movement through telephone interviews conducted on weekday mornings is haphazard and is likely to include an unduly large proportion of housewives and other people who are not employed outside the home. A simple random sample must ensure that each member of the population has the same probability of selection.

The Stratified Random Sample

It is often possible to identify distinguishable **strata** (i.e., *homogeneous groups*) within the population under consideration. For various reasons it might be more efficient and thus desirable to *give unequal emphasis in the sample to the various population strata*—that is, to take a **stratified random sample.**

One such reason might be a lack of homogeneity among the population strata. Suppose we have a list of 1,000 commercial banks in a given population and we wish to select 50 of them for a sample. The 1,000 banks naturally vary in size. Although large banks may make up only a small percentage of the population, we may wish to give them a greater probability of being selected, because if more large banks are chosen, we will obtain better coverage of the total deposits. We could define two strata: (1) "large" banks (those with $100 million or more in deposits) and (2) all the others. If only 30 of the 1,000 banks are defined as large, we may decide to allocate our sample in such a way that 20 banks will be selected among the large banks and the rest will be selected among the others. So, although only 3 percent of the banks in the population are large (30 out of 1,000), the large banks will make up 40 percent (20 out of 50) of the observations in the sample. This procedure of sample selection will better meet our need to obtain even coverage of the deposits in the 1,000 banks.

There are other reasons why unproportional representation of population strata might lead to greater efficiency, such as unequal variance of the population strata. It is often advantageous to obtain a greater proportion of sample observations from the strata with greater variance.

Stratified sampling is primarily important if the overall sample size is small. If only a small number of observations is taken and if distinguishable groups in the population exist, it is quite possible that not all of them will be well represented in a random sample. As the sample size increases, simple rules of probability are likely to bring about a more even and well-balanced representation of all the population's groups in a simple random sample, and no special steps will need to be taken to ensure balanced representation.

The Cluster Random Sample

A population stratum is one that is homogeneous with respect to the characteristic with which we are concerned. Sometimes we define *groups within homogeneous population strata on the basis of their accessibility.* We refer to such groups as **clusters. Cluster random samples** are used almost exclusively when groups are defined by geographical location. A personal interview with 100 people across the country could require 100 trips to various locations. To minimize our costs, we might wish to divide the country into states, counties, or cities, and choose—by a simple random process—a limited number of such geographical areas. We might, for example, choose five cities and hold our personal interviews there, thus making only 5 trips instead of 100. Cluster samples result in greater sampling errors than do simple random samples (which yield the smallest sampling errors), because people from one cluster may be more similar to one another than they are to people in other locations. But the money we save by focusing on a limited number of clusters can be used to increase the sample size and offset at least part of the otherwise relatively large sampling errors. Sometimes the savings from using a cluster sample rather than a random sample may be substantial.

9.5 The Distribution of the Sample Average ($\overline{X}$)

We have discussed several types of samples and the circumstances under which they might be relevant. In what follows, however, we shall assume that the samples are simple random samples.

Intuitively we know that the greater the sample size (that is, the larger the number of observations taken), the closer the estimated average will tend to be to the true mean of the entire population. By averaging the heights of 20 U.S. citizens, we more accurately portray the mean U.S. citizen's height than if we merely average the heights of 5 of them. Likewise, the average height of 100 citizens will probably be even closer to the "true" mean height (the mean height of *all* U.S. citizens).

This point is so crucial to a good understanding of sampling and estimation that it deserves a detailed illustration. Suppose that the store manager of the Gorgeous Lamps Company is considering the frequency of his inventory orders. One of his objectives is to estimate the mean delivery period (that is, the average number of days that elapse from the time an order is placed until the time the merchandise actually comes into the store). Let us assume that the delivery period—which is a random variable—has the probability distribution shown in Table 9.1.

TABLE 9.1
Probability Distribution of Delivery Periods

Delivery period (X)	Probability
4	0.01
5	0.02
6	0.05
7	0.08
8	0.12
9	0.14
10	0.16
11	0.14
12	0.12
13	0.08
14	0.05
15	0.02
16	0.01
	1.00

The mean delivery period is 10 days:

$$E(X) = 4 \cdot 0.01 + 5 \cdot 0.02 + 6 \cdot 0.05 + \cdots + 14 \cdot 0.05 + 15 \cdot 0.02$$
$$+ 16 \cdot 0.01 = 10.00$$

The probability distribution in Table 9.1 may be interpreted in the following way: on the average, one would expect to find a delivery period of 4 days occurring once in 100 times, a delivery period of 5 days twice in 100 times, a period of 6 days 5 times in 100, and so on.

Suppose the probability distribution is not known to the manager, perhaps because past records were never kept. If he wants to estimate the mean, he will have to take a sample of delivery periods, whose average ($\overline{X}$) will serve as a proxy to (estimate of) the population mean (μ). How good the proxy will be depends on the number of observations taken. If only one observation is taken—in other words, if only one delivery period is observed and used as an estimate—then the likelihood is fairly good that the estimated delivery period will differ substantially from the true mean delivery period. For example, out of 100 delivery periods, 56 (on the average) are either less than or equal to 8 days or greater than or equal to 12 days—in other words, there are 56 periods in 100 that deviate from the mean by 2 or more days. Similarly, 84 periods in 100 (on the average) are either less than or equal to 9 days or greater than or equal to 11 days—in other words, 84 periods in 100 deviate from the mean by 1 or more days. Thus, by taking a sample of only one observation, one faces the likelihood of coming up with an inaccurate estimate. If, however, one bases the estimate on the average of, say, 5 observations, *the probability that the average will miss the true average by 2 or more days is far below 56 percent, and the probability that it will miss the mean by one or more days is far below 84 percent. This is the major reason that we use samples of several observations rather than one observation to estimate population parameters.* Technically, we state that the variance of the average of *n* observations is smaller than the variance of one observation.

To illustrate our argument, we can take a close look at the average of samples of various sizes. We begin by constructing Table 9.2, in which 100

TABLE 9.2
One Hundred Delivery Periods Implied "on the Average" by the Probability Distribution of X

4	7	8	9	9	10	11	11	12	13
5	7	8	9	9	10	11	11	12	13
5	7	8	9	10	10	11	12	12	14
6	7	8	9	10	10	11	12	12	14
6	7	8	9	10	10	11	12	13	14
6	7	8	9	10	10	11	12	13	14
6	8	8	9	10	10	11	12	13	14
6	8	8	9	10	10	11	12	13	15
7	8	9	9	10	11	11	12	13	15
7	8	9	9	10	11	11	12	13	16

delivery periods are listed in frequencies implied by the probability distribution. The number 4 appears once, the number 5 appears twice, the number 6 appears five times, and so on, according to their probability of occurrence. After assigning a serial number to each of the numbers in Table 9.2, we draw random samples of various sizes out of the table, using the random numbers table (see Appendix Table A.11). Note that any of the numbers in Table 9.2 can be selected more than once, depending on the number of times its serial number appears in the random numbers table. This procedure guarantees that the samples are chosen by a random selection process out of the specified probability distribution.

First we select 20 samples of one observation each. These samples are listed in Table 9.3, and as we can see, if only one observation were to be used as an estimate for the mean, our error might be large: quite a few of the observations listed in Table 9.3 vary substantially from the true mean, 10. In Table 9.4, 20 random samples of 5 observations each and their respective averages are listed. An examination of the sample averages shows, for example, that *not even one average deviates from 10 by two or more days.* (Contrast the observations in Table 9.3: a relatively large percentage of these *individual* observations deviate from 10 by two or more days.) When we take *sample averages,* large deviations from the true average are possible, but they are far less probable than when we take individual observations. The greater the sample size, the greater the tendency of the sample average to cluster around the true mean of the distribution, and the rarer large deviations become. Table 9.5 presents a set of 20 randomly selected samples of 12 observations each. We obtained these observations by choosing them from Table 9.3, using the random numbers table (Appendix Table A.11). Here, as expected, the averages of the various samples cluster even more closely around the mean. Only 4 of the sample averages in Table 9.5 deviate by more than one day from the distribution mean, as compared to 6 for samples of size 5 (Table 9.4) and 13 for samples of size 1 (Table 9.3). This tendency of the sample averages to cluster more closely around the distri-

TABLE 9.3
Twenty Randomly Selected Samples of One Observation Each

12	12	6	9	7	14	7	10	8	10
11	14	12	10	11	5	12	13	9	7

TABLE 9.4
Twenty Randomly Selected Samples of 5 Observations Each, and Their Averages

1	2	3	4	5	6	7	8	9	10
14.0	11.0	9.0	8.0	11.0	10.0	10.0	10.0	14.0	10.0
10.0	7.0	13.0	9.0	9.0	8.0	7.0	10.0	7.0	9.0
9.0	9.0	9.0	9.0	11.0	10.0	9.0	10.0	10.0	10.0
10.0	6.0	11.0	8.0	7.0	8.0	11.0	6.0	10.0	7.0
12.0	12.0	8.0	7.0	10.0	12.0	7.0	13.0	8.0	14.0
$\overline{X}_1 = 11.0$	$\overline{X}_2 = 9.0$	$\overline{X}_3 = 10.0$	$\overline{X}_4 = 8.2$	$\overline{X}_5 = 9.6$	$\overline{X}_6 = 9.6$	$\overline{X}_7 = 8.8$	$\overline{X}_8 = 9.8$	$\overline{X}_9 = 9.8$	$\overline{X}_{10} = 10.0$

11	12	13	14	15	16	17	18	19	20
9.0	14.0	10.0	10.0	11.0	7.0	12.0	12.0	9.0	8.0
7.0	5.0	9.0	9.0	11.0	11.0	14.0	11.0	8.0	9.0
7.0	13.0	11.0	9.0	6.0	11.0	10.0	12.0	10.0	13.0
12.0	15.0	11.0	13.0	11.0	8.0	8.0	9.0	9.0	10.0
9.0	11.0	8.0	10.0	7.0	12.0	12.0	8.0	5.0	15.0
$\overline{X}_{11} = 8.8$	$\overline{X}_{12} = 11.6$	$\overline{X}_{13} = 9.8$	$\overline{X}_{14} = 10.2$	$\overline{X}_{15} = 9.2$	$\overline{X}_{16} = 9.8$	$\overline{X}_{17} = 11.2$	$\overline{X}_{18} = 10.4$	$\overline{X}_{19} = 8.2$	$\overline{X}_{20} = 11.0$

TABLE 9.5
Twenty Randomly Selected Samples of 12 Observations Each, and Their Averages

1	2	3	4	5	6	7	8	9	10
14.0	6.0	11.0	8.0	10.0	9.0	9.0	11.0	14.0	9.0
8.0	14.0	10.0	10.0	13.0	8.0	10.0	11.0	10.0	11.0
11.0	6.0	12.0	14.0	10.0	10.0	14.0	10.0	9.0	8.0
9.0	15.0	10.0	12.0	12.0	4.0	9.0	11.0	10.0	8.0
7.0	8.0	11.0	12.0	8.0	8.0	13.0	9.0	12.0	9.0
9.0	14.0	7.0	10.0	12.0	7.0	14.0	9.0	11.0	9.0
10.0	16.0	12.0	15.0	7.0	11.0	15.0	12.0	7.0	8.0
13.0	11.0	10.0	13.0	11.0	8.0	14.0	11.0	9.0	7.0
10.0	12.0	10.0	8.0	10.0	12.0	9.0	5.0	6.0	15.0
14.0	8.0	14.0	10.0	7.0	8.0	11.0	8.0	12.0	9.0
7.0	10.0	11.0	8.0	14.0	11.0	11.0	13.0	9.0	13.0
9.0	11.0	8.0	10.0	6.0	11.0	11.0	9.0	13.0	10.0
$\overline{X}_1 = 10.1$	$\overline{X}_2 = 10.9$	$\overline{X}_3 = 10.5$	$\overline{X}_4 = 10.8$	$\overline{X}_5 = 10.0$	$\overline{X}_6 = 8.9$	$\overline{X}_7 = 11.7$	$\overline{X}_8 = 9.9$	$\overline{X}_9 = 10.2$	$\overline{X}_{10} = 9.7$

11	12	13	14	15	16	17	18	19	20
11.0	9.0	9.0	9.0	8.0	9.0	10.0	11.0	8.0	9.0
9.0	11.0	10.0	6.0	10.0	7.0	10.0	10.0	11.0	8.0
11.0	7.0	7.0	14.0	13.0	7.0	11.0	14.0	6.0	8.0
7.0	10.0	14.0	11.0	13.0	12.0	12.0	12.0	10.0	11.0
9.0	10.0	6.0	8.0	13.0	9.0	5.0	11.0	12.0	12.0
10.0	10.0	11.0	12.0	8.0	14.0	4.0	9.0	11.0	14.0
8.0	6.0	11.0	8.0	12.0	7.0	10.0	5.0	5.0	4.0
10.0	15.0	6.0	9.0	10.0	13.0	12.0	9.0	9.0	14.0
8.0	14.0	11.0	7.0	12.0	15.0	7.0	9.0	10.0	13.0
12.0	7.0	11.0	11.0	12.0	11.0	8.0	11.0	11.0	8.0
10.0	10.0	9.0	9.0	9.0	10.0	8.0	10.0	8.0	12.0
7.0	10.0	13.0	8.0	11.0	9.0	10.0	10.0	5.0	8.0
$\overline{X}_{11} = 9.3$	$\overline{X}_{12} = 9.9$	$\overline{X}_{13} = 9.8$	$\overline{X}_{14} = 9.3$	$\overline{X}_{15} = 10.9$	$\overline{X}_{16} = 10.2$	$\overline{X}_{17} = 8.9$	$\overline{X}_{18} = 10.1$	$\overline{X}_{19} = 8.8$	$\overline{X}_{20} = 10.1$

bution mean as the sample size increases is seen also in Figure 9.1, in which the averages of the samples of Tables 9.3, 9.4, and 9.5 are presented together.

It is extremely important to realize that *in most sampling situations, only one sample is taken.* However, from the tendency of sample averages to cluster more closely around the distribution mean as the sample size increases, it follows that *the larger the sample size, the greater the probability that the average of the one sample taken will be close to the distribution mean,* where "close" is defined as some arbitrary maximum deviation, such as one day in the delivery-period example.

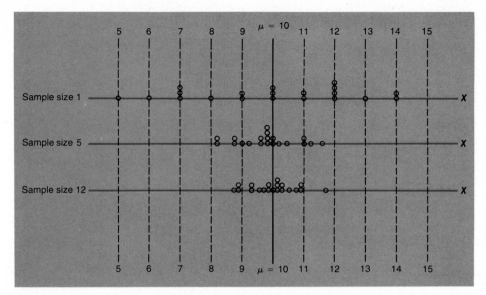

Figure 9.1

The dispersion of averages of samples of various sizes taken from the distribution of delivery periods

THE STANDARD DEVIATION OF THE SAMPLE AVERAGE ($\overline{X}$)

If the nature of the distribution of $\overline{X}$ is understood, the mathematics of it should be readily understood. We have already explained that $\overline{X}$ is a random variable. The delivery-period example illustrated another important characteristic of $\overline{X}$: the average of many samples' averages tends toward the mean of the distribution, or $E(\overline{X}) = \mu$. Let us see how we can get this result by using simple equations involving random variables.

Since by definition $\overline{X} = \dfrac{\Sigma X}{n} = \dfrac{X_1 + X_2 + \cdots + X_n}{n}$

we may write

$$E(\overline{X}) = E\left(\frac{X_1 + X_2 + \cdots + X_n}{n}\right) = E\left[\frac{1}{n}(X_1 + X_2 + \cdots + X_n)\right]$$

$$= \frac{1}{n}E(X_1 + X_2 + \cdots + X_n) = \frac{1}{n}[E(X_1) + E(X_2) + \cdots + E(X_n)]$$

Since $X_1, X_2, \cdots, X_n$ are all taken from the same population, they all have the same mean, μ:

$$E(X_1) = E(X_2) = \cdots = E(X_n) = \mu$$

so that

$$E(\overline{X}) = \frac{1}{n}\underbrace{[\mu + \mu + \cdots + \mu]}_{n \text{ times}} = \frac{1}{n} \cdot n\mu = \mu$$

In brief, we may write this:

$$E(\overline{X}) = \mu_{\overline{x}} = \mu \tag{9.1}$$

where $\mu_{\overline{x}}$ denotes the expected value of $\overline{X}$.

We have shown that the variability of $\overline{X}$ decreases with sample size. Now let us take a closer look at variability as measured by the variance. Denoting the variance of $\overline{X}$ by $\sigma_{\overline{X}}^2$ we write

$$\begin{aligned} \mathrm{var}(\overline{X}) \equiv \sigma_{\overline{X}}^2 = \mathrm{var}\left(\frac{\Sigma X}{n}\right) &= \mathrm{var}\left(\frac{X_1 + X_2 + \cdots + X_n}{n}\right) \\ &= \mathrm{var}\left[\frac{1}{n}(X_1 + X_2 + \cdots + X_n)\right] \\ &= \frac{1}{n^2}\mathrm{var}(X_1 + X_2 + \cdots + X_n) \end{aligned} \tag{9.2}$$

Assuming that $X_1, X_2, \cdots, X_n$ are independent observations, we get

$$\mathrm{var}(X_1 + X_2 \cdots + X_n) = \mathrm{var}(X_1) + \mathrm{var}(X_2) + \cdots + \mathrm{var}(X_n)$$

and since the variances of all the X_is are equal to one another (different observations of the same distribution) and are denoted by σ^2, we get

$$\mathrm{var}(X_1) + \mathrm{var}(X_2) + \cdots + \mathrm{var}(X_n) = \underbrace{\sigma^2 + \sigma^2 + \cdots + \sigma^2}_{n \text{ times}} = n\sigma^2 \tag{9.3}$$

Substituting Equation 9.3 into Equation 9.2 yields

$$\sigma_{\overline{X}}^2 = \frac{1}{n^2}n\sigma^2 = \frac{\sigma^2}{n} \tag{9.4}$$

and as a direct result we find

$$\sigma_{\overline{x}} = \sqrt{\frac{\sigma^2}{n}} = \frac{\sigma}{\sqrt{n}} \tag{9.5}$$

where $\sigma_{\overline{x}}$ stands for "**standard error** of the mean," which is simply the standard deviation of $\overline{X}$.

$$\sigma_{\overline{X}}^2 = \frac{\sigma^2}{n}$$

$$\sigma_{\overline{x}} = \frac{\sigma}{\sqrt{n}} \tag{9.6}$$

THE SAMPLING DISTRIBUTION OF $\overline{X}$ WHEN THE POPULATION IS FINITE

We have seen that the standard error of the mean is given by the expression $\sigma_{\overline{X}} = \dfrac{\sigma}{\sqrt{n}}$. The estimated standard error is given by $S_{\overline{X}} = \dfrac{S}{\sqrt{n}}$ where S is the standard deviation in the sample. These expressions are correct for infinite populations or for finite populations when sampling is done with replacement. When the population is finite and sampling is done with no replacement, the above expressions overstate the standard error of $\overline{X}$. To illustrate, suppose the population consists of the annual income of 36 students in a statistics class. Since the population is *defined* as the income of the particular 36 students in the class, it is clear that if all 36 students have been sampled, the sample average $\overline{X}$ will *precisely* equal the population mean. This does not mean that there is no variability in the income of the 36 students but only that the population mean has been estimated with full precision. For example, if σ = \$600 (i.e., the average deviation of the income among the 36 students from the class average is \$600), the expression $\sigma_{\overline{X}} = \dfrac{\sigma}{\sqrt{n}}$ incorrectly indicates that the standard deviation of $\overline{X}$ for n = 36 is $\sigma_{\overline{X}} = \dfrac{600}{\sqrt{36}}$ = \$100. In fact, $\dfrac{\sigma}{\sqrt{n}}$ should be multiplied by a finite population correction factor, $\sqrt{\dfrac{N-n}{N-1}}$, where N is the population size and n is the sample size. More specifically, for a finite population and sampling without replacement, the standard error of the mean is given by the equation:

$$\sigma_{\overline{X}} = \frac{\sigma}{\sqrt{n}} \sqrt{\frac{N-n}{N-1}} \qquad (9.7)$$

Its estimator is given by the equation

$$S_{\overline{X}} = \frac{S}{\sqrt{n}} \sqrt{\frac{N-n}{N-1}} \qquad (9.8)$$

Clearly, when $N = n$, we get $\sigma_{\overline{X}} = S_{\overline{X}} = 0$ as in the above example. On the other hand, if N is very large compared to n, the term $\sqrt{\dfrac{N-n}{N-1}}$ is very close to 1, and $\sigma_{\overline{X}} = \dfrac{\sigma}{\sqrt{n}}$ and $S_X = \dfrac{S}{\sqrt{n}}$ are good proxies. For example, suppose N = 11,000, n = 1,000, and σ = 100. When disregarding the correction factor we get

$$\sigma_{\overline{X}} = \frac{\sigma}{\sqrt{n}} = \frac{100}{\sqrt{1,000}} \frac{100}{31.6} = \$3.16$$

After correcting for the finite population we obtain

$$\sigma_{\bar{X}} = \frac{\sigma}{\sqrt{n}}\sqrt{\frac{N-n}{N-1}} = \frac{100}{\sqrt{1,000}}\sqrt{\frac{11,000-1,000}{11,000-1}}$$

$$= 3.16\sqrt{0.909} = \$3.01$$

Thus the approximate standard error overestimated $\sigma_{\bar{X}}$ by about 5 percent in this particular case.

THE SAMPLING DISTRIBUTION OF $\bar{X}$ WHEN X IS NORMALLY DISTRIBUTED

Our main concern in the remainder of this chapter is the distribution of $\bar{X}$ when the random variable X is normally distributed. The main rule to remember here is that when X is normally distributed, then $\bar{X}$ is also normally distributed with the same mean, μ. Thus, if

$$X \sim N(\mu, \sigma)$$

then

$$\bar{X} \sim N(\mu, \sigma_{\bar{X}}) \tag{9.9}$$

and $\sigma_{\bar{X}}$ is equal to $\dfrac{\sigma}{\sqrt{n}}$ if the population is infinite or if it is finite and the sampling is done with replacement. If the population is finite and the sampling is without replacement, it is equal to $\dfrac{\sigma}{\sqrt{n}}\sqrt{\dfrac{N-n}{N-1}}$. In the rest of this chapter we will assume $\sigma_{\bar{X}} = \dfrac{\sigma}{\sqrt{n}}$

In Chapter 8 we saw that if

$$X \sim N(\mu, \sigma)$$

then

$$\frac{X-\mu}{\sigma} \sim N(0,1)$$

and it follows that

$$Z = \frac{X-\mu}{\sigma} \quad \text{and} \quad X = \mu + \sigma Z$$

Similar relationships exist in the distribution of $\bar{X}$, so that ZA is obtained when we subtract the mean of $\bar{X}$ (μ) and divide by the standard error of the mean (σ_x). Assuming $\sigma_{\bar{X}} = \dfrac{\sigma}{\sqrt{n}}$, we get

If

$$\overline{X} \sim N\left(\mu, \frac{\sigma}{\sqrt{n}}\right)$$

then

$$\left(\frac{\overline{X} - \mu}{\sigma/\sqrt{n}}\right) \sim N(0,1) \qquad (9.10)$$

or

$$Z = \frac{\overline{X} - \mu}{\sigma/\sqrt{n}} \qquad \text{and} \qquad \overline{X} = \mu + \frac{\sigma}{\sqrt{n}} \cdot Z$$

Example 9.1 illustrates an application of the relationships just established.

EXAMPLE 9.1

Suppose the daily revenue of a laundromat chain is normally distributed with a mean of $20,000 and standard deviation of $4,000. Find the probability that the average daily revenue for 25 days will be between $19,000 and $21,000.

Our approach is to convert the question from one concerning the probability of $\overline{X}$, for which $\mu_{\overline{X}} = 20$ (thousand dollars) and $\sigma_{\overline{X}} = \frac{4}{\sqrt{25}}$, to one concerning the corresponding distribution of Z, so we proceed as follows:

$$P(19 \leq \overline{X} \leq 21) = P\left(\frac{19 - 20}{4/\sqrt{25}} \leq \frac{\overline{X} - 20}{4/\sqrt{25}} \leq \frac{21 - 20}{4/\sqrt{25}}\right)$$

$$= P\left(\frac{-1.0}{0.8} \leq Z \leq \frac{1.0}{0.8}\right) = P(-1.25 \leq Z \leq 1.25) = 0.7888$$

Any other problem concerning the probability of $\overline{X}$ may be tackled in a similar way. For example, to find the probability that the average daily revenue of 36 days will exceed $20,400, we calculate the following:

$$P(\overline{X} \geq 20.4) = P\left(\frac{\overline{X} - 20}{4/\sqrt{36}} \geq \frac{20.4 - 20.0}{4/\sqrt{36}}\right) = P\left(Z \geq \frac{0.40}{4/6}\right)$$

$$= P(Z \geq 0.6) = 0.2743$$

To calculate the probability that the average revenue of 64 days will be less than $19,100, we proceed as follows:

$$P(\overline{X} \leq 19.1) = P\left(\frac{\overline{X} - 20}{4/\sqrt{64}} \leq \frac{19.1 - 20.0}{4/\sqrt{64}}\right) = P\left(Z \leq \frac{-0.9}{0.5}\right)$$

$$= P(Z \leq -1.8) = 0.0359$$

9.6 Interval Estimate vs. Point Estimate

Consider the following types of statements concerning the estimation of the average height of adults in a given population:

(a) The sample average is equal to 5'7"; thus 5'7" is our estimate of the mean height in the population considered.

(b) The sample shows that we should have 95 percent confidence that the interval from 5'5" to 5'9" contains the true mean population height.

Statement *a* provides us with a **point estimate** while statement *b* gives us an **interval estimate.** In statement *a*, only one value is given as the estimated average height; thus the term "point estimate." Statement *b* is a probabilistic one concerning the chance that a given interval includes the true mean of the population. The way in which we derive such probabilistic interval estimates is our concern here.

Consider the area accumulated under the standard normal distribution between the values $Z = 0$ and $Z = 1.645$. According to the normal distribution table printed inside the back cover of this book, this area is (approximately) equal to 0.4500, so that the tail remaining to the right of the value $Z = 1.645$ is (approximately) equal to 0.05 (= 0.5000 − 0.4500). We shall denote the value of Z that leaves a 5 percent right tail by $Z_{0.05}$. Similarly, $Z_{0.10}$ refers to the Z score that leaves a 10 percent tail (or area of 0.10) to its right. To determine the value of $Z_{0.10}$, one must realize that the Z score that leaves an area of 0.10 on its right is the same Z score that accumulates an area of 0.40 when the accumulation starts at $Z = 0$ and continues toward the right. This becomes clear in Figure 9.2: $Z_{0.10}$ is that Z score which divides the right side of the standard normal distribution into areas of 0.40 and 0.10 as shown. The standard normal distribution table shows that the value up to which there is an accumulation of 0.40 is equal to 1.28, and thus $Z_{0.10} = 1.28$, as illustrated in Figure 9.2.

Figure 9.2

The location of $Z_{0.10}$ on the standard normal scale

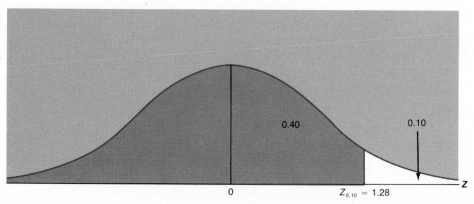

In general, Z_α is that Z score which leaves a right tail equal to α, as seen in Figure 9.3,[1] and it follows that $Z_{\alpha/2}$ is that Z score which leaves a right tail equal to $\alpha/2$.

---

[1] In fact, Z_α is the $1 - \alpha$ percentile of the standard normal distribution.

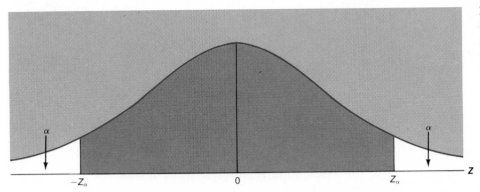

Figure 9.3

The location of the Z_α score on the standard normal scale

Notice that if Z_α leaves a right tail of α, the negative value $-Z_\alpha$ is the one that leaves a left tail equal to α (see Figure 9.3). Similarly, the area to the left of $-Z_{\alpha/2}$ is equal to $\alpha/2$. Figure 9.4 shows a standard normal distribution with both $Z_{\alpha/2}$ and $-Z_{\alpha/2}$ plotted along the horizontal axis. It is clear from the graph and from our earlier definitions that the sum of the right-tail area ($\alpha/2$) and the left-tail area ($\alpha/2$ as well) is equal to α, implying that the area between $-Z_{\alpha/2}$ and $Z_{\alpha/2}$ equals $1 - \alpha$. Note that $\alpha/2 + \alpha/2 + (1 - \alpha) = 1$, the entire area under the density function.

Since the area over the range from $-Z_{\alpha/2}$ to $Z_{\alpha/2}$ is equal to $1 - \alpha$, and since the area under the curve represents probability, we may write as a direct conclusion

$$P(-Z_{\alpha/2} \leq Z \leq Z_{\alpha/2}) = 1 - \alpha \qquad \text{(9.11)}$$

and since $\dfrac{\overline{X} - \mu}{\sigma/\sqrt{n}} = Z$ by Equation 9.10, we may write

$$P\left(-Z_{\alpha/2} \leq \frac{\overline{X} - \mu}{\sigma/\sqrt{n}} \leq Z_{\alpha/2}\right) = 1 - \alpha \qquad \text{(9.12)}$$

Multiplying within the parentheses by $\sigma/\sqrt{n}$ and adding the constant μ to each term in the parentheses gives

$$P\left(\mu - \frac{\sigma}{\sqrt{n}}\, Z_{\alpha/2} \leq \overline{X} \leq \mu + \frac{\sigma}{\sqrt{n}}\, Z_{\alpha/2}\right) = 1 - \alpha \qquad \text{(9.13)}$$

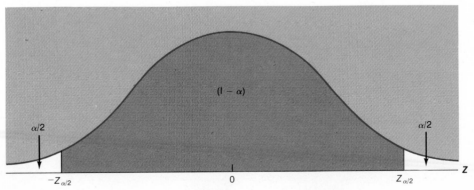

Figure 9.4

The area between $-Z_{\alpha/2}$ and $Z_{\alpha/2}$ and in the two tails

Since $\sigma_{\bar{X}} = \dfrac{\sigma}{\sqrt{n}}$ by Equation 9.6, we may write Equation 9.13 as

$$P(\mu - \sigma_{\bar{X}}Z_{\alpha/2} \leq \overline{X} \leq \mu + \sigma_{\bar{X}}Z_{\alpha/2}) = 1 - \alpha \tag{9.14}$$

Note that Equation 9.14 holds for infinite as well as finite populations, where for infinite populations we have $\sigma_{\bar{X}} = \dfrac{\sigma}{\sqrt{n}}$ and for finite populations we have $\sigma_{\bar{X}} = \dfrac{\sigma}{\sqrt{n}} \sqrt{\dfrac{N - n}{N - 1}}$. Finally, if we wish to denote

$$e = \sigma_{\bar{X}}Z_{\alpha/2} \tag{9.15}$$

where e stands for "error," we can derive from Equation 9.13

$$P(\mu - e \leq \overline{X} \leq \mu + e) = 1 - \alpha \tag{9.16}$$

Equation 9.16 may be interpreted to mean that if the sample average is normally distributed, the probability is $1 - \alpha$ that it will be within the range from $\mu - e$ to $\mu + e$, and thus will not miss the population mean (μ) by more than e—the **tolerable error.** Consequently, there is a probability α—called the **risk probability**—that $\overline{X}$ will miss the mean by more than the tolerable error, e. Figure 9.5 helps to clarify this point.

Figure 9.5

The relationship between the risk probability, α, and the tolerable error, e

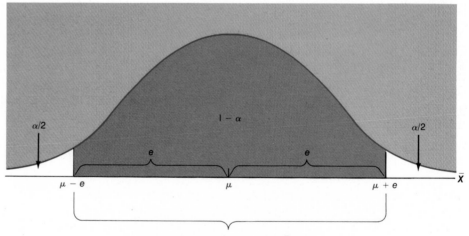

There is a probability of $1 - \alpha$ that $\overline{X}$ will fall within this range, and thus will miss μ by no more than e. There is a risk probability equal to α that $\overline{X}$ will miss by more than e.

EXAMPLE 9.2

Suppose we plan to draw a random sample to determine the average amount invested in common stock by individuals in the age group 18–23 in the United States. For the sake of simplicity, assume normal distribution and that the standard deviation of the investment in common stock is known to equal $100. If 81 individuals will be surveyed, determine a symmetrical interval around the true mean common stock investment within which the sample average is expected with 90 percent probability.

When we use Equation 9.15 for the case of infinite populations, the solution to the problem becomes a simple matter indeed. The data given in the problem are

$$\sigma = 100$$

$$n = 81$$

$$1 - \alpha = 0.90$$

It follows that $\alpha = 0.10$, $\alpha/2 = 0.05$, and from the normal distribution table, we get $Z_{0.05} = 1.645$. Substituting in Equation 9.15 yields

$$e = \frac{100}{\sqrt{81}} \cdot 1.645 = 18.28$$

We thus conclude that

$$P(\mu - 18.28 \leq \overline{X} \leq \mu + 18.28) = 0.90$$

meaning that while we do not know the value of the population average, μ, there is 90 percent probability that the sample average will fall within the range from $\mu - 18.28$ to $\mu + 18.28$. *Thus we have 90 percent confidence that the sample average will be within $18.28 of the true investment mean.*

Note that while the total width of the interval from $\mu - e$ to $\mu + e$ is equal to $2e$ (see Figure 9.5), the half-interval width e is often of more concern.

THE FACTORS DETERMINING e—AND WHAT HAPPENS WHEN THEY CHANGE

It is important to note here that e is determined by three factors: the population's standard deviation (σ), the sample size (n), and the risk probability, α. Let us examine the effect of each of these factors on the width of the interval around μ.

The Effect of a Change in Standard Deviation

Suppose that in Example 9.2, we assume $\sigma = \$150$ rather than 100. This increase of 50 percent in σ will cause a 50 percent increase in e. Using $\sigma = 150$, we get $e = \dfrac{150}{\sqrt{81}} \cdot 1.645 = 27.42$, which is exactly 50 percent greater than 18.28. Thus the interval is wider for the same α.

The Effect of a Change in Sample Size

Returning to the original data of Example 9.2, suppose we now increase the sample size from 81 to 324 (a fourfold increase: $81 \cdot 4 = 324$). We shall see that e is decreased by a factor of 2 (which is equal to the square root of 4, since there is a square-root operator on n).

$$e = \frac{100}{\sqrt{324}} \cdot 1.645 = 9.14$$

Indeed, 9.14 is half of the original value of e, 18.28. Thus the interval is narrower for the same α.

The Effect of a Change in $1 - \alpha$

Since the value of Z is not a linear function of α or $1 - \alpha$, we cannot generally determine the exact magnitude of the effect of a change in $1 - \alpha$ on e. We can, however, state the direction of the change: an increase in $1 - \alpha$ causes a widening of the interval (in other words, an increase in e), and a decrease in $1 - \alpha$ causes a narrowing of the interval (a decrease in e). If $1 - \alpha$ is increased from 0.90 to 0.95, we get

$$1 - \alpha = 0.95 \qquad \alpha = 0.05 \qquad \alpha/2 = 0.025 \qquad Z_{\alpha/2} = 1.96$$

Thus $e = \dfrac{100}{\sqrt{81}} \cdot 1.96 = 21.78$, which is greater than 18.28.

RELATIONSHIPS AMONG THE FACTORS DETERMINING e

Now let us take a close look at the usefulness of Equation 9.16 in conjunction with Equation 9.15.

Determining the Sample Size

Consider a real-estate firm that is interested in building an apartment complex in a given location and wants to estimate the mean rent paid for existing two-bedroom apartments. If we know that the standard deviation of the rent is $40, how many observations have to be taken so that we will have a 99 percent probability that the average of *all* the rents (for similar apartments) in the area is not missed by more than $5?

This problem may be easily solved if we make use of Equations 9.15 and 9.16. Since we are maintaining a 99 percent probability that the mean is not missed by more than $5, we know that $1 - \alpha$ should equal 0.99 and e should equal 5 (dollars). Note that since $1 - \alpha = 0.99$, α must equal 0.01, which implies that $\alpha/2 = 0.005$ and (by the normal distribution table) $Z_{\alpha/2} = 2.57$. We can now proceed directly to the solution. Consider Equation 9.15 applied to an infinite population once again:

$$e = \frac{\sigma}{\sqrt{n}} Z_{\alpha/2}$$

Multiplying both sides of the equation by $\sqrt{n}$ and dividing by e gives

$$\sqrt{n} = \frac{\sigma Z_{\alpha/2}}{e} \qquad\qquad \textbf{(9.17)}$$

and after squaring both sides of Equation 9.17 we get

$$n = \frac{\sigma^2 Z_{\alpha/2}^2}{e^2} \qquad\qquad \textbf{(9.18)}$$

We substitute $\sigma = 40$, $Z_{\alpha/2} = 2.57$, and $e = 5$ in Equation 9.18 and get

$$n = \frac{40^2 \cdot 2.57^2}{5^2} = \frac{1,600 \cdot 6.6049}{25} = 422.71$$

Since n must be an integer, the smallest sample size that will guarantee at least 99 percent confidence that $\overline{X}$ does not miss the true mean by more than \$5 is 423.[2] Figure 9.6 provides a diagrammatic interpretation of the

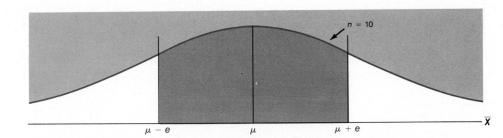

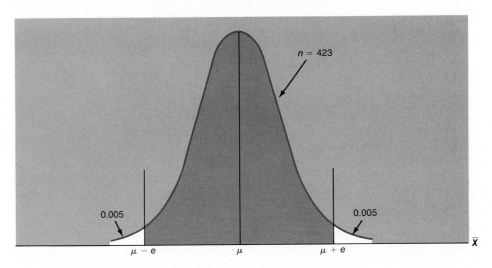

The risk probability is represented by the white tails.

Figure 9.6

Change in risk probability (α) with a change in sample size (n).

[2] Note that we should round off fractions in an upward direction even when the first decimal digit is less than 5. Our answer would have been 423, then, even if Equation 9.18 had yielded 422.1, say.

solution. The assumption that σ is known determines the degree of dispersion of the distribution of rents. The assumption that $1 - \alpha = 0.99$ determines a tail of 0.005 percent on each side of the distribution and the predetermined value of e (5) fixes an interval around the population mean, μ. The probability distribution of $\overline{X}$ becomes less and less dispersed as the sample size increases. Our problem was to determine how much we should condense the distribution of $\overline{X}$ to force 99 percent of its area into the interval from $\mu - e$ to $\mu + e$.

Determining the Probability that the Sample Average Will Fall within a Given Interval of Error

Let us turn now to a different type of situation. Suppose the number of observations to be taken is fixed (perhaps because of a budget constraint or time limitation). In this kind of situation we can either solve for e when $1 - \alpha$ is given or solve for $1 - \alpha$ when e is given. Suppose a consumer protection organization is concerned with the average difference between estimates given on car repairs and the actual bills. To evaluate the average discrepancy, 100 discrepancies are observed. If the standard deviation of the discrepancies is $15, what is the probability that the average discrepancy as estimated from the sample does not deviate from the true average discrepancy by more than $3? Assume a normal distribution. Given $\sigma = 15$, $n = 100$, and $e = 3$, we simply have to determine the area under the normal curve (having a standard deviation of $\sigma_{\overline{x}} = 15/\sqrt{100} = 1.5$) over the interval from $\mu - 3$ to $\mu + 3$. The solution may easily be derived if we use Equation 9.15 again. Remember that Equation 9.15 is $e = \dfrac{\sigma}{\sqrt{n}} Z_{\alpha/2}$. Multiplying both sides by $\dfrac{\sqrt{n}}{\sigma}$ gives us

$$Z_{\alpha/2} = \frac{e\sqrt{n}}{\sigma} \tag{9.19}$$

In this case we get

$$Z_{\alpha/2} = \frac{3 \cdot \sqrt{100}}{15} = \frac{30}{15} = 2$$

From the normal distribution table we find that the accumulated area under the standard normal distribution over the range from 0 to 2 is equal to 0.4772, so that the tail to the right of 2 must be equal to 0.0228 ($= 0.5 - 0.4772$). If the tail to the right of $Z_{\alpha/2}$ is equal to 0.0228, then by definition we get $\alpha/2 = 0.0228$, implying that $\alpha = 0.0456$ and that $1 - \alpha = 0.9544$. Thus we would have a probability of 95.44 percent that the sample's average will not deviate from the mean by more than $3.

Determining the Value of e, Given the Risk Probability and the Sample Size

If the number of observations taken in a sample is, say, 64, what is the maximum error we can get with a 90 percent probability if the population is infinite?

Here we apply Equation 9.15 directly:

$$e = \frac{\sigma}{\sqrt{n}} Z_{\alpha/2} = \frac{15}{\sqrt{64}} \cdot 1.645 = 3.084$$

There is thus a 90 percent chance that our estimate will not miss the true mean by more than $3.084.

One may wonder at this point whether the applicability of the approach presented in this section is hampered by the fact that it hinges on the assumption of a normal distribution. In the following section we shall present the well-known central limit theorem, which will show that the assumption of distribution normality is not always vital.

9.7 The Central Limit Theorem

Let X be a random variable having *any distribution at all* with mean μ and standard deviation σ, and let $\overline{X}$ be the average of a random sample taken from this distribution. We can then state the following theorem:

CENTRAL LIMIT THEOREM

The probability distribution of $\overline{X}$ will approach a normal distribution as the number of observations increases.

The **central limit theorem** has far-reaching implications. After all, when we are conducting a study on the time needed to transport merchandise from warehouses to department stores, or when we are studying late flight arrivals and departures, the profit distribution across firms in a given industry, the distribution of hourly earnings in a given occupation, the effect of counter display position on the marketing of a product, or similar questions, we generally do not know whether the variables under study are normally distributed. But we do know from the central limit theorem that the *average* of a sample of observations approximates normal distribution—provided that the observations are independent of one another and the number of observations is large. The larger the sample size, other things being equal, the closer the distribution to normal. Twenty-five independent observations will generally produce an average with a distribution pretty close to normal.[3]

Consider Figure 9.7. Here we present several distributions that vary from the normal and examine the distribution of their sample averages. Three columns of diagrams are shown, each column presenting a different distribution. The left column, for example, shows the distribution of sample averages of various sizes taken from a uniform distribution. The average of 2 observations taken independently from the uniform distribution has a triangular distribution, while the average of 10 observations is similar to the normal. With a sample of 25 observations, the distribution of the average is close to normal in all three cases, even though the original distributions (shown in the top line of diagrams in Figure 9.7) vary considerably from the normal. As the sample size increases (other things held constant), the distributions of the averages tend to be more and more like the normal distribution.

---

[3] Note that the normal approximation of the binomial distribution, presented in Chapter 8, makes use of the fact that a sample's average tends to be normally distributed when the observations are independent.

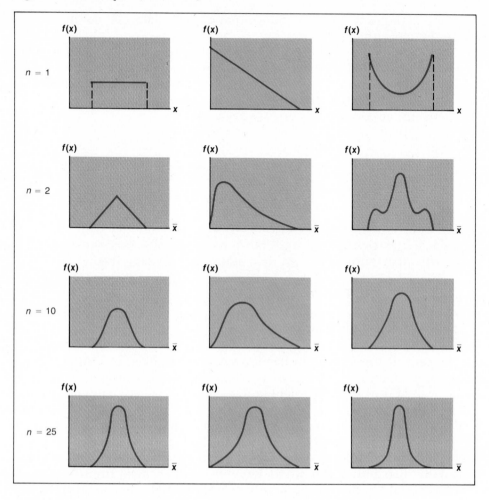

9.8 Confidence Intervals for the Population Mean

In our discussion of point estimation we noted that if we are to estimate the mean of a given population by using a *single (numerical) value,* we will obtain the best results by using that value which we find to be the average of a sample chosen at random from the population under consideration. The advantage of a point estimate lies in the fact that it provides us with a single value to focus on. Its disadvantage is the likelihood that it will miss the population mean, perhaps by a substantial amount. Because of this disadvantage, we developed an equation (9.13) and its variations (9.14 and 9.16) to evaluate the probability that $\overline{X}$ may miss μ by a given magnitude. We further learned about the trade-off among the error (e), the sample size (n), and the confidence probability ($1 - \alpha$), and how to solve for one of these three variables when we know the value of the other two.

An equation similar (but not identical) to 9.13 may be developed by rearranging 9.13:

CONFIDENCE INTERVAL FOR A POPULATION'S MEAN

$$P(\overline{X} - \sigma_{\overline{X}}Z_{\alpha/2} \leq \mu \leq \overline{X} + \sigma_{\overline{X}}Z_{\alpha/2}) = 1 - \alpha \qquad \textbf{(9.20)}$$

---

or using $e = \sigma_{\overline{X}}Z_{\alpha/2}$:

$$P(\overline{X} - e \leq \mu \leq \overline{X} + e) = 1 - \alpha \qquad \textbf{(9.21)}$$

which holds for infinite as well as finite populations.

Equations 9.20 and 9.21 determine a *symmetrical interval around the sample average* $\overline{X}$ and assert that the probability that the interval established includes the true mean of the distribution (μ) is equal to $1 - \alpha$. The interval established, from $\overline{X} - e$ to $\overline{X} + e$, is called the **confidence interval,** and $1 - \alpha$ is known as the **confidence level.** The quantity α, as we have seen, is often termed the *risk probability*.

It must be understood that the confidence interval may not actually contain the mean. The interval is centered on $\overline{X}$, whose value varies from one sample to another. Thus, the interval ends are random variables, and when $\overline{X}$ differs considerably from μ, the interval may very well not contain the mean. In Figure 9.8, where a few confidence intervals are shown, all but two contain the value μ. $(1 - \alpha)$ percent of all intervals of size n formed will contain μ.

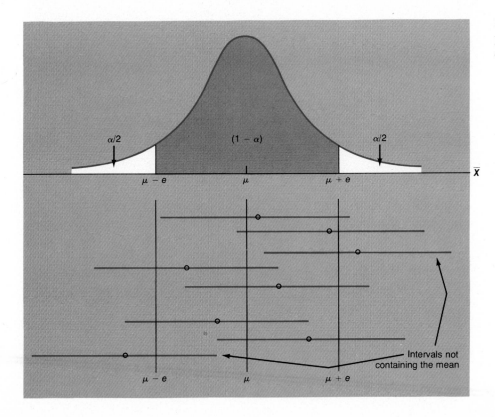

Figure 9.8

Interval estimation

EXAMPLE 9.3

Florida Paper, Inc., has agreed to buy paper to recycle from the Central Florida Waste Collection Company (CFWCC). CFWCC is to supply the waste paper in packages of 300 pounds each, for which Florida Paper is to pay by the package. To speed up waste packing, CFWCC is packaging 300 pounds by *approximation*. Florida Paper does not object to this procedure as long as it gets 300 pounds per package *on the average*. To estimate the average weight of raw material in a package, Florida Paper weighed 75 randomly selected packages and found that the average weight was 290 pounds. Assuming that the standard deviation of weight in the packages is 15 pounds, let us determine whether we can indeed assume that there are 300 pounds per package.

Let us first identify the values given above. The sample average is 290 pounds, so $\bar{X} = 290$. The number of observations taken is 75, so $n = 75$. The population's standard deviation (σ) is assumed to equal 15 pounds. To summarize:

$$\bar{X} = 290$$

$$\sigma = 15$$

$$n = 75$$

The point estimate produced by the sample ($\bar{X} = 290$) is less than the desired mean weight of raw material per package in the population. We know, however, that the sample average almost always differs from the population's mean. Suppose we want to construct a confidence interval with a 99 percent confidence level. According to the central limit theorem, the distribution of $\bar{X}$ is approximately normal, so we let $1 - \alpha$ equal 0.99, which implies that $\alpha = 0.01$, $\alpha/2 = 0.005$, and $Z_{\alpha/2} = 2.57$.

Substituting our data in Equation 9.20 yields

$$P\left(290 - \frac{15}{\sqrt{75}} \cdot 2.57 \leq \mu \leq 290 + \frac{15}{\sqrt{75}} \cdot 2.57\right) = 0.99$$

$$P(290 - 4.45 \leq \mu \leq 290 + 4.45) = 0.99$$

or

$$P(285.55 \leq \mu \leq 294.45) = 0.99$$

The confidence interval is thus 285.55 to 294.45, indicating 99 percent confidence that the interval from 285.55 to 294.45 covers the mean weight per package. It also indicates that there is only a 1 percent chance that the true average weight per package (μ) is not covered by this interval. It is quite likely, then, that the 300-pound average weight per package claimed by CFWCC is exaggerated.

We have basically tested whether or not the mean weight per package is 300 pounds. Tests of a similar nature are more directly dealt with in Chapter 10. The methodology there is somewhat different but the essence of the problem is the same.

THE TRADE-OFF BETWEEN CONFIDENCE LEVEL AND INTERVAL WIDTH

In a textbook problem the confidence level is often indicated, so we are relieved of the need to determine it. In real life, however, the researcher has to determine the confidence level before the confidence interval may be established. While a high confidence level is desirable, one should be aware of the trade-off between the confidence level and the confidence interval width. For any given sample size, the higher the confidence level, the wider the interval derived. Suppose a restaurant manager wants to know how many waiters to employ and how much food to prepare for dinner. By constructing confidence intervals, she might find that there is a confidence level of 0.80 that the average number of people who come to dine is between 90 and 110, a confidence level of 0.90 that the average number of people who come to dine is between 80 and 120, and a 99 percent confidence level that on the average between 50 and 150 people dine in the restaurant. There is a trade-off, then, between precision in terms of the confidence level and accuracy in terms of the interval width within which the population mean is believed to lie.

Choosing a very high level of confidence is likely to make the interval too wide to be meaningful. The interval 50 to 150 diners is probably too broad to enable the restaurant's management to make any kind of decision. An interval such as 90 to 110 people is more specific and of greater assistance to management. There is a good chance, however (a 20 percent chance), that the interval 90 to 110 does not really cover the true average of the population. Hence a trade-off between a high probability statement and a wider interval exists, as long as the sample size is held constant.

In most cases a confidence level between 0.90 and 0.99 is chosen, and the decision of which level to choose within that range depends on the type of question under consideration. At all times, however, one should be aware of the trade-off between precision in the confidence level and accuracy in the width of the interval.

9.9 A Confidence Interval for μ When σ Is Unknown

The extensive use we have made of the normal distribution is well justified. Many variables in business and economics have normal or close-to-normal distributions. Furthermore, we have pointed out that even in those cases in which the population distribution differs from the normal, the sample's average is approximately normally distributed as long as we have a sufficient number of independent observations in the sample (the central limit theorem). We have been assuming, however, that while the population's mean (μ) is unknown and has to be estimated, the population's standard deviation (σ) is known and may be used to provide an interval estimate for the population's mean. We shall now assume instead that σ is unknown and attempt to establish a confidence interval for μ. As a standard procedure in statistics, we shall substitute for σ its sample estimate, S. S is defined as follows:

$$S = \sqrt{\frac{\Sigma(X - \overline{X})^2}{n - 1}} = \sqrt{\frac{\Sigma X^2 - n\overline{X}^2}{n - 1}} \qquad \text{(9.22)}$$

Instead of using the statistic $\dfrac{\overline{X} - \mu}{\sigma/\sqrt{n}}$, we shall have to use $\dfrac{\overline{X} - \mu}{S/\sqrt{n}}$, but while the former has a normal distribution, the latter is distributed according to the t distribution, provided X is normally distributed. We shall therefore discuss the t distribution before proceeding.

THE STUDENT t DISTRIBUTION

The **Student t distribution** (or, in short, the t distribution), is similar in its general shape to the standard normal distribution. Like the standard normal distribution, the t distribution is bell-shaped, with its mean, median, and mode coinciding at zero. The difference between the two, as illustrated in Figure 9.9, is that the t distribution is more dispersed and thus has thicker tails. While there is only one standard normal distribution, there is a whole family of t distributions, distinguishable by a parameter called the *degrees of freedom*. A t distribution with one degree of freedom (denoted by $t^{(1)}$) is somewhat different from a t distribution with 2, 5, or 10 degrees of freedom. Figure 9.9 shows two t distributions, one with 5 degrees of freedom ($t^{(5)}$) and the other with 30 degrees of freedom ($t^{(30)}$). While the shape of the $t^{(5)}$ distribution differs considerably from that of the standard normal distribution, the $t^{(30)}$ distribution is in fact very much like the standard normal distribution. As the number of degrees of freedom increases beyond 30, the t distribution resembles even more closely the shape of the standard normal distribution.

Since the t distribution is a function of the number of degrees of freedom, only a limited number of the more frequently used critical t scores are presented in the t distribution table, which is printed inside the back cover of this book. We shall use the notation $t_\alpha^{(n)}$ to denote that t score which leaves a right-tail area equal to α under the $t^{(n)}$ distribution. It follows directly that $t_{\alpha/2}^{(n)}$ is that t score which leaves a right tail equal to $\alpha/2$ on the $t^{(n)}$ distribution. Figure 9.10 helps clarify the notation $t_\alpha^{(n)}$: the t score 2.110,

Figure 9.9

The t density function compared with the normal curve

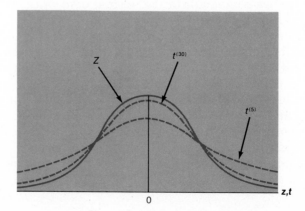

which is $t_{0.025}^{(17)}$, leaves a right-tail area of 2.5 percent under the $t^{(17)}$ distribution. Figure 9.11 describes the way we find $t_\alpha^{(n)}$ on the t distribution table: the value $t_{0.025}^{(17)}$, for example, may be found at the intersection of 17 df (degrees of freedom) indicated in the left-hand column of the table and the 0.025 value indicated at the top of the table.

As additional examples, let us find $t_\alpha^{(8)}$, $t_{\alpha/2}^{(8)}$, and $t_{\alpha/2}^{(\infty)}$, assuming that $\alpha = 0.05$. Since $\alpha = 0.05$, obviously $\alpha/2 = 0.025$. Focusing on $t^{(8)}$ first, we find that $t_{0.05}^{(8)} = 1.860$, as indicated on the table at the intersection of $8df$ and 0.050, and similarly we find that $t_{0.025}^{(8)} = 2.306$. We find that $t_{0.05}^{(\infty)}$ is equal to 1.645, a number we are well acquainted with from the normal distribution table. More specifically, we realize that $t_{0.05}^{(\infty)} = Z_{0.05}$, and in general for any given α we get $t_\alpha^{(\infty)} = Z_\alpha$. When the number of degrees of freedom approaches infinity, the t distribution approaches the standard normal distribution, and as a result all the individual t scores approach their respective scores on the standard normal distribution.

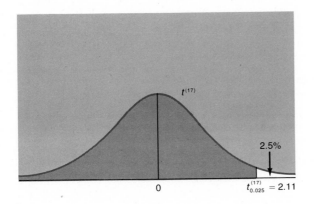

Figure 9.10

A t distribution ($t^{(17)}$) showing the value $t_{0.025}^{(17)}$

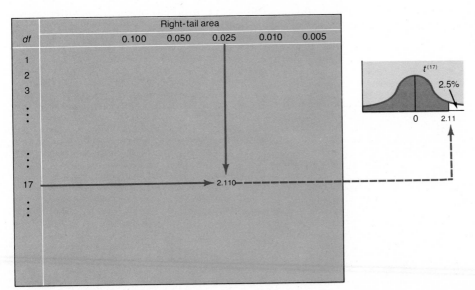

Figure 9.11

The t-distribution table

CONSTRUCTING THE CONFIDENCE INTERVAL

While $\sigma_{\bar{x}} = \dfrac{\sigma}{\sqrt{n}}$ for an infinite population and $\sigma_{\bar{x}} = \dfrac{\sigma}{\sqrt{n}} \sqrt{\dfrac{N-n}{N-1}}$ for a finite population (with no replacement), the estimated value of $\sigma_{\bar{x}}$ is given by:

$$S_{\bar{X}} = \frac{S}{\sqrt{n}} \qquad (9.23)$$

for an infinite population, and

$$S_{\bar{X}} = \frac{S}{\sqrt{n}} \sqrt{\frac{N-n}{N-1}} \qquad (9.24)$$

for a finite population (with no replacement). Unless otherwise stated, we shall assume that the population is infinite.

Whereas $\dfrac{\bar{X} - \mu}{\sigma/\sqrt{n}}$ has a standard normal distribution, $\dfrac{\bar{X} - \mu}{S/\sqrt{n}}$ has a t distribution with $(n - 1)$ degrees of freedom.

$$\frac{\bar{X} - \mu}{S/\sqrt{n}} \sim t^{(n-1)} \qquad (9.25)$$

Note that when the sample involves more than 30 observations, the t distribution in Equation 9.25 has at least 30 degrees of freedom and thus will closely approximate the normal distribution.

Earlier, we have seen how a confidence interval for the mean may be derived when σ is known. Following similar derivations, one may obtain the following confidence interval, which involves the t distribution:

$$P\left(\bar{X} - \frac{S}{\sqrt{n}} \cdot t_{\alpha/2}^{(n-1)} \leq \mu \leq \bar{X} + \frac{S}{\sqrt{n}} \cdot t_{\alpha/2}^{(n-1)}\right) = 1 - \alpha \qquad (9.26)$$

Equation 9.26 implies that a confidence interval with a confidence level of $(1 - \alpha)$, when the standard deviation is unknown, is

$$\bar{X} - \frac{S}{\sqrt{n}} \cdot t_{\alpha/2}^{(n-1)} \leq \mu \leq \bar{X} + \frac{S}{\sqrt{n}} \cdot t_{\alpha/2}^{(n-1)}$$

EXAMPLE 9.4

The Consumer Protection Agency of Wyoming is advocating truth in advertising and in labeling. As part of its regular activities, the agency conducts market surveys involving estimation by samples of products' weight. One of the products sampled consists of bags of charcoal briquettes advertised to contain 5 pounds of briquettes. A sample of 16 bags had the following weights (in pounds): 4.8, 4.7, 5.0, 5.2, 4.7, 4.9, 5.0, 5.0, 4.6, 4.7, 5.0, 5.1, 4.7, 4.5, 4.9, 4.9.

The Consumer Protection Agency needs to estimate and derive from these data the confidence interval for the mean weight of the bags of charcoal briquettes at a confidence level of 95 percent.

The point estimate may simply be obtained by averaging the 16 weights to obtain

$$\Sigma X = 4.8 + 4.7 + 5.0 + 5.2 + \cdots + 4.9 = 77.7$$

$$\overline{X} = \frac{77.7}{16} = 4.856 \text{ pounds}$$

To obtain an interval estimate for the mean, we first estimate the standard deviation of the weights by means of Equation 9.22.

$$\Sigma X^2 = 4.8^2 + 4.7^2 + 5.0^2 + 5.2^2 + \cdots + 4.9^2 = 377.89$$

$$n\overline{X}^2 = 16 \cdot 4.856^2 = 16 \cdot 23.581 = 377.29$$

$$S = \sqrt{\frac{377.89 - 377.29}{16 - 1}} = \sqrt{\frac{0.60}{15}} = \sqrt{0.04} = 0.20$$

Since $\dfrac{\overline{X} - \mu}{S/\sqrt{n}} \sim t^{(15)}$ and since $\alpha/2 = 0.025$, the t value to be used here is $t^{(15)}_{0.025} = 2.131$. Using Equation 9.27, we determine that the confidence interval is from $4.856 - \dfrac{0.20}{\sqrt{16}} \cdot 2.131 < \mu < 4.856 + \dfrac{0.20}{\sqrt{16}} \cdot 2.131$, or $4.749 < \mu < 4.963$.

The interval obtained implies that there is a 95 percent chance that the range between 4.749 and 4.963 includes the true mean weight, and thus also that there is at most 5 percent confidence that the mean weight is really 5 pounds.

The confidence interval constructed when the standard deviation of the population is unknown is wider (other things being equal) than that constructed when it is known. If we definitely knew that $\sigma = 0.20$ (Example 9.4), the interval would have been from $\overline{X} - \dfrac{\sigma}{\sqrt{n}} Z_{\alpha/2} < \mu < \overline{X} + \dfrac{\sigma}{\sqrt{n}} Z_{\alpha/2}$, which for our example is $4.856 - \dfrac{0.20}{\sqrt{16}} \cdot 1.96 < \mu < 4.856 + \dfrac{0.20}{\sqrt{16}} \cdot 1.96$, or $4.758 < \mu < 4.954$. Indeed, this interval is narrower than the interval between 4.749 and 4.963, which we obtained for the unknown σ. The wider interval reflects the uncertainty concerning the value of σ: if σ is estimated, we need to allow a wider range to be sure that the unknown μ will fall within it.

9.10 Estimation of Proportions

Just as when we estimated the mean of a distribution, when we wish to estimate a population's proportion by means of a random sample our result is subject to a sampling error. To describe the sample's results more meaningfully, we must provide both point and interval estimates for the proportion.

In Chapter 7 we expressed the number of successes in a binomial experiment in terms of the proportion of the trials resulting in success, by simply dividing the number of successes (X) by the total number of trials (n) to get $\dfrac{X}{n}$. We denoted $\dfrac{X}{n}$ by $\hat{p}$ and demonstrated there that

$$E(\hat{p}) = p$$

$$\sigma_{\hat{p}}^2 = \frac{pq}{n} \qquad \text{(9.28)}$$

$$\sigma_{\hat{p}} = \sqrt{\frac{pq}{n}}$$

Our point estimate for the true population proportion is the statistic $\hat{p} = \dfrac{X}{n}$. To obtain an interval estimate, we recall that X—and thus also $\dfrac{X}{n}$—have approximately normal distribution (provided that $np \geq 5$ and $nq \geq 5$). When the normal approximation applies, we may write

$$\hat{p} \sim N\left(p, \sqrt{\frac{pq}{n}}\right) \qquad \text{(9.29)}$$

so that

$$\frac{\hat{p} - p}{\sqrt{\dfrac{pq}{n}}} \sim N(0,1) \qquad \text{(9.30)}$$

However, since the standard deviation is equal to $\sqrt{\dfrac{pq}{n}}$ and p is unknown (if it were known, we would not have needed the estimation to begin with), we shall use the estimated standard deviation $S_{\hat{p}} = \sqrt{\dfrac{\hat{p}(1 - \hat{p})}{n}}$, and obtain

$$\frac{\hat{p} - p}{S_{\hat{p}}} \cong N(0,1) \qquad \text{(9.31)}$$

Following an approach similar to the one we used to derive a confidence interval for the mean, we may construct a confidence interval for the true proportion and obtain the following:

$$P(\hat{p} - S_{\hat{p}} \cdot Z_{\alpha/2} \leq p \leq \hat{p} + S_{\hat{p}} \cdot Z_{\alpha/2}) = 1 - \alpha \qquad \text{(9.32)}$$

Note that the widest interval is obtained when $\hat{p} = \frac{1}{2}$ (other things being held constant).

EXAMPLE 9.5

A new program for a youth club is planned in a small city. To determine whether or not the program will get the city government's support, it is necessary to estimate the proportion of young people who plan to use the club's facilities. A survey of 100 randomly selected young people has shown that 22 will use the facilities if they become available.

To obtain a point estimate of the proportion of young people who plan to use the facilities, we merely divide the number of successes (22) by the total number of trials (100) to get: $\dfrac{X}{n} = \dfrac{22}{100} = 0.22$. We then estimate the standard deviation $S_{\hat{p}}$:

$$S_{\hat{p}} = \sqrt{\frac{\hat{p}(1 - \hat{p})}{n}} = \sqrt{\frac{0.22 \cdot 0.78}{100}} = 0.0414$$

To construct the confidence interval for the true proportion with a confidence level of 99 percent, we apply Equation 9.32:

$$P(0.22 - 0.0414 \cdot 2.57 \leq p \leq 0.22 + 0.0414 \cdot 2.57)$$

$$= P(0.1136 \leq p \leq 0.3264) = 0.99$$

The sample indicates that there is a 99 percent chance that the interval between 11.36 and 32.64 percent contains the true population proportion.

9.11 APPLICATION 1:
THE ELECTION PROBLEM

Every presidential election campaign is accompanied by a flood of forecasts assessing each candidate's chances of winning. The forecasts are based on polls that estimate the proportion of voters who are expected to vote for each candidate on election day. The sampling is carried out on a state-by-state basis, with a separate estimate derived for each state. Depending on the poll results in each state, the candidates must decide whether to increase their campaign efforts in the state or to concede it to the rival candidate and concentrate their efforts on campaigns in other states.

Suppose that one week before the election, a sample of 900 voters in New York reveals that 400 would vote for the Democrat and 500 would vote for the Republican. Should the Democratic candidate concede New York State? Could he still win New York's electoral seats?

The point estimate of the proportion of Democratic votes in New York is $\hat{p} = \dfrac{X}{n} = \dfrac{400}{900} = 0.444$. If $\frac{4}{9}$ of the voters do vote for the Democratic candidate and $\frac{5}{9}$ of them vote for the Republican, the Democrat will lose the New York State electors. In view of the poll's results, however, is it possible that the Democrat may get 51 percent of the votes in New York? By constructing a 0.95 confidence interval for the true proportion, p, we see that there is a 95 percent chance that the true proportion is covered by the interval between 41.2 and 47.6 percent—an interval that does *not* cover the value of 0.51.

$$P(\hat{p} - S_{\hat{p}} \cdot Z_{\alpha/2} \le p \le \hat{p} + S_{\hat{p}} \cdot Z_{\alpha/2}) = 1 - \alpha$$

Using $\alpha = 0.05$, $\hat{p} = 0.444$, and $n = 900$, we get

$$P\left(0.444 - \sqrt{\frac{0.444 \cdot 0.556}{900}} \cdot 1.96 \le p \le 0.444 + \sqrt{\frac{0.444 \cdot 0.556}{900}} \cdot 1.96\right) = 0.95$$

$$P(0.412 \le p \le 0.476) = 0.95$$

Note that we should not be too concerned by the fact that the standard deviation of $\hat{p}$ used to construct the above confidence interval is itself an estimate. Taking the largest possible value of the standard deviation (which is obtained when $p = 0.5$), we get

$$\sigma_{\hat{p}} = \sqrt{\frac{0.5 \cdot 0.5}{900}} = 0.01667$$

and the confidence interval becomes

$$0.444 - 0.01667 \cdot 1.96 < p < 0.444 + 0.01667 \cdot 1.96$$

or

$$0.411 < p < 0.477$$

—an interval very close to that obtained above. This interval doesn't cover the value 0.51 either. Should the Democratic candidate stop campaigning in New York and concentrate on other states? No, because a week still remains before election day and New York is a large state with a large number of electors. Our advice would be to campaign heavily in New York, since a gain of about 5 percent of the voters would give the Democrat a reasonable chance of winning the state.

Now assume that out of a sample of 2,500 voters in Kentucky only 500—20 percent—support the Republican candidate. Should the Republican keep on campaigning in Kentucky or should he focus his efforts elsewhere?

A 95 percent confidence interval at the highest possible value of the standard deviation of p is

$$0.20 - \sqrt{\frac{0.5 \cdot 0.5}{2,500}} \cdot 1.96 < p < 0.20 + \sqrt{\frac{0.5 \cdot 0.5}{2,500}} \cdot 1.96$$

$$= 0.20 - 0.0196 < p < 0.20 + 0.0196$$

or

$$0.1804 < p < 0.2196$$

If we used $\hat{p} = 0.20$ instead of 0.50 for the standard deviation, we would have gotten an even narrower range; thus it is clear that the true p is most probably substantially below 51 percent. Our advice to the Republican would be to concentrate his effort elsewhere.

9.12 APPLICATION 2:
QUALITY CONTROL

Before the 1930s, American firms were relatively small and quality control had not been instituted. During the 1930s the concept of quality control was developed, thanks mainly to Walter A. Stewhart of Bell Telephone Laboratories. Modern industry is characterized by mass production, and quality control departments exist in many industrial firms. Today consumers are so conscious of the quality and safety of the products they buy that a defective or unsafe product may greatly hurt a firm's reputation. Firms are therefore concerned about keeping their products at a specified level of quality.

Take a bottling company. Soft-drink bottles are filled by automatic machines that are set to seal the bottles with an interior pressure of 2 atmospheres. The pressure may vary from one bottle to another, however, owing to improper setting of the filling machine or other reasons. A pressure greater than 2.5 atmospheres may cause an explosion; in fact, many soft-drink buyers were injured by such explosions. One way to avoid explosions is to decrease the pressure all the way down to, say, 1 atmosphere. At such low pressure, however, the soft drink loses its taste and value.

The quality-control process that's important here is a checking procedure through which management tries to eliminate systematic change of the degree of pressure. If such a systematic change occurs, production should be stopped while the machine is adjusted. With the pressure in the bottles a random variable, when should we stop the machine? Suppose that the quality control department takes a sample of 10 bottles every hour and the average pressure, $\overline{X}$, is measured. Past experience shows that

$$\overline{X} \sim N(\mu, \sigma_{\overline{X}})$$

where $\mu = 2$ and $\sigma_{\overline{X}} = 0.15$.

The quality control department established an upper control limit (UCL) and lower control limit (LCL) such that the probability that the average pressure of 10 bottles ($\overline{X}$) will be greater than UCL is one-half of 1 percent (that is, 0.005) and the probability that $\overline{X}$ will be lower than LCL is also 0.005. To determine UCL we write

$$P(\overline{X} > UCL) = P\left(\frac{\overline{X} - \mu}{\sigma_{\overline{X}}} > \frac{UCL - \mu}{\sigma_{\overline{X}}}\right) = P\left(Z > \frac{UCL - 2}{0.15}\right) = 0.005$$

From the normal distribution table (inside back cover) we find that the value above which there is a 0.005 right-tail area is 2.576, so that

$$\frac{UCL - 2}{0.15} = 2.576$$

and it follows that

$$UCL = 2 + 0.15 \cdot 2.576 = 2.386 \text{ atmospheres}$$

Similarly, the *LCL* is determined as follows:

$$P(\overline{X} < LCL) = P\left(\frac{\overline{X} - \mu}{\sigma_{\overline{x}}} < \frac{LCL - \mu}{\sigma_{\overline{x}}}\right) = P\left(Z < \frac{LCL - 2}{0.15}\right) = 0.005$$

so that

$$\frac{LCL - 2}{0.15} = -2.576$$

and

$$LCL = 2 - 0.15 \cdot 2.576 = 1.614 \text{ atmospheres}$$

Figure 9.12 is a quality control chart showing the *UCL* and *LCL*. The average pressure of 10 bottles as measured every hour is plotted on the chart. As long as the pressure is represented between the *UCL* and the *LCL*, production continues. Such results as those seen in Figure 9.12, however, show very clearly that the machine has shown a tendency to move out of adjustment, and at 3 P.M. the average was out of the acceptable limits. The bottles produced at and around 3 P.M. are of unacceptable quality, and the machine should be stopped and adjusted.

More sophisticated quality control methods exist, of course, and many of them are based on the confidence-interval approach.

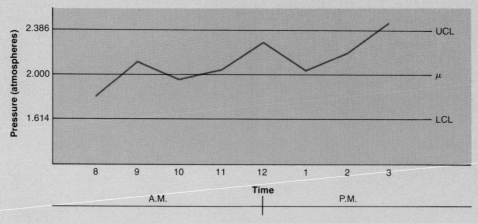

Figure 9.12
Quality control chart

Problems

9.1. Why do we use samples rather than censuses in many studies of populations?

9.2. Distinguish between sampling errors and nonsampling errors.

9.3. Explain the difference between a biased and an unbiased statistic.

9.4. List and briefly explain the major types of samples, indicating in each case whether it is a probability sample and under what circumstances it is most appropriately used.

9.5. Suppose the cost of sampling is $10 per observation. What is the optimal sample size if the population's variance is zero? Explain.

9.6. (*a*) Consider a probability distribution consisting of the integers 1 through 100, each having an equal probability of $\frac{1}{100}$. Using the random digit table (Appendix Table A.11), select at random 10 samples of 10 digits each, 10 samples of 20 digits each, and 10 samples of 30 digits each. Record the averages in the table below.

(*b*) For each sample size, compute the average of the 10 sample averages and the standard deviation of the sample averages. The standard deviation is calculated by means of the equation $S = \sqrt{\frac{\Sigma(\bar{X}_i - \bar{\bar{X}})^2}{n-1}}$. The value $\bar{X}_i$ is the average of the *i*th sample, $\bar{\bar{X}}$ is the average of all the samples of a given size, and *n* is the number of sample averages of each sample size. In this problem $n = 10$. Draw the conclusion.

Sample	Sample size		
number	10	20	30
1			
2			
3			
4			
5			
6			
7			
8			
9			
10			

9.7. *X* is a random variable distributed as follows: $X \sim N(150, 25)$. A sample of 25 observations is taken. What is the probability that the sample's average is greater than 140 but less than 147.5?

9.8. *X* is a random variable distributed as follows: $X \sim N(200, 20)$. A sample is taken and the following probability concerning the sample average holds:

$$P(\bar{X} \geq 203.92) = 0.025.$$

What is the sample size?

9.9. Table P9.9 gives the time deposits as a percentage of all deposits and the return on average assets of 20 banks on December 31, 1980. Assume that 20 banks constitute the entire industry (i.e., population). Suppose now that the time deposit as a percentage of total deposits is not known, and you are trying to estimate the industry's mean percentage by taking random samples of various sizes.

Using the random digit table (Table A.11) for the selection, choose samples of sizes 5, 10, 15, and 20.

Compare your estimates with the population's mean. What is the relationship between the sample size and the deviation of the sample average from the population mean? Explain.

9.10. "If $X \sim N(\mu, \sigma)$ and the population is infinite, then the average of a sample of *n* independent observations, $\bar{X}$, is given by $\bar{X} \sim N\left(\mu, \frac{\sigma}{\sqrt{n}}\right)$." Show graphically the relationship between the distributions of *X* and $\bar{X}$.

9.11. Suppose $X \sim N(\mu, \sigma)$, where $\mu = 0$ and $\sigma = 9$. Calculate the probability $P(\bar{X} \leq 3)$ if $\bar{X}$ is the average of a sample of size *n* and *n* is alternately equal to 9, 64, 81, and 100.

9.12. Repeat Problem 9.11, but this time calculate the probability $P(-3 \leq \bar{X} \leq 3)$.

9.13. The 1980 return on average assets for 20 banks is given in Table P9.9. Choose a random sample of 10 observations (using Appendix Table A.11 for the selection) and construct a 95 percent confidence interval for the industry mean, assuming that σ is unknown.

9.14. Rework problem 9.13, this time assuming that the standard deviation of the population is known to be 0.2928.

TABLE P9.9

Time Deposits as a Percentage of All Deposits and Return on Average Assets of 20 Banks, December 31, 1980

Bank	Time deposits (percent of all deposits)	Return on average assets
Florida National Bank of Florida (Jacksonville)	55%	0.78%
United Bank Corp. of New York	67	1.32
Equitable Bancorp (Baltimore)	65	0.49
Old Kent Financial (Grand Rapids)	78	1.06
First American (Nashville)	63	0.80
Zions Utah Bancorp (Salt Lake City)	72	1.11
Heritage Bancorp (Cherry Hill, N.J.)	67	1.03
Colorado National Bancshares	58	1.29
North East Bancorp	62	1.08
First Tulsa Bancorp	46	1.04
Indiana National (Indianapolis)	58	0.79
Cullen First Bankers (San Antonio)	63	0.85
First United Bancorp (Fort Worth)	62	1.04
Central Fidelity (Richmond)	72	0.87
Whitney Holding (New Orleans)	48	1.48
City National (Beverly Hills)	57	1.33
Northern States Bancorp (Detroit)	71	0.37
Colonial Bancorp (Waterbury, Conn.)	70	0.61
General Bancshares (St. Louis)	71	0.89
LITCO Bancorp of New York (Garden City)	62	0.50

Data from *Business Week*, April 13, 1981, pp. 100–101.

9.15. Compare the confidence intervals you obtained in Problems 9.13 and 9.14. Which is wider? Can you explain why?

9.16. The daily return per $100 investment in a given stock is a random variable whose distribution is normal. The following is a random sample of five daily returns on the stock:

Day	Return
1	−0.2
2	0.1
3	0.1
4	−0.6
5	−0.9

(*a*) Construct a 95 percent confidence interval for the mean daily rate of return on the stock, assuming that the variance is 0.1.

(*b*) Rework part *a*, this time assuming that the variance is unknown.

9.17. The amount of oil piped into a barrel by Quality Oil Corporation is a random variable normally distributed with a known standard deviation of 3 gallons. Thirty-six barrels are randomly and independently selected, and the average number of gallons in a barrel is found to be 102.

(*a*) What is the probability that the 102-gallon average of the sample is within 2 gallons of the actual population mean?

(*b*) Construct a confidence interval around the mean of the above population (using a 99 percent confidence level).

9.18. A health insurance company is conducting periodic surveys on the cost of health services in order to update its health insurance premiums. Suppose you head a research team whose responsibility is to estimate the cost of hospitalization (excluuding the cost of special services) per day in California. If the standard deviation of the cost of hospitalization per day is $40, how many observations will you need in order to determine the cost with only a 2 percent probability of making an error in excess of $4?

9.19. Still referring to Problem 9.18, imagine now that as a result of a recent recession, your office is subject to serious budget cuts, and consequently you can sample only 25 observations. The sample shows that the daily hospitalization cost is $170 (on the average) and that the standard deviation is $30. Determine a 99 percent confidence interval and briefly explain the meaning of the interval.

9.20. The number of dollars collected in tolls per day (Monday through Friday) on a given bridge is a random variable with a known standard deviation of $4,000. A sample of 16 observations is taken to estimate the mean toll collection. What is the probability that the estimate will be off by more than $2,570 from the true mean collection?

9.21. The number of miles per gallon obtained by a given type of car is a random variable normally distributed with a mean of 20 miles and standard deviation of 1 mile. Four such cars are randomly selected and the average miles per gallon are measured. Find the probability that the sample's average is beteen 20.5 and 21.0 miles.

9.22. (*a*) Assume that the actual expenditures on drugs per household of three in a given location is normally distributed, and its average has to be estimated with an accuracy of $2 and 95 percent probability (i.e., there should be 95 percent probability that the estimated mean is not off by more than $2 from the true mean of the distribution). If the standard deviation of drug expenditures is $14 and each observation costs $3 to make, determine the budget for the sample.

(*b*) If the sample size is 64 observations, what is the probability that the error will not exceed $3?

9.23. To determine the optimal number of items a company should hold in inventory, it is necessary to estimate the average weekly sales of that particular item. A random sample is taken, and the number of items sold per week is as follows:

$$64, 57, 49, 81, 76, 70, 59$$

Assume that the number of items sold is normally distributed.

Give a point as well as an interval estimate for the mean number of items sold per week, assuming a 95 percent confidence level. Explain the meaning of your interval estimate.

9.24. The proportion of a certain brand of tires that will become flat within the first 1,000 miles is to be estimated by a sample. To ensure randomness and independence, each tire examined is mounted on a different kind of car. One hundred tires are examined and 5 of them become flat within the first 1,000 miles.

(*a*) Construct an interval estimate for the true proportion at a 99 percent confidence level.

(*b*) What is the sample size required to determine the proportion in such a way that the probability of making an error in excess of 0.01 is 0.05?

9.25. A sample of 14 out of 48 firms in an industry is observed. The sample shows that 6 firms use a given technology known as Technology *A*. Give an interval estimate for the proportion of firms using Technology *A* in the industry. If all 48 firms are observed, what, then, is the standard deviation of the estimated proportion? Explain.

9.26. A firm is considering producing and promoting a new product. It is estimated that the product will be profitable if at least 5 percent of the population buy the product within the first year of its introduction. A survey of 200 people has shown that 6 would buy the product.

(*a*) Estimate the proportion of buyers in the population.

(*b*) Construct a 99 percent confidence interval for the proportion of buyers in the population.

(*c*) If the true proportion of buyers is exactly 0.05 (i.e., 5 percent), what is the probability that 200 observations selected at random will include 6 or fewer buyers?

(*d*) If the true proportion of buyers is exactly 0.05, what is the probability that 10,000 observations selected at random will include 300 or fewer buyers?

(*e*) Suppose the true proportion of buyers is exactly 0.05. What is the sample size required so that the probability of missing the true proportion by more than 0.005 will not exceed 5 percent?

9.27. One month before the election of the U.S. president, a survey of 10,000 randomly selected voters in a certain state showed that 4,500 favored the Democratic candidate.

(*a*) Estimate the proportion of voters favoring the Democratic candidate in this state.

(*b*) Give a 95 percent confidence interval for the above proportion.

(*c*) Rework parts *a* and *b*, this time assuming that the sample size was 100, of whom 45 favored the Democratic candidate.

9.28. "Constructing a confidence interval for a proportion, p, is somewhat problematic. The problem is that if p is unknown, the standard deviation is also unknown and it is impossible to construct the interval. On the other hand, we can always be conservative and estimate the standard deviation by using $p = \frac{1}{2}$." Evaluate this statement.

9.29. Suppose that when estimating a proportion, p, we want to make sure we have a 95 percent chance that the true proportion will not be missed by more than 0.01. How many observations are needed to achieve this result?

9.30. When a 95 percent confidence interval for the mean of an infinite population is constructed, is it true that the interval constructed is *necessarily* wider when σ is unknown than when it is known? Explain.

9.31. Table P9.31 provides data on long-term debt as a percentage of total invested capital in 1980

TABLE P9.31

Long-Term Debt as a Percentage of Total Invested Capital for 10 Firms of the Beverage Industry

Anheuser-Busch	32.7%
Brown-Forman Distillers	24.9
Coca-Cola	6.0
Coca-Cola Bottling Co. of N.Y.	46.9
Heublein	29.7
Pabst Brewing	6.8
PepsiCo	32.6
Royal Crown	30.2
Schlitz (Joseph) Brewing	28.3
Wometco Enterprises	41.2

Data from *Business Week*, October 13, 1980, p. 70.

in 10 firms of the beverage industry. Assume that the 10 firms represent a sample taken from a very large population of firms, whose distribution is normal.

(*a*) Give the point estimate of the long-term debt as a percentage of the invested capital in the industry.

(*b*) Give a 95 percent confidence interval for your estimate, assuming that σ is known to be 10.

(*c*) Again give a 95 percent confidence interval for your estimate, this time assuming that σ is unknown.

CHAPTER TEN OUTLINE

10.1 Basic Hypothesis-Testing Concepts: An Example Concerning a Discrete Distribution

Establishing the Decision Rule
The Null Hypothesis and the Alternative Hypothesis
Errors in Hypothesis Testing: Type I and Type II

10.2 Testing Hypotheses Concerning the Mean of a Distribution

The Acceptance Region and the Rejection Region
Using a Test Statistic Having a Standard Normal Distribution
Upper-Tail and Lower-Tail Tests

10.3 Using a Two-Tailed Test Concerning the Mean

10.4 Type I Error, Type II Error, and the Power of the Test

Calculating the Probability of Type II Error

10.5 The Trade-Off between α and β

10.6 The Relationship among α, β, and n

10.7 Testing Hypotheses Concerning the Mean of the Distribution When the Standard Deviation Is Unknown

10.8 Tests of Hypotheses Concerning the Difference between the Means of Two Distributions

10.9 Testing Hypotheses Concerning a Single Proportion

10.10 Testing Hypotheses Concerning the Difference between Two Proportions

10.11 Limitations

Appendix 10A: Type II Error and the Power Function

The Power of a One-Tailed Test: An Example

Key Words

decision rule	test statistic
null hypothesis	significance level
alternative hypothesis	upper-tail test
Type I error	lower-tail test
Type II error	simple hypothesis
critical value	composite hypothesis
acceptance region	power of a test
rejection region	power function

10
HYPOTHESIS TESTING

10.1 Basic Hypothesis-Testing Concepts: An Example Concerning a Discrete Distribution

Chapter 9 was concerned with the estimation of parameters by the use of sample data. In this chapter we shall show more directly how the results of a sample may be used as a guide in decision-making processes. In fact, we often rely heavily on sample data in decision-making. Public opinion polls may help a presidential candidate to decide whether to keep running or get out of a primary race. A market survey may help a firm to estimate consumers' interest in a given product and thus to determine whether or not funds should be allocated to research and development of that product. A sample of items produced by a machine may determine whether the machine should be stopped for adjustment or allowed to keep running. Sample results may be used as either the sole determinant of a decision or as a major input in the decision-making process.

In this chapter we shall develop a systematic method of decision-making based on sample results. We shall begin by introducing a simple example that makes no assumptions with regard to the shape of the probability distribution of the random variable. Our example will clearly illustrate the main idea and procedures involved in hypothesis testing.

A large hospital is using a given medicine that is effective for 50 percent of the patients treated with it. The cost of the medicine is $100 a dose, and one dose is enough to treat a patient. After years of laboratory research, Merca Company has announced the discovery of a superior medicine. Merca claims that the new medicine is effective for 70 percent of the patients treated. One dose of the new medicine is enough to treat a patient, but the cost per dose is $1,100. Suppose we must decide on behalf of the hospital whether or not to buy the Merca product. Our first logical step will be to find out whether the new medicine really constitutes a significant improvement over the old one. In other words, we must look for evidence to support or reject Merca's claim.

The hospital's management decides to buy eight doses of the new medicine and try it on eight patients. Denoting the number of patients *cured* by the new medicine by X, we realize that the management's "experiment" may result in the following values of X: 0, 1, 2, 3, 4, 5, 6, 7, or 8. It is important to realize that Merca's claim refers to the *probability* of curing a patient, and in any experiment involving a finite number of patients, X can take on any of the values 0, 1, 2, 3, 4, 5, 6, 7, and 8 *even if Merca's claim is correct*. In order to make a decision, however, we must have a **decision rule** that will prudently take into consideration the probabilistic nature of the problem.

ESTABLISHING THE DECISION RULE

The first step in any decision-making process of hypothesis testing is the division of the sample space into two mutually exclusive and collectively exhaustive sets, in such a way that any of the experiment's results will lead to a clear decision. For example, we may decide that if $X = 7$ or $X = 8$, there is a high enough percentage of successes to support Merca's claim and to warrant the adoption of the new medicine, while if $X = 0, 1, 2, 3, 4, 5,$ or 6, the frequency of successes is not high enough to support Merca's claim and to warrant the adoption of the new medicine.

THE NULL HYPOTHESIS AND THE ALTERNATIVE HYPOTHESIS

Hypothesis-testing procedures are generally conservative. In our example, the conservatism is built into the test, since the old medicine will be used unless the new one is proved to be better: we "disbelieve" the new medicine's claim unless we can substantiate improvement. We formulate two distinct hypotheses, which are called the *null hypothesis* and the *alternative hypothesis*, and take an initial stand favoring the null hypothesis.

The **null hypothesis,** denoted by H_0, states that the new medicine does not increase the proportion of patients cured. Formally we write

$$H_0: \quad p = 0.50$$

where p stands for the proportion of patients cured.

The **alternative hypothesis,** denoted by H_1, states that the new medicine increases the proportion of patients cured. In our example the alternative is rather specific: the proportion of patients cured must be equal to 70 percent. Formally we write

$$H_1: \quad p = 0.70$$

Recall that our decision rule divides the possible outcomes of the experiment into two mutually exclusive and collectively exhaustive sets:

Accept H_0 if $X = 0, 1, 2, 3, 4, 5, 6.$
Reject H_0 if $X = 7, 8.$

ERRORS IN HYPOTHESIS TESTING: TYPE I AND TYPE II

Because of the probabilistic nature of the experiment's results, our acceptance or rejection of H_0 may be either "correct" or "erroneous." The possibility (and the probability) of errors in the decision-making process is at the core of hypothesis testing.

What kinds of errors can we commit? First, it is possible that in truth the new medicine is no better than the old one, that on the average it really cures no more than 50 percent of patients treated; in other words, that $p = 0.50$ and H_0 is really correct. In the experiment conducted by the hospital we may by chance get the result $X = 7$ or $X = 8$, and thus, following the decision rule, decide to adopt the new medicine even though it is no better than the old. This type of error—namely, accepting H_1 when in fact H_0 is correct—is known as a **Type I error.** The probability of committing a Type I error can be easily calculated by the binomial distribution equation:

$$P\{X = 7, 8\} = P(X = 7) + P(X = 8) = \binom{8}{7}\left(\frac{1}{2}\right)^7\left(\frac{1}{2}\right)^1 + \binom{8}{8}\left(\frac{1}{2}\right)^8\left(\frac{1}{2}\right)^0 = 0.035$$

or

$$P\{X = 7, 8\} = 3.5\%$$

The reason we use $\frac{1}{2}$ as the success probability in this binomial experiment is that we assumed that the probability is *in truth* equal to 0.50.

At this point, have we considered all possible errors? Not really. We still have to consider the possibility of deciding against adoption of the new medicine when in truth it does provide an improvement, and cures, on the average, 70 percent of patients, as the Merca Company claims. The error of accepting H_0 when in fact H_1 is correct is known as **Type II error.** This error can occur in our example if in truth $p = 0.70$ when the experiment provides results with X equal to one of the values 0, 1, 2, 3, 4, 5, 6. This probability is equal to 74 percent:

$$P\{X = 0, 1, 2, 3, 4, 5, 6\} = \sum_{X=0}^{6}\binom{8}{X}(0.70)^X(0.30)^{8-X} = 0.74$$

or

$$P\{X = 0, 1, 2, 3, 4, 5, 6\} = 74\%$$

Note that the probability of success used here is 0.70 because it is assumed that this is *in truth* the probability of curing a patient.

Given the two possible errors and their respective probabilities, can we consider this statistical test a good one? Almost everyone is sure to agree that the 74 percent chance of committing the second type of error is much too high. Such a high probability gives very little chance for the new medicine to prove itself, since even if the new medicine is an improvement, under the procedure we have adopted there is a 74 percent chance that its merits will not be recognized.

This high probability of 74 percent for a Type II error is such an absurdity that Merca's management would be better off if the hospital's administrator made the decision by flipping a coin rather than by carrying out the experiment. That way the new medicine would have a 50 percent chance of being adopted, whereas under the procedure we have described, the chances were only 26 percent $(1.00 - 0.74 = 0.26)$.

One possible way to give the new medicine a better chance is to modify the decision rule. Suppose we establish the following decision rule:

> Accept H_0 if $X = 0, 1, 2, 3, 4, 5.$
> Reject H_0 if $X = 6, 7, 8.$

Under this decision rule, the probability of Type II error—that of deciding against the new medicine if in truth it is superior—drops from 74 percent to 45 percent:

$$P\{X = 0, 1, 2, 3, 4, 5\} = \sum_{X=0}^{5} \binom{8}{X} (0.7)^X (0.3)^{8-X} = 0.45$$

At the same time, however, the probability of the Type I error—that of deciding to adopt the new medicine when in truth it does not provide any improvement over the old—increases to 15 percent, up from 3.5 percent under the previous decision rule:

$$P\{X = 6, 7, 8\} = \sum_{X=6}^{8} \binom{8}{X} \left(\frac{1}{2}\right)^X \left(\frac{1}{2}\right)^{8-X} = 0.15$$

We thus conclude that *by changing the decision rule, we can decrease the probability of committing one type of error only at the expense of increasing the probability of committing the other.*

The Effect of Sample Size on Type I and Type II Errors

Fortunately, there is a way to decrease the probability of both types of errors simultaneously: by *increasing the sample size.* Suppose we try the new medicine on 60 (rather than just 8) patients. Suppose also that we adopt the following decision rule:

> Accept H_0 if $X = 0, 1, 2, 3, \ldots, 36, 37.$
> Reject H_0 if $X = 38, 39, 40, \ldots, 59, 60.$

In this case, the probability of favoring the new medicine when in truth it is no better than the old one (and thus committing a Type I error) is

$$P\{X = 38, 39, \ldots, 59, 60\} = \sum_{X=38}^{60} \binom{60}{X} \left(\frac{1}{2}\right)^X \left(\frac{1}{2}\right)^{60-X} = 0.046$$

or 4.6 percent. The probability of committing the Type II error (that is, rejecting the new medicine even though it is better than the old) is

$$P\{X = 0, 1, 2, \ldots, 36, 37\} = \sum_{X=0}^{37} \binom{60}{X} (0.70)^X (0.30)^{60-X} = 0.06$$

or 6 percent.

It is evident, then, that by increasing the sample size we may decrease the probabilities of both errors. Yet while this analysis may lead to the conclusion that the larger the sample size, the lower the error probabilities, and thus that we should always seek to increase the sample size as much as possible, we must remember that the sample size is often limited due to cost, limited time, or other considerations (see Chapter 9).

To summarize, the hospital must consider three types of costs involved in the statistical test:

1. If the new medicine is *accepted* when in truth it is no better than the old, there will be $1,000 extra cost per patient with no extra benefit. The loss will continue until the hospital realizes that the new medicine is indeed no improvement over the old.
2. If the new medicine is *rejected* when in truth it is better than the old, there will be a social cost (more sick people who can in fact be cured) as well as an economic cost, since the hospital may lose patients to other hospitals that are using the new medicine.
3. Each sample observation has its cost. In our case, this cost amounts to $1,000.

To conclude our example of the Merca Company's medicine, let us touch upon one more point. Suppose Merca claims that the new medicine is 100 percent effective. What would be our decision rule in this case, and how would the error probabilities be affected? The decision rule becomes rather simple: we shall accept Merca's claim only if all patients treated are cured by the new medicine, and reject it if one or more patients are not cured. In such a situation, the probability of committing the Type II error (rejecting the new medicine when it really cures 100 percent of patients) is equal to zero. Nevertheless, the probability of the Type I error (accepting the new medicine when in truth it is no better than the old) is greater than zero. It will depend on the number of patients treated in the experiment, since occasionally all patients in the sample also recover with the old medicine.

10.2 Testing Hypotheses Concerning the Mean of a Distribution

Now that we have been exposed to the basic concepts and trade-offs involved in hypothesis testing, we continue with some specific hypothesis-testing procedures. Here, we shall look at some examples involving continuous probability distributions. We shall assume that the *average* of the sample is normally distributed; although this is not really the case under all circumstances, the central limit theorem suggests that the assumption is acceptable in many situations. (When the assumption of normality cannot be accepted, we should use nonparametric hypothesis-testing procedures, some of which are detailed in Chapter 18.)

First, let us consider the following situation. The management of a movie house is searching for ways to boost profits, which have been declining lately as a result of soaring costs. In the past the management has shown only movies rated *G* ("suitable for general audiences"). The average weekly revenue has been $3,000. It has now been suggested that movies rated *PG* and *R* ("parental guidance suggested" and "restricted") be shown. The theater management has decided to have a 16-week trial period during which *PG*- and *R*-rated movies are to be run, and to permit these movies to continue after the trial period only if average weekly revenue increases significantly.

We shall simplify matters here by dichotomizing the possible results of the trial period. We assume that management believes that expected weekly revenue will either stay at $3,000 or rise to $3,250 (any other values being ruled out at this stage for the sake of simplicity). We shall assume further that management has agreed not to change the existing *G* policy unless

Figure 10.1

The probability distribution of $\overline{X}$ under two alternative hypotheses

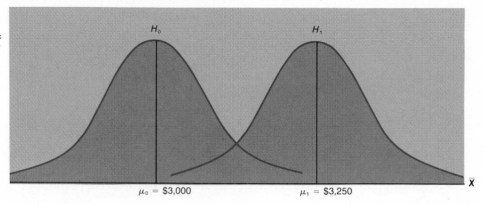

significant evidence exists that the new average weekly revenue is $3,250. We shall also assume that weekly revenues are normally distributed, with standard deviation equal to $400. Note that this assumption implies that the average revenue of the 16 weeks of the trial period is normally distributed with $400/\sqrt{16} = \$100$ standard error.

Insight into the possible events during the trial period and into the decision reached on the basis of the trial may be gained from Figure 10.1. Here the distribution of the average revenue of 16 weeks is graphed in two alternative ways: the first, headed H_0 (the null hypothesis), assumes that the mean weekly revenue is $3,000, and thus the 16-week average revenue also has a mean of $3,000 with standard error of $400/\sqrt{16} = \$100$. The second, headed H_1 (the alternative hypothesis), assumes that the mean weekly revenue is $3,250, and thus the 16-week average revenue also has a mean of $3,250 with $100 standard error. Notationally, we state the null and the alternative hypotheses as

$$H_0: \quad \mu = \mu_0 = \$3,000$$
$$H_1: \quad \mu = \mu_1 = \$3,250$$

Obviously, the actual 16-week average revenue of the trial period ($\overline{X}$) is not likely to equal exactly $3,000 or precisely $3,250. After the trial period is over and $\overline{X}$ is calculated, management may either conclude that the null hypothesis (H_0) is more likely to be correct *or* that the alternative (H_1) is more likely to be correct.[1] Clearly if $\overline{X}$ is in the neighborhood of $3,000, H_0 will seem to be more likely, so that if $2,990 or $3,015 is the average revenue of the 16-week period, it will be reasonable to accept H_0 and reject H_1—to conclude that $3,000 is still the weekly average revenue and that no increase in average weekly revenue has occurred as a result of the policy change. But if the 16-week trial period produces an average of $3,230, $3,246, or $3,270, say, management will have good reason to accept H_1 and reject H_0—to believe that an increase from $3,000 to $3,250 weekly revenue actually occurs as a result of the new policy. The question, of course, is which hypothesis the management should accept if the sample average (the weekly average in the trial period) falls somewhere between $3,000 and $3,250—around $3,125, say.

---

[1] In earlier chapters we used a lower-case letter to denote specific values assumed by a random variable. We feel that a distinction between upper-case and lower-case letters in this chapter will make notations cumbersome. Thus, only upper-case letters will be used.

THE ACCEPTANCE REGION AND THE REJECTION REGION

To enable us to choose between the two alternatives efficiently and objectively, we need to establish a cutoff point on the $\overline{X}$ axis. (Recall that in the previous example we also had to choose the cutoff point that represented the minimum number of patients cured by the new medicine.) Suppose we choose to select $3,125, the midpoint between $3,000 and $3,250, as a cutoff point. This selection will imply the following decision rule:

Accept H_0 if the sample average ($\overline{X}$) is less than or equal to $3,125.

Reject H_0 (and thus accept H_1) if $\overline{X}$ is greater than $3,125.

While we have selected the midpoint between the two means to serve here as the cutoff point for decision making, the midpoint will not, in general, be a good cutoff point, for two reasons:

1. Only in a limited number of cases is the distribution's mean under the alternative hypothesis (H_1) so well defined as in our (admittedly simplified) example. When the mean under H_1 is not numerically specified, no midpoint such as ours exists, and a different approach to the determination of the cutoff point must be found.
2. The midpoint in our example, $3,125, is 1.25 standard errors (of $100 each) to the right of $3,000. The right-tail area of H_0 to the right of $3,125 is the same as the right-tail area of the Z distribution to the right of $Z = 1.25$ and is equal to 0.1056. The probability of rejecting the null hypothesis when in fact it is correct (that is, committing a Type I error) is thus equal in our case to 10.56 percent, which may be either too high or too low in management's opinion. Depending on management's attitude toward the proposed policy change, they may decide to set this probability at 5 percent. In such a case we merely have to set the cutoff point—or, as it is often called, the **critical value**—equal to that value which leaves a 5 percent right tail under the distribution, assuming that H_0 is in truth correct.

Recalling that $Z_{0.05} = 1.645$ is that Z score which leaves a 5 percent right tail under the standard normal distribution, we may easily get the critical value:

$$\overline{X}^* = \mu_0 + Z_{0.05}\sigma_{\overline{X}} = \mu_0 + Z_{0.05}\frac{\sigma}{\sqrt{n}} \tag{10.1}$$

where $\overline{X}^*$ is the critical value and μ_0 is the distribution mean, assuming H_0 is correct. In our case we get

$$\overline{X}^* = 3,000 + 1.645 \cdot \frac{400}{\sqrt{16}} = 3,000 + 164.5 = \$3,164.5$$

Figure 10.2 clearly illustrates the location of $\overline{X}^*$: it leaves a 5 percent right tail of the H_0 distribution. The importance of the critical value is that it leads to a decision rule by defining the **acceptance region** (of the null hypothesis) and the **rejection region** (of the null hypothesis). If the average of the 16-week trial period, $\overline{X}$, is less than or equal to $3,164.5—the null

Figure 10.2

The location of the critical value, the acceptance region, and the rejection region

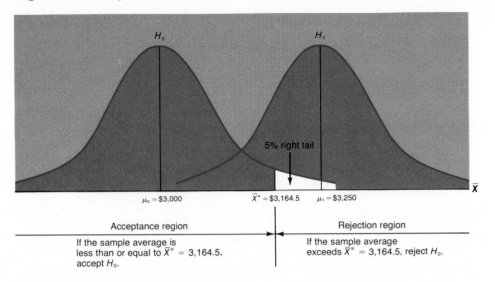

hypothesis should be accepted. If the sample average exceeds $3,164.5, however, the null hypothesis should be rejected.

USING A TEST STATISTIC HAVING A STANDARD NORMAL DISTRIBUTION

As you recall, we have defined a statistic as a function of the observations in the sample. A **test statistic** is a statistic *used as a vehicle to test hypotheses*.

Thus far we have been using $\overline{X}$ as our test statistic. Before proceeding, let us suggest a different (though similar) procedure to test the hypothesis under consideration. Instead of using $\overline{X}$ as our test statistic, we shall use $Z = \dfrac{\overline{X} - \mu_0}{\sigma/\sqrt{n}}$, which has a standard normal distribution. Let us set the critical value Z^* at $Z_{0.05}$, so that all Z values less than or equal to $Z_{0.05} = 1.645$ are in the acceptance region, and all Z values that are greater than $Z_{0.05}$ are in the rejection region. Now suppose the sample average is equal to $3,140. To find the Z value that corresponds to the sample average, we write

$$Z = \frac{\overline{X} - \mu_0}{\sigma/\sqrt{n}} = \frac{3,140 - 3,000}{400/\sqrt{16}} = 1.40$$

Since 1.40 is in the acceptance region, the null hypothesis should be accepted. If the Z value corresponding to $\overline{X}$ had exceeded 1.645, the null hypothesis should have been rejected. Obviously, both approaches (the one using $\overline{X}$ and the one using Z) *lead to the same acceptance or rejection decision*, and the approach one chooses is largely a matter of taste.

The acceptance or rejection of the null hypothesis depends, among other things, on the exact location of $\overline{X}^*$, the critical value, in relation to μ_0. The location of $\overline{X}^*$, in turn, is a function of the probability we set for rejecting the null hypothesis when in fact it is correct. This probability is equal to 5 percent in our example and is represented by the tail of the null hypothesis distribution over the rejection region (see Figure 10.3). In general, we shall denote this probability by α and call it the test's **significance level**. The significance level of the test, denoted by α, is *the probability of rejecting the* *null hypothesis when it is in truth correct* (and thus should be accepted). It is thus equal to the probability of committing a Type I error.

UPPER-TAIL AND LOWER-TAIL TESTS

There is one thing you should note at this point. In our example, we have positioned the rejection region to the right of the null hypothesis mean (see Figure 10.2) because the alternative mean (μ_1) is *greater* than the null hypothesis mean (μ_0). This type of test is known as an **upper-tail test.** In the case in which $\mu_1 < \mu_0$, the test is known as a **lower-tail test,** and the rejection region will be located to the left of μ_0. Let us consider an example that will specifically illustrate this kind of situation.

A greeting-card printer has been told of the availability of a new machine that will produce the same output (from the point of view of both quality and quantity) as his present machine, but is claimed to bring savings in operating costs because it can be adjusted more easily when card design is changed, it has a lower breakdown rate, and so on. The existing machine costs an average of $100 a week to operate. Assume that operating cost is normally distributed with a weekly standard deviation of $25 for both machines, and that a sample of 9 weeks has shown an average operating cost of $75 per week for the new machine. Assume also that $\alpha = 0.01$ and test the following hypotheses:

$$H_0: \quad \mu = \mu_0 = 100$$
$$H_1: \quad \mu = \mu_1 < 100$$

The null hypothesis may also be written in the following way:

$$H_0: \quad \mu = \mu_0 \geq 100$$

The testing procedure in this case is precisely as when H_0 is $\mu = 100$.

This example differs from our earlier example in two ways:

1. The alternative hypothesis assumes that μ_1 is *less than* μ_0.
2. No specific value is assumed for μ_1, and H_1 only indicates that μ_1 is assumed to be less than 100.

Let us use $Z = \dfrac{\overline{X} - \mu_0}{\sigma / \sqrt{n}}$ as our test statistic. Since $\alpha = 0.01$ and since the

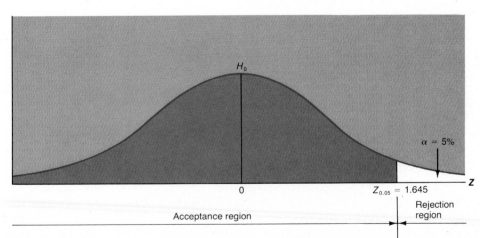

Figure 10.3

The acceptance and rejection regions and the test significance level, α, when the test statistic is $Z = \dfrac{\overline{X} - \mu_0}{\sigma / \sqrt{n}}$ and $\alpha = 0.05$

H_0

$\alpha = 5\%$

0

$Z_{0.05} = 1.645$

Z

Acceptance region

Rejection region

Figure 10.4

The location of the critical value in a lower-tail test when the test statistic is

$$Z = \frac{\overline{X} - \mu_0}{\sigma/\sqrt{n}}$$

and $\alpha = 0.01$

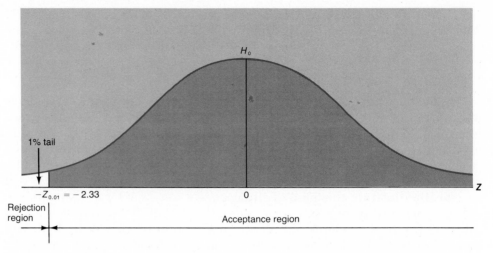

alternative hypothesis suggests that μ_1 is *less than* μ_0, we set the critical value (Z^*) equal to $-Z_{0.01}$ (see Figure 10.4). The acceptance region includes all the Z values that are greater than $-Z_{0.01} = -2.33$. Our decision rule is as follows:

Accept H_0 if the test statistic appears within the acceptance region.

Reject H_0 if the test statistic appears within the rejection region.

All we have to do now is to compute the test statistic value in our particular sample:

$$Z = \frac{\overline{X} - \mu_0}{\sigma/\sqrt{n}} = \frac{75 - 100}{25/\sqrt{9}} = \frac{-25}{25/3} = -3$$

A glance at Figure 10.4 will reveal that -3 is in the rejection region, so we reject the null hypothesis.

The same problem may be solved with $\overline{X}$ as the test statistic (see Figure 10.5). We first determine the critical value ($\overline{X}^*$) as follows:

$$\overline{X}^* = \mu_0 + (-Z_{0.01})\frac{\sigma}{\sqrt{n}} = 100 + (-2.33) \cdot \frac{25}{\sqrt{9}}$$

$$= 100 - 19.42 = 80.58$$

Our decision rule is

Accept H_0 if $\overline{X} \geq 80.58$.
Reject H_0 if $\overline{X} < 80.58$.

Since $\overline{X} = 75$ and it is in the rejection region, we reject the null hypothesis.

10.3 Using a Two-Tailed Test Concerning the Mean

So far we have considered only one-tailed tests: the lower-tail test and the upper-tail test. Sometimes the test may be two-tailed. For example, a baker may wish to test the hypothesis that his loaves of bread weigh exactly 2

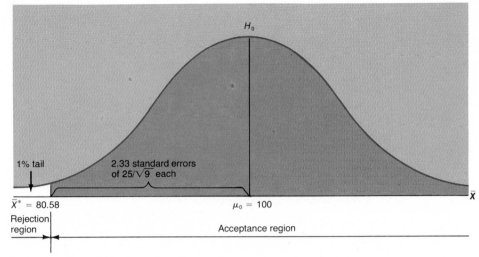

Figure 10.5

The location of the critical value in a lower-tail test when the test statistic is $\overline{X}$ and $\alpha = 0.01$

pounds (32 ounces). He may wish to take some corrective measures both in the case of overweight (because of unnecessary cost) and in the case of underweight (because this might alienate customers). Suppose the loaves' weight is normally distributed with standard deviation of 2 ounces. Periodically the baker takes a random sample of 20 loaves and decides (on the basis of the sample result) whether or not corrective measures should be undertaken. If the significance level (α) is 5 percent and the sample average is 31 ounces, what course of action should the baker take?

In this two-tailed test we have the following hypotheses:

$$H_0: \quad \mu = \mu_0 = 32$$
$$H_1: \quad \mu \neq 32$$

Since a significant deviation above or below 32 ounces would be consistent with the alternative hypothesis, the rejection region should be split between the two tails of the distribution. Thus we allocate the probability α to both tails of the Z distribution in equal amounts to determine two critical values, $-Z_{0.025} = -1.96$ and $Z_{0.025} = 1.96$, as illustrated in Figure 10.6,

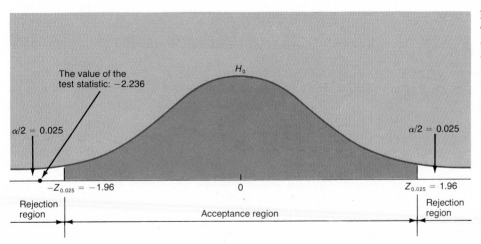

Figure 10.6

The location of the critical values in a two-tailed test when the test statistic is $Z = \dfrac{\overline{X} - \mu_0}{\sigma/\sqrt{n}}$ and $\alpha = 0.05$

where the test statistic is $Z = \dfrac{\overline{X} - \mu_0}{\sigma/\sqrt{n}}$. If the value of the test statistic falls between the two critical values (that is, in the acceptance region), H_0 is accepted; otherwise it is rejected. In our case we get

$$Z = \frac{\overline{X} - \mu_0}{\sigma/\sqrt{n}} = \frac{31 - 32}{2/\sqrt{20}} = \frac{-1}{2/4.472} = -2.236$$

This value, -2.236, is well outside the acceptance region and thus the null hypothesis is rejected: corrective measures should be undertaken.

As before, one may decide instead to use $\overline{X}$ as the test statistic and determine two critical values, $\overline{X}_1^*$ and $\overline{X}_2^*$, as follows:

$$\overline{X}_1^* = \mu_0 + (-Z_{0.025}) \frac{\sigma}{\sqrt{n}} = 32 - 1.96 \cdot \frac{2}{\sqrt{20}} = 32 - 0.877 = 31.123$$

$$\overline{X}_2^* = \mu_0 + Z_{0.025} \frac{\sigma}{\sqrt{n}} = 32 + 1.96 \cdot \frac{2}{\sqrt{20}} = 32 + 0.877 = 32.877$$

Since 31, the sample average, is outside the acceptance region—the interval between 31.123 and 32.877—we reject the null hypothesis. We wish to emphasize once again that the decision concerning the acceptance or rejection of the null hypothesis is reached whether $\overline{X}$ or $Z = \dfrac{\overline{X} - \mu_0}{\sigma/\sqrt{n}}$ is chosen as the test statistic. It should also be clear that *using one test statistic is sufficient to draw the proper conclusion.* We have presented the procedures involving $\overline{X}$ and $Z = \dfrac{\overline{X} - \mu_0}{\sigma/\sqrt{n}}$ side by side only for the purpose of providing further insight.

Note the similarity between the two-tailed test and the confidence interval. If a confidence interval is constructed around $\overline{X}$ at a given confidence level $(1 - \alpha)$ and μ_0 is included in it, then a two-tailed test of hypotheses with a significance level α will lead to the acceptance of H_0. If the confidence interval does not cover the value μ_0, however, the corresponding test will require the rejection of H_0. In the above example, we rejected H_0 because the sample average ($\overline{X} = 31$) was outside the acceptance region (from 31.123 to 32.877). The rejection of H_0 can be seen in another way, too. A 0.95 confidence interval is $31 \pm 1.96 \cdot \dfrac{2}{\sqrt{20}} = 31 \pm 0.877$, or the interval from 30.123 to 31.877. This interval does not include $\mu_0 = 32$, and thus H_0 is rejected.

We may now summarize step by step the hypothesis-testing procedures for upper-tail, lower-tail, and two-tailed tests as follows:

PROCEDURE FOR UPPER-TAIL TEST

Step 1: Clearly identify the null and alternative hypotheses:

$$H_0: \quad \mu \leq \mu_0$$
$$H_1: \quad \mu > \mu_0$$

Step 2: Determine the significance level, α, that will be suitable for the test.

Step 3: Choose your test statistic, Z or $\overline{X}$, and proceed accordingly. Determine the critical value and the decision rule.

If $Z = \dfrac{\overline{X} - \mu_0}{\sigma/\sqrt{n}}$ is the test statistic

the critical value is Z_α and the decision rule is

Accept H_0 if $Z \leq Z_\alpha$.
Reject H_0 if $Z > Z_\alpha$.

If $\overline{X}$ is the test statistic

the critical value is

$$\overline{X}^* = \mu_0 + Z_\alpha \cdot \frac{\sigma}{\sqrt{n}}$$

and the decision rule is

Accept H_0 if $\overline{X} \leq \overline{X}^*$.

Reject H_0 if $\overline{X} > \overline{X}^*$.

Step 4: Calculate the value of the test statistic using the sample data.
Step 5: Apply the decision rule established in Step 3, and accept or reject H_0 accordingly.

PROCEDURE FOR LOWER-TAIL TEST

Step 1: Clearly identify the null and alternative hypotheses:

$$H_0: \quad \mu \geq \mu_0$$
$$H_1: \quad \mu < \mu_0$$

Step 2: Determine the significance level, α, that will be suitable for the test.

Step 3: Choose your test statistic, Z or $\overline{X}$, and proceed accordingly. Determine the critical value and the decision rule.

If $Z = \dfrac{\overline{X} - \mu_0}{\sigma/\sqrt{n}}$ is the test statistic

the critical value is $-Z_\alpha$, and the decision rule is

Accept H_0 if $Z \geq -Z_\alpha$.
Reject H_0 if $Z < -Z_\alpha$.

If $\overline{X}$ is the test statistic

the critical value is

$$\overline{X}^* = \mu_0 - Z_\alpha \cdot \frac{\sigma}{\sqrt{n}}$$

and the decision rule is

Accept H_0 if $\overline{X} \geq \overline{X}^*$.

Reject H_0 if $\overline{X} < \overline{X}^*$.

Step 4: Calculate the value of the test statistic using the sample data.
Step 5: Apply the decision rule established in Step 3, and accept or reject H_0 accordingly.

PROCEDURE FOR TWO-TAILED TEST

Step 1: Clearly identify the null and alternative hypotheses:

$$H_0: \quad \mu = \mu_0$$
$$H_1: \quad \mu \neq \mu_0$$

Step 2: Determine the significance level, α, that will be suitable for the test.

Step 3: Choose your test statistic, Z or $\overline{X}$, and proceed accordingly. Determine the two critical values and the decision rule.

$\boxed{\text{If } Z = \dfrac{\overline{X} - \mu_0}{\sigma/\sqrt{n}} \text{ is the test statistic}}$	$\boxed{\text{If } \overline{X} \text{ is the test statistic}}$

the critical values are $-Z_{\alpha/2}$ and $Z_{\alpha/2}$, and the decision rule is

Accept H_0 if $-Z_{\alpha/2} \leq Z \leq Z_{\alpha/2}$.
Reject H_0 if $Z < -Z_{\alpha/2}$ or if $Z > Z_{\alpha/2}$.

the critical values are

$$\overline{X}_1^* = \mu_0 - Z_{\alpha/2} \cdot \frac{\sigma}{\sqrt{n}}$$

$$\overline{X}_2^* = \mu_0 + Z_{\alpha/2} \cdot \frac{\sigma}{\sqrt{n}}$$

and the decision rule is

Accept H_0 if $\overline{X}_1^* \leq \overline{X} \leq \overline{X}_2^*$.
Reject H_0 if $\overline{X} < \overline{X}_1^*$ or if $\overline{X} > \overline{X}_2^*$.

Step 4: Calculate the value of the test statistic using the sample data.
Step 5: Apply the decision rule established in Step 3, and accept or reject H_0 accordingly.

10.4 Type I Error, Type II Error, and the Power of the Test

In the example given in Section 10.1, we indicated the possibility that one may commit an error by accepting a hypothesis that is not correct. Here we shall consider the probabilities of error in testing hypotheses concerning a mean of distribution, assuming X is normally distributed, as in Section 10.2.

The decision rule illustrated in Section 10.2 may lead to either correct or incorrect decisions. If H_0 is right and it is accepted, a correct decision has been made; if H_0 is wrong and it is rejected, again a correct decision has been made. Because of sampling errors, however, the test statistic may appear in the rejection region even though H_0 may be correct, leading us to (incorrectly) reject H_0 (Type I error). It is also possible that H_1 is in truth correct, but the test statistic appears in the acceptance region, leading us to (incorrectly) accept H_0 (Type II error). These four possibilities are summarized in the following table:

POSSIBLE DECISIONS IN HYPOTHESIS TESTING			
		True Situation	
		H_0 is correct	H_1 is correct
Decision	Accept H_0	Correct decision.	Incorrect decision. Type II error.
	Reject H_0	Incorrect decision. Type I error.	Correct decision.

As the table indicates, and as we saw in Section 10.1, our decision is subject to two possible errors. A **Type I error** occurs when H_0 is correct but sampling errors lead us to reject it. A **Type II error** occurs when H_1 is correct but we reject it in favor of H_0, again because of sampling errors. Note that neither *ex ante* (that is, before the sample is selected) nor *ex post* (that is, after the sample data become available) is it possible to determine whether an error of either type has been committed. In order to make a prudent decision on the basis of the sample data, one should be aware of the *probability* of committing each type of error. The probabilities can be understood with the help of Figure 10.7.

Consider first the probability of committing a Type I error. By definition, this error occurs when $\overline{X}$ falls somewhere within the rejection region while H_0 is in fact the correct hypothesis. This probability is represented in Figure 10.7 by that part of the right-tail area of the H_0 distribution which lies in the rejection region. Recall that this tail was set equal to α (the chosen significance level of the test) by positioning $\overline{X}^*$ in such a location that it leaves a tail area equal to α. We conclude, then, that the significance level of a test, α, is also the probability of committing Type I error.

The probability of committing Type II error is represented in Figure 10.7 by that part of the tail of the H_1 distribution which extends into the acceptance region. This probability, denoted by β, is thus generally defined as follows:

$$\beta = P(\text{committing Type II error})$$

$$= P(\overline{X} \text{ appears in the acceptance region when } H_1 \text{ is correct})$$

(10.2)

In other words, the probability of Type II error (β) is equal to the probability that $\overline{X}$ will fall within the acceptance region *when H_1 is correct*. We shall often write the probability on the right-hand side of Equation 10.2 in a slightly more compact form:

$$\beta = P_{H_1}(\overline{X} \text{ shows in the acceptance region})$$

(10.3)

The subscript H_1 in Equation 10.3 means that the probability is conditional upon H_1's being correct.

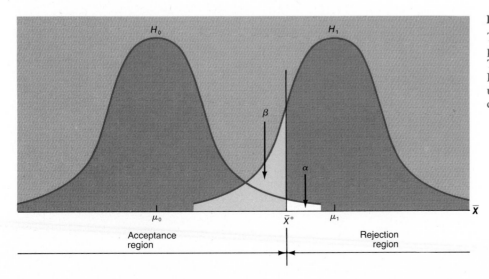

Figure 10.7

The probabilities of Type I and Type II errors: the upper-tail test case

CALCULATING THE PROBABILITY OF TYPE II ERROR

When H_1 is a **simple hypothesis**—that is, its parameter μ_1 has a single specific value (for example, H_1: $\mu = 10$)—β is also a single probability value. Frequently, however, the alternative hypothesis is a **composite** rather than a **simple hypothesis**—that is, its parameter has a value within a given range of values (for example, H_1: $\mu > 10$). In such cases, the probability of a Type II error is specified as a function rather than a single value. Discussion of this function is presented in Appendix 10A. Let us consider here the calculation of β in the case of a simple alternative hypothesis.

When the alternative hypothesis is simple, we test H_0: $\mu = \mu_0$ against H_1: $\mu = \mu_1$; the test is a lower-tail test if $\mu_1 < \mu_0$ and an upper-tail test if $\mu_1 > \mu_0$. In an upper-tail test situation (like the one described in Figure 10.7), the acceptance region is to the left of $\overline{X}^*$. In this case we proceed in the following manner:

$$\beta = P_{H_1}(\overline{X} \text{ shows in the acceptance region}) = P_{H_1}(\overline{X} \leq \overline{X}^*) \qquad \textbf{(10.4)}$$

To find the probability in Equation 10.4, which is conditional upon H_1's being correct, one has to transform $\overline{X}$ into a Z variable. Assuming that μ_1 is the true mean and that σ is known, we can transform $\overline{X}$ into Z by subtracting μ_1 and dividing by $\sigma/\sqrt{n}$:

$$\frac{\overline{X} - \mu_1}{\sigma/\sqrt{n}} = Z \qquad \textbf{(10.5)}$$

Substituting Equation 10.5 in Equation 10.4 and performing a similar transformation on $\overline{X}^*$, we obtain

$$\beta = P_{H_1}(\overline{X} \leq \overline{X}^*) = P_{H_1}\left(\frac{\overline{X} - \mu_1}{\sigma/\sqrt{n}} \leq \frac{\overline{X}^* - \mu_1}{\sigma/\sqrt{n}}\right) = P\left(Z \leq \frac{\overline{X}^* - \mu_1}{\sigma/\sqrt{n}}\right) \qquad \textbf{(10.6)}$$

In the case of a lower tail, after performing similar derivations we get

$$\beta = P\left(Z \geq \frac{\overline{X}^* - \mu_1}{\sigma/\sqrt{n}}\right) \qquad \textbf{(10.7)}$$

To summarize, the probability of Type II error when H_1 is a simple hypothesis is as follows:

PROBABILITY OF TYPE II ERROR FOR LOWER-TAIL ALTERNATIVE

$$(\mu_0 > \mu_1)$$

$$\beta = P\left(Z \geq \frac{\overline{X}^* - \mu_1}{\sigma/\sqrt{n}}\right) \qquad \textbf{(10.8)}$$

PROBABILITY OF TYPE II ERROR FOR UPPER-TAIL ALTERNATIVE

$$(\mu_0 < \mu_1)$$

$$\beta = P\left(Z \leq \frac{\overline{X}^* - \mu_1}{\sigma/\sqrt{n}}\right) \qquad \textbf{(10.9)}$$

The **power of a test** is the probability of accepting H_1 when H_1 is correct. This probability is equal to $1 - \beta$. Obviously, other things being equal, the greater the power of the test, the better the test.

EXAMPLE 10.1

To test $H_0: \mu = \mu_0 = 130$ against $H_1: \mu = \mu_1 = 125$, a sample of 64 observations is used. Assume that σ is equal to 25 and that the level of significance used is $\alpha = 0.05$. Determine the probability of committing a Type II error and the power of the test.

Following the procedure for determining $\overline{X}^*$ in the case of a lower-tail test like the one at hand, we write

$$\overline{X}^* = \mu_0 - Z_\alpha \frac{\sigma}{\sqrt{n}} = 130 - 1.645 \cdot \frac{25}{\sqrt{64}} = 130 - 5.14 = 124.86$$

Once $\overline{X}^*$ is determined, we can calculate β by using Equation 10.9:

$$\beta = P\left(Z \geq \frac{124.86 - 125.00}{25/\sqrt{64}}\right) = P(Z \geq -0.04) = 0.5160$$

The power of this test is equal to $1 - \beta = 1 - 0.5160 = 0.4840$, indicating that there is only a 48.4 percent chance that if H_1 is correct it will be accepted.

To illustrate the procedure for an upper-tail test, let us change only one parameter in the previous set of numbers. Let us change $H_1: \mu = 125$ to $H_1: \mu = 137$. Since now we are dealing with an upper-tail test, we have to find the appropriate critical point by using the formula for an upper-tail test:

$$\overline{X}^* = \mu_0 + Z_\alpha \frac{\sigma}{\sqrt{n}} = 130 + 1.645 \cdot \frac{25}{\sqrt{64}} = 135.14$$

We now use Equation 10.8 for the upper-tail test:

$$\beta = P\left(Z \leq \frac{135.14 - 137}{25/\sqrt{64}}\right) = P(Z \leq -0.60) = 0.2743$$

The power of this test is equal to $1 - \beta = 1 - 0.2743 = 0.7257$.

10.5 The Trade-Off between α and β

Since both α and β are probabilities of committing errors in hypothesis testing, we are naturally concerned with ways to minimize them. In our introductory example (Section 10.1) we discussed the trade-off between α and β. We shall now discuss that same trade-off in testing hypotheses of the mean when normal distributions are assumed. We shall show that for any given sample size there is a trade-off between α and β, so that when α is lowered, β increases, and vice versa. Figure 10.8 helps to clarify the type of trade-off that exists between α and β. In Figure 10.8a, α is equal to 5 percent and the resulting probability of a Type II error is 25 percent. In Figure 10.8*b*, α is 1 percent. We make α equal to 1 percent by shifting $\overline{X}^*$ to the right of its position in 10.8*a*. While this shift lowers α, it simultaneously increases β from 25 percent to 45 percent. The same sort of trade-off, of course, exists in the case of a lower-tail test, as illustrated in Figure 10.9. While it is more difficult to see the trade-off between α and β in a diagram of a two-tailed test, it does exist in this case as well.

Figure 10.8

The trade-off between α and β: the upper-tail case

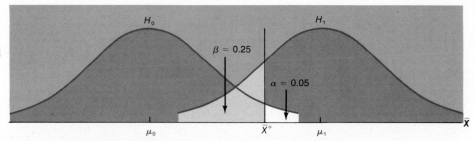

(a) The significance level (α) is set to equal 5%, and β is equal to 25%.

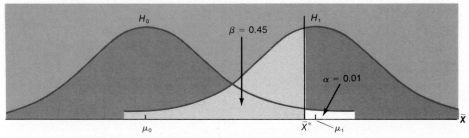

(b) When the significance level (α) is changed to 1% by the shifting of $\overline{X}^*$ to the right, β increases from 25% to 45%.

Figure 10.9

The trade-off between α and β: the lower-tail case

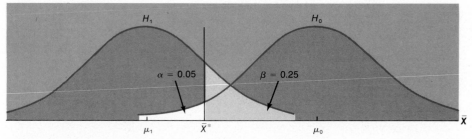

(a) The significance level (α) is set to equal 5%, and β is equal to 25%.

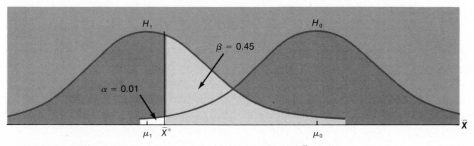

(b) When the significance level (α) is changed to 1% by the shifting of $\overline{X}^*$ to the left, β increases from 25% to 45%.

10.6 The Relationship among α, β, and n

For any given sample size, a trade-off exists between the probability of a Type I error (α) and the probability of a Type II error (β). If the sample size (n) is allowed to increase, however, it is possible for both α and β to decrease simultaneously, as we saw in Section 10.1. Figure 10.10 shows the effect of increased sample size on both α and β in the case of a lower-tail test, assuming $\overline{X}$ to be normally distributed.

In Figure 10.10a, the distributions are rather dispersed, giving rise to the following error probabilities: $\alpha = 0.05$, $\beta = 0.38$. When the sample size increases (from n to some greater n'), the distributions become more condensed around the (true) mean. If we hold $\overline{X}^*$ in Figure 10.10b in the same position as in 10.10a, both α and β will decrease. In the case illustrated, α is decreased to 0.005 and β to 0.08. The effect of increasing the sample size is positive: it decreases the probability of both Type I and Type II errors. Since the power of the test is equal to $(1 - \beta)$, it is also clear that since β decreases, $(1 - \beta)$ increases, so that an increase in n has a positive effect on the power of the test as well. The only problems we run into when the sample size is increased are related to availability of observations and cost of sampling (see Section 10.1). The question of deriving optimal sample size when one takes into consideration possible losses resulting from acceptance of the wrong hypothesis, as well as from the cost of sampling, is left for more advanced texts in decision-making.

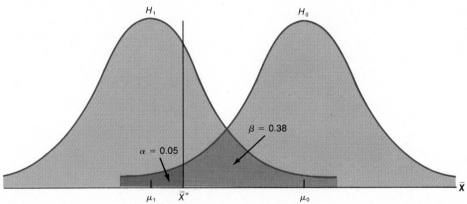

(a) For sample size n, the probabilities of error are $\alpha = 0.05$, $\beta = 0.38$.

Figure 10.10

The effect of sample size on α and β

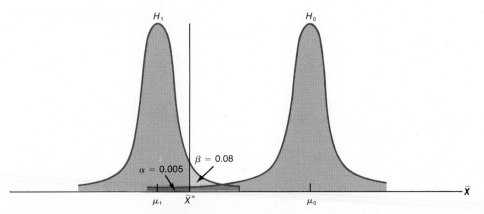

(b) For sample size n' $(n'>n)$, the probabilities of errors decrease to $\alpha = 0.005$, $\beta = 0.08$.

EXAMPLE 10.2

Consider testing $H_0: \mu = \mu_0 = 300$ against $H_1: \mu = \mu_1 = 310$, assuming that the standard deviation of X is equal to 30. At first, let us suppose that the significance level (and thus also the probability of committing Type I error) is equal to 5 percent: $\alpha = 0.05$. Determine the probability of committing Type II error under two alternative assumptions regarding sample size: $(a) n = 36$, $(b) n = 144$.

Assuming $n = 36$ first, we determine the critical value, $\overline{X}^*$, as follows:

$$\overline{X}^* = \mu_0 + Z_\alpha \frac{\sigma}{\sqrt{n}} = 300 + 1.645 \cdot \frac{30}{\sqrt{36}} = 308.23$$

Given this critical value, the probability of committing Type II error is easily calculated:

$$P(\text{Type II error}) \equiv \beta = P_{H_1}(\overline{X} \leq 308.23)$$

$$= P\left(\frac{\overline{X} - 310}{30/\sqrt{36}} \leq \frac{308.23 - 310.0}{30/\sqrt{36}} \right)$$

$$= P(Z \leq -0.35) = 0.3632$$

Assuming now that $n = 144$, our $\overline{X}^*$ is

$$\overline{X}^* = \mu_0 + Z_\alpha \frac{\sigma}{\sqrt{n}} = 300 + 1.645 \cdot \frac{30}{12} = 304.11$$

The probability of committing Type II error under this assumption is

$$\beta = P\left(\frac{\overline{X} - 310}{30/\sqrt{144}} \leq \frac{304.11 - 310.0}{30/\sqrt{144}} \right) = P(Z \leq -2.36) = 0.0091$$

Although in this example we only saw that β decreased as a result of an increase in sample size, it is quite possible to demonstrate a simultaneous reduction in both α and β. With a sample of 144 observations, we can decrease α from 5 percent to 1 percent and get a new critical value:

$$\overline{X}^* = \mu_0 + Z_\alpha \frac{\sigma}{\sqrt{n}} = 300 + \frac{30}{\sqrt{144}} \cdot 2.33 = 305.83$$

The probability of Type I error is thus reduced from 5 percent to 1 percent, and as we can see, the Type II error probability is also decreased from its original 36.32 percent all the way to 4.75 percent:

$$\beta = P\left(\frac{\overline{X} - 310}{30/\sqrt{144}} \leq \frac{305.83 - 310.0}{30/\sqrt{144}} \right) = P(Z \leq -1.67) = 0.0475$$

10.7 Testing Hypotheses Concerning the Mean of the Distribution When the Standard Deviation Is Unknown

So far in our hypothesis-testing procedures we have focused on cases in which the test concerns the mean of the distribution when the distribution's standard deviation is known. When the hypotheses concern the mean of the distribution and the standard deviation is unknown, we shall proceed as in Chapter 9 and use test statistics that have t distributions. We substitute S for σ in the test statistic to get $T = \dfrac{\overline{X} - \mu_0}{S/\sqrt{n}}$, where the S statistic is defined as:

$$S = \sqrt{\frac{\Sigma(X - \overline{X})^2}{n - 1}} = \sqrt{\frac{\Sigma X^2 - n\overline{X}^2}{n - 1}}$$

The test statistic T has a t distribution with $(n - 1)$ degrees of freedom. The critical values that determine the acceptance and rejection regions are therefore derived from the t distribution table. The prescribed steps for the testing of the population mean when σ is unknown can be summarized as follows:

PROCEDURE FOR TESTING HYPOTHESES CONCERNING THE MEAN WHEN THE STANDARD DEVIATION IS UNKNOWN

Step 1: Clearly identify the null and alternative hypotheses.
Step 2: Determine the significance level, α, that will be suitable for the test.
Step 3: Determine the critical value(s) and the decision rule. Step 3 varies slightly from one test to another, depending on whether we are dealing with an upper-tail, lower-tail, or two-tailed test.
    (*a*) For a lower-tail test, the critical value is $-t_\alpha^{(n-1)}$. Accept H_0 if the test statistic is greater than the critical value; otherwise reject.
    (*b*) For an upper-tail test, the critical value is $t_\alpha^{(n-1)}$. Accept H_0 if the test statistic is less than the critical value; otherwise reject.
    (*c*) For a two-tailed test, the critical values are $-t_{\alpha/2}^{(n-1)}$ and $t_{\alpha/2}^{(n-1)}$. Accept H_0 if the test statistic is between the two values; otherwise reject.
Step 4: Calculate the value of the test statistic by use of the sample data:
$$T = \frac{\overline{X} - \mu_0}{S/\sqrt{n}}.$$
Step 5: Apply the decision rule established in Step 3 and accept or reject H_0 accordingly.

EXAMPLE 10.3

The watches produced by Exact Time, Inc., are said to tell the correct time to within, on the average, 5 seconds per month. A random sample of 9 watches is observed for a month and the time discrepancy in seconds is as follows: 2, 7, 6, 1, 10, 10, 7, 5, 5. Assuming that the time discrepancy is normally distributed, would you accept or reject the company's claim? Assume a significance level of 5 percent.

Following the steps established for the testing of hypotheses concerning the distribution's mean when σ is unknown, we first identify the hypotheses involved:

$$H_0: \quad \mu \le 5$$
$$H_1: \quad \mu > 5$$

The significance level is 5 percent, so $\alpha = 0.05$. Since σ is unknown and 9 observations are available, our test statistic has a t distribution with $8 \,(= n - 1)$ degrees of freedom. Furthermore, since we are dealing with an upper-tail test, the critical value is $t_{0.05}^{(8)} = 1.86$. Our decision rule will therefore be: Accept H_0 if $T = \dfrac{\bar{X} - \mu_0}{S/\sqrt{n}} \le 1.86$ and reject H_0 if $T > 1.86$.

Our next step is to calculate the value of the test statistic. This is done in Table 10.1.

TABLE 10.1

Calculating the Value of the Test Statistic

i	X	X^2	*Calculation of test statistic, T*
1	2	4	
2	7	49	$\bar{X} = \dfrac{\Sigma X}{n} = \dfrac{53}{9} = 5.889$
3	6	36	
4	1	1	
5	10	100	$S = \sqrt{\dfrac{\Sigma X^2 - n\bar{X}^2}{n - 1}} = \sqrt{\dfrac{389 - 9 \cdot 5.889^2}{9 - 1}} = \sqrt{9.61} = 3.100$
6	10	100	
7	7	49	
8	5	25	$T = \dfrac{\bar{X} - \mu_0}{S/\sqrt{n}} = \dfrac{5.889 - 5.000}{3.100/\sqrt{9}} = \dfrac{0.889}{1.033} = 0.861$
9	5	25	
$\Sigma X = 53$	$\Sigma X^2 = 389$		

The last step is to determine whether the value of the test statistic falls within the acceptance or rejection region and to make our decision accordingly. Following the decision rule established for this problem, we ought to accept H_0, since 0.861 (the value of the test statistic) is less than 1.86 (the critical value). We thus conclude that while the sample shows an average time discrepancy greater than 5 seconds, the sample average is not significantly greater than 5 and the null hypothesis cannot be rejected.

We have already mentioned several times that when the number of observations available exceeds 30, the t distribution's shape approaches that of the standard normal distribution, and the various t scores closely approximate the corresponding Z scores. It is therefore conventional to use critical values taken from the Z distribution table when the number of observations exceeds 30 even when σ is unknown. For example, if 31 watches or more had been observed, we could have used $Z_{0.05} = 1.645$ for a critical value.

10.8 Tests of Hypotheses Concerning the Difference between the Means of Two Distributions

We shall often be interested in testing hypotheses concerning the difference between the means of two distributions. A special case of this type of hypothesis testing occurs when we wish to determine whether the two means are equal to or different from each other. Denoting the mean of one distribution by μ_1 and the mean of the other by μ_2, we state the null hypothesis as

$$H_0: \quad \mu_1 = \mu_2$$

or

$$H_0: \quad \mu_1 - \mu_2 = 0$$

It is convenient to denote the difference between the means by D, so that $D = \mu_1 - \mu_2$. Using this notation, we may write the null hypothesis as

$$H_0: \quad D = D_0 = 0$$

The alternative hypothesis, H_1, may take on any of three familiar forms:

$$H_1: \quad D = D_1 < 0 \qquad \text{(meaning } \mu_1 < \mu_2\text{; lower-tail test)}$$

or

$$H_1: \quad D = D_1 > 0 \qquad \text{(meaning } \mu_1 > \mu_2\text{; upper-tail test)}$$

or

$$H_1: \quad D = D_1 \neq 0 \qquad \text{(meaning } \mu_1 \neq \mu_2\text{; two-tailed test)}$$

We shall be addressing ourselves to the third form of H_1, assuming that you are capable of adjusting the relevant equations for the other two forms.

The only difference between tests involving the difference between two means and a test involving only one distribution's mean is the form of the test statistic and its distribution.

Since the sample average is our point estimate for a distribution's mean, it is only natural for us to estimate the difference $D = \mu_1 - \mu_2$ by using the difference of the averages of the samples taken from the two distributions under consideration. When we denote the sample average of the first and second distributions by $\overline{X}_1$ and $\overline{X}_2$, respectively, our point estimate for $D = \mu_1 - \mu_2$ is $\hat{D} = \overline{X}_1 - \overline{X}_2$, where $\hat{D}$, then, is the estimator of D. We note that the expected value of $\hat{D} = \overline{X}_1 - \overline{X}_2$ is equal to $D = \mu_1 - \mu_2$:

$$E(\hat{D}) = E(\overline{X}_1 - \overline{X}_2) = E(\overline{X}_1) - E(\overline{X}_2) = \mu_1 - \mu_2 = D \qquad \textbf{(10.10)}$$

Assuming that the observations from the two distributions are independent, we derive the variance of $\hat{D} = \overline{X}_1 - \overline{X}_2$ as follows:[2]

$$\operatorname{var}(\hat{D}) = \operatorname{var}(\overline{X}_1 - \overline{X}_2) = \operatorname{var}(\overline{X}_1) + \operatorname{var}(\overline{X}_2) = \frac{\sigma_1^2}{n_1} + \frac{\sigma_2^2}{n_2} \qquad \textbf{(10.11)}$$

---

[2] Formula 10.11 holds *only* when $\overline{X}_1$ and $\overline{X}_2$ are independent. Recalling the discussion of linear transformation of random variables (Chapter 6), we note that:

$$\begin{aligned} \operatorname{var}(\overline{X}_1 - \overline{X}_2) &= \operatorname{var}(\overline{X}_1) + \operatorname{var}(-\overline{X}_2) \\ &= \operatorname{var}(\overline{X}_1) + (-1)^2 \operatorname{var}(\overline{X}_2) \\ &= \operatorname{var}(\overline{X}_1) + \operatorname{var}(\overline{X}_2). \end{aligned}$$

where n_1 and n_2 are the sample sizes taken from the first and second distributions, respectively, and σ_1^2 and σ_2^2 are the variances of the first and second distributions, respectively. The standard deviation of $\hat{D} = \overline{X}_1 - \overline{X}_2$, $\sigma_{\hat{D}}$, is

$$\sigma_{\hat{D}} = \sigma_{\overline{X}_1 - \overline{X}_2} = \sqrt{\frac{\sigma_1^2}{n_1} + \frac{\sigma_2^2}{n_2}} \qquad \textbf{(10.12)}$$

When $\overline{X}_1$ and $\overline{X}_2$ are normally distributed, the difference $\hat{D} = \overline{X}_1 - \overline{X}_2$ is also normally distributed. Assuming that H_0 is correct, we get

$$\frac{\hat{D} - D_0}{\sigma_{\hat{D}}} \sim N(0,1) \qquad \textbf{(10.13)}$$

or

$$Z = \frac{\hat{D} - D_0}{\sigma_{\hat{D}}} \qquad \textbf{(10.14)}$$

If the variances σ_1^2 and σ_2^2 are unknown, then $\sigma_{\hat{D}}$ as given in Equation 10.13 is also unknown, and the test statistic Z from Equation 10.14 cannot be used to test the hypothesis. In this case, we use the t distribution. A more detailed explanation is provided following Example 10.4.

EXAMPLE 10.4

A department-store chain has opened two new stores in two locations. After a period of operations, management has decided to test whether the two stores have the same mean daily sales. Suppose the sales are normally distributed and the variances are known to equal 8,000 (dollars squared) for the first store ($\sigma_1^2 = 8{,}000$) and 10,000 (dollars squared) for the second store ($\sigma_2^2 = 10{,}000$). Suppose independent random samples of sizes 10 and 14 are taken from the first and second stores, respectively (that is, $n_1 = 10$ and $n_2 = 14$), and these samples produce averages of $\overline{X}_1 = \$15{,}000$ and $\overline{X}_2 = \$14{,}800$.

With these data we can use the test statistic given in Equation 10.14 to test hypotheses such as the following:

$$H_0: \quad D = 0$$
$$H_1: \quad D \neq 0$$

If we use $\alpha = 0.01$, our acceptance region is the range between $-Z_{0.005}$ and $Z_{0.005}$ (that is, the range from -2.57 to 2.57), and the rejection region is all other values of Z. In our example we find that

$$\sigma_{\hat{D}} = \sqrt{\frac{\sigma_1^2}{n_1} + \frac{\sigma_2^2}{n_2}} = \sqrt{\frac{8{,}000}{10} + \frac{10{,}000}{14}} = \sqrt{800 + 714} = 38.9$$

$$Z = \frac{\hat{D} - D_0}{\sigma_{\hat{D}}} = \frac{(15{,}000 - 14{,}800) - 0}{38.9} = \frac{200.0}{38.9} = 5.14$$

and the Z value we compute is well within the rejection region. We conclude, then, that the stores do not have the same sales volume. It is worth noting that our test suggests *different* sales volumes for the two stores: since in the sample we obtained $\overline{X}_1 > \overline{X}_2$, we are led to conclude that the first store has a greater sales volume.

Example 10.4 is largely hypothetical. It is very unlikely that a situation might arise in which the sales variances were known while the means were not; in reality, if the means are unknown, the variances are likely to be unknown as well. We want a statistic that will be useful in this kind of situation. Fortunately, if we can assume that the variances of the two distributions are equal, such a statistic is available:

$$T = \frac{(\overline{X}_1 - \overline{X}_2) - (\mu_1 - \mu_2)}{\sqrt{\left[\frac{(n_1 - 1)S_1^2 + (n_2 - 1)S_2^2}{(n_1 + n_2 - 2)}\right]\left(\frac{n_1 + n_2}{n_1 \cdot n_2}\right)}} \sim t^{(n_1 + n_2 - 2)} \qquad \textbf{(10.15)}$$

Where S_1^2 and S_2^2 are the estimated variances of the first and second samples respectively.

As shown in Equation 10.15, the statistic has a t distribution with degrees of freedom equal to $n_1 + n_2 - 2$.

EXAMPLE 10.5

Nan Rowell, a student of business administration, is deciding which of two processing-by-mail services to use to process her film. She considers two, Quickie and Fastie, which charge the same price and provide similar quality. Nan selects the one that provides faster service. Her records show the waiting period (in days) between the mailing of the film and the receipt of prints from the two firms:

Waiting period for prints (days)

Quickie	Fastie
4	5
8	6
10	6
10	4
9	7
8	

Is there a significant difference in the waiting period for the color prints between the two companies? Assume a 20 percent significance level and that the waiting period is normally distributed.

There is an obvious difference between the sample averages of the waiting periods. The difference, however, does not necessarily imply that one service is *significantly* faster than the other, and could very well result from sampling errors only. The hypotheses are

$$H_0: \quad D = \mu_1 - \mu_2 = 0$$
$$H_1: \quad D = \mu_1 - \mu_2 \neq 0$$

Since $n_1 = 6$, $n_2 = 5$, and $\alpha = 0.20$, our critical points are

$$-t_{0.10}^{(6+5-2)} = -t_{0.10}^{(9)} = -1.383$$

and

$$t_{0.10}^{(6+5-2)} = t_{0.10}^{(9)} = 1.383$$

The test statistic is

$$T = \frac{(\overline{X}_1 - \overline{X}_2) - (\mu_1 - \mu_2)}{\sqrt{\left[\frac{(n_1 - 1)S_1^2 + (n_2 - 1)S_2^2}{(n_1 + n_2 - 2)}\right]\left(\frac{n_1 + n_2}{n_1 \cdot n_2}\right)}}$$

If the value of the test statistic falls within the acceptance region—that is, between -1.383 and 1.383—Nan should accept the null hypothesis; otherwise she should reject it. The calculation of the value of the test statistic in the sample, assuming that H_0 is correct, is presented in Table 10.2, and is equal to 2.32. Since 2.32 is outside the range from -1.383 to 1.383, we conclude that we have 80 percent confidence that the two mean waiting periods are not the same and we reject H_0. Obviously, since $\overline{X}_1 = 8.167 > 5.600 = \overline{X}_2$, the result implies that the faster service is provided by Fastie.

TABLE 10.2
Calculation of the Value of the Test Statistic

i	X_1	X_2	X_1^2	X_2^2
1	4	5	16	25
2	8	6	64	36
3	10	6	100	36
4	10	4	100	16
5	9	7	81	49
6	8	—	64	—
	$\Sigma X_1 = 49$	$\Sigma X_2 = 28$	$\Sigma X_1^2 = 425$	$\Sigma X_2^2 = 162$

Calculation of the test statistic, T

$$\overline{X}_1 = \frac{\Sigma X_1}{n_1} = \frac{49}{6} = 8.167 \qquad S_1^2 = \frac{\Sigma X_1^2 - n_1 \overline{X}_1^2}{n_1 - 1} = \frac{425 - 6 \cdot 8.167^2}{5} = \frac{24.8}{5} = 4.96$$

$$\overline{X}_2 = \frac{\Sigma X_2}{n_2} = \frac{28}{5} = 5.600 \qquad S_2^2 = \frac{\Sigma X_2^2 - n_2 \overline{X}_2^2}{n_2 - 1} = \frac{162 - 5 \cdot 5.600^2}{4} = \frac{5.2}{4} = 1.3$$

$$T = \frac{(\overline{X}_1 - \overline{X}_2) - (\mu_1 - \mu_2)}{\sqrt{\left[\dfrac{(n_1 - 1)S_1^2 + (n_2 - 1)S_2^2}{(n_1 + n_2 - 2)}\right]\left(\dfrac{n_1 + n_2}{n_1 \cdot n_2}\right)}} = \frac{(8.167 - 5.600) - 0}{\sqrt{\left(\dfrac{5 \cdot 4.96 + 4 \cdot 1.30}{9}\right)\left(\dfrac{6 + 5}{6 \cdot 5}\right)}}$$

$$= \frac{2.5670}{1.1055} = 2.32$$

Finally, let us note that the null hypothesis does not necessarily have to state that the difference between the means is equal to zero. Any other hypothesized difference between μ_1 and μ_2 can be tested by the procedure established in this section. For example, if we want to test

$$H_0: \quad \mu_1 - \mu_2 = 1$$
$$H_1: \quad \mu_1 - \mu_2 \neq 1$$

the value of the test statistic (see Table 10.2) will be

$$\frac{(8.167 - 5.600) - 1.000}{1.1055} = \frac{1.5670}{1.1055} = 1.417$$

The critical values are still -1.383 and 1.383, and H_0 is rejected.

10.9 Testing Hypotheses Concerning a Single Proportion

In Section 10.1 we presented a test of a single proportion and used the binomial probability distribution. Here we shall be using the normal approximation to the binomial distribution.

In principle, all tests of hypotheses are alike in rationale and procedure. They differ only in the test statistic that is used, and in its distribution. Let us, then, consider the test statistic for the testing of proportions.

In our discussion of the binomial distribution in Chapter 7, we noted that the number of successes of a binomial experiment (X) may also be expressed as a proportion of successes if we just divide X by the number of trials in the experiment (n). Thus $\hat{p} \equiv \dfrac{X}{n}$ is the sample proportion of successes, which serves as a statistic for estimating the distribution's probability of success (p), as discussed in Chapter 9. We also recall that $E(\hat{p}) = p$, that the standard deviation of $\hat{p}$ is $\sigma_{\hat{p}} = \sqrt{\dfrac{pq}{n}}$, and that if both np and nq are greater than or equal to 5, $\hat{p}$ is approximately normally distributed. We shall assume throughout this section that np and nq meet this requirement.

Suppose the distribution's probability of success is unknown, and we wish to test hypotheses concerning its value. The null hypothesis will take the form

$$H_0: \quad p = p_0$$

The alternative hypothesis may take on one of three familiar forms:

$$H_1: \quad p = p_1 < p_0 \qquad \text{(lower-tail test)}$$

or

$$H_1: \quad p = p_1 > p_0 \qquad \text{(upper-tail test)}$$

or

$$H_1: \quad p = p_1 \neq p_0 \qquad \text{(two-tailed test)}$$

In testing any of these forms, we can proceed along one of two avenues: we may use either $\hat{p}$ or $\dfrac{\hat{p} - p_0}{\sigma_{\hat{p}}}$ as our test statistic. If we use $\hat{p}$ as our test statistic, the critical values (p^*) and the decision rules are

---

LOWER-TAIL TEST

$$p^* = p_0 - Z_\alpha \, \sigma_{\hat{p}}$$

Accept H_0 if $\hat{p} \geq p^*$; otherwise reject H_0.

UPPER-TAIL TEST

$$p^* = p_0 + Z_\alpha \, \sigma_{\hat{p}}$$

Accept H_0 if $\hat{p} \leq p^*$; otherwise reject H_0.

TWO-TAILED TEST

$$p_1^* = p_0 - Z_{\alpha/2}\, \sigma_{\hat{p}}$$

$$p_2^* = p_0 + Z_{\alpha/2}\, \sigma_{\hat{p}}$$

Accept H_0 if $p_1^* \leq \hat{p} \leq p_2^*$; otherwise reject H_0.

Note that in all three cases we obtain the value of $\sigma_{\hat{p}}$ by assuming that H_0 is correct and using the formula $\sigma_{\hat{p}} = \sqrt{\dfrac{p_0 q_0}{n}}$, where $q_0 = (1 - p_0)$.

EXAMPLE 10.6

The annexation to a city of an unincorporated area is on the ballot in an upcoming election. Voters may vote either for or against the annexation. A local newspaper is predicting that 60 percent of the votes will favor annexation and 40 percent will oppose it. A random sample of 100 people reveals that 58 are for annexation and 42 are against. Would you accept the newspaper's prediction? The hypotheses to be tested are

$$H_0:\quad p = 0.60$$
$$H_1:\quad p \neq 0.60$$

Since a significance level is not given in the problem, we have to choose one. Let us choose a significance level of 1 percent. Next we must choose the test statistic. Although only one statistic is necessary to reach a decision, we shall work out the problem with both $\hat{p}$ and $\dfrac{\hat{p} - p_0}{\sigma_{\hat{p}}}$ as test statistics. Note that $\sigma_{\hat{p}} = \sqrt{\dfrac{pq}{n}}$; assuming that H_0 is correct we get

$\sigma_{\hat{p}} = \sqrt{\dfrac{0.6 \cdot 0.4}{100}} = \sqrt{0.0024} = 0.049$. The derivation of the critical values, and the decision rules, are shown in Table 10.3.

TABLE 10.3
Critical Values and Decision Rules Derived for Two Alternative Test Statistics

Test statistic: $\hat{p}$	*Test statistic: $Z = \dfrac{\hat{p} - p_0}{\sigma_{\hat{p}}}$*
Critical values:	Critical values:
$p_1^* = p_0 - Z_{\alpha/2}\, \sigma_{\hat{p}}$	$Z_1^* = -Z_{\alpha/2} = -2.57$
$\quad = 0.6 - 2.57 \cdot 0.049 = 0.474$	$Z_2^* = Z_{\alpha/2} = 2.57$
$p_2^* = p_0 + Z_{\alpha/2}\, \sigma_{\hat{p}}$	
$\quad = 0.6 + 2.57 \cdot 0.049 = 0.726$	
Accept H_0 if $\hat{p}$ falls between 0.474 and 0.726; otherwise reject H_0.	Accept H_0 if test statistic falls between -2.57 and 2.57; otherwise reject H_0.

In our example we have $\hat{p} = \dfrac{58}{100} = 0.58$, so that $\hat{p}$ falls in the acceptance region and H_0 is not rejected. If we use the test statistic Z we calculate

$$Z = \frac{\hat{p} - p_0}{\sigma_{\hat{p}}} = \frac{0.58 - 0.60}{0.049} = -0.408$$

Again we do not reject H_0.

10.10 Testing Hypotheses Concerning the Difference between Two Proportions

We are sometimes interested in testing hypotheses concerning the difference between proportions. For example, we might wish to compare the proportions of credit-card holders in two population strata, or the proportions of adult population members owning at least $1,000 worth of common stock in two locations (such as urban versus rural areas). Suppose we hypothesize that the proportions in both populations are equal to each other and to some value p_0:

$$H_0: \quad p_1 = p_2 = p_0$$
$$H_1: \quad p_1 \neq p_2$$

Alternatively, we may define $\Delta p = p_1 - p_2$ and rewrite the hypotheses as follows:

$$H_0: \quad \Delta p = 0$$
$$H_1: \quad \Delta p \neq 0$$

As before, we shall use the sample proportion $\hat{p} = \dfrac{X}{n}$ as our estimate of the population proportion, p. Since in the current problem we have two populations, we will actually obtain two estimates of proportions ($\hat{p}_1 = \dfrac{X_1}{n_1}$ and $\hat{p}_2 = \dfrac{X_2}{n_2}$, where the subscripts 1 and 2 denote variables pertaining to the first and second distributions, respectively), one for each population, and denote the difference between the estimated proportion by $\widehat{\Delta p}$. The expected value of the sample difference between the proportions is

$$E(\widehat{\Delta p}) = E(\hat{p}_1 - \hat{p}_2) = E(\hat{p}_1) - E(\hat{p}_2) = p_1 - p_2 = \Delta p \qquad \textbf{(10.16)}$$

Assuming that the two samples are independent, so that $\hat{p}_1$ and $\hat{p}_2$ are independent as well, we get

$$\mathrm{var}(\widehat{\Delta p}) = \mathrm{var}(\hat{p}_1 - \hat{p}_2) = \mathrm{var}(\hat{p}_1) + \mathrm{var}(\hat{p}_2) = \frac{p_1 q_1}{n_1} + \frac{p_2 q_2}{n_2}$$

and

$$\sigma_{\widehat{\Delta p}} = \sqrt{\frac{p_1 q_1}{n_1} + \frac{p_2 q_2}{n_2}} \qquad \textbf{(10.17)}$$

If the null hypothesis is correct, then $p_1 = p_2 = p_0$ (implying that $q_1 = q_2 = q_0$) and

$$\left. \begin{array}{l} E(\widehat{\Delta p}) = \Delta p = 0 \\[2ex] \sigma_{\widehat{\Delta p}} = \sqrt{p_0 q_0 \left(\dfrac{1}{n_1} + \dfrac{1}{n_2} \right)} \end{array} \right\} \quad \text{Holds if } H_0 \text{ is correct} \tag{10.18}$$

The test statistic we can use to test H_0 against H_1 is

$$P = \frac{\widehat{\Delta p} - \Delta p}{\sigma_{\widehat{\Delta p}}} = \frac{(\hat{p}_1 - \hat{p}_2) - (p_1 - p_2)}{\sigma_{(\hat{p}_1 - \hat{p}_2)}} \tag{10.19}$$

If the proportion p_0 is not specified under H_0, then we evaluate $\sigma_{\widehat{\Delta p}}$ by using $\bar{p}$ instead of p_0, where $\bar{p}$ is the weighted mean of the observed sample proportion:

$$\bar{p} = \frac{n_1 \hat{p}_1 + n_2 \hat{p}_2}{n_1 + n_2} \tag{10.20}$$

For example, if we test $H_0: p_1 = p_2$ against $H_1: p_1 \neq p_2$ using $\alpha = 0.05$ and the first sample (with $n_1 = 6$) shows $\hat{p}_1 = 0.60$ and the second (with $n_2 = 9$) shows $\hat{p}_2 = 0.68$, we estimate $\bar{p}$ to be:

$$\bar{p} = \frac{6 \cdot 0.60 + 9 \cdot 0.68}{6 + 9} = 0.648$$

thus,

$$\sigma_{\widehat{\Delta p}} = \sqrt{(0.648)(1 - 0.648) \left(\frac{1}{6} + \frac{1}{9} \right)} = \sqrt{0.06336} = 0.252$$

and using Equation 10.19 we get:

$$P = \frac{\Delta \hat{p} - \Delta p}{\sigma_{\widehat{\Delta p}}} = \frac{(0.60 - 0.68) - 0}{0.252} = -0.317$$

Since $-0.317 < 1.96 = Z_{0.025}$ and also $-0.317 > -1.96 = -Z_{0.025}$, we cannot reject the null hypothesis which asserts that p_1 equals p_2.

10.11 Limitations

In this chapter we have presented the basic concepts of classical hypothesis-testing procedures along with a discussion of specific types of hypotheses and the ways they should be tested. We have indicated several times that some procedures are based on the *assumption* that the observations in our samples are drawn from normal distributions. The central limit theorem allows us to apply those procedures when the sample sizes are large and the observations are independent, even if the original distributions are other than normal.

One should keep in mind, however, that when these assumptions do not hold, and when the central limit theorem does not justify use of the procedures described in this chapter, one should take extra care in interpreting the test's results and consider using other statistical methods, such as nonparametric methods, which we discuss in Chapter 18. Much too often people use statistical methods without first ascertaining that the tests and procedures used are applicable. We urge you to bear these limitations in mind.

Appendix 10A:
Type II Error and the Power Function

The meaning of Type I error, Type II error, and the power of a test have been explained in Chapter 10. We saw that the power of a test, denoted by $1 - \beta$, is the probability of rejecting H_0 and accepting H_1 when H_1 is in fact correct.

THE POWER OF A ONE-TAILED TEST: AN EXAMPLE

Suppose that the average weekly wage in a large company in the past year was \$300. In wage negotiations, the union claims that the workers' wage is below the average wage of all workers in the same industry, while the firm's management claims that the opposite is true. Assume that it is known that the wage is normally distributed with a standard deviation of $\sigma = \$120$.

To test the workers' claim, the following hypotheses need to be tested:

$$H_0: \quad \mu = \mu_0 = \$300$$
$$H_1: \quad \mu = \mu_1 > \$300$$

since they claim that the mean industry income is higher than their average wage.

The hypothesis H_1 is one-sided and composite. It is composite because under H_1 no single value of income is specified for μ_1. The alternative H_1: $\mu > \$300$, for example, means that μ can take any value above \$300. For a given significance level, the power of the test is a function—called the **power function**—of the specific value of μ under H_1. Since μ is not specified by a single value, a distinct power is obtained for each value assumed for μ.

In order to calculate the power, assume that a sample of 64 observations is taken from the population of all workers in the industry and that we choose a significance level $\alpha = 0.01$. We determine the critical value

$$\overline{X}^* = \mu_0 + \frac{\sigma}{\sqrt{n}} Z_{0.01}$$

$$= 300 + \frac{120}{\sqrt{64}} \cdot 2.33 = \$334.95$$

and momentarily change the alternative hypothesis, H_1, so that μ is specified as a single value. Suppose $H_1: \mu = \$350$ is now the alternative hypothesis. What is the power of the test? The power is given by P (rejection of H_0 under the condition that H_1 is correct), that is,

$$P(\overline{X} > 334.95 \mid H_1 \text{ is correct}) \equiv P_{H_1}(\overline{X} > 334.95)$$

or

$$P_{H_1}\left(\frac{\overline{X} - \mu}{\sigma/\sqrt{n}} > \frac{334.95 - 350}{120/\sqrt{64}}\right) = P\left(Z > \frac{-15.05}{15}\right) = P(Z > -1.00) = 0.8413$$

Thus for H_1: $\mu = \$350$, the power is 84.13 percent. Note that in the calculation we use $\mu = \$350$, that is, our mean wage under the alternative hypothesis. Suppose now that the H_1 is given by H_1: $\mu = \$320$. Through similar calculations we derive the power

$$P_{H_1}\left(\frac{\overline{X} - \mu}{\sigma/\sqrt{n}} > \frac{334.95 - 320}{15}\right) = P(Z > 0.997) = 0.1587$$

Thus the power diminishes when μ_1 is shifted toward μ_0.

Graphically, the power of the test is measured by the area under the alternative hypothesis (H_1) curve over the rejection region. Since that area changes with the location of the curve's mean, so does the power of the test. We can see in Figure 10A.1 that as μ_1 approaches μ_0, the area that represents the power decreases. Indeed, the closer μ_1 is to μ_0, the more difficult it becomes to favor μ_1 rather than μ_0 on the basis of a sample average that is subject to random variations.

When μ_1 is close to $\$300$, the power of the test is very poor, meaning that the workers have very little chance to prove that their claim is correct even if in fact it is. What can the workers do in such a situation? They are almost sure that the average weekly pay in the industry is $\$320$, but they cannot prove it. One way to get out of the trap is to finance a larger sample. On the basis of a sample of 10,000 observations, for example, it can be easily and

Figure 10A.1

The power of the test under various assumptions concerning the value of the mean by alternative hypothesis H_1

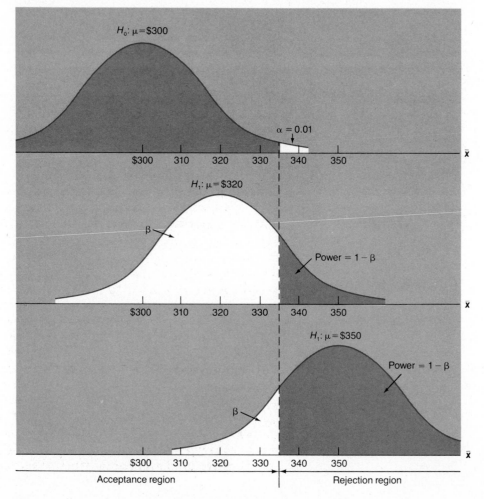

powerfully determined whether H_0 or H_1 should be favored. Assuming H_0: $\mu = \mu_0 = \$300$ and H_1: $\mu = \mu_1 = \$320$, we first determine the rejection area. We reject the null hypothesis (with $\alpha = 1$ percent as before) if

$$\frac{\overline{X} - \mu_0}{\sigma/\sqrt{n}} = \frac{\overline{X} - 300}{250/\sqrt{10,000}} = \frac{\overline{X} - 300}{2.5} > Z_{0.01} = 2.33$$

or if

$$\overline{X} > 300 + 2.33 \cdot 2.5 = 305.83$$

Thus the power is given by

$$P_{H_1}\left(\frac{\overline{X} - \mu_1}{\sigma/\sqrt{n}} > \frac{305.83 - 320}{2.5}\right) = P_{H_1}(Z > -5.7) \cong 1.00$$

It seems that the union should spend some money on sampling, since the larger sample size increases the power of the test from about 4 percent to almost 100 percent! If the workers are right and indeed their wage is below the average in the industry, the cost of sampling will probably pay off.

To summarize the one-tailed test, Figure 10A.2 illustrates the power function, as well as the probability of committing a Type II error, β. As μ_1 slides to the right, the power increases and β decreases.

Figure 10A.3 illustrates the power function of the test and the probability of Type II error in the case of a lower-tail test where the hypotheses are:

$$H_0: \quad \mu = \mu_0 = \$300$$

$$H_1: \quad \mu = \mu_1 < \$300$$

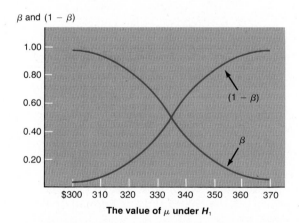

Figure 10A.2

The power function and the probability of committing a Type II error for a sample size of 64 observations and a significance level, α, of 1 percent: the upper-tail test

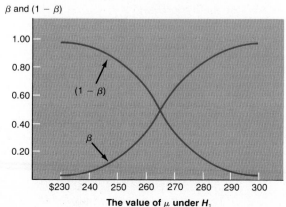

Figure 10A.3

The power function and the probability of committing a Type II error for a sample size of 64 observations and a significance level, α, of 1 percent: the lower-tail test

Problems

10.1. Explain the following terms:

   (*a*) Acceptance region.
   (*b*) Rejection region.
   (*c*) Significance level.
   (*d*) Type I error.
   (*e*) Type II error.

10.2. Use a graphic presentation to explain and evaluate the following statements:

   (*a*) "The probability of committing Type I error is always smaller than that of committing Type II error."
   (*b*) "It is absolutely impossible to have a statistical test and choose between hypotheses without running the risk of committing either Type I or Type II error."
   (*c*) "No matter what the alternative hypothesis, a two-tailed test is always better than a one-tailed test."
   (*d*) "When testing hypotheses we should pay very little attention to the power of the test. Rather, we should concentrate on minimizing the probability of committing Type II error for a given level of significance."

10.3. Evaluate and graphically explain the following statement:

   When $H_0: \mu = \mu_0$ is tested against $H_1: \mu > \mu_0$, the power of the test is always no smaller than the significance level, α."

10.4. "The testing of hypotheses with a sample of an infinite number of observations is characterized by zero probability of Type I and Type II errors." Do you agree with this statement? In your answer distinguish among:

   (*a*) The testing of hypotheses concerning one mean.
   (*b*) The testing of hypotheses concerning two means when one sample is infinite and the other is finite.
   (*c*) The testing of hypotheses concerning two means when both samples are infinite.

10.5. "When one decides to reject a hypothesis and accept another, at least one type of error must be committed." Do you agree? Explain.

10.6. "While we do not know ahead of time if an error will be committed, it becomes apparent after the decision is made whether an error has been committed or not." Do you agree? Explain.

10.7. Suppose we test the hypothesis $H_0: \mu = 10$ against $H_1: \mu > 10$. If $n = 10, \overline{X} = 10.1$, and $S = 0$, what is the lowest significance level under which H_0 would be rejected?

10.8. The weekly income of the residents of the city of Buka is normally distributed: $X \sim N(\mu, \sigma)$, where σ is known to equal 10. A sample of 100 residents is chosen (i.e., $n = 100$) and their income is measured. Suppose we test the hypothesis

$$H_0: \mu = \$200.00$$

against

$$H_1: \mu = \$202.00$$

   (*a*) If a significance level $\alpha = 0.05$ is used to test the above hypothesis, what is the critical value $\overline{X}^*$ such that if the observed sample average, $\overline{X}$, is greater than $\overline{X}^*$, H_0 will be rejected?
   (*b*) Calculate the probability of committing Type II error. Are you satisfied with this test? What is the probability of accepting H_1 when indeed $\mu = \$202.00$?
   (*c*) Suppose that in order to avoid such a high probability of Type II error, we decide to

increase α. What should α be in order that the probability of Type II error will be 20 percent?

(d) As we have seen, the initial level of the probability of Type II error is very high. Reducing this error involves an increase in the probability of Type I error. Suppose now that you want neither Type I error nor Type II error to have a probability exceeding 5 percent. What is the minimum sample size, n, that will achieve this goal?

(e) Illustrate your answer to parts a, b, c, and d on a chart.

10.9. A certain plant runs a machine whose daily output is a random variable (X) and whose distribution is normal with expected value 600 and standard deviation of 91: $X \sim N(600, 91)$. A suggestion has been made that the machine be replaced by another whose output is also normally distributed, with a standard deviation of 91 but a higher mean. In order to test whether the mean is indeed higher, the new machine's daily output is checked for 40 days. The 40 days' output provides the sample data for testing whether the new machine has a higher average production rate. Assume $\alpha = 0.05$.

(a) What is the rejection region when the alternative hypothesis is that the output of the new machine is higher?

(b) What is the rejection region when the alternative hypothesis is that the output of the new machine is different from that of the old machine?

(c) What is the power of the test if the alternative hypothesis is H_1: $\mu = 635$?

(d) What is the power of the test if the alternative hypothesis is H_1: $\mu = 560$?

10.10. The monthly output of a plant producing plywood was measured in eight randomly selected months. The results obtained (in tons) are 110, 120, 100, 102, 130, 140, 150, and 140. Test the hypothesis that the average monthly output of the plant is 140 tons against the alternative hypothesis that the average monthly output is 120 tons. Assume that output is a random variable (X) with a normal distribution, and choose a significance level $\alpha = 0.05$.

10.11. The price-earnings ratio (P/E) measures the number of times greater than the earnings per share (E) a share's price (P) is. Table P10.11 presents the price-earnings ratio for firms in the container industry. Assume that the companies shown are randomly selected from an infinitely large industry and the population is normally distributed (of course, this is only an approximation), and test the following hypotheses:

$$H_0: \mu = 10.0$$

against

$$H_1: \mu < 10.0$$

Use a significance level $\alpha = 0.05$, and assume that σ is known to equal 3.0.

TABLE P10.11

Firm	Price–earnings ratio
Container manufacturers	
American Can	10
Anchor Hocking	7
Ball	6
Brockway Glass	8
Continental Group	6
Crown Cork & Seal	7
Dorsey	9
Kerr Glass Mfg.	8
National Can	4
Owens-Illinois	6
Stone Container	9

TABLE P10.11 (continued)

Firm	Price–earnings ratio
Drug companies	
Abbott Laboratories	16
American Home Products	11
American Hospital Supply	15
Baxter Travenol Laboratories	15
Becton, Dickinson	14
Johnson & Johnson	16
Lilly (Eli)	14
Mallinckrodt	13
Merck	16
Pfizer	15
Richardson-Merrell	9
Robins (A. H.)	11
Schering-Plough	9
Searle (G. D.)	15
Smith Kline	17
Squibb	14
Sterling Drug	12
Syntex	12
Upjohn	11

Data from *Business Week*, May 18, 1981, pp. 80–83.

10.12. A common measure for a firm's liquidity is the "current ratio"—the ratio of current assets (i.e., cash and other assets that can be easily liquidated) to current liabilities (i.e., debts that the firm must pay off within one year's time). Table P10.12 presents the 1980 current ratio for 11 companies in the food and lodgings industry.

(a) Assuming that the current ratio is normally distributed with $\sigma = 0.5$, and that the 11 companies are a random sample of an infinite population, test the hypotheses

$$H_0: \mu = 0.9$$

against

$$H_1: \mu = 1.2$$

Use a significance level $\alpha = 0.05$.

(b) What is the power of the above test?

(c) Assume now that σ is unknown and retest the hypotheses. Compare your results to those of part *a*.

TABLE P10.12

The Current Ratio for 11 Companies of the Food and Lodgings Industry, 1980

ARA Services	1.2
Caesars World	1.2
Denny's	1.0
Hilton Hotels	2.3
Holiday Inns	0.9
Marriott	1.1
McDonald's	0.7
Ramada Inns	0.9
Saga	1.0
Sambo's Restaurants	0.4
Webb (Del E.)	1.6

Data from *Business Week*, October 13, 1980, p. 78.

10.13. Test the hypothesis that the mean current ratio of the food and lodgings industry (shown in Table P10.12) is equal to the mean current ratio of the instruments industry (shown in Table P10.13). Use $\alpha = 0.05$ and make the following two alternative assumptions:

    (*a*) The variances of the current ratio are known to be $\sigma = 0.5$ for the food and lodgings industry, and $\sigma = 1.0$ for the instruments industry.

    (*b*) The variances are unknown, but it is known that they are equal to one another.

TABLE P10.13

The Current Ratio for 12 Companies of the Instruments Industry, 1980

Ametek	3.0
Bausch & Lomb	2.5
Beckman Instruments	2.9
Foxboro	3.5
General Signal	2.3
Itek	2.5
Johnson Controls	1.8
Perkin-Elmer	3.0
Robertshaw Controls	2.5
Sybron	2.4
Talley Industries	2.8
Tektronix	2.8

Data from *Business Week*, October 13, 1980, p. 80.

10.14. Even though it was the oil-producing countries that quadrupled prices in 1973, many Americans believed that the real villain in the case was the oil industry. Table P10.14 shows the 1980 return on equity (i.e., the percentage of profit earned by shareholders on their invested capital) of eight oil companies. It also shows that the average for all manufacturing companies was 12.9 percent. Use $\alpha = 0.01$.

    (*a*) Assuming that the eight companies represent a sample taken from an infinite population, test the hypothesis that the average rate of return on the equity of oil companies is 12.9 percent against the alternative that it is less than 12.9 percent.

    (*b*) Repeat part *a*, this time assuming that the eight firms were selected at random from a population of 30 firms.

TABLE P10.14

1980 Return on Equity of Eight Oil Companies

Company	Revenues* (billions of dollars)	Profits (billions of dollars)	Profit margin (percent)	Return on stockholders' equity (percent)
Exxon	$110.4	$5.6	4.5	24.3
Mobil	62.8	2.8	3.3	24.7
Texaco	52.5	2.2	3.6	20.0
Standard Oil of California	42.9	2.4	5.4	24.6
Gulf Oil	28.4	1.4	4.3	15.8
Standard Oil of Indiana	27.8	1.9	5.4	21.9
Atlantic Richfield	24.2	1.7	5.4	25.3
Shell Oil	20.0	1.5	8.0	21.1
Average for all manufacturing companies	—	—	4.9	12.9

* Excludes excise and sales taxes.

Source: *Business Week*, March 16, 1981, and May 4, 1981.

10.15. In order to increase efficiency and facilitate performance evaluation, corporations have executive-level managers responsible for cost minimization and other such managers responsible for profit maximization. A hypothetical study designed to examine variations in opportunity for independent thought among such executives included 46 executives, 23 of each type. They were asked to indicate their feelings concerning their opportunities for independent thought on a seven-point response scale, where 1 indicated "extremely dissatisfied" and 7 "extremely satisfied." Hence, a high score indicated a high degree of satisfaction with respect to the opportunity for independent thought. The sample results are shown in Table P10.15. Test the hypothesis that the mean scores of both types of executive manager are the same. Use $\alpha = 0.05$. Specify the assumptions needed for the test.

TABLE P10.15

Cost managers	Profit managers
5.63	5.28
5.25	6.00
5.79	6.01
5.44	5.79
5.93	5.72
5.31	6.32
5.31	5.28
4.75	5.24
5.54	5.83
5.78	6.18
5.97	5.94
5.42	6.22
5.97	6.26
5.97	6.15
5.75	6.17
5.61	6.20
5.90	6.27
5.89	6.09
5.88	5.89
5.49	4.80
5.17	6.30
4.72	5.02
4.76	5.03

10.16. During the planning stages of a new airline route, management assumed that the variance of monthly passenger traffic would be 30 (squared thousands of passengers). The number of monthly passengers (in thousands) during the first fifteen months was as follows: 55, 51, 49, 50, 48, 54, 52, 54, 54, 55, 55, 56, 46, 47, 48. Assuming that the number of passengers is normally distributed, test the hypothesis that the variance of the monthly passenger traffic is 30, at $\alpha = 0.05$, against the alternative that the variance is other than 30.

10.17. Use Table P10.11 to test the hypothesis that the mean P/E ratio in the container industry is equal to that of the drug industry. Use $\alpha = 0.05$ and a two-tailed test. What assumptions do you need to make to carry out your test?

10.18. The California Citrus Groves Corporation wished to compare the efficiency of two fertilization treatments for raising yields. It chose 10 trees to undergo Treatment *A* and 6 trees to undergo Treatment *B*. The trees' yields (in tens of pounds) was as follows:

Treatment *A:* 80, 82, 84, 79, 77, 80, 78, 76, 83, 85
Treatment *B:* 80, 78, 74, 82, 79, 80

Assume that the weight of the yield is a random variable having a normal distribution, and test whether, on the average, the two fertilization treatments give the same yields. Assume identical variances of the weights under both treatments. Use $\alpha = 0.05$.

10.19. Investor A has invested in 5 stocks that were randomly selected from those traded on the New York Stock Exchange. Investor B has invested in 5 bonds. Assume that the rates of return on stocks and on bonds are normally distributed. Here are the rates of return on those securities in 1982 (in percent):

Stocks: 4.0, 3.0, 2.0, 5.0, 6.0
Bonds: 2.0, 6.0, 2.0, 2.0, 6.0

Assume that the variance of the rate of return on stocks is equal to that on bonds, and test the null hypothesis that the mean rate of return on stocks is equal to the mean rate of return on bonds. Use $\alpha = 0.05$.

10.20. Suppose we have two samples with equal numbers of observations ($n_1 = n_2 = n_0$), and when we test the hypothesis

$$H_0: \mu_1 = \mu_2$$

we find that the estimated variances are equal ($S_1^2 = S_2^2 = S_0^2$). Show how the t statistic of Equation 10.15 in the text can be simplified in this specific case.

CHAPTER ELEVEN OUTLINE

Key Terms
chi-square distribution
two-by-two (2 × 2)
  contingency table
random walk theory

11

CHI-SQUARE: TESTS FOR INDEPENDENCE AND GOODNESS OF FIT

In this chapter we deal with two types of statistical tests: tests for independence of variables and tests of goodness of fit. The test statistics of these tests follow the **chi-square distribution,** so we first turn to the description of this distribution.

11.1 The Chi-Square Distribution

The chi-square distribution is an important continuous probability distribution with wide applications. Given a normally distributed random variable $X \sim N(\mu, \sigma)$, the *probability distribution of the variable from its mean* can be analyzed by the chi-square distribution. Also, the sum of the squared deviations of each of a number of independent normal variables from their respective means can be analyzed by the chi-square distribution.

The shape of the chi-square function depends on one parameter alone. The parameter that specifies the chi-square distribution is known as the *degrees of freedom* of the distribution (often denoted df), and it represents the number of random variables inherent in the chi-square variable.[1] A chi-square variable with n degrees of freedom is denoted by $\chi^{2(n)}$. As the chi-square variable is

---

[1] If Z is a standard normal variable, then Z^2 has a chi-square distribution with *one* degree of freedom. If Z_1 and Z_2 are two *independent* standard normal variables, then the variable $Z_1^2 + Z_2^2$ has a chi-square distribution with two degrees of freedom. Denoting a chi-square variable with n degrees of freedom by $\chi^{2(n)}$, we may write

$$\chi^{2(n)} = Z_1^2 + Z_2^2 + \cdots + Z_n^2 = \sum_{i=1}^{n} Z_i^2$$

where all the Z variables are *independent* standard normal variables.

squared, it can assume only positive values, and thus it is located over the positive range of real numbers. The distribution is unimodal and skewed to the right. Figure 11.1 depicts some chi-square distributions. As one can easily see, the higher the number of degrees of freedom, the farther the probability distribution stretches toward the right, and the more symmetrical it becomes.

As in the case of the normal distribution, a $\chi^{2(n)}$ symbol with a subscript α ($\chi_\alpha^{2(n)}$) indicates the $\chi^{2(n)}$ value (or score) that leaves the right-tail area equal to α. For example, $\chi_{0.05}^{2(10)}$ is that value which bounds a 5 percent right-hand tail under the $\chi^{2(10)}$ distribution. Since the total area under any chi-square distribution is equal to 1, it follows that the area under the curve to the left of $\chi_\alpha^{2(n)}$ must equal $1 - \alpha$. Figure 11.2 illustrates the location of another chi-square score: $\chi_{0.10}^{2(8)}$. It equals 13.362, and on the $\chi^{2(8)}$ distribution, it bounds a 10 percent right-tail area.

USING THE CHI-SQUARE TABLE

Since the chi-square distribution is a function of its degrees of freedom, it is impractical to provide a separate table comparable to the normal distribution table for each chi-square distribution. Instead, only a limited number of more frequently used chi-square scores are provided for each distribution. The chi-square table is presented in Table A.4, Appendix A, and a schematic description of it is provided in Figure 11.3. When we want to find a given chi-square score, we first go down the left-hand column (headed *df*, for "Degrees of freedom") to the desired degrees of freedom and then

Figure 11.1

Some chi-square density functions

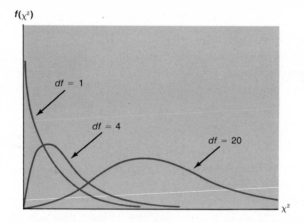

Figure 11.2

A chi-square distribution: the location of $\chi_{0.10}^{2(8)}$

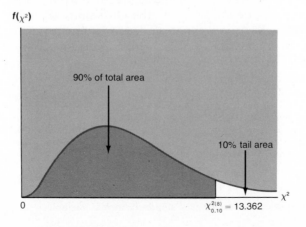

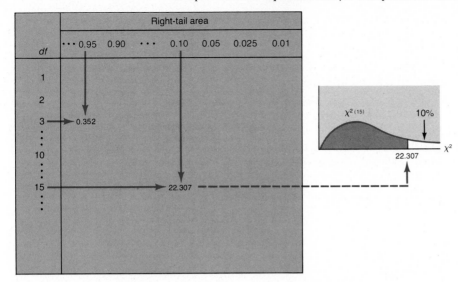

Figure 11.3

Chi-square distribution table

move across the table horizontally to the right until we reach the column denoting the desired right-tail area. For example, if we are looking for the chi-square score that leaves a right-tail area of 10 percent under the $\chi^{2(15)}$ distribution, we go down the left-hand column to "15" (see Figure 11.3), then cross to the right column headed "0.10." The number indicated in the table, 22.307, is the chi-square value that leaves a 10 percent tail on the right under $\chi^{2(15)}$. To find a chi-square value that leaves a given left-tail area, we follow across the table to the column denoting the complement (to 1) of that area. For example, to find the $\chi^{2(3)}$ value that delimits a 5 percent *left-tail* area, we go down the left-hand column to the appropriate degrees-of-freedom row and then move to the right to the column headed "0.95" (since 0.95 is the complement of 0.05 for 1.0, and a left-tail area of 5 percent is complemented by an area of 95 percent on the right). In the case at hand, the chi-square value is equal to 0.352, as illustrated in Figure 11.3. We urge you to verify the following values:

$$\chi^{2(8)}_{0.01} = 20.090$$

$$\chi^{2(12)}_{0.05} = 21.026$$

$$\chi^{2(30)}_{0.10} = 40.256$$

$$\chi^{2(20)}_{0.95} = 10.851$$

$$\chi^{2(20)}_{0.05} = 31.410$$

$$\chi^{2(3)}_{0.95} = 0.352$$

11.2 Chi-Square Test for Independence

The question of whether two variables are dependent or independent often has significant implications in business. For example, it may be important to find out if the type of community (say, urban versus rural) and the media used for advertising in it (TV, radio, newspapers) are independent in terms of the effectiveness of their advertising. It is important for a firm's management to know about this type of dependence or independence, since such knowledge can help in the planning of an effective advertising campaign.

Take another example. The Watchdog Company is a wholesaler of burglar and fire-alarm systems for private homes and public buildings. Watchdog operates in 31 states and sells to numerous retailers. Most of Watchdog's sales are made on credit and the company uses a credit-rating service to obtain updated information about the financial status of existing and potential customers. The credit-rating service classifies customers into the categories very good, good, average, and poor. In reviewing their records, Watchdog's accountants suspect that there is little relationship between these classifications and customers' bill-paying behavior. Since the credit information provided to Watchdog is not costless, Watchdog wants to find out whether the information is in fact valuable. One solution is to test for independence between the categories provided by the credit service and the actual credit standing of Watchdog's customers. If a statistical test substantiates independence between the two variables, the company may very well consider terminating its employment of the credit-rating service.

THE TWO-BY-TWO CONTINGENCY TABLE AND THE TEST HYPOTHESES

Table 11.1 presents what is known as a **two-by-two (2 × 2) contingency table,** which we will use to explain the test for independence. The following notation will be used in our discussion:

i Index of row categories. Here, i equals either 1 or 2. We will use I as a general notation for the largest value of i, so that for Table 11.1, $I = 2$.

j Index of column categories. Here, j equals either 1 or 2. J will be used as a general notation for the largest value of j, so that for Table 11.1, $J = 2$.

o_{ij} The observed sample frequency in cell i,j.

e_{ij} The expected sample frequency in cell i,j.

r_i Sum of cell frequencies in row i.

c_j Sum of cell frequencies in column j.

n Total number of sample observations: $n = \sum_i r_i = \sum_j c_j$.

The objective of the chi-square test is to determine whether it is likely or unlikely that the two sets of classifications (that is, the row and column classifications) are independent. In fact, we want to test the null hypothesis

TABLE 11.1
Two-by-Two Contingency Table

i	j		
	$j = 1$	$j = 2$	*Total*
$i = 1$	o_{11}	o_{12}	r_1
$i = 2$	o_{21}	o_{22}	r_2
Total	c_1	c_2	n

that the two variables (used for the classification) are independent, and the alternative hypothesis that the two variables are dependent. In short,

H_0: the two variables are independent

H_1: the two variables are dependent

In virtually all the hypothesis-testing procedures presented in Chapter 10, acceptance or rejection of H_0 depends on how the *actually observed* sample value compares with what is *expected to be observed* in the sample under H_0. For example, when we test the hypothesis H_0: $\mu = \mu_0 = 100$ versus H_1: $\mu = \mu_1 = 100$, we basically compare the sample average $\overline{X}$ to its hypothesized value under H_0. For any given sample size, it is the deviation of $\overline{X}$ from μ_0 that determines whether H_0 is accepted or rejected. When σ and n are given, the value of the test statistic $\dfrac{\overline{X} - \mu_0}{\sigma/\sqrt{n}}$ depends on the deviation $\overline{X} - \mu_0$. In the test for independence of two variables, we also have to compare the observed sample value with the value expected if H_0 is correct. If the deviation between the observed and expected frequencies is significantly large, we reject H_0. Otherwise H_0 is not rejected.

We must now take two steps. First, we determine the expected frequencies in the contingency table that would appear *if H_0 were correct*. These expected frequencies will provide the basis for comparison in the second step. Second, given the expected as well as observed frequencies, we determine whether the deviations of the observed from the expected are small enough to be interpreted as mere sampling variations or large enough to lead us to the conclusion that H_0 should be rejected.

DETERMINING THE EXPECTED FREQUENCIES

Let us now take the first step and determine the expected frequencies. Recall our discussion of independent events from Chapter 5. We defined statistical independence between events A and B as a situation in which the probability of the intersection of the two events is equal to the product of the two probabilities: if A and B are independent, then $P(A \cap B) = P(A)P(B)$.

In terms of our contingency table, each cell may be viewed as an intersection of its respective row and column. Thus, if H_0 is correct—that is, if the row and column classifications are independent—then the probability that an observation will belong to a given cell is the product of the probabilities that it will belong to the respective row and column. The probability that an observation will belong to row i (P_i) can be *estimated* by the following sample proportion:

$$\hat{P}_i = \frac{r_i}{n} \tag{11.1}$$

Similarly the probability that an observation will belong to column j (P_j) may be *estimated* by this sample proportion:

$$\hat{P}_j = \frac{c_j}{n} \tag{11.2}$$

If H_0 is correct, then the probability that any observation will belong to cell ij (that is, P_{ij}) may be *estimated* in the following way:

$$\hat{P}_{ij} = \hat{P}_i \cdot \hat{P}_j \tag{11.3}$$

or

$$\hat{P}_{ij} = \left(\frac{r_i}{n}\right)\left(\frac{c_j}{n}\right) \tag{11.4}$$

To estimate the expected *frequency* in each cell—that is, e_{ij}—we simply multiply $\hat{P}_{ij}$ by the total number of observations in the sample (n), which yields

$$e_{ij} = \hat{P}_{ij} \cdot n = \left(\frac{r_i}{n}\right)\left(\frac{c_j}{n}\right)n \tag{11.5}$$

This can be simply stated in the following way:

$$e_{ij} = \frac{r_i c_j}{n} \tag{11.6}$$

THE TEST STATISTIC

The observed and expected frequencies must now be compared throughout the table's cells and a decision rule must be established to allow acceptance or rejection of the null hypothesis. Before doing this, however, we must make sure that the *expected* frequency in each cell is at least 5. If this is not the case, cells must be grouped together so that the redefined cells will have at least 5 expected observations in each. The reason for this is that the test statistic we present below is approximately chi-square distributed. The approximation is not good enough if there are not at least 5 expected observations in each cell.

The test statistic to be used is the sum over all the table's cells of the expression $(e_{ij} - o_{ij})^2/e_{ij}$. Denoting the test statistic by χ^2 and ignoring indices for simplification, we can write the following equation:

$$\chi^2 = \Sigma \frac{(e - o)^2}{e} \tag{11.7}$$

If the sample observed frequencies are precisely equal to the expected frequencies in all the table's cells, we have a "perfect fit" situation and the statistic in Equation 11.7 is equal to zero. The greater the deviations (in either direction) of the observed frequencies, o, from the expected frequencies, e, the greater the value of the test statistic, χ^2. (Thus the test is always an upper-tail test, never lower-tail or two-tailed.) The test statistic approximates the χ^2 distribution, and the larger the sample, the better the approximation.

DEGREES OF FREEDOM OF THE CHI-SQUARE DISTRIBUTION

We have yet to determine the number of degrees of freedom of the χ^2 distribution. Given I rows and J columns, there are IJ cells altogether. Suppose the joint probabilities P_{ij} are somehow known and need not be estimated. In this case the number of degrees of freedom of our test statistic is $IJ - 1$. Why is one degree of freedom lost? Given any sample size, n, it is possible to determine only the frequencies of the number of cells minus one. The frequency of one of the cells can be computed by subtracting the total of the frequencies of all other cells from n. Thus $IJ - 1$ frequencies can be determined independently, but once they are determined, the last expected frequency is left with no degree of freedom—it must equal n minus the sum of all the other frequencies.

In most cases, we determine e_{ij} by first estimating the joint probability P_{ij}, and then applying Equation 11.5. To get an estimate of P_{ij}, however, we must estimate the marginal probabilities of P_i and P_j, as found in Equation 11.3. There are I different P_i's and J different P_j's to estimate, but once $I - 1$ of the P_i's are estimated, the last P_i is by necessity the complement of their sum to one. Similarly, once $J - 1$ of the P_j's are estimated, the last P_j is by necessity the complement of their sum to one. As a rule, each estimated parameter absorbs one degree of freedom, so that by estimating all the P_i's and P_j's, we "lose" $(I - 1) + (J - 1)$ degrees of freedom. We can now summarize as follows: if the joint probabilities are known on the basis of outside information and need not be estimated, the number of degrees of freedom for the test statistic of a contingency table is equal to $IJ - 1$. If, however, the various P_{ij}'s have to be estimated to arrive at the e_{ij} values, as they usually must, then the degrees of freedom are

$$(IJ - 1) - [(I - 1) + (J - 1)] = IJ - 1 - I + 1 - J + 1$$

$$= IJ - I - J + 1 = (I - 1) \cdot (J - 1)$$

This can be clearly stated as follows:

$$\text{The number of degrees of freedom of } \chi^2 \text{ is} \begin{cases} IJ - 1 & \text{if all } P_{ij}\text{'s are known in advance} \\ (I - 1) \cdot (J - 1) & \text{if all } P_{ij}\text{'s are to be estimated} \end{cases} \tag{11.8}$$

This can be stated more generally as:

$$\begin{pmatrix} \text{The number of degrees} \\ \text{of freedom of } \chi^2 \end{pmatrix} = IJ - 1 - \begin{pmatrix} \text{the number of parameters} \\ \text{that need to be estimated} \end{pmatrix} \tag{11.9}$$

In a two-by-two contingency table, for example, where $I = J = 2$, the number of degrees of freedom is $IJ - 1 = 2 \cdot 2 - 1 = 3$ if the probabilities P_{ij} are known; the number of degrees of freedom is $(I - 1)(J - 1) = (2 - 1)(2 - 1) = 1$ if they must be estimated.

Now, let us proceed with two examples in which we carry out the test of independence.

EXAMPLE 11.1

A theory in finance known as the **random walk theory** suggests that short-term changes in stock prices follow a random pattern. According to this theory, yesterday's prices can tell us virtually nothing of value about today's prices. Let us denote the *change* in price of a stock in time t—which refers to a given trading day—by ΔP_t and the *change* in price in the next trading day by ΔP_{t+1}. Suppose we observe price changes of

TABLE 11.2

Two-by-Two Contingency Table of Observed Daily Stock Price Changes

Price changes in day $t+1$	Price changes in day t		Total
	$\Delta P_t > 0$ $(j = 1)$	$\Delta P_t \leq 0$ $(j = 2)$	
$\Delta P_{t+1} > 0$	47	53	100
$\Delta P_{t+1} \leq 0$	63	77	140
Total	110	130	240

240 stocks that have been randomly selected and obtain the results shown in Table 11.2. To test the hypotheses

H_0: price changes in day $t + 1$ are independent of changes in day t
H_1: the two variables are not independent

we first have to determine the expected frequencies. The estimated joint probabilities are obtained by means of Equation 11.6:

$$e_{11} = \frac{100 \cdot 110}{240} = 45.83$$

$$e_{12} = \frac{100 \cdot 130}{240} = 54.17$$

$$e_{21} = \frac{140 \cdot 110}{240} = 64.17$$

$$e_{22} = \frac{140 \cdot 130}{240} = 75.83$$

Given the observed as well as the expected frequencies, we may go right ahead and compute the test-statistic value. Since the joint probabilities P_{ij} are not given in this example, the number of degrees of freedom of the test statistic is $(I - 1)(J - 1) = (2 - 1)(2 - 1) = 1$. Assuming a 5 percent significance level ($\alpha = 0.05$), the critical value is $\chi^2_{0.05}{}^{(1)} = 3.841$, and so all χ^2 values that are less than 3.841 lead to the acceptance of H_0. Table 11.3, which is a work sheet for computing the test statistic χ^2,

TABLE 11.3
Work Sheet for Computation of Test Statistic χ^2

Row and column	e	o	$e - o$	$(e - o)^2$	$\dfrac{(e - o)^2}{e}$
(1,1)	45.83	47.00	−1.17	1.3689	0.02987
(1,2)	54.17	53.00	1.17	1.3689	0.02527
(2,1)	64.17	63.00	1.17	1.3689	0.02133
(2,2)	75.83	77.00	−1.17	1.3689	0.01805
Total	240.00	240.00	0		$\chi^2 = 0.09452$

shows that in our example we have $\chi^2 = 0.09452 < 3.841$, so we accept the null hypothesis and view our results as supporting the random walk theory.

EXAMPLE 11.2

Let us modify Example 11.1 and retest a hypothesis concerning the independence of two successive price changes. This time we shall add the assumptions that the stock price is in equilibrium and the probability of a price increase on the next trading day (day t) is the same as the probability of a price decrease. Also we assume that the probability of a price increase on the following day (day $t + 1$) is equal to the probability of a price decrease, *independently* of the rise or fall of the price on day t.

Suppose we observe changes in the price of a stock over 80 trading days and obtain the results shown in Table 11.4. The null hypothesis

TABLE 11.4
Two-by-Two Contingency Table of Observed Daily Stock Price Changes

Price changes in day $t + 1$	Price changes in day t		Total
	$\Delta P_t > 0$	$\Delta P_t \leq 0$	
$\Delta P_{t+1} > 0$	18	17	35
$\Delta P_{t+1} \leq 0$	19	26	45
Total	37	43	80

in this example entails more than one that concerns only the independence of price changes on day t and price changes on day $t + 1$. Under H_0, specific marginal probabilities are assumed. The specific hypotheses to be tested here are:

H_0: The directions of price changes on two successive trading days are independent; furthermore, $P(\Delta P_t > 0) = P(\Delta P_t \leq 0)$ $= P(\Delta P_{t+1} > 0) = P(\Delta P_{t+1} \leq 0) = \frac{1}{2}$

H_1: H_0' (i.e., anything but H_0)

If H_0 is correct, then the joint probabilities P_{ij} are equal to $\frac{1}{4}$, since $P_i = \frac{1}{2}$ and $P_j = \frac{1}{2}$:

$$P_{ij} = P_i \cdot P_j = \tfrac{1}{4}$$

Substituting this result in Equation 11.5 but using P_{ij} rather than $\hat{P}_{ij}$, we find $e_{ij} = \frac{1}{4} \cdot 80 = 20$. The test statistic is computed in Table 11.5.

Since the various P_i and P_j values were assumed rather than estimated, the number of degrees of freedom of the test statistic is $IJ - 1 = 2 \cdot 2 - 1 = 3$. Assuming a significance level $\alpha = 0.05$, we determine

TABLE 11.5
Work Sheet for Computation of Test Statistic χ^2

Row and column	e	o	$e - o$	$(e - o)^2$	$\dfrac{(e - o)^2}{e}$
(1,1)	20	18	2	4	0.20
(1,2)	20	17	3	9	0.45
(2,1)	20	19	1	1	0.05
(2,2)	20	26	−6	36	1.80
Total	80	80	0		$\chi^2 = 2.50$

the critical point to be $\chi^{2(3)}_{0.05} = 7.815$. Since $\chi^2 = 2.50$ is *less than* 7.815, our observed statistic falls within the acceptance region and consequently we do not reject H_0.

11.3 The Chi-Square Test for Goodness of Fit

Occasionally we need to test two probability distributions (either discrete or continuous) to determine whether or not they are identical. In other words, we are not testing hypotheses concerning a parameter of a given distribution; rather, we are questioning the entire shape of the distribution, to see how well it fits that of another distribution.

WHEN DO WE NEED TO CONSIDER GOODNESS OF FIT?

Let us consider some situations in which the form of the distribution may be of interest. Suppose a nationwide public-opinion poll on a given issue is conducted at a given time and a frequency distribution over the various categories of classification is obtained. (For example, suppose the issue of interest is abortion. The classifications could be "strongly object," "object in general but condone under certain circumstances," and so on.) Suppose after a few years a similar sample using the same categories is taken. Because of sampling errors, we are bound to find some differences between the two frequency distributions. Do the differences indicate a shift in public opinion on the issue? Can we say that the newly observed distribution is *significantly different* from what we would have expected had there been no change in public opinion? In such a situation we may want to test hypotheses concerning the form of the entire distribution rather than on a specific parameter.

Consider another example. When a company places an order for its inventories, a certain time period is usually allowed before it expects to actually receive its order. How early should the order be placed? That depends on the number of items in stock and on the future demand for those items. Future demand is usually unknown, but we may wish to estimate its probability distribution in order to make better decisions with regard to the optimal time for the ordering of new stock. If we have reason to believe (perhaps because of some theoretical considerations) that the probability distribution is of Poisson form, we may test the following hypotheses:

H_0: the probability distribution is of Poisson form
H_1: the probability distribution is not of Poisson form

In another situation, we might wish to test the hypothesis that a given distribution is normal. Mutual funds, for example, are investment companies that pool funds from many individuals and institutions and invest them in diversified portfolios of bonds and stocks. Whether the distribution of the rates of return on such portfolios is normal or not is important for several reasons. One reason is that a normal distribution is fully specified by only two parameters—the mean and the variance. If indeed a portfolio's rate of return is normally distributed, theoretical as well as practical risk analysis of the returns' distribution can be greatly simplified. In this case one would have to focus on the mean and the variance only, and one need not deal with the skewness or any other moments of the distribution. Thus, we might want to test the following hypotheses:

H_0: the rate of returns' probability distribution is normal
H_1: the rate of returns' probability distribution is not normal

CARRYING OUT TESTS FOR GOODNESS OF FIT

In principle as well as in practice, goodness-of-fit tests are straightforward. We draw a random sample and compare the sample frequency distribution with the frequency we would have expected to see had the null hypothesis really been correct. If the fit between the hypothesized frequency under H_0 and the observed sample frequency is good, we favor the null hypothesis, and if it is not good we favor the alternative hypothesis.

The Test Statistic

As usual, we need to develop a test statistic and a decision rule so that our conclusion will not be arbitrary. The test statistic we develop will resemble the one used for the test of independence in Section 11.1. The statistic is

$$\chi^2 = \Sigma \frac{(e-o)^2}{e} \qquad \text{(11.10)}$$

where e and o are the expected and observed frequencies in various intervals or categories of the relevant variable.

Categorizing the Variable

Evidently, then, the variable to be considered must first be classified into categories or intervals. There is no specific rule for this classification; when the variable is continuous, the classification is somewhat arbitrary. We should try to have many intervals, but no interval should include fewer than 5 *expected* observations. If an interval has fewer than 5 expected observations, it should be combined with an adjacent interval.

Determining Expected Frequencies

We find the expected frequency in interval j (e_j) by multiplying P_j (the expected proportion of the population in interval j) by n (the sample size):

$$e_j = nP_j \qquad (11.11)$$

Degrees of Freedom

The number of degrees of freedom of the test statistic is equal to the number of categories (or intervals) minus one. If, however, the expected frequencies can be derived only after some distribution parameters have been estimated from the sample, then one degree of freedom per estimated parameter will be lost. For example, if the null hypothesis is

H_0:   the density function·is normal with parameters $\mu = 100$ and $\sigma = 10$

then the expected frequencies in the various intervals can be arrived at *without estimating* μ and σ. If, however, the null hypothesis is

H_0:   the density function is normal with mean $\mu = 100$

then the expected frequencies in the various intervals can be computed only after σ has been estimated. By estimating σ we lose one degree of freedom. Finally, if H_0 is

H_0:   the density function is normal

we must first estimate both μ and σ and thereby lose two degrees of freedom. Our rule, then, is

$$\begin{pmatrix} \text{The number of} \\ \text{degrees of freedom} \\ \text{of } \chi^2 \end{pmatrix} = \begin{pmatrix} \text{number of} \\ \text{intervals} \\ \text{or} \\ \text{categories} \end{pmatrix} - 1 - \begin{pmatrix} \text{number of} \\ \text{parameters} \\ \text{to be} \\ \text{estimated} \end{pmatrix} \qquad (11.12)$$

EXAMPLE 11.3

Truck Rental Service, Inc. (TRS), is a large firm operating nationwide and organized in many regional offices. Consider one of these offices, located in a large metropolitan area. It is the responsibility of the regional office to replace trucks that have any kind of mechanical or electrical problem if it cannot be fixed within a very short time. As a result, the office must maintain a standby truck fleet at all times. How many trucks should be standing by? This depends, of course, on the probability distribution of the number of trucks that must be replaced on any given day. Consider the following hypotheses:

H_0:   the probability distribution of the number of replacements per day is of Poisson form

H_1:   the probability distribution is not of Poisson form

To test such hypotheses, we need to take a sample of days and record the number of replacements that were needed each day. Suppose a sample of 250 days was taken. The data gathered are presented in Table 11.6.

TABLE 11.6
Frequency Distribution of Truck Replacements

Number of replacements	Frequency of days
0	2
1	8
2	21
3	31
4	44
5	48
6	39
7	22
8	17
9	13
10	5
	250

Using $\alpha = 0.05$, should we accept or reject H_0 on the basis of this sample? As we recall (from Chapter 7), the Poisson distribution is determined by the parameter λ, which in our example is the mean daily number of truck replacements that the office must make. Since this figure is neither provided nor hypothesized, we must use the sample to estimate it before the expected frequencies can be obtained. The average of daily replacements is equal to the total number of replacements made within the sample period divided by the number of days in the period. In our case, 1,250 replacements took place within the 250 days. Note that to calculate the total number of replacements we simply multiply each replacement value by its frequency and add:

$$0 \cdot 2 + 1 \cdot 8 + 2 \cdot 21 + 3 \cdot 31 + \cdots + 10 \cdot 5 = 1,250$$

TABLE 11.7
Probability Distribution of Truck Replacements

x	$P(x) = \dfrac{5^x e^{-5}}{x!}$
0	0.0067
1	0.0337
2	0.0842
3	0.1404
4	0.1755
5	0.1755
6	0.1462
7	0.1044
8	0.0653
9	0.0363
10	0.0181
11 or more $1.0 - 0.9863 =$	0.0137
Total	1.0000

and to get the average daily replacements, we divide the total by the number of days: $\frac{1,250}{250} = 5$. Using the Poisson equation, $P(X = x)$ $= \frac{\lambda^x e^{-\lambda}}{x!}$ where 5 is used for λ, we derive the probabilities for the various number of replacements shown in Table 11.7.

Given 250 observations (days), we determine the expected frequencies by multiplying the probability of each category by 250 and then proceed to calculate the test statistic. The calculations are provided in Table 11.8. Note that those categories in which the expected frequency, e, is less than 5 have been grouped together with their adjacent categories.

TABLE 11.8[a]

Work Sheet for Computation of Test Statistic χ^2

j	P_j	$e = 250 \cdot P(x)$	o	$e - o$	$(e - o)^2$	$\frac{(e - o)^2}{e}$
0	0.0067	1.675 }10.100	2 }10	0.100	0.010	0.0010
1	0.0337	8.425	8			
2	0.0842	21.050	21	0.050	0.002	0.0001
3	0.1404	35.100	31	4.100	16.810	0.4789
4	0.1755	43.875	44	−0.125	0.016	0.0004
5	0.1755	43.875	48	−4.125	17.016	0.3878
6	0.1462	36.550	39	−2.450	6.002	0.1642
7	0.1044	26.100	22	4.100	16.810	0.6441
8	0.0653	16.325	17	−0.675	0.456	0.0279
9	0.0363	9.075	13	−3.925	15.406	1.6976
10	0.0181	4.525 }7.950	5 }5	2.950	8.702	1.0946
11+	0.0137	3.425	0			
	1.0000	250.000				$\chi^2 = 4.4966$

[a] An additional category (11+) was added so that the expected frequencies as well as the actual frequencies would add up to 250. The probability of this category was calculated by subtracting the sum of all the other categories from 1.

With 10 categories and one estimated parameter (λ), the number of degrees of freedom of the test statistic is $10 - 1 - 1 = 8$, and the critical value is $\chi^{2(8)}_{0.05} = 15.507$. Since $\chi^2 = 4.4966 < 15.507$, we accept the null hypothesis.

EXAMPLE 11.4

The management of a large pension fund is interested in studying the probability distribution of monthly rates of return on large, well-diversified portfolios of common stock. In particular, the management is interested in finding out whether such probability distributions can be reasonably assumed to be normal. Intensive gathering of data has yielded 90 monthly returns on various large and well-diversified portfolios. These returns (expressed as gross returns per \$1,000 invested) had a sample average $\overline{X} = \$1,010$ and sample standard deviation $S = \$20$. The data were classified into intervals and had the frequency distribution presented in Table 11.9. In the table each interval includes

TABLE 11.9
Frequency Distribution of Gross Returns

Range[a] (dollars)	Frequency
Less than $970	4
$ 970–980	4
980–990	10
990–1,000	13
1,000–1,010	16
1,010–1,020	15
1,020–1,030	13
1,030–1,040	8
1,040–1,050	5
1,050 or more	2

[a] Each closed interval includes the lower limit
but excludes the upper limit.

the lower limit but excludes the upper limit. To determine the expected frequencies, we first standardize the range of gross returns by expressing each nonstandardized range limit X as $Z = \dfrac{X - \overline{X}}{S}$. For example, the limit 1,020 will be standardized as follows: $\dfrac{1,020 - 1,010}{20} = 0.5$. Once the conversion into standardized ranges has been made, we can determine the relative frequency directly from the normal distribution table (inside back cover) and then multiply it by 90 (the number of observations) to obtain the frequency in the category. When the expected frequencies are determined and categories with expected frequency of less than 5 are combined, we continue as usual with the calculation of the test statistic. The calculations are provided in Table 11.10.

TABLE 11.10
Work Sheet for Computation of Test Statistic χ^2

(1) Nonstandardized range X	(2) Standardized range $Z = \dfrac{X-\mu}{S}$	(3) Relative frequency	(4) = (3) · 90 Expected frequency e	(5) o	(6) e − o	(7) (e − o)²	(8) $\dfrac{(e-o)^2}{e}$
Less than $970	Less than −2.0	0.023	2.07 ⎱ 6.03	4 ⎱ 8	−1.97	3.881	0.644
$ 970–980	−2.0 to −1.5	0.044	3.96 ⎰	4 ⎰			
980–990	−1.5 to −1.0	0.092	8.28	10	−1.72	2.958	0.357
990–1,000	−1.0 to −0.5	0.150	13.50	13	0.50	0.250	0.019
1,000–1,010	−0.5 to 0.0	0.191	17.19	16	1.19	1.416	0.082
1,010–1,020	0.0 to 0.5	0.191	17.19	15	2.19	4.796	0.279
1,020–1,030	0.5 to 1.0	0.150	13.50	13	0.50	0.250	0.019
1,030–1,040	1.0 to 1.5	0.092	8.28	8	0.28	0.078	0.009
1,040–1,050	1.5 to 2.0	0.044	3.96 ⎱ 6.03	5 ⎱ 7	−0.97	0.941	0.156
1,050 or more	2.0 and up	0.023	2.07 ⎰	2 ⎰			
		1.000	90.00	90			$\chi^2 = 1.565$

Since the mean and standard deviation of the returns are calculated on the basis of the sample data, two degrees of freedom have been "used up" and so our χ^2 statistic has $8 - 1 - 2 = 5$ degrees of freedom. The

> critical value, if $\alpha = 0.05$, is $\chi^2_{0.05(5)} = 11.070$, and as $\chi^2 = 1.565 < 11.070$, we do not reject H_0 and conclude that the returns' distribution is normal. More accurately, we do not have statistical evidence that will dispute the hypothesis that the distribution is normal.

A CONCLUDING NOTE REGARDING CLASSIFICATION

Tests of goodness of fit to distributions—uniform, exponential, binomial, and so on—follow the same logic and procedure as the tests that have been described in our examples. The difference between one test and another is in the derivation of the expected frequency for which the relevant probability distribution must be used. We would like to note once again that in the case of a continuous distribution, the classification of categories is rather arbitrary, and unless a sufficient number of categories is defined, the test may be meaningless. As an extreme case, suppose we define only two categories in Example 11.4: "less than \$1,010" and "1,010 or more." The expected relative frequency will be 50 percent in the first category and 50 percent in the second category.[1] Even if the null hypothesis is accepted in this case, it would be absolutely wrong to interpret the result as a confirmation of normality. The test would indicate only that normality cannot be rejected on the basis of the test results. But the very same data will lead to the conclusion that we cannot reject a hypothesis that the distribution is uniform if only these two broad categories are used. If the distribution is assumed to be uniform, the expected relative frequency will also be 50 percent in each category, and will result in the same value of the test statistic and the same number of degrees of freedom $(2 - 1 = 1)$.[2] Consequently, the decision rule will be the same as for the test of a normal distribution.

We conclude, then, that the categories should be determined in such a way that there will be enough categories to make the expected frequencies resemble the probability distribution hypothesized under H_0, because when other things are held constant,[3] the greater the number of categories, the higher the power of the test (and thus the lower the probability of committing a Type II error).

Problems

11.1. Explain why the chi-square tests of independence and of goodness of fit are always upper-tail tests.

11.2. "In a chi-square test, whenever the observed frequency is less than 5, categories must be combined before the test can be carried out." Do you agree? Why?

11.3. Is it true that the number of degrees of freedom of the test statistic of a chi-square test of independence is equal to the number of cells in the contingency table minus one? Explain your answer.

---

[1] Assuming that the mean and standard deviation of the normal distribution are somehow exogenously known. Otherwise there will not be enough degrees of freedom to test such hypotheses.

[2] Assuming the distribution parameters are exogenously known.

[3] In particular one should remember that a minimum of 5 *expected* observations should be included in each category.

11.4. Students who have taken two courses in statistics and economics on a pass–fail basis have the following distribution:

		Statistics		
		Pass	*Fail*	*Total*
	Pass	583	147	730
Economics	*Fail*	95	29	124
	Total	678	176	854

Are passing economics courses and passing statistics courses independent? Show your calculations, and use $\alpha = 0.05$.

11.5. The following table shows the frequency of pass–fail grades by grade and by the relation of the sex of the teacher to that of the student.

		Grade		
		Pass	*Fail*	*Total*
Sex of teacher	*Opposite sex*	156	8	164
vs.	*Same sex*	284	99	383
Sex of student	*Total*	440	107	547

Is there any evidence of sex discrimination in the grades? Use a significance level of 5 percent.

11.6. The following table shows the distributions of real estate investments of an insurance company by location and type of investment.

Class	*West*	*South*	*Midwest*	*East*
Offices	10	12	7	9
Industrial	96	26	86	7
Commercial	14	6	7	3
Hotels and motels	0	7	0	6
Apartments	2	1	2	2
Land	1	0	0	0

Source: C. M. Ballard and B. V. Strum, "Pension Funds in Real Estate: New Challenges/ Opportunities for Professionals," *Appraisal Journal,* October 1978, pp. 551–569.

Test the hypothesis that the class and geographic location of the property in which the insurance company has invested are independent. Use $\alpha = 0.01$.

11.7. The home ownership of a certain sample is distributed as follows, by income:

	Home ownership		
Income	*Yes*	*No*	*Total*
$ 0–4,999	20	347	367
5,000–9,999	60	124	184
10,000–14,999	272	50	322
15,000 +	498	66	564
Total	850	587	1,437

(*a*) Explain why home ownership and income are likely to show dependence in a chi-square test applied to the above table.

(*b*) Test the hypothesis that home ownership and income are independent at a 5 percent significance level.

TABLE P11.8

Demographic characteristics	Paid all charges	Paid part of charges
Age of head of household		
Under 31 years	200	200
31–40 years	210	190
41–50 years	240	160
51–60 years	300	100
Over 60	240	160
Education of head of household		
Under 12 years	150	250
High school graduate	180	220
Some college	190	210
College graduate	250	150
Postcollege	300	100

11.8. Table P11.8, based on a hypothetical survey, shows the number of households that paid all or part of their charge-card bills for the month prior to that in which the survey was made, by demographic characteristics.

(a) Test the hypothesis that age of head of household is independent of whether all or part of the charges are paid. Use $\alpha = 0.01$.

(b) Test the hypothesis that education of head of household is independent of whether all or part of the charges are paid. Use $\alpha = 0.01$.

11.9. There are financial services that rate bonds according to their probability of default (that is, inability to meet interest or repayment of principal). Standard and Poor's is one such service. The service rates bonds as AAA ("triple A") if they are judged to have a negligible risk of default and thus to be of highest quality. AA ("double A") are bonds of high quality but are judged to be not quite so free of default risk as AAA bonds. Bonds rated BBB are of medium quality.

The table below presents the results of a 300-firm sample: it shows the number of firms that defaulted on their bonds last year, by bond rating.

Bond rating	Number of defaulting firms	Number of firms not defaulting	Total
AAA	5	95	100
AA	5	95	100
BBB	20	80	100
Total	30	270	300

(a) Calculate the expected frequency in each cell, assuming that default and bond rating are independent.

(b) Test the hypothesis that the rating service does not provide valuable information with regard to the chances of default. Formulate the null and alternative hypotheses, and use $\alpha = 0.05$ to test the hypotheses.

11.10. Rework Problem 11.9, this time assuming that the sample results were as in the table below:

Bond rating	Number of defaulting firms	Number of firms not defaulting	Total
AAA	5	95	100
AA	5	95	100
BBB	80	20	100
Total	90	210	300

11.11. Rework Problem 11.9 again, assuming that the sample showed the following results:

Bond rating	Number of defaulting firms	Number of firms not defaulting	Total
AAA	20	80	100
AA	10	90	100
BBB	0	100	100
Total	30	270	300

Considering the test result, evaluate the bond rating service: is it worthwhile? (Do not forget to use common sense!)

11.12. The following table shows the observed frequency of a sample of 100 people classified by wealth and habit of drinking Coca-Cola. The people were randomly chosen from the general adult population.

Coca-Cola habit	Rich	Poor	Total
Drinking	25	25	50
Nondrinking	25	25	50
Total	50	50	100

Can you say without carrying out any calculations whether a null hypothesis of independence between the two variables should be accepted or rejected? Explain.

11.13. Consider Problem 11.12 again. Suppose now the sample shows the following result:

Coca-Cola habit	Rich	Poor	Total
Drinking	X	$50 - X$	50
Nondrinking	$50 - X$	X	50
Total	50	50	100

What is the maximum value of X such that $X \leq 25$ and the hypothesis of independence between wealth and Coca-Cola drinking will be rejected at $\alpha = 0.05$?

11.14. A large company employing tens of thousands of employees is considering offering its employees low-cost life insurance rather than a pay raise. A sample of 200 administrators, production workers, and salespeople has shown the following distribution of preference for the plan:

Employees	Pay raise	Life insurance	Indifferent
Administrators	15	20	10
Production workers	18	30	15
Salespeople	25	45	12

Test the hypothesis that the preference for life insurance or pay raise is independent of the workers' classification; in other words, the preference of the administrators is the same as that of the production workers and the salespeople. Use $\alpha = 0.01$.

11.15. An insurance firm is reexamining its auto insurance premiums. One question currently considered is whether young people and old people are involved in more accidents than others. A sample of 400 files has been randomly chosen, 100 from each age group, and the involvement in accidents has been recorded as follows:

Age group	Was involved in accidents	Was not involved in accidents
Under 26	30	70
26–40	5	95
41–55	10	90
56+	25	75

Test the null hypothesis, which would hold that involvement in accidents is not dependent on age, against the alternative hypothesis that the involvement in accidents is dependent on age. Do you think the company should charge higher auto insurance premiums for young and old drivers?

11.16. Cash turnover is defined as company sales divided by cash. Table P11.16 presents the cash turnover of 59 firms. Test the hypothesis that the cash turnover is normally distributed with mean of $\mu = 13.0$ and standard deviation of $\sigma = 15.0$. When deriving the test statistic, use symmetrical intervals of 7.5 around the mean 13.0. Use $\alpha = 0.01$.

TABLE P11.16

Firm	Cash turnover	Firm	Cash turnover
1	13.9	31	2.0
2	10.9	32	17.1
3	8.9	33	9.7
4	10.0	34	2.6
5	28.9	35	62.0
6	3.6	36	2.0
7	64.1	37	36.6
8	48.9	38	3.2
9	36.2	39	9.1
10	16.2	40	16.0
11	10.8	41	9.2
12	13.4	42	30.3
13	16.7	43	45.6
14	5.1	44	31.5
15	12.2	45	2.7
16	15.2	46	32.5
17	20.2	47	6.5
18	8.5	48	19.1
19	3.8	49	27.3
20	23.2	50	12.6
21	29.9	51	11.8
22	12.8	52	20.7
23	9.3	53	25.7
24	2.9	54	6.1
25	11.5	55	24.9
26	11.8	56	5.5
27	11.7	57	12.1
28	22.7	58	14.9
29	27.4	59	27.5
30	16.3		

11.17. Rework Problem 11.16, this time assuming that while the mean is known to equal 13.0, the standard deviation is unknown.

11.18. Rework Problem 11.16 once more, assuming that both the mean and the standard deviation of the cash turnover are unknown.

11.19. How many degrees of freedom does the test statistic for testing goodness of fit to the normal distribution have when the number of intervals is 12 and the mean and standard deviation are unknown?

11.20. Suppose you test the goodness of fit of sample data to the normal distribution when the mean is known to equal zero. Suppose also you decide to use two intervals—one for values less than or equal to zero and the other for values greater than zero.

(a) How many degrees of freedom does the test statistic have?

(b) Is it a good test? Explain.

11.21. "If $o_i = e_i$ for each class in a goodness-of-fit test, we would not reject the null hypothesis, no matter what the distribution assumed under that hypothesis." Do you agree with the statement? Would you necessarily have much confidence that the null hypothesis is indeed correct?

11.22. The price of Zoom Corporation's common stock has risen (or stayed unchanged) in 36 of a sample period of 50 days. It has declined in 14 of the 50 days. Use the chi-square goodness-of-fit test to test the null hypothesis that the daily price change is binomially distributed with equal probability of "success" (i.e., price increase or no change) and "failure" (i.e., price decrease). Use $\alpha = 0.01$. Hint: Recall that the expected value of the binomial distribution is np.

11.23. Using the sample data of Problem 11.22, can you test the null hypothesis that the distribution is binomial with $p = \frac{1}{2}$ without using the chi-square goodness-of-fit test? If yes, show your calculations. What is your decision if you use $\alpha = 0.01$?

11.24. Compare the results of Problems 11.22 and 11.23. Which is more accurate?

11.25. The number of tourists at Florida's Disney World was recorded for 90 weeks (excluding such busy seasons as Christmas and Easter). The records show the following distribution:

Number of tourists per week (thousands)	Frequency of weeks
0.00–10.99	0
11.00–13.99	12
14.00–16.99	10
17.00–19.99	15
20.00–22.99	19
23.00–25.99	12
26.00–28.99	14
29.00–31.99	3
32.00–34.99	0
35.00–37.99	4
38.00–40.99	1

(*a*) Calculate the average and standard deviation of the sample data.

(*b*) Test the hypothesis that the distribution is normal. Use $\alpha = 0.05$.

11.26. A carnival roulette wheel is marked 1 through 10. A sample of 100 spins shows the following frequency of numbers:

Number	Frequency
1	7
2	11
3	10
4	10
5	8
6	12
7	11
8	11
9	13
10	7
	100

Test the hypothesis that the roulette wheel is balanced (i.e., that there is an equal probability for each of the numbers 1 through 10).

11.27. The period of time from the day a customer is billed to the day he actually pays is referred to as a collection period. A power company's records show that frequencies of collection periods (in days) are as given in Table P11.27.

TABLE P11.27

Collection period (days)	Frequency
0	0
1	0
2	20
3	20
4	40
5	70
6	110
7	100
8	175
9	170
10	165
11	110
12	20
13	0
14	0
	1,000

(a) Test the hypothesis that the collection period is Poisson-distributed with $\lambda = 5.0$. Use $\alpha = 0.05$.

(b) Test the hypothesis that the collection period is Poisson-distributed, but do not assume any value for λ. Rather, estimate λ from Table P11.27. Use $\alpha = 0.05$ again.

CHAPTER TWELVE OUTLINE

12.1 Introduction to Analysis of Variance

12.2 The *F* Distribution

12.3 One-Way Analysis of Variance
Beginning the Analysis: Basic Procedures
Total Variation
Within-Group Variations
Between-Group Variations
Mean Squares within Groups and between Groups
The Test Statistic
Shortcut Equations for One-Way Analysis of Variance

12.4 Two-Way Analysis of Variance
How Two-Way ANOVA Works: The Fundamental Approach

12.5 Interpreting Computer Printouts of Two-Way Analysis of Variance

Key Terms

analysis of variance (ANOVA)
one-way analysis of variance
average within the group
grand average
total sum of squares
sum of squares within the groups
unexplained sum of squares
sum of squares between the groups

mean squared deviations within the groups
mean squared deviations between the groups
two-way analysis of variance
column effect
row effect
interaction effect

12
ANALYSIS OF VARIANCE

12.1 Introduction to Analysis of Variance

The statistical inference discussed so far has concerned parameters of either one or two populations. In this chapter we shall discuss a technique known as **analysis of variance (ANOVA),** which *allows for inference concerning the means of more than two populations.* Experience shows that analysis of variance is better understood if we consider the general framework first and the detailed equations later. Computer packages with easy-to-use ANOVA programs make the computation of analysis of variance of secondary importance. It is far more important to understand the purpose and approach of ANOVA, which in turn will enable you to set up correct experimental designs and draw correct conclusions from the analysis. We shall describe the computational part of ANOVA primarily to provide deeper insight into the analysis.

We perform analysis of variance when we want to test the hypothesis that the means of several distributions are the same against the hypothesis that not all those means are equal to one another. Figure 12.1*a* shows three density functions, each with a separate mean. Each of these density functions can be thought of as representing a separate population or group. The density functions could, for example, represent the grade distributions of students in three schools, or transportation time from city *A* to city *B* by three methods of transportation. A sample is observed from each of these populations, the values of which are drawn below the population density functions and marked o. The samples' averages are shown by solid squares. We wish to compare the range of the entire set of observations with that

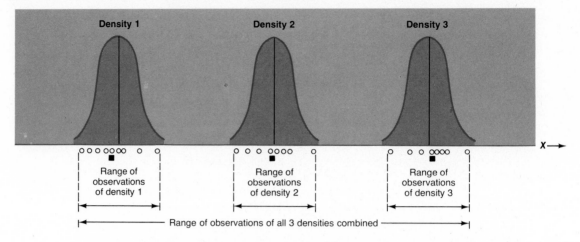

(a) The means of the population are far apart

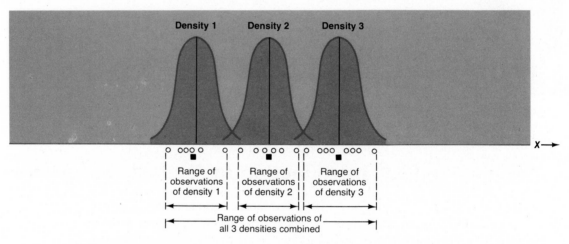

(b) The means are closer together

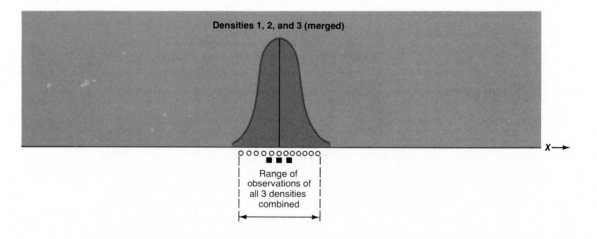

(c) The means are equal to one another

Figure 12.1
Three sets of populations and the relationship among their means

drawn from the individual populations. In Figure 12.1*a*, the range of the entire set of observations (that is, those drawn from all three populations) is far wider than the range of any one sample considered individually, because the means of the populations are widely dispersed.

In Figure 12.1*b* the means of the three populations are closer together. The sample diversity *within* each population remains the same as in 12.1*a*. Here, however, because the population means are closer together, the range of the observations in total is narrower than in 12.1*a*.

Finally, Figure 12.1*c* illustrates the extreme case: the three population density functions are assumed to have the *same* mean, which merges all three densities exactly on top of each other. Here, the diversity of the sample observations in total is therefore of the same magnitude as the diversity of any one of the individual samples, and the sample averages are close to one another.

We can now draw the conclusion that the *reason* the total diverisity of the three samples in Figures 12.1*a* and 12.1*b* is greater than the diversity of each separate sample is that the samples are selected from density functions with *unequal means*.

From assumed known structured patterns of three populations we have made deductions about the relationships among sample observations drawn from those populations. Reversing our approach, we shall now infer from observed samples the structures of the populations. In Figure 12.2, observations of three samples, each drawn from a different population, are

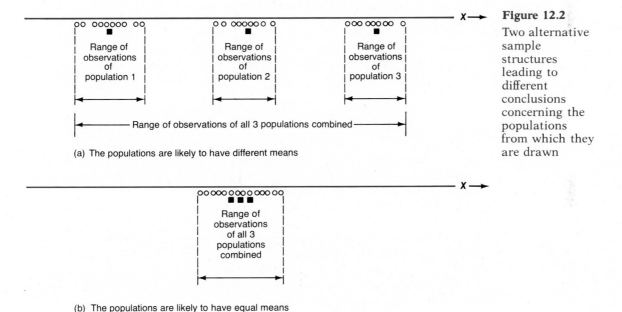

Figure 12.2

Two alternative sample structures leading to different conclusions concerning the populations from which they are drawn

presented in two alternative ways. In Figure 12.2*a* the observations are shown in three separate clusters, leading us to conclude that there is more to this separation than mere chance. Apparently we get three clusters because the means of the three populations are different. In Figure 12.2*b* the observations of all three samples are clustered together, leading us to conclude that we do not have statistical evidence to reject a hypothesis that all three population means are equal. Formally, we test the null hypothesis

$$H_0: \quad \mu_1 = \mu_2 = \mu_3$$

where μ_1, μ_2, and μ_3 are the means of the three distributions, against the alternative:[1]

$$H_1: \quad H_0' \text{ (or: not all the } \mu_i \text{ are equal)}$$

Acceptance or rejection of H_0 will depend on the structure of the observations drawn from the three populations. If the structure is like that of Figure 12.2*a*, almost everyone will agree that H_0 should be rejected. If it is like that of Figure 12.2*b*, almost everyone will agree to accept H_0. In real life, of course, the observations are seldom as neatly separated and clustered as in Figure 12.2*a* or as close together as in 12.2*b*. They are usually scattered, like those in Figure 12.3, where intuitive conclusions are difficult to reach. Statistical

Figure 12.3

A hypothesized relationship among the observations of three samples

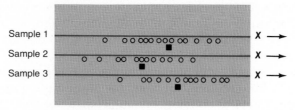

tools must be used to decide which of the alternative hypotheses we should accept.

We shall now explain the inductive argument (deriving the population structure from the sample structure) using a numerical example.

EXAMPLE 12.1

A cosmetics company is considering four container designs for its new antiperspirant: a metal spray container, a see-through glass spray container, a metal roll-on container, and a see-through glass roll-on container. All containers hold 5 ounces and cost the same to produce. It has been decided that, at least for the time being, only one container design is to be chosen, and the question, of course, is which of the four is most appealing to consumers. The marketing department has decided to set a trial period during which containers of different designs will be sold in randomly selected stores across the country. The results of the trial period are shown in Table 12.1.

Let us first make a few comments on Table 12.1. First, the data are arranged in four groups. (As you recall, our explanation of Figures 12.1 and 12.2 may be extended to include more than three population groups.) Second, analysis of variance can be carried out when the number of observations in the various groups is either equal or unequal. In Table 12.1, not all the groups have the same number of sample observations. Third, the store number is simply a number chosen for convenience. The order in which the observations are listed within each group is arbitrary.

---

[1] The symbol H_0' is used to denote the "complement of H_0," which means "any relationship between μ_1, μ_2, and μ_3 but the one hypothesized in H_0." In H_0' the following are included:

$$\mu_1 = \mu_2 \neq \mu_3$$

$$\mu_1 \neq \mu_2 = \mu_3$$

$$\mu_1 \neq \mu_2 \neq \mu_3$$

$$\mu_1 = \mu_3 \neq \mu_2$$

TABLE 12.1
Number of Containers Sold, by Design

Store	Metal spray	Glass spray	Metal roll-on	Glass roll-on
1	95	149	220	362
2	106	153	219	376
3	100	154	217	353
4	98	151	225	350
5	110	145	221	356
6	103	150	218	362
7	—	148	220	348
8	—	150	—	349
Average	102	150	220	357

The difference between these two figures is due to chance fluctuations.	*The difference between these two figures is primarily due to reasons other than chance.*

One does not have to be an experienced statistician to conclude that the company will be better off selling its antiperspirant in glass roll-on containers. Is it possible that the average of 357 containers sold in the glass roll-on design is only *by chance* greater than the average of the sales of the other designs? While the possibility exists, the probability is so remote that it may be considered nil. The diversity of sales across stores *within* any of the four design categories is of far lesser magnitude than the diversity of sales between the four different design groups. Specifically, we attribute the variation in sales of the glass roll-on design among individual stores to chance fluctuations. We also attribute the diversity among sales of individual stores in any of the other three designs to chance fluctuations. But the difference between the 95 metal spray containers sold by the first store (and for that matter any other sales figure taken from the same group) and the 356 glass roll-on containers sold by the fifth store (and for that matter any other sales figure taken from the same group) is primarily attributable to the fact that the glass roll-on design is overwhelmingly preferred by consumers.

The data in Example 12.1 are unusual in that the variations in sales *within* each group are relatively small, while the variations *between* the groups are relatively large. (We set up these hypothetical data in such a way as to make our point as clear as possible.) When actual rather than hypothetical data are involved, the difference between diversity *within* the groups and diversity *between* the groups is often less clear, and the use of a statistical procedure is required for objective decision-making. It is clear, however, that if the variance between the groups is relatively large in comparison to the variances within the groups, we tend to believe that the samples are taken from populations with different means. If the variances within the groups are relatively large, we tend to believe that the difference in the means of the samples merely reflects sampling variations. That is to say, if we repeated the sampling of the various groups many times, the differences between the sample averages would tend toward zero. Since we only have one sample from each group, however, statistical tools are needed to interpret the differences. Before turning to another example we need to consider the **F distribution,** which we will use for the analysis.

12.2 The *F* Distribution

The **F distribution** is located over the *range of positive real numbers*. It is a unimodal and positively skewed distribution, as illustrated in Figure 12.4. The *F* distribution has two parameters, v_1, and v_2, which are degrees of freedom. Briefly, if two independent chi-square variables have v_1 and v_2 degrees of freedom respectively, then the ratio $\dfrac{[\chi^{2(v_1)}/v_1]}{[\chi^{2(v_2)}/v_2]}$ has an *F* distribution with degrees of freedom v_1 and v_2, or

$$\frac{[\chi^{2(v_1)}/v_1]}{[\chi^{2(v_2)}/v_2]} \sim F^{(v_1, v_2)}$$

The function's exact shape depends solely on the values of the parameters v_1 and v_2. In Figure 12.4 the parameters shown are 4 and 10, and $F_{0.05}^{(4,10)} = 3.48$ is the *F* score that leaves a 5 percent right-tail area.

Let us turn to a description of the *F* distribution table. Three *F* distribution tables are given in this book: one for the upper 5 percent points, one for the 2.5 percent, and one for the 1 percent. Consider, for example, the table with the 5 percent points. The table (schematically reproduced in Figure 12.5)

Figure 12.4

An *F* distribution: $F^{(4,10)}$ and the location of the value $F_{0.05}^{(4,10)}$

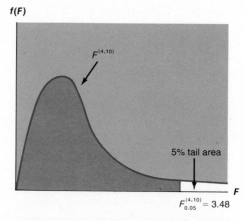

Figure 12.5

The *F* distribution table: upper 5 percent point

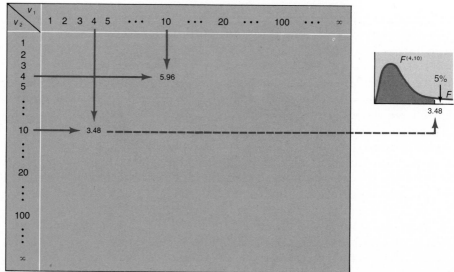

gives us the F scores that leave a right-tail area of 5 percent for various combinations of v_1 and v_2. One should be careful not to mistake v_1 for v_2. As Figure 12.5 shows, $F_{0.05}^{(4,10)}$ is not the same as $F_{0.05}^{(10,4)}$. Note that the degrees of freedom listed first are those of the numerator of the F ratio, and the degrees of freedom listed second are those of the denominator.

12.3 One-Way Analysis of Variance

Let us consider another example where analysis of variance is to be carried out. Here, our analysis will be a **one-way analysis of variance.**

Suppose we want to test the hypothesis that the time required to replace old mufflers with new ones in compact cars is the same at three large muffler shops. The time from check-in to check-out is measured at the three shops for a few randomly selected cars. The results (in minutes) are reported in Table 12.2. Naturally, the average number of minutes spent by the sampled cars in the three shops is unequal. This variation is to be expected, whether the group's means are equal to one another or not. (That is to say, even if the means were equal, we would expect unequal sample averages as a result of sampling errors.) What we want to find out is whether these differences should lead us to conclude that the group's *means* are unequal. More formally, we want to test

$$H_0: \quad \mu_1 = \mu_2 = \mu_3$$

against

$$H_1: \quad H_0' \text{ (i.e., } H_0 \text{ does not hold)}$$

TABLE 12.2

Muffler Replacement Time in Three Muffler Shops

(in minutes)

Observation (i)	Shop (j)		
	1	2	3
1	120	200	160
2	100	160	80
3	174	110	110
4	140	180	100
5	90	100	180
6	90	160	—
7	—	140	—
$\sum_i X_{ij}$	714	1,050	630
n_j	6	7	5
$\overline{X}_{.j}$	$\dfrac{714}{6} = 119$	$\dfrac{1,050}{7} = 150$	$\dfrac{630}{5} = 126$
$\overline{\overline{X}}$	$\dfrac{714 + 1,050 + 630}{6 + 7 + 5} = \dfrac{2,394}{18} = 133$		

BEGINNING THE ANALYSIS: BASIC PROCEDURES

At this point, we can no longer avoid using new notations and equations to develop the test statistic and decision rule. Let us briefly consider these new items.

Summing Observations within Groups

The sample data in Table 12.2 are organized in three columns representing three shops (or groups, in more general terms), and the various rows represent the sequential observations in the groups. To facilitate reference to the various sample observations, we shall use the notation X_{ij} to denote the ith observation in the jth group. For example, X_{11} is equal to 120, the value of the first observation in Shop 1. Similarly, $X_{21} = 100$, $X_{42} = 180$, and so on. The number of observations in Group j is denoted by n_j. Thus $n_1 = 6$, $n_2 = 7$, and $n_3 = 5$. The sum of all the observations in Group j is given by $\sum_i X_{ij}$, so that for $j = 1$ we obtain $\sum_i X_{i1}$, meaning the sum of all the observations within Group 1. Similarly, $\sum_i X_{i2}$ and $\sum_i X_{i3}$ are the sums of all the observations in Groups 2 and 3, respectively.

Finding the Averages within Groups

To find the **average value within a given group,** we simply divide the total values of all the observations in the group by the number of observations in that particular group. The group average is denoted by $\overline{X}_{\cdot j}$, where the dot symbolizes the fact that the average was carried out across the index i. The following formula applies:

AVERAGE WITHIN THE GROUP

$$\overline{X}_{\cdot j} = \frac{\sum_i X_{ij}}{n_j} \tag{12.1}$$

For example, if $j = 3$, we get $\sum_i X_{i3} = 630$, $n_3 = 5$, and thus $\overline{X}_{\cdot 3} = \frac{630}{5} = 126$.

The Grand Average

Next we want to introduce the notations n and $\overline{\overline{X}}$ ("X double bar"):

$$n = \sum_j n_j \tag{12.2}$$

and

GRAND AVERAGE

$$\overline{\overline{X}} = \frac{\sum_i \sum_j X_{ij}}{n} \tag{12.3}$$

---

where $\sum_i \sum_j X_{ij}$ means summation of all the X_{ij} in all groups. The notation n then stands for the total number of observations in all the available groups.

Since in our case $n_1 = 6$, $n_2 = 7$, and $n_3 = 5$, we get

$$n = 6 + 7 + 5 = 18$$

The **grand average**, $\overline{\overline{X}}$, is simply the *average of all the observations in the sample.* As one can easily verify, the total of all the available observations $\sum\sum X_{ij}$ is equal to 2,394, so that $\overline{\overline{X}} = \dfrac{2,394}{18} = 133$.

TOTAL VARIATION

Suppose we ignore the grouping of the sample observations and treat them all as observations drawn from the same population. We may then compute the **total sum of squares** by adding up the squared deviations of all of the sample observations from the grand average, $\overline{\overline{X}}$. Denoting the total sum of squares by SS_t, we find

---

TOTAL SUM OF SQUARES

$$SS_t = \sum_i\sum_j (X_{ij} - \overline{\overline{X}})^2 \tag{12.4}$$

Using the data of Table 12.2, we compute

$$SS_t = (120 - 133)^2 + (100 - 133)^2 + (174 - 133)^2 + \cdots$$
$$+ (180 - 133)^2 = 23,874$$

Next, as we shall see, the total sum of squares, SS_t, can be separated into two components: the sum of the squares *within* each group and the sum of squares *between* the groups.

WITHIN-GROUP VARIATIONS

To obtain the sum of squares within the groups, which we denote by SS_w, we add up the squared deviations of all observations from each group's average:[4]

SUM OF SQUARES WITHIN THE GROUPS

$$SS_w = \sum_j\sum_i (X_{ij} - \overline{X}_{\cdot j})^2 \tag{12.5}$$

In our example we obtain

for $j = 1$: $(120 - 119)^2 + (100 - 119)^2 + (174 - 119)^2 + (140 - 119)^2$
$+ (90 - 119)^2 + (90 - 119)^2 = 5,510$

for $j = 2$: $(200 - 150)^2 + (160 - 150)^2 + (110 - 150)^2 + (180 - 150)^2$
$+ (100 - 150)^2 + (160 - 150)^2 + (140 - 150)^2 = 7,800$

for $j = 3$: $(160 - 126)^2 + (80 - 126)^2 + (110 - 126)^2 + (100 - 126)^2$
$+ (180 - 126)^2 = 7,120$

and

$$SS_w = 5,510 + 7,800 + 7,120 = 20,430$$

The sum of squares within the groups is sometimes referred to as the **unexplained sum of squares;** since we do not identify any particular cause for this type of variation, we say that it is due to chance.

BETWEEN-GROUP VARIATIONS

The **sum of squares between the groups,** SS_b, is a measure that reflects the variability of the sample averages of the groups' means. We add up the squared deviation of each group average ($\overline{X}._j$) from the grand average ($\overline{\overline{X}}$) after it is multiplied by the number of observations in each group (n_j) to get

---

SUM OF SQUARES BETWEEN THE GROUPS

$$SS_b = \sum_j n_j(\overline{X}._j - \overline{\overline{X}})^2 \qquad (12.6)$$

We note that when a deviation $(\overline{X}._j - \overline{\overline{X}})^2$ is multiplied by n_j, the group size attains its appropriate weight relative to the sample size n_j. In our example we find

$$SS_b = 6 \cdot (119 - 133)^2 + 7 \cdot (150 - 133)^2 + 5 \cdot (126 - 133)^2$$
$$= 6 \cdot 196 + 7 \cdot 289 + 5 \cdot 49 = 3{,}444$$

Adding up SS_w and SS_b, we compute $20{,}430 + 3{,}444 = 23{,}874$, which is equal to the total sum of squares. Since this result holds in general, we come up with the following formula:

TOTAL SUM OF SQUARES

$$SS_t = SS_w + SS_b \qquad (12.7)$$

The **total sum of squares,** SS_t, can be broken down into the sum of the squares within the groups, SS_w, and the sum of the squares between the groups, SS_b. By this means the total variability is decomposed or analyzed into its components. Hence the name "analysis of variance."

MEAN SQUARES WITHIN GROUPS AND BETWEEN GROUPS

We are approaching the main purpose of this exercise. We must remember that our principal goal in conducting the analysis of variance is to find out whether the primary source of the variance is variations within the groups or between the groups. To that end we analyze SS_t into its components SS_w and SS_b, and compare the two components. Before we can compare the magnitude of these two components, however, we have to convert them into *average squared deviations*; thus both SS_w and SS_b must be divided by their respective degrees of freedom.

There are n deviations built into the SS_w formula. Each deviation is measured around its respective group average. Each group average is *estimated* by the sample observations and thus uses up one degree of freedom out of the n independent observations provided by the sample. Denoting the number of groups by J, we conclude that the number of degrees of freedom in SS_w is $n - J$. From this, we can derive the following equation:

MEAN SQUARED DEVIATIONS WITHIN THE GROUPS

$$MS_w = \frac{SS_w}{n - J} \tag{12.8}$$

Turning now to the SS_b, we realize that there are J group averages and J deviations of the averages around the grand average, $\overline{\overline{X}}$. Since by estimating $\overline{\overline{X}}$ we used up one degree of freedom, the number of degrees of freedom in SS_b is $(J - 1)$. The following equation can thus be derived:

MEAN SQUARED DEVIATIONS BETWEEN THE GROUPS

$$MS_b = \frac{SS_b}{J - 1} \tag{12.9}$$

In our example, where $n = 18$ and $J = 3$, we obtain

$$MS_w = \frac{SS_w}{n - J} = \frac{20,430}{18 - 3} = 1,362$$

and

$$MS_b = \frac{SS_b}{J - 1} = \frac{3,444}{3 - 1} = 1,722$$

THE TEST STATISTIC

Assuming that the sample observations are drawn independently from a normal distribution and that the null hypothesis is correct,[2] it can be shown that the ratio of MS_b to MS_w has an F distribution with $(J - 1)$ and $(n - J)$ degrees of freedom.[3]

---

[2] We assume in addition that the variances in the various groups are equal. Thus if H_0 does not hold, the groups differ in their means but are equal in variance.

[3] Assuming normality and that the null hypothesis is correct, and recalling our assumption that the variances in the various groups (σ^2) are equal to one another, the quantity $\dfrac{SS_b}{\sigma^2}$ can be shown to have a χ^2 distribution with $J - 1$ degrees of freedom. It follows that $\dfrac{SS_b}{(J - 1)\sigma^2} = \dfrac{MS_b}{\sigma^2} \sim \dfrac{\chi^{2(J-1)}}{J - 1}$. Similarly it can be shown that the quantity $\dfrac{SS_w}{\sigma^2}$ has a χ^2 distribution with $n - J$ degrees of freedom, so that $\dfrac{SS_w}{(n-J)\sigma^2} = \dfrac{MS_w}{\sigma^2} \sim \dfrac{\chi^{2(n-J)}}{n - J}$. From the definition of the F variable, it follows that

$$\frac{MS_b}{\sigma^2} \bigg/ \frac{MS_w}{\sigma^2} = \frac{MS_b}{MS_w} \sim F^{(J-1,n-J)}$$

$$\frac{MS_b}{MS_w} \sim F^{(J-1,n-J)} \tag{12.10}$$

Thus the ratio in Equation 12.10 serves as a test statistic. A ratio in the neighborhood of 1 indicates that MS_b is about the same magnitude as MS_w. In this case, the null hypothesis, which states that all the groups' means are equal, cannot be rejected. A computed ratio substantially greater than 1, on the other hand, indicates that the mean square deviations between the groups is large in relation to the mean square deviations within the groups, and we will tend to reject this null hypothesis.

Given a significance level α, we can easily determine the acceptance and rejection regions and accept or reject H_0 accordingly. Since the numerator of the test statistic is squared, large deviations between the samples' means imply that we obtain a large value for the test statistic. Thus the test is necessarily an upper-tail test; the critical value is given by $F_\alpha^{(J-1,n-J)}$ and the decision rule is

Do not reject H_0 if

$$\frac{MS_b}{MS_w} \leq F_\alpha^{(J-1,n-J)}$$

Reject H_0 if

$$\frac{MS_b}{MS_w} > F_\alpha^{(J-1,n-J)}$$

Returning to our example (Table 12.2) and assuming $\alpha = 0.05$, we find that

$$F_\alpha^{(J-1,n-J)} = F_{0.05}^{(2,15)} = 3.68$$

so that our decision rule is (see Figure 12.6)

$$\text{Do not reject } H_0 \text{ if } \frac{MS_b}{MS_w} \leq 3.68$$

$$\text{Reject } H_0 \text{ if } \frac{MS_b}{MS_w} > 3.68$$

Figure 12.6

Acceptance and rejection regions for a one-way ANOVA

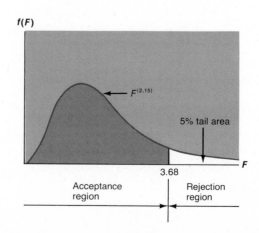

Substituting the figures derived earlier in the test statistic, we get

$$\frac{MS_b}{MS_w} = \frac{1,722}{1,362} = 1.264$$

and we cannot reject H_0. It is customary and convenient to summarize the sum of squares, the degrees of freedom, the mean squares, and the test statistic in a table such as Table 12.3. The numerical example is summarized in Table 12.4.

TABLE 12.3

One-Way Analysis of Variance Summary Table

Source of variation	Sum of squares	df	Mean squares	F
Between groups	$SS_b = \sum_j n_j(\bar{X}_{\cdot j} - \bar{\bar{X}})^2$	$J - 1$	$MS_b = \dfrac{SS_b}{J-1}$	
				$\dfrac{MS_b}{MS_w} \sim F^{(J-1,\,n-J)}$
Within groups	$SS_w = \sum_j \sum_i (X_{ij} - \bar{X}_{\cdot j})^2$	$n - J$	$MS_w = \dfrac{SS_w}{n-J}$	
Total	$SS_t = \sum_i \sum_j (X_{ij} - \bar{\bar{X}})^2 = SS_w + SS_b$			

TABLE 12.4

One-Way Analysis-of-Variance Summary Table: The Numerical Example

Source of variation	Sum of squares	df	Mean squares	F
Between groups	3,444	2	1,722	
				$\dfrac{1,722}{1,362} = 1.264$
Within groups	20,430	15	1,362	
Total	23,874	17		

Since the value of the test statistic falls within the acceptance region, we conclude that although MS_b is greater than MS_w, the difference may be due to sampling variation. Therefore we do not reject the null hypothesis.

SHORTCUT EQUATIONS FOR ONE-WAY ANALYSIS OF VARIANCE

The equations for SS_b and SS_w may be written in shortened form which is easier to use with hand calculators. Since some students may find these shortcut equations easier to work with, we will present them here. Let us define the quantities A, B, and C as follows:

$$A = \sum_i \sum_j X_{ij}^2 \qquad\qquad (12.11)$$

$$B = \frac{1}{n}\left(\sum_i \sum_j X_{ij}\right)^2 \qquad (12.12)$$

$$C = \sum_j \left(\frac{1}{n_j}\right)\left(\sum_i X_{ij}\right)^2 \qquad (12.13)$$

Quantity A is simply the sum of all the *squared* sample observations. Quantity B is the squared value of the sum of all the sample observations. Quantity C is the sum of the squares of the totals of all the groups, each group's total divided by the number of observations in the group. The computations of A, B, and C can be best understood by referring back to the data of Table 12.2:

$$A = 120^2 + 100^2 + 174^2 + \cdots + 180^2 = 342{,}276$$

$$B = \frac{1}{18}(120 + 100 + 174 + \cdots + 180)^2 = \frac{5{,}731{,}236}{18} = 318{,}402$$

$$C = \frac{1}{6}(120 + 100 + \cdots + 90)^2 + \frac{1}{7}(200 + 160 + \cdots + 140)^2$$

$$+ \frac{1}{5}(160 + 80 + \cdots + 180)^2 = 84{,}966 + 157{,}500 + 79{,}380 = 321{,}846$$

You can check to see (compare with Table 12.4) that

$$SS_b = C - B \qquad (12.14)$$

and

$$SS_w = A - C \qquad (12.15)$$

12.4 Two-Way Analysis of Variance

One-way analysis of variance is conducted to test the simultaneous equality of means of different probability distributions or groups. The groups are defined by a single classification—location, design, religion, company, and so on. But **two-way analysis of variance,** as its name suggests, deals with *the analysis of probability distributions or groups defined by dual classification.* In principle, one-way and two-way analyses of variance are similar: the purpose is to find out whether or not the means of all the populations considered are equal to one another. But in two-way analysis of variance we use sample data to test more than one hypothesis.

HOW TWO-WAY ANOVA WORKS: THE FUNDAMENTAL APPROACH

Explanation of two-way ANOVA will be greatly facilitated by a numerical example. Table 12.5 presents information on the production level of various crews working on two types of machines. The data measure the results of five days of production by three crews, measured in units per day. Thus five observations are given for each combination of machine and crew. The purpose of the two-way analysis of variance is threefold:

1. To determine whether a **column effect** exists, namely, that the mean production level varies among the three crews.
2. To determine whether a **row effect** exists, namely, that the mean production level varies between the two types of machine.
3. To determine whether an **interaction effect** exists, namely, that there is an interaction between the type of machine and the crew that uses it, in terms of productivity; that is, we wish to determine whether some specific combination of machine and crew results in greater productivity.

As in one-way ANOVA, we assign the index i to the table's rows and the index j to its columns, so that for Table 12.5 we have the values 1 and 2 for i ($i = 1, 2$) and the values 1, 2, or 3 for j ($j = 1, 2, 3$).

TABLE 12.5
Production (Units Completed) by Type of Machine and Crew

Type of machine	Crew		
	A ($j = 1$)	*B* ($j = 2$)	*C* ($j = 3$)
Type 1 ($i = 1$)	90 95 80 85 89	75 86 80 83 90	72 88 83 82 78
Type 2 ($i = 2$)	88 85 79 75 78	81 83 74 69 74	69 63 72 80 85

Why do we need to examine the differences in productivity? If a significant column effect, i.e., difference in productivity among the crews, is found, it may have implications for wage determination, employee promotion, and so on. If a significant row effect, i.e., difference in the machines' productivity, is found, it may lead management to use just one type of machine instead of two. Finally, there is a much more delicate question: is it possible that there is an interaction between crews and machines? It may be that Crew *A* produces much more with the first type of machine than with the second type, while the opposite may be true of Crew *B*. If there is evidence to support such a hypothesis, namely, that the interaction effect is found to be statistically significant, then the manager can use it in assigning crews to machines. If such an interaction exists, it is possible that having two types of machine may prove to be the best policy. Thus, a table of data such as Table 12.5 provides the basic input for two-way ANOVA.

To test the significance of the column effect we compute a statistic SS_{column} which is the sum of squared deviations of the average of the columns around the total sample average. Then we obtain the mean squared deviations for the columns, MS_{column}, by dividing SS_{column} by $J - 1$ where J is the number of columns in the sample. Thus $MS_{column} = SS_{column}/(J - 1)$. For testing the significance of the row effect we calculate the statistic SS_{row} which is the sum of squared deviations of the rows' averages around the total sample average. To obtain the mean squared deviation for the rows (MS_{row}) we devide SS_{row} by $I - 1$ where I is the number of rows in the sample. To test for interaction we calculate the statistic $SS_{\text{interaction}}$ and then derive the mean squares for interaction ($MS_{\text{interaction}}$) by dividing $SS_{\text{interaction}}$ by $(I-1)(J-1)$: $MS_{\text{interaction}} = SS_{\text{interaction}}/(I-1)(J-1)$.

The sum of squares not explained by the columns, rows, or interaction is referred to as the residual sum of squared: SS_{residual} and to obtain the mean squares of the residuals we divide $SS_{\text{residuals}}$ by $IJ(K-1)$, where K is the number of observations within each cell.

Testing the existence of a significant column effect is done by using the ratio $MS_{\text{column}}/MS_{\text{residual}}$ which has an $F^{[(J-1),IJ(K-1)]}$ distribution. Testing the existence of a significant row effect is done by using the ratio $MS_{\text{row}}/MS_{\text{residual}}$ which has an $F^{[(I-1),IJ(K-1)]}$ distribution and testing for significant interaction effect is done by the ratio $MS_{\text{interaction}}/MS_{\text{residual}}$ which has an $F^{[(I-1)(J-1),IJ(K-1)]}$ distribution. Although the computations involved in two-way ANOVA are not too difficult to handle with hand calculators, particularly when the sample is not large, it is undoubtedly more convenient to omit the computational step and leave the burden to the computer. Many computer programs are readily available and can be executed without extensive knowledge of computer programming. We turn now to the interpretation of computer outputs.

12.5 Interpreting Computer Printouts of Two-Way Analysis of Variance

This section is not intended to be a substitute for a computer programming manual, or to provide a specific description of the way ANOVA programs are executed. Rather, it is intended to encourage you to inquire about such programs by showing you how simple it is to run ANOVA programs by means of computer services, and to provide an explanation about the correct interpretation of the results.

Figure 12.7 shows results from a computer printout of the two-way ANOVA for the data of Table 12.5.[4] On the left-hand side we have added a sequential number to each line to facilitate reference throughout our explanation.

Figure 12.7

Results from computer printout of a two-way ANOVA

```
1. TWO  WAY  EXAMPLE
   FILE   NONAME   (CREATION  DATE  =  11/05/80)
   /***********A N A L Y S I S   O F   V A R I A N C E***************
                     PROD
2. {        BY  CREW
                  MACHINE
   \***************************************************************

                              SUM OF                 MEAN            SIGNIF
3. SOURCE  OF  VARIATION      SQUARES      DF        SQUARE     F     OF  F
4. MAIN  EFFECTS              610.496       3        203.499   5.128  0.007
5.     CREW                   270.463       2        135.231   3.408  0.049
6.     MACHINE                340.033       1        340.033   8.569  0.007
7. 2-WAY  INTERACTIONS          0.067       2          0.033   0.001  0.999
8.   CREW      MACHINE          0.067       2          0.033   0.001  0.999
9. EXPLAINED                  610.562       5        122.112   3.077  0.028
10. RESIDUAL                  952.402      24         39.683
11. TOTAL                    1562.964      29         53.895
12.    30 CASES  WERE  PROCESSED.
13.     0 CASES  (   0.0  PCT)  WERE  MISSING.
```

[4] The computer printout used is that of the Statistical Package for Social Sciences (SPSS), a computer package available in most computer centers.

To generate the ANOVA table, 44 computer cards were used—14 program cards and 30 data cards (one per observation). The 14 cards were used to inform the computer about the package to be used, the number and names of the variables involved, the number of observations, and so on. The specifics of how to set up these cards are spelled out in a manual for the computer package used—the Statistical Package for Social Sciences (SPSS). Most universities provide assistance to interested students in setting up their computer programs.

Let us interpret the output line by line.

Line 1   A job name provided by the user. We called our job "two-way example."

Line 2   A title indicating that two-way analysis of variance is to follow, with data classified by crew and type of machine.

Line 3   Titles of the various columns.

Line 4   The sum of "main effects" (row and column effects).

Line 5   Statistics on the columns: the sum of the squares is 270.463, the degrees of freedom are 2, and the mean squares is $270.467/2 = 135.231$. Under the column headed "F," the F statistic for testing for the existence of column effects is given. Specifically, the F value 3.408 is equal to $MS_{column}/MS_{residual} = 135.231/39.683$. In the next column, headed "Signif. of F," we read "0.049." This means that the computed value (that is, 3.408) leaves a right-hand tail equal to 4.9 percent on the $F^{(2,24)}$ distribution. It is thus clear that if we test the hypothesis that there is no column effect (H_0) against the hypothesis that there is a column effect (H_1), the value 0.049 is the maximum α value under which H_0 will not be rejected. For example, if $\alpha = 0.050$ (that is, greater than 0.049), H_0 will be rejected, while if $\alpha = 0.048$ or lower, H_0 will be accepted.

Line 6   Similar to line 5, except that here the statistics apply to the rows.

Line 7   The sum of all two-way interactions. Since in our analysis there is only one kind of two-way interaction, line 7 is identical to line 8. When an analysis of variance involving three or more dimensions is studied, the interaction between each pair of variables is listed in line 8. Line 7 shows the sum of all the interactions.

Line 8   Statistics on the interaction effect. The F statistic is 0.001, and the existence of interaction should be rejected in all tests of hypotheses when an α value of less than 0.999 is used.

Line 9   The sum of squares here is the total of the sums of squares in lines 5, 6, and 7. Similarly, the degrees of freedom are the sum of the degrees of freedom in those lines. This line provides a sort of simultaneous test for rows, columns, and interaction. In our example the F statistic is 3.077 ($= 122.113/39.683$) and the significance of F is 0.028. This means that for all α greater than 2.8 percent we should reject a hypothesis that no effect (either row or column or interaction) exists.

Line 10   The sum of squares, degrees of freedom, and mean squares for the residuals.

Line 11   The total sum of squares and the respective degrees of freedom and mean squares. The sum of squares and degrees of freedom in this line is the sum of those in lines 5, 6, 7, and 10.

Line 12   The number of observations in the sample.

Line 13   The number of missing values that the computer has come across.

We shall give one more example in which a computer output will be given and interpreted.

EXAMPLE 12.2

In a study on inflation, it is necessary to determine whether price increases of U.S. companies vary according to the size of the company or according to the degree of concentration in the industry in which the company operates. (If there is perfect competition in the industry, concentration is minimal; if there is a monopoly, concentration is at its highest level.) Assume that five levels of concentration and three company sizes are distinguished, and that four observations (four companies) for each combination of size and concentration are available, as summarized in Table 12.6. Note that the numbers in the table are rates of price increases by each company sampled in the past year. The ANOVA table is given in Figure 12.8. The conclusions from the ANOVA table are as follows (we assume $\alpha = 0.05$ for all tests):

1. Row effect is *not* significant ($0.172 > 0.05$). We do not reject H_0, which claims that the rate of price increase is not a function of the size of the firm.
2. Column effect *is* significant ($0.001 < 0.05$). The rate of price increase varies among the various concentration levels.
3. There is no significant interaction effect.

TABLE 12.6
Rates of Price Increases of Sampled Companies, by Concentration and Size

Size	Concentration 1	2	3	4	5
1	3	4	5	9	12
	5	7	7	10	11
	7	7	9	10	9
	4	7	6	7	9
2	12	5	8	10	10
	10	9	8	14	9
	3	7	9	10	11
	5	7	10	11	8
3	5	6	7	11	6
	3	10	11	20	15
	4	9	8	9	9
	2	8	11	8	9

This example shows that once the *meanings* of the various effects are understood, one can easily use the analysis of variance without having to confront any of the equations presented earlier.

A word or two should be said here on the relationship between one-way and two-way analysis of variance. In testing the significance of a possible interaction effect, one must employ a two-way analysis of variance with at least two observations in each cell. In fact, testing the significance of an interaction effect is the main purpose of two-way analysis. One can, however, use two-way ANOVA even when only one observation is available in each cell. In this case it is impossible to test hypotheses regarding interaction effects, but we can test hypotheses

Figure 12.8
Results from computer printout of a two-way ANOVA

```
ADDITIONAL EXAMPLE
FILE   NONAME   (CREATION  DATE  =  11/05/80)
**************A N A L Y S I S   O F   V A R I A N C E***************
          INFLATN
     BY  SIZE
         CONCENT
*********************************************************************
                       SUM OF              MEAN            SIGNIF
SOURCE OF VARIATION     SQUARES    DF      SQUARE    F      OF F
MAIN EFFECTS            249.467     6      41.578   6.785   0.001
   SIZE                  22.300     2      11.150   1.820   0.172
   CONCENT              227.167     4      56.792   9.268   0.001
2-WAY INTERACTIONS       54.033     8       6.754   1.102   0.380
   SIZE      CONCENT     54.033     8       6.754   1.102   0.380
EXPLAINED               303.500    14      21.679   3.538   0.001
RESIDUAL                275.747    45       6.128
TOTAL                   579.247    59       9.818
     60 CASES WERE PROCESSED.
      0 CASES (   0.0 PCT) WERE MISSING.
```

concerning column and row effects. A two-way analysis of variance is not a must here, since one-way analysis can be employed twice instead: one analysis can be conducted for row effect, and one for column effect. Employing a two-way ANOVA is more economical, however, since it requires a smaller sample size for a given power of the test than that required to perform one-way ANOVA twice.

Problems

12.1. Explain the meaning of the sum of squares *within* groups and the sum of squares *between* groups in analysis of variance.

12.2. The following table shows advertising expenditures as a percentage of sales for three firms in the same industry over five years.

Year	Firm A	Firm B	Firm C
1976	1.1	1.2	1.3
1977	1.1	1.2	1.3
1978	1.1	1.2	1.3
1979	1.1	1.2	1.3
1980	1.1	1.2	1.3

Test the hypothesis that the average advertising expenditure as a percentage of sales is equal for the three firms. Reach a conclusion without consulting any statistical tables. Are the shortcut equations of ANOVA helpful for this problem? Explain.

12.3. The productivity of three workers, measured by number of units of output per day, is reported in the table below:

Day	Worker A	Worker B	Worker C
1	5	10	20
2	10	10	5
3	15	10	5

Use ANOVA to test the null hypothesis that the productivity of the three workers is equal. Use $\alpha = 0.05$. How would you change your conclusion for a significance level of $\alpha = 0.50$? Reach a decision without using any statistical tables. It is recommended that you not use the shortcut equations. Can you explain why?

12.4. Is it possible to get a negative MS_b in a sample? A negative MS_w? Explain.

12.5. Suppose the total number of observations in a one-way analysis of variance is $n = 25$. Find the critical value F_α ($\alpha = 0.05$) for J groups, where J takes on the following alternative values:

(a) $J = 2$.
(b) $J = 3$.
(c) $J = 4$.
(d) $J = 5$.

For a given n, what is the relationship between the critical value, F_α, and the number of groups, J?

12.6. What is the source of the name "analysis of variance"? Use a simple numerical example to demonstrate that

$$SS_t = SS_w + SS_b$$

12.7. Explain the difference between one-way and two-way analysis of variance.

12.8. Explain the meaning of *interaction*. Give an example.

12.9. Suppose the cost per unit of items produced by four production processes is examined in order to find out whether the mean cost per unit is the same for all four processes. Suppose we carry out the analysis twice: once when the cost is expressed in dollars and once when it is expressed in tens of dollars. What is the impact of the units used on SS_t, SS_w, SS_b? Do the units have an impact on the calculated F value? Explain.

12.10. The table below shows the market value of property owned by three individuals in New York and three individuals in Chicago (in thousands of dollars).

New York	Chicago
40	20
50	30
60	40

Test the hypothesis that the average market value of property held by individuals in both cities is the same against the alternative that the averages are not the same:

$$H_0: \mu_1 = \mu_2$$

$$H_1: \mu_1 \neq \mu_2$$

Test the hypothesis twice: once by using the t distribution and once by using the F distribution ($\alpha = 0.05$). What is the relationship between the results? Is it possible to reach opposite conclusions by the two procedures? Explain.

12.11. The monthly sales of a given product in three stores in January, February, and March have been recorded and are listed below in thousands of dollars.

Month	Store 1	Store 2	Store 3
January	5	3	1
February	6	10	10
March	7	8	13
Average monthly sales	6	7	8

Suppose the assumptions required for analysis of variance indeed hold, and suppose we tested the null hypothesis (under which the average monthly sales are the same for all three stores) and found that it is accepted at a given significance level.

Change the sample data in such a way that the changes will be in the *direction* that will make the null hypothesis more likely to be *rejected*. When making the changes, be sure to keep the average monthly sales of the three stores unchanged. Explain your answer.

12.12. Four new drugs are tested for effectiveness by two hospitals. In each hospital a team of doctors (independent of the team at the other hospital) uses a one-way analysis of variance to test for the equality of the drugs' effectiveness. Suppose the two teams obtain the very same value for the test statistic: $F = \dfrac{MS_b}{MS_w}$. Nevertheless, and despite the fact that both teams have used the same significance level for their tests, they have reached different conclusions. One team has accepted the hypothesis that the four drugs have the same effectiveness while the other team has rejected this hypothesis. Is it possible? Why? Explain by means of a numerical example.

12.13. U.S. companies publish annual reports in which they provide balance sheets, income statements, and other information for their stockholders. Since the annual report provides a comprehensive summary of a firm's operations and financial position, the firm needs time to prepare it. Therefore, there is a time lag between the end of the year covered by the annual report and the publication date of the report. A sample of the lag (in days) for three firms in the last three years is given below.

Firm	First year	Second year	Third year	Average lag
Bank	10	20	30	20
Computer manufacturer	60	50	40	50
Food chain	80	90	70	80

(*a*) Apply a one-way ANOVA to test the hypothesis that the mean time lag (in days) of the three firms is equal. Set $\alpha = 0.01$.

(*b*) Retest the same hypothesis, this time using the following data:

Firm	First year	Second year	Third year	Average lag
Bank	5	20	35	20
Computer manufacturer	100	30	20	50
Food chain	80	40	120	80

(*c*) Since the average time lag is the same in parts *a* and *b*, how do you account for the difference in the answers obtained?

(*d*) Draw a chart showing the sample range of the time lag for each of the three firms in parts *a* and *b*. Does the relationship between the ranges help to explain the accept–reject decision as obtained in *a* and *b*?

12.14. The Dow Jones Industrial Average is one of the most widely quoted stock market indexes. This index is based on a sample of 30 industrial stocks, the so-called Dow Jones industrials (DJI), and the average is arrived at by adding up the prices of the 30 stocks and dividing the total by the so-called Dow Jones divisor to obtain an average stock price adjusted for past stock splits and stock dividends. (The Dow Jones divisor in 1977 was 1.443, so that when the prices of the 30 DJI stocks added up to $1,214.66, the Dow Jones Industrial Average stood at $1,214.66/1.443 = $841.76.)

A similar method is used to calculate the earnings on the Dow Jones Industrial Average: the reported quarterly earnings per share for the 30 DJI stocks are added up and the sum is divided by the DJI divisor. The result reflects average earnings per share adjusted for the cumulative effect of stock splits and stock dividends. The following table gives the quarterly earnings or DJI average for the period 1972–78.

Quarter	1978	1977	1976	1975	1974	1973	1972
1st	22.04	21.91	23.12	16.91	22.48	19.19	14.32
2nd	29.66	27.52	25.85	17.04	26.68	22.88	17.30
3rd	26.40	16.18	23.50	18.37	26.73	20.26	15.73
4th		23.49	24.25	23.34	23.15	23.84	19.76
Total		89.10	96.72	75.66	99.04	86.17	67.11

Source: *Barron's*, December 25, 1978.

In 1977, for instance, the fourth-quarter earnings figure of 23.49 obviously translates into $23.49 · 1.443 = $33.90 as the total of the reported quarterly earnings of the 30 DJI stocks. Thus the average dollar earnings for the fourth quarter were $33.90/30 = $1.13.

A glance at the earnings table reveals a fairly pronounced cyclical pattern: second- and fourth-quarter earnings are on the whole higher than first- and third-quarter earnings. This may be due to the effect of seasonal shopping patterns: Easter in the second quarter and Christmas in the fourth quarter are usually associated with increased sales and hence increased quarterly earnings.

Use one-way ANOVA to check whether our impression of a seasonal pattern is statistically significant. Note that on December 25, 1978, we did not yet have the earnings of the fourth quarter of 1978, and hence we have groups with unequal numbers of observations.

12.15. Table P12.15 shows yields on U.S. Treasury bills with various maturities, i.e., 7 days, 1 month, 2 months, 3 months, 6 months, and 1 year, as recorded on three days in May 1981, three days in June 1981, three days in July 1981, and three days in August 1981.

(a) Test the hypothesis that the yields on the Treasury bills were the same in the four months shown.

(b) Test the hypothesis that the yields on the Treasury bills were the same for the various maturities.

(c) Test the hypothesis that there is an interaction between the maturity and the month.

Use $\alpha = 0.01$ for the above tests.

TABLE P12.15

Month	Maturity					
	7 days	1 month	2 months	3 months	6 months	1 year
May 1981	12.3	13.7	15.0	15.4	15.4	15.2
	17.5	16.3	17.3	17.5	17.1	16.5
	16.7	16.2	16.2	16.1	15.6	15.2
June 1981	16.1	16.2	16.0	15.6	15.4	14.7
	17.7	15.9	15.2	15.0	14.8	14.7
	14.0	14.2	14.4	14.6	14.8	14.6
July 1981	14.1	14.3	14.6	14.9	14.6	14.8
	16.6	14.9	15.2	15.2	15.2	15.1
	13.8	14.7	15.3	15.7	16.3	16.1
August 1981	14.7	14.9	15.2	15.6	16.2	16.2
	16.5	15.6	15.8	16.0	16.9	16.6
	14.5	15.0	15.9	16.1	16.9	16.8

Source: *Wall Street Journal*, various issues.

12.16. Table P12.16 represents research and development (R&D) spending as a percentage of sales of a few large U.S. companies in 1977. The data are divided into three industry groups (paper, tires and rubber, and information processing).

(*a*) Calculate the sample average of spending on R&D as a percentage for each industry group.

(*b*) In view of the fact that the data provided are only a sample, do you hold the view that the mean percentage spending of the industries is the same? Use a 1 percent significance level in a test that will validate your answer. Use the shortcut equation and present your calculation.

(*c*) What assumptions do you have to make in order to be able to employ ANOVA?

TABLE P12.16

Company	R&D spending as percent of sales	Company	R&D spending as percent of sales
Paper		Tires, Rubber (*cont.*)	
Bemis	1.8	Firestone Tire & Rubber	1.7
Boise Cascade	0.2	Goodyear Tire & Rubber	2.1
Consolidated Paper	0.7	Uniroyal	1.6
International Paper	0.7		
Lydall	1.9	Information Processing	
Rexham	1.3	Amdahl	15.8
Union Camp	1.0	Apple Computer	6.2
		Control Data	6.6
Tires, Rubber		Digital Equipment	7.9
Amerace	2.5	Honeywell	6.0
Armstrong Rubber	1.5	IBM	5.8
Carlisle	0.4	Prime Computer	7.6

Data from *Business Week*, July 6, 1981.

12.17. Table P12.17 lists the annual rate of return on assets for a sample of seven telephone companies for the ten-year period 1965–74. The telephone industry is regulated: the rates are fixed by government agencies so as to provide an adequate compensation to stockholders. If regulation is effective, the rates of return they earn should be roughly the same.

TABLE P12.17

Year	American Telephone & Telegraph (AT&T)	General Telephone & Utilities	Cincinnati Bell	Mid-Continent Telephone	Mountain States Telephone & Telegraph	Rochester Telephone & Telegraph	United Tele-communi-cation
1965	7.8	5.6	7.5	6.8	7.1	6.4	7.3
1966	7.9	7.0	7.6	6.7	7.2	6.4	7.8
1967	7.8	6.5	7.7	6.5	7.2	6.8	8.0
1968	7.5	6.7	7.3	7.5	6.9	6.8	8.2
1969	7.6	7.1	6.8	7.4	7.7	7.1	8.7
1970	7.5	7.4	7.3	7.4	7.6	7.4	7.4
1971	7.5	7.4	7.4	7.9	7.6	7.9	7.4
1972	7.7	7.9	7.1	7.8	7.5	8.1	7.5
1973	8.0	7.9	7.9	8.3	7.9	8.5	7.6
1974	8.0	7.9	8.6	8.5	7.4	7.3	7.7

Source: Derived from the companies' annual reports.

Use one-way analysis of variance to test the null hypothesis that the average profitability (as measured by the rate of return on assets over the sample period) is the same for the seven telephone companies. In your analysis assume $\alpha = 5$ percent level of significance.

Do you think that government regulation is effective?

12.18. Business firms generally use two sources of funds to finance new activities: equity (shareholders' capital, including common stock, accrued earnings, and so on) and debt (in the form of bonds issued to the public, long-term loans from banks and government, and the like). The ratio of debt to total invested capital is known in finance literature as *financial leverage*. A firm's financial leverage is of great importance and is closely watched by bankers, financial analysts, and investors. It is generally assumed that companies in the same industry should have roughly the same financial leverage, and managers will attempt to correct any obvious deviation from the industry mean. Across industries, on the other hand, financial leverage is expected to vary in line with the overall financing needs of each industry.

Table P12.18 lists the 1977 ratio of debt to total invested capital for a sample of 48 companies drawn from four industry groups.

Calculate the average financial leverage for each group and use one-way ANOVA to test whether the group means are equal. Discuss your findings.

Arrange your calculations on detailed worksheets and keep them for the next problem.

Hint: Use the shortcut equations for one-way ANOVA (12.11 through 12.15).

TABLE P12.18

Company	Ratio of debt to total invested capital (percent)	Company	Ratio of debt to total invested capital (percent)
Airlines		Instruments	
American Airlines	55.3	Bausch & Lomb	15.8
Braniff International	59.4	Beckman Instruments	17.8
Continental Air Lines	45.5	Bell & Howell	22.4
Delta Air Lines	18.3	Foxboro	11.4
Eastern Air Lines	66.1	General Signal	12.0
National Airlines	23.3	Hewlett-Packard	1.0
Northwest Airlines	11.4	Johnson Controls	18.1
Pan Am	70.7	Perkin-Elmer	19.1
Tiger International	74.5	Sybron	28.4
TWA	65.0	Talley Industries	42.6
UAL	53.0	Technicon	21.4
Western Airlines	43.9	Tektronix	9.9
Food and lodgings		Office equipmt., computers	
ARA Services	24.4	Addressograph/Multigraph	26.3
Caesars World	65.0	Burroughs	8.5
Denny's	40.0	Control Data	22.0
Hilton Hotels	43.7	Data General	25.4
Holiday Inns	36.7	Digital Equipment	10.6
Howard Johnson	4.0	Honeywell	15.3
Hyatt	73.1	IBM	2.0
Marriott	48.3	Memorex	51.5
McDonald's	49.3	NCR	29.5
Ramada Inns	63.8	Pitney Bowes	33.1
Sambo's Restaurants	63.2	Sperry Rand	21.8
Webb	42.5	Xerox	24.8

Data from *Business Week*, October 16, 1978.

PART 4

REGRESSION AND CORRELATION (INDUCTION CONTINUED)

Part 3 of this book deals with statistical inference; Part 4 continues this topic but focuses on the relationship between two or more variables, using regression analysis, which is one of the analytical tools that are most frequently used in many areas of business and economics. Chapter 13 and most of Chapter 14 focus on simple regression analysis: the former presents the techniques of fitting a regression line, and the latter presents the theory and applications. An extension of the analysis to multiple regression is presented in Chapter 14, and Chapter 15 is devoted to simple and multiple correlation coefficients.

Part 4 includes applications, examples, and problems in cost accounting (allocation of indirect costs), finance (measuring security risk), marketing (investment in advertising), economics (income and consumption), the relationship between return on investments and their risk, the relationship between interest rates and inflation, and many other areas.

CHAPTER THIRTEEN OUTLINE

13.1 Introduction
Why Study Regression Analysis?
Simple and Multiple Regression Analysis

**13.2 Fitting a Regression Line: Preliminary
Considerations**
The Fitted Line Basic Equation
Fitting the Line: Methods and Criteria

13.3 How the Ordinary Least-Squares Method Works
Using the OLS Method
What the Regression Line May Indicate

Key Terms

simple regression analysis
multiple regression analysis
dependent variable
independent (explanatory)
  variable
scatter diagram

ordinary least-squares method
normal equations
direct relationships
inverse relationships
uncorrelated variables

13
SIMPLE LINEAR REGRESSION: THE TECHNIQUE

13.1 Introduction

Our exposure to statistical inference in previous chapters has ranged from inference concerning a single parameter of a distribution to inference concerning parameters of several distributions (ANOVA). Thus far, however, we have not fully treated the relationship between two or more variables. Regression analysis, which deals with the way one variable tends to change as another variable changes (or other variables change), is one of the ways of dealing with this question. By employing regression analysis, we can find out not only whether a relationship between variables exists, but also the direction and amount of change that can be expected in one variable when another variable changes by a given amount. To illustrate, suppose we want to analyze the impact on sales of a TV campaign. By employing regression analysis we can estimate the amount of dollar change in sales resulting from a given incremental expenditure on TV commercial advertising. In addition, we can find out the simultaneous impact on sales of other variables and assess the impact of sampling errors on our estimates. Regression analysis is perhaps the most frequently used statistical procedure in business and economics; it is so important that we will devote not only this chapter, but the next two as well, to considering it.

WHY STUDY REGRESSION ANALYSIS?

The relationship between variables needs to be studied for several reasons. First, it must be determined whether such relationships exist or not. For example, at one time it was questionable whether a relationship existed between people's smoking habits and the incidence of lung cancer. This relationship has been established by various studies in recent years, and the findings have discouraged many people from smoking. Other studies have tried to determine whether music played in industrial plants tends to increase productivity, and if so, what is the "optimal" volume (from the firm's point of view).

Second, the direction, the strength, and the type (linear, nonlinear, and so on) of relationship is also of great importance in many studies and deserves careful investigation. Specific knowledge of the relationships between variables may help us to understand complex phenomena and improve our predictions concerning the future values of some variables. For example, the quantity of agricultural products available in the summer is, among other things, a function of weather conditions during the preceding spring. The more favorable the weather conditions, other things being equal, the more agricultural products in the market. Basic rules of supply and demand show that the greater the supply, the lower the price. Since spring weather conditions become known before the summer, it is important to study the relationship between the spring weather conditions and the quantity of supply in the following summer in order to predict prices of agricultural products. Or consider the way the federal government determines its monetary and fiscal policies with the objective of stabilizing the national economy in future months at a low unemployment level, stable price level, and rising gross national product (GNP). It is quite difficult to achieve these goals simultaneously, and good predictions of future economic events are necessary. There are some economic variables whose current behavior today serves as an indicator of future unemployment, inflation, and GNP trends. Among these variables are current investment trends, current levels of corporate inventory, current prices on the stock exchange, and many more. The more we know about the relationship between the *current* levels of such variables and *future* unemployment, inflation, and GNP trends, the better and more efficient will be the monetary and fiscal actions that the federal government can adopt today.

SIMPLE AND MULTIPLE REGRESSION ANALYSIS

In certain cases we deal with relationships between two variables only; the regression analysis we perform in these cases is called **simple regression analysis.** In business statistics, however, we often deal with the relationship between one variable (customarily denoted by Y) and other variables (customarily denoted by $X_1, X_2, X_3, \ldots, X_K$). This type of regression analysis is called **multiple regression analysis.** For example, the time it takes for a truck to deliver goods from point A to point B is a function of the distance between the two points as well as of a host of other variables—the condition of the roads, the density of traffic, the type of truck driven, and so on. If we use regression analysis to determine the relationship between time and distance only, we perform a simple regression analysis, while if we relate the time to other variables as well, we perform a multiple regression analysis.

13.2 Fitting a Regression Line: Preliminary Considerations

When we study the relationship between two variables by means of simple regression analysis, we distinguish between the dependent variable, denoted by Y, and the independent variable, denoted by X. The **dependent variable** is usually *one whose value depends on the value of the other variable*. The **independent variable** is also known as the **explanatory variable**, since its *value normally gives at least a partial explanation of the behavior of the dependent variable*. When regression analysis is applied to the impact of advertising expenditures on sales, for example, "sales" is the dependent variable and "advertising expenditures" is the independent, or explanatory, variable.

Let us look now at Figure 13.1, which shows the relationship between annual household income (the independent variable) and household consumption (the dependent variable). We present only households with annual incomes of $10,000, $20,000, and $30,000. For each income level, nine to ten observations are marked (though of course many more belong to the population), showing various levels of consumption. The diversity of consumption levels among households of the same income level is due to variations in household size and wealth (assets), wealth and income of relatives, habits, philosophy of life, and so on. For each household, both income and consumption levels are represented by one dot in the diagram. The diagram shows that the higher the income, the higher the *average* level of consumption of the household. Diversity exists in the consumption of individual households, however, and thus for each level of income, various consumption levels are observed. The dependent variable is in fact a random variable, though the independent variable is not. In other words, given a certain level of income, it is impossible to determine the consumption level of a household before that household's consumption is observed.

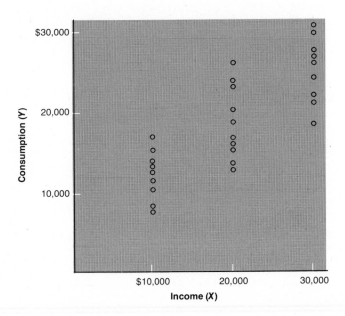

Figure 13.1

Relationship between income (X) and consumption (Y)

The data presented in Figure 13.1 are atypical in the sense that quite a few observations are shown for each of the selected values of the independent variable. The presentation is intended to demonstrate the point that the Y values obtained in a sample for a given value of X differ from one another: they are in fact random. In a more typical case, there will be few observations—often only one—for each of the observed values of the independent variable, X. When these observations are plotted in the (X, Y) plane, a **scatter diagram** is obtained, as in Figure 13.2, the data for which are given in Table 13.1. Note that each observation consists of an income value as well as a consumption value. Ten observations, then, consist of twenty pieces of data. A glance at Figure 13.2 reveals two important facts. First, there appears to be a direct relationship between X and Y: consumption tends to rise as income rises. Second, the observations are scattered and do not lie along a straight line. In order to characterize the relationship between the dependent and independent variables, we wish to draw a straight line through the scattered data points in such a way that it will best "fit" the data. The data points will, as we have mentioned, deviate from the fitted line. The objective in fitting a line is to minimize any errors involved.

Figure 13.2

Scatter diagram based on the data from Table 13.1 and a fitted line $\hat{Y} = a + bX$

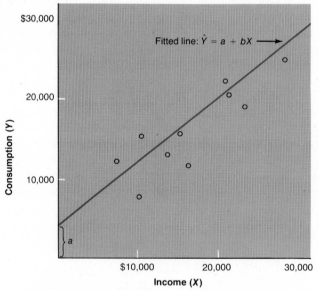

TABLE 13.1

Ten Observations of Annual Income and Household Consumption

(thousands of dollars)

Observation i	Income X_i	Consumption Y_i
1	$10.0	$15.0
2	18.0	23.5
3	28.0	25.0
4	7.0	12.5
5	13.0	13.0
6	16.0	12.0
7	22.0	19.0
8	15.0	16.0
9	20.0	21.0
10	10.0	8.0

THE FITTED LINE BASIC EQUATION

A straight line relating X to Y is given by

$$\hat{Y} = a + bX \tag{13.1}$$

where a is the vertical intercept (or Y intercept), b is the slope of the line, and $\hat{Y}$ is the fitted value of X (see Figure 13.2). When searching for a line that fits the data, we want to find the values a and b that minimize total sampling errors involved.

FITTING THE LINE: METHODS AND CRITERIA

One way of fitting a line through the scattered data is to try a freehand fitting: we simply do our best by approximation. This method is certainly unacceptable, however, because of the large potential errors in determining the line and the resulting inability to quantify the sampling errors. Clearly, a quantitative method must be developed which will uniquely determine the best line for given sample data. Let us examine some of the possible methods.

One possibility is to find the line that minimizes the sum of *horizontal* deviations of each point from the line (see Figure 13.3). This method is also unacceptable, however, since the horizontal distance between a point and the line simply does not measure any sampling error. As we explained earlier, the dependent variable, Y, is a random variable, while the independent variable is not; therefore minimizing deviation in X is meaningless. For a given value of X, X_0, the fitted Y value is $\hat{Y}_0 = a + bX_0$. If the sample observation Y_0 is equal to $\hat{Y}_0$, the observation lies on the fitted line, and there is no deviation of the observation from the line. If the observation Y_0 is greater or less than $\hat{Y}_0$, it lies above or below the fitted line, and a positive or negative deviation exists. It is the deviation in Y that we need to minimize, not that in X.

Given, then, that the vertical rather than the horizontal deviations from the fitted line demand our attention, we may wish to fit to the data the line that minimizes the sum of all the vertical deviations:

$$\sum_{i=1}^{n} (Y_i - \hat{Y}_i)$$

where n is the sample size, Y_i is the Y value of the ith observation, and $\hat{Y}_i$ is

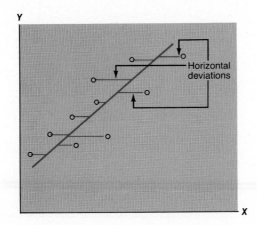

Figure 13.3

Horizontal deviations from the fitted line

the corresponding Y value on the fitted line, as shown in Figure 13.4. While the minimizing of $\sum_{i=1}^{n}(Y_i - \hat{Y}_i)$ is possibly a very intuitively appealing choice of a criterion for fitting a line through the scattered points, it is perhaps the worst criterion one could choose. If we follow this criterion, we will get a line that does not even get close to the sample points. The reason becomes obvious when we study Figure 13.4. Here two lines are presented, one that passes through the sample points and another that passes far above them. The line passing through the points (clearly a better choice for our purpose) gives a sum of deviations of +0.4, while the other line, which obviously is far from being the line we are looking for, gives a much *lower* sum of deviations: −115.6. It is clear that the higher the line above the sample points, the lower the quantity $\sum_{i=1}^{n}(Y_i - \hat{Y}_i)$; moreover, the line that minimizes this quantity passes at an infinite distance above the data points. Obviously this criterion is inappropriate.

Instead of minimizing the quantity $\sum_{i=1}^{n}(Y_i - \hat{Y}_i)$ and sending the line off to infinity, we can force the line through the sample points by setting this quantity at zero. In other words, the criterion we want to examine now is the selection of a line that brings the sum of all the deviations to zero: $\sum_{i=1}^{n}(Y_i - \hat{Y}_i) = 0$. This criterion, however, also suffers from a serious drawback. Consider a small sample of five observations, as follows:

Observation	Y	X
1	3	3
2	5	2
3	6	5
4	4	7
5	7	8

Figure 13.4

The sum of the deviations from two arbitrary (parallel) lines

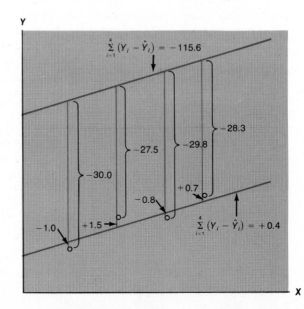

The scatter diagram is presented in Figure 13.5. In Figure 13.5a the line $\hat{Y}$ = 3.46155 + 0.30769X is drawn and the vertical deviation of each observation from the line is marked. A simple check shows that the sum of the deviations is equal to zero, so that the line meets the criterion. Note that the line passes through the point of averages $(\bar{X}, \bar{Y})$. In Figure 13.5b the same sample points are shown, but this time a different line is drawn through the points of averages: $\hat{Y}$ = 7.5 − 0.5X. The vertical deviation of the observations from the line are marked here too, and as we can see, they add up to zero again. This line, then, also meets the criterion. In fact, an infinite number of lines can be drawn through the point $\bar{X}$, $\bar{Y}$ and all of them meet the criterion. Most of these lines fall far short of characterizing the relationship between X and Y; consequently this criterion is faulty as well.

Having dismissed the previous criteria for fitting a line as inappropriate, let us examine the criterion of bringing the magnitude $\sum_{i=1}^{n} |Y_i - \hat{Y}_i|$, the sum of the absolute deviations, to a minimum. This method is significantly better than the ones presented earlier, but it suffers from being complex mathematically. It also results in a line whose statistical properties are inferior to those of the line produced by the criterion we propose next.

The criterion we propose here is the one we shall use: the selection of the line that minimizes the sum of squared deviations from the line. If you think that minimizing the absolute deviations from the line and minimizing the squared deviations result in the same regression line, let us consider a simple example to show the contrary. Suppose a small sample of three observations provides the following data:

X	Y
2	6
4	14
6	10
$\bar{X} = 4$	$\bar{Y} = 10$

Figure 13.6 shows the scatter diagram with two alternative lines fitted through the data. In Figure 13.6a the line $\hat{Y} = 6 + X$ is drawn through the

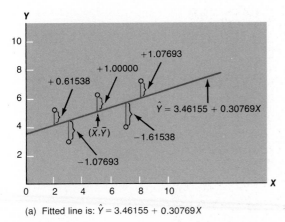

(a) Fitted line is: $\hat{Y} = 3.46155 + 0.30769X$

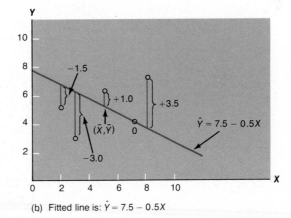

(b) Fitted line is: $\hat{Y} = 7.5 − 0.5X$

Figure 13.5

Two lines fitted by the criterion $\Sigma(Y_i - \hat{Y}_i) = 0$

data and in 13.6b the line $\hat{Y} = 4 + X$ is drawn through the same data. By methods that will be shown later in the chapter, we can definitely say that the line in Figure 13.6a—that is, $\hat{Y} = 6 + X$—brings the sum of the squared deviations from the line to a minimum; no other straight line can be drawn that will bring the sum of squared deviations below the sum obtained by the line $\hat{Y} = 6 + X$. The sum of squared deviations from this line is

$$\Sigma(Y_i - \hat{Y})^2 = (4)^2 + (-2)^2 + (-2)^2 = 16 + 4 + 4 = 24$$

The sum of the absolute deviations is

$$\Sigma |Y_i - \hat{Y}| = |4| + |-2| + |-2| = 4 + 2 + 2 = 8$$

Now consider the line in Figure 13.6b: $\hat{Y} = 4 + X$. The sum of the squared deviations, as we could have expected, rises above 24, but at the same time the sum of absolute deviations falls below 8:

$$\Sigma(Y_i - \hat{Y})^2 = (0)^2 + (6)^2 + (0)^2 = 36 > 24$$

and

$$\Sigma |Y_i - \hat{Y}| = |0| + |6| + |0| = 6 < 8$$

This example shows that the two criteria result in two different lines. As we argued earlier, the sum of squared deviations is preferable because of its convenience in mathematical treatment, and even more so because the resulting regression line has some superior qualities from a statistical standpoint.

Minimizing the sum of squared deviations, $\Sigma(Y_i - \hat{Y})^2$, is a method called the **ordinary least-squares method.** It is the method used in most regression analyses; we devote this chapter and the next two to its description and its interpretation. The advantages of the method are these:

1. It identifies the vertical (rather than the horizontal) deviation between each point and the line as the relevant deviation magnitude.
2. By squaring each deviation, the criterion overcomes the sign disadvantage of the method that uses the deviation itself.
3. It is much easier to handle mathematically than the method that sums the absolute deviations.

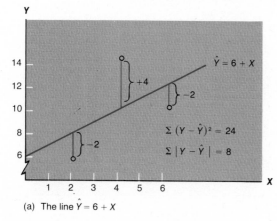

(a) The line $\hat{Y} = 6 + X$

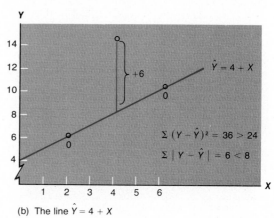

(b) The line $\hat{Y} = 4 + X$

Figure 13.6

Comparing alternative criteria for fitting a regression line: the minimum sum of squared deviations and the minimum sum of absolute deviations

4. There is a theoretical argument in favor of the ordinary least-squares method, known as the Gauss-Markov theorem. We will discuss it briefly in Chapter 14.

13.3 How the Ordinary Least-Squares Method Works

The principle of the ordinary least-squares (OLS) method is as follows: out of all of the possible regression lines that we can draw through the scatter diagram, we choose the one that minimizes the sum of the squared deviations from the line. Figure 13.7 helps to explain this principle. Here a straight line drawn through sample observations in an arbitrary manner is shown. We shall refer to this line as the "arbitrary regression line." From each point representing an observation, we drop or raise a vertical line toward the arbitrary regression line. In this way we obtain for each observation, Y_i, a corresponding point on the arbitrary regression line, which we denote by $\hat{Y}_i$. For each Y_i we now define a deviation, or an "error" term, which we denote by e_i and which is defined as

$$e_i = Y_i - \hat{Y}_i \tag{13.2}$$

Figure 13.7 illustrates the relationship between Y_i and $\hat{Y}_i$ and clearly identifies the "error" term, e_i, for a given observation, which may be positive, zero, or negative. The OLS method leads us to choose the particular regression line (out of an infinite number of possible lines) which *brings the sum of squared errors to a minimum*. Thus, if we denote by a and b the parameters of the line, we locate the points $\hat{Y}_i$ on the line by means of our equation

$$\hat{Y}_i = a + bX_i \tag{13.3}$$

and the OLS method leads us to derive the particular a and b that bring the expression Σe_i^2 to a minimum. By definition we get

$$\Sigma e_i^2 = \Sigma (Y_i - \hat{Y}_i)^2 \tag{13.4}$$

From Equation 13.3 we know that $\hat{Y}_i = a + bX_i$. Substituting this relationship in Equation 13.4 yields

$$\Sigma e_i^2 = \Sigma (Y_i - a - bX_i)^2 \tag{13.5}$$

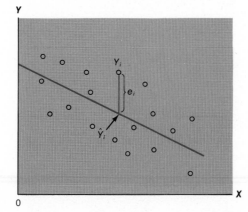

Figure 13.7

An arbitrary line drawn through sample data

To bring Equation 13.5 to a minimum by varying a and b, we take the partial derivatives of Σe_i^2 with respect to both a and b, equate the derivatives to zero, and solve. Following this procedure, we get

$$\frac{\partial \Sigma (Y_i - a - bX_i)^2}{\partial a} = \Sigma (2)(Y_i - a - bX_i)(-1) = 0 \tag{13.6}$$

$$\frac{\partial \Sigma (Y_i - a - bX_i)^2}{\partial b} = \Sigma (2)(Y_i - a - bX_i)(-X_i) = 0 \tag{13.7}$$

From Equations 13.6 and 13.7 we derive Equations 13.8 and 13.9, which are known as the **normal equations:**

$$\Sigma Y_i - na - b\Sigma X_i = 0 \tag{13.8}$$

$$\Sigma X_i Y_i - a\Sigma X_i - b\Sigma X_i^2 = 0 \tag{13.9}$$

where n stands for the number of observations available. The two unknowns in Equations 13.8 and 13.9 are the constants a and b. All the other quantities are available from the sample.

After dividing Equation 13.8 by n throughout, we get

$$\frac{\Sigma Y_i}{n} - a - b\frac{\Sigma X_i}{n} = 0 \tag{13.10}$$

and since $\dfrac{\Sigma Y_i}{n} = \overline{Y}$ and $\dfrac{\Sigma X_i}{n} = \overline{X}$, we may now rewrite Equation 13.10 as follows:

$$a = \overline{Y} - b\overline{X} \tag{13.11}$$

Substituting Equation 13.11 in 13.9, we obtain

$$\Sigma X_i Y_i - (\overline{Y} - b\overline{X})\Sigma X_i - b\Sigma X_i^2 = 0 \tag{13.12}$$

or

$$b(\overline{X}\Sigma X_i - \Sigma X_i^2) = -\Sigma X_i Y_i + \overline{Y}\Sigma X_i \tag{13.13}$$

Since $\Sigma X_i = n\overline{X}$, we get $\overline{X}\Sigma X_i = n\overline{X}^2$ and $\overline{Y}\Sigma X_i = n\overline{X}\overline{Y}$. By substituting these expressions in Equation 13.13 and multiplying both sides by -1, we obtain

$$b(\Sigma X_i^2 - n\overline{X}^2) = \Sigma X_i Y_i - n\overline{X}\overline{Y} \tag{13.14}$$

and finally

$$b = \frac{\Sigma X_i Y_i - n\overline{X}\overline{Y}}{\Sigma X_i^2 - n\overline{X}^2}$$

Summarizing, therefore, we can state that the values of a and b that bring $\Sigma (Y_i - \hat{Y}_i)^2$ to a minimum are given by the following equations:

$$b = \frac{\Sigma X_i Y_i - n\overline{X}\,\overline{Y}}{\Sigma X_i^2 - n\overline{X}^2} \qquad\qquad (13.15)$$

$$a = \overline{Y} - b\overline{X} \qquad\qquad (13.16)$$

We use the available data to obtain b from Equation 13.15, then use b along with $\overline{Y}$ and $\overline{X}$ to derive the value of a from Equation 13.16.

USING THE OLS METHOD

Table 13.2 is basically a work sheet for the calculation of the coefficients a and b in our income and consumption example (Table 13.1). Once the sample data are compiled as in Table 13.2, we may proceed directly to calculate the desired coefficients of the estimated regression line. Substituting in Equation 13.15, we get

$$b = \frac{\Sigma X_i Y_i - n\overline{X}\,\overline{Y}}{\Sigma X_i^2 - n\overline{X}^2} = \frac{2{,}879.5 - (10)(15.9)(16.5)}{2{,}891.0 - (10)(15.9)^2} = \frac{256.0}{362.9} = 0.705$$

and using this result we can also derive the coefficient a:

$$a = \overline{Y} - b\overline{X} = 16.5 - (0.705)(15.9) = 16.5 - 11.2 = 5.3$$

The regression equation (13.1) is

$$\hat{Y} = 5.3 + 0.705X_i$$

TABLE 13.2
Work Sheet for Calculation of the Coefficients a and b

Observation i	Annual income (thousands of dollars) X_i	Annual consumption (thousands of dollars) Y_i	X_i^2	Y_i^2	$X_i Y_i$
1	10.00	15.00	100.00	225.00	150.00
2	18.00	23.50	324.00	552.25	423.00
3	28.00	25.00	784.00	625.00	700.00
4	7.00	12.50	49.00	156.25	87.50
5	13.00	13.00	169.00	169.00	169.00
6	16.00	12.00	256.00	144.00	192.00
7	22.00	19.00	484.00	361.00	418.00
8	15.00	16.00	225.00	256.00	240.00
9	20.00	21.00	400.00	441.00	420.00
10	10.00	8.00	100.00	64.00	80.00
Total	$\Sigma X_i = 159.00$ $\overline{X} = 15.90$	$\Sigma Y_i = 165.00$ $\overline{Y} = 16.50$	$\Sigma X_i^2 = 2{,}891.00$	$\Sigma Y_i^2 = 2{,}993.50$	$\Sigma X_i Y_i = 2{,}879.50$

The interpretation of the regression line is as follows: for each $1 increase in annual income, there is an increase of $0.705 in average consumption. For households with no income ($X = 0$), we estimate average consumption to be $5,300. This relationship between X and Y is shown in Figure 13.8.

Figure 13.8

The ordinary least-squares regression line in the income-consumption example

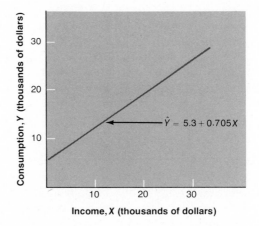

$\hat{Y} = 5.3 + 0.705X$

WHAT THE REGRESSION LINE MAY INDICATE

Several comments seem to be in order at this point. First, the *Y intercept* (that is, the coefficient a) as well as the slope of the regression (that is, the coefficient b) are positive in our income and consumption example. This is not necessarily the case, of course, under other circumstances. The value of a and b may be positive, negative, or zero. Figure 13.9 shows alternative

Figure 13.9

Alternative strengths of relationship between variables

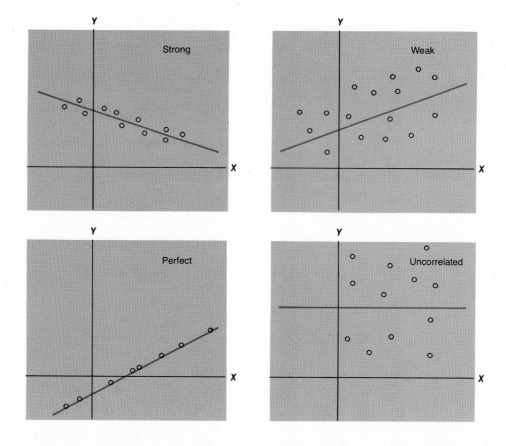

regression lines. If the slope of the regression line is positive, we say that a **direct relationship** exists between X and Y. If the slope is negative, an **inverse relationship** is said to exist between the variables. If the slope is equal to 0, X and Y are **uncorrelated.**

Second, the relationship between X and Y may be either strong or weak. When the error terms e_i are relatively small, the relationship is strong, while when the observations are more dispersed around the regression line and the error terms are of relatively large magnitude, the relationship is weak. The strongest relationship between X and Y is found when the error terms e_i are all equal to zero. When this is the case, all the observations are actually lying on the regression line and the variables are said to be perfectly correlated. In such cases we have a basically deterministic relationship between the variables. The other extreme case exists when the two variables are uncorrelated: as one variable varies, the other is not affected. Figure 13.10 shows alternative sample diagrams of variables having various strengths of relationship.

In this chapter we have dealt exclusively with the techniques of fitting the best regression line to sample data. Though the chapter deals with simple regression analysis only, the OLS method can be extended to multiple regression analysis (more than one independent variable). The next chapter deals with the theoretical aspects of simple regression analysis, and a significant part of it is devoted to practical applications.

Problems

13.1. Assume that the average of X in the regression $\hat{Y} = a + bX$ is equal to zero: $\overline{X} = 0$. Show that

$$a = \overline{Y}$$

and

$$b = \frac{\Sigma X_i Y_i}{\Sigma X_i^2}$$

13.2. A sample consists of 10 observations: $n = 10$. It is given that $\overline{X} = \overline{Y} = 1$. Show that in the regression $\hat{Y} = a + bX$,

$$a = 1 - \frac{\Sigma X_i Y_i - 10}{\Sigma X_i^2 - 10}$$

$$b = \frac{\Sigma X_i Y_i - 10}{\Sigma X_i^2 - 10}$$

13.3. The sample again consists of 10 observations. Show that if $\overline{X} = \overline{Y} = 0$, then

$$a = 0$$

and

$$b = \frac{\Sigma X_i Y_i}{\Sigma X_i^2}$$

13.4. Each of three regression lines of the form $\hat{Y} = a + bX$ has a slope equal to 2: $b = 2$. Additional information about the regressions is as follows:

Regression A	Regression B	Regression C
$\overline{X} = 1$	$\overline{X} = 2$	$\overline{X} = 0.5$
$\overline{Y} = 0$	$\overline{Y} = 4$	$\overline{Y} = 10.0$

Which of the regressions has the highest intercept, a?

13.5. Suppose we run the following two regression lines on the same set of data:

$$\hat{Y} = a_1 + b_1 X$$

$$\hat{X} = a_2 + b_2 Y$$

Do you agree with the following statements? Prove your answers.

(a) "If $\Sigma(X - \bar{X})^2 = \Sigma(Y - \bar{Y})^2$, then $b_1 = b_2$."

(b) "If $\bar{X} = \bar{Y}$, then $a_1 = a_2$."

13.6. The regression line $\hat{Y} = a + bX$ was estimated from a sample of 10 observations. Find the value of ΣXY, given the following information:

$$a = 1 \qquad \bar{X} = 2$$

$$b = 1 \qquad \Sigma X^2 = 100$$

13.7. Net profit before tax of ABC Corporation is denoted by X. The after-tax profit, Y, is (precisely) equal to $(1 - T)(X)$, so that $Y = (1 - T)(X)$, where T is the corporate tax rate (assume that $T = 0.50$).

(a) What are the intercept and the slope of the regression line relating Y (dependent) to X (explanatory)?

(b) What are the intercept and the slope of the regression line relating X (dependent) to Y (explanatory)?

(c) Draw both regression lines.

(d) Given that Y is precisely equal to $(1 - T)(X)$, are the intercept and slope of the line functions of the specific data collected in a sample? Would you get a different intercept and slope for another company that is subject to the same corporate tax rate?

(e) Suppose the corporate tax rate is reduced to $T = 0.30$. What are now the intercepts and slopes of the regression lines of parts a and b?

13.8. A regression of annual consumption (Y) on annual income (X) of households produces the following estimated equation:

$$\hat{Y} = 3{,}000 + 0.7X$$

(a) Suppose *each* of the households in the sample spends exactly \$1,000 annually on entertainment. Denote the spending on entertainment by E. Suppose we ran the regression $\widehat{Y - E} = a + bX$, using the same sample data as before. Could you say what the intercept and slope of this regression line will be?

(b) Suppose now that *each* of the households in the sample has exactly \$500 interest income (denoted by I). Suppose we ran the regression $\hat{Y} = a + b(X - I)$. Could you say what the regression coefficients will be?

(c) What are the intercept and slope of the following regression?

$$\widehat{Y - E} = a + b(X - I)$$

13.9. Suppose that 40 percent of the consumption of *each* household expenditure in Problem 13.8 is spent on food and *each* household makes 90 percent of its income from wages. Denote food consumption by F and wage income by W. Recall that the regression of Y on X is given by $\hat{Y} = 3{,}000 + 0.7X$.

(a) What are the regression coefficients of $\hat{F} = a + bX$?

(b) What are the regression coefficients of $\hat{Y} = a + bW$?

(c) What are the regression coefficients of $\hat{F} = a + bW$?

13.10. Give your own numerical example in which you show that more than one regression equation satisfies the criterion $\Sigma(Y_i - \hat{Y}_i) = 0$. Choose your own data and draw at least two lines that satisfy this criterion.

13.11. The annual savings of individuals at savings banks and the interest rates paid on those savings during four years are given below:

Year	Savings	Interest rate
1977	$1,000	9%
1978	800	10
1979	1,200	8
1980	2,000	11

(*a*) Draw a scatter diagram of savings (dependent) and interest rate (explanatory). Without performing any calculations, draw a regression line through the scattered data points and determine its intercept and slope.

(*b*) Calculate the OLS regression coefficients and then draw on the same diagram the OLS regression line. Compare the line with the one you drew earlier.

13.12. Using a given randomly selected sample, one can run the regression

$$\hat{Y} = a_1 + b_1 X$$

or, alternatively, use X for the dependent variable and Y for the explanatory variable and run

$$\hat{X} = a_2 + b_2 Y$$

Comment on the following statements:

(*a*) "If $b_1 = 0$, $b_2 = 0$."
(*b*) "If b_1 is positive, so is b_2."
(*c*) "If a_1 is positive, so is a_2."

Prove your answers, making use of the equations presented in this chapter.

13.13. The following are three sample points to which a regression line must be fitted:

Y	X
100	70
200	12
600	25

(*a*) Find the OLS regression equation for the data, and compute the value of $\Sigma(Y - \hat{Y})^2$.

(*b*) For the line you estimated in part *a*, calculate the magnitude $\Sigma|Y - \hat{Y}|$.

(*c*) Fit another line through the data points so that the sum of the absolute deviations (that is, $\Sigma|Y - \hat{Y}|$) for the line will be smaller than the value you calculated in part *b*.

(*d*) Find the sum of squared deviations [that is, $\Sigma(Y - \hat{Y})^2$] from the line you have drawn in part *c*. What is your conclusion from this problem?

13.14. The interest on bonds and the time left for their maturity (in years) is given below:

Interest rate (Y) (percent)	Bond maturity (X) (years)
3	0.25
4	0.50
7	1.00
8	5.00
9	10.00
10	20.00

Find the ordinary least-squares line $\hat{Y} = a + bX$, and explain its meaning.

13.15. For each of the pairs of variables below, indicate which is the dependent and which is the independent variable:

    (*a*) Education and personal income.
    (*b*) Density of population (that is, number of inhabitants per square mile) and average rent per room, when measured across various regions around the country.
    (*c*) Number of accidents per 1,000 miles of road and the quality of road measured numerically by some code.
    (*d*) The number of questionnaires answered and the number of questionnaires distributed.
    (*e*) Time traveled and distance traveled.

13.16. Write down the normal equations for the regression $\hat{Y} = a + bX$, using the following sample data:

Y	X
5	10
8	12
14	10

13.17. For the following sample data calculate the ordinary least squares line, using Y as the dependent variable and X as the independent variable:

Y	X
10	5
10	10
10	15
10	20
10	25

Explain your results.

13.18. Table P13.18 presents the annual sales, the return on sales (the net profit as percentage of sales), and the return on common equity (the net profit as a percentage of the firm's capital that belongs to the common shareholders) for the largest American textile companies. Note

TABLE P13.18
The 10 Largest Publicly Held U.S. Textile Companies

Name	Annual sales (millions of dollars)	Return on sales (percent)	Return on common equity (percent)
Burlington Industries	$2,456	3.0%	7.3%
J. P. Stevens	1,736	2.2	8.2
West Point Pepperell	909	3.4	11.0
Springs Mills	685*	4.2*	9.6*
United Merchants & Manufacturers	642**	1.4**	NM**
Cone Mills	618	5.8	15.0
M. Lowenstein & Sons	605	1.8	7.9
Collins & Aikman	557	4.0	12.9
Cannon Mills	547	4.5	8.9
Dan River	530	2.5	8.2

Data: Company annual reports: figures based on four most recent quarters reported.
  * Includes nontextile business amounting to 22% of sales.
** Affected by Chapter XI bankruptcy. NM = not meaningful.

Source: Data from *Business Week*, April 9, 1979.

that United Merchants & Manufacturers has gone bankrupt, so do not include this company in your calculations.

(*a*) Find the regression line when sales is the explanatory variable and return on sales is the dependent variable. Explain the meaning of the regression coefficients.

(*b*) Find the regression line when sales is the explanatory variable and return on common equity is the dependent variable.

(*c*) In view of the results of parts *a* and *b*, does profitability increase with sales? Explain.

13.19. The capacity and production of oil by the oil-producing and -exporting countries (OPEC) during the first quarter of 1979 are shown in Table P13.19.

TABLE P13.19
Who Produces OPEC's Oil

Country	Estimated sustainable production capacity	Estimated production first quarter 1979
	Millions of bbl per day	
Saudi Arabia	10.7	9.8
Iran	6.5	1.0
Iraq	3.0	3.2
Kuwait	2.9	2.6
Venezuela	2.6	2.3
Libya	2.3	2.1
Nigeria	2.3	2.4
United Arab Emirates	2.3	1.8
Indonesia	1.7	1.6
Algeria	1.3	1.2
Qatar	0.6	0.5
Ecuador	0.2	0.2
Gabon	0.2	0.2

Source: Data from *Business Week*, April 9, 1979.

(*a*) Treat capacity as the explanatory variable and draw a scatter diagram of the data.

(*b*) Calculate and draw the regression line through your scatter diagram.

(*c*) Rework parts *a* and *b*, this time eliminating Iran. Can you tell that Iran was undergoing a political change during the first quarter of 1979 by comparing the two regression lines you have obtained? Explain.

CHAPTER FOURTEEN OUTLINE

14.1 **Estimated vs. "Real" Regression Lines**
How the Value u Affects the Value Y

14.2 **Analysis of the Error Term, u**
The expected Value and Variance of u

14.3 **The Relationship between the Estimated Regression Line and the Distribution of the Error Term, u**

14.4 **Properties of the Ordinary Least-Squares Estimators**

14.5 **The Estimated Standard Deviation of the Dependent Variable Y and of the Error Term, e**
The Estimated Standard Deviation of Y
The Estimated Standard Deviation of the Error Term, e
Computing the Statistics S_Y and S_e

14.6 **The Probability Distribution of b and the Testing of Hypotheses**
The Probability Distribution of b
Testing Hypotheses Concerning β

14.7 **Confidence Intervals for the Mean μ_0 and the Individual Observation Y_0**
Using $\hat{Y}_0$ as an Estimator of μ_0
Deriving a Confidence Interval for μ_0
Deriving a Confidence Interval for an Individual Observation, Y_0

14.8 **Application 1: Cost Accounting: Allocation of Indirect Cost**

14.9 **Application 2: Finance: Measuring Security Risk**

14.10 **Application 3: Marketing: Investment in Advertising**

14.11 **Multiple Regression Analysis**
Multiple Regression with Two Explanatory Variables

14.12 **Interpretation of the Regression Coefficients**

14.13 **Multiple Regression with Two, Three, or More Explanatory Variables**

14.14 **Association between Explanatory Variables**

Key Terms

disturbance term
statistical relationship
Gauss-Markov theorem
best linear unbiased estimator (BLUE)
standard error of estimate
confidence bands

indirect variable cost
fixed costs
indirect cost
multiple regression analysis
normal equations

14

LINEAR REGRESSION: THEORY AND APPLICATIONS

The previous chapter was devoted exclusively to the technique of fitting a straight line to sample data. In this chapter we analyze the meaning of this line, the properties of its parameters a and b, and provide a test of significance for the sample estimates. We conclude with several applications, and extend the analysis to multiple regression.

As a necessary first step we explain the concept of statistical linear relationship between variables. When two variables X and Y relate to one another as in Equation 14.1, we say that the relationship is *linear and deterministic.*

$$Y = \alpha + \beta X \tag{14.1}$$

In Equation 14.1, α and β are two parameters (that is, *constants*). Given a value for the variable X, an accurate corresponding value of Y may be derived. For example, assuming $\alpha = 10$ and $\beta = 2$, the value of Y that corresponds to $X = 5$ is $Y = 10 + 2 \cdot 5 = 20$. The relationship in Equation 14.1 is *not* the typical relationship between two variables in business or economics. We are more likely to observe a *statistical relationship* in these areas, the nature of which becomes clear from Equation 14.2:

$$Y = \alpha + \beta X + u \tag{14.2}$$

Here α and β are two parameters, X is the independent variable, and u, the **disturbance term,** is a *random variable.* The variable Y is the dependent variable, for its value "depends" on the values of the parameters and variables appearing on the right-hand side of Equation 14.2. Since u is a random variable, so is the variable Y, which has u as one of its components. Thus, while the variable Y in Equation 14.1 is not a random variable (and 14.1 represents a deterministic model of relationships between X and Y), the variable Y in Equation 14.2 is a random variable (and 14.2 represents a statistical model of relationships). Note that in both models the variable X is not random. A **statistical relationship**, then, is *one in which the dependent variable is not totally determined by the independent variable but is determined in part by the independent variable and in part by the value of the random disturbance term.*

14.1 Estimated vs. "Real" Regression Lines

Just as we distinguish between a distribution's mean (μ) and its sample estimate ($\overline{X}$) and between a distribution's standard deviation (σ) and its sample estimate (S), we distinguish between "real" or "true" regression parameters, α and β, and their sample counterparts, a and b, which are merely point estimators of α and β. (If the true line were known, we would have no need to estimate the regression line.)

Generally speaking, the estimates a and b differ from the true regression line's intercept and slope and vary from one sample to another, since they are subject to sampling errors. (This was demonstrated in Chapter 13.) While there is only one true regression line (which is not observed), many regression lines may be estimated, each with its own value of the intercept a and the slope b, as shown in Figure 14.1.

Figure 14.1

A "true" regression line and three estimated lines

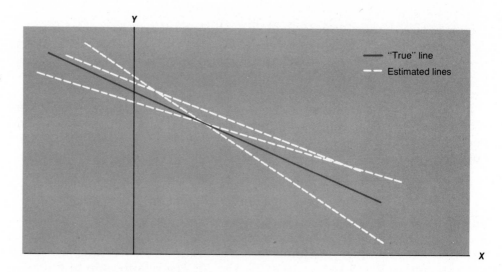

HOW THE VALUE u AFFECTS THE VALUE Y

To see more clearly why the estimated and true regression lines do not generally coincide, recall that the true regression line is given by

$$Y = \alpha + \beta X + u$$

Since u is a random variable, a whole probability distribution of Y corresponds to any *single* value of X. We saw earlier that if $Y = \alpha + \beta X$, $\alpha = 10$,

$\beta = 2$, and $X = 5$, we get a single value of Y, namely, $Y = 10 + 2 \cdot 5 = 20$. If, however, $Y = \alpha + \beta X + u$ and u takes on various values from one observation to the next, so does Y. Table 14.1 shows how the value of the dependent variable Y varies with u *for a given level of* X.

TABLE 14.1
Alternative Values of Y for a Given Value of X, When the Relationship Is
$Y = 10 + 2X + u$, **and u Is a Random Variable**

Observation i	Value of independent variable X_i	α	β	Value of random factor u_i	Value of dependent variable Y_i
1	5	10	2	1.5	21.5
2	5	10	2	0.0	20.0
3	5	10	2	0.2	20.2
4	5	10	2	−1.6	18.4
5	5	10	2	0.9	20.9
6	5	10	2	−2.0	18.0
7	5	10	2	3.0	23.0
8	5	10	2	−1.0	19.0
9	5	10	2	4.0	24.0
10	5	10	2	5.0	25.0

The consequence of Y's being a random variable for any value of X is shown in Figure 14.2, where a probability distribution of Y is drawn for each of the three given levels of X. The result is that samples taken differ in their Y values, even for the same values of X; consequently the estimated regression line varies from one sample to the next. Figure 14.2 also shows that the probability distributions of Y vary from one value of X to the next (in Figure 14.2, the higher the value of X, the higher the mean of Y). In fact, the probability distribution of Y is a conditional probability distribution (conditional, that is, on the respective value of $X : X_i$) and is denoted by $f(Y \mid X_i)$.

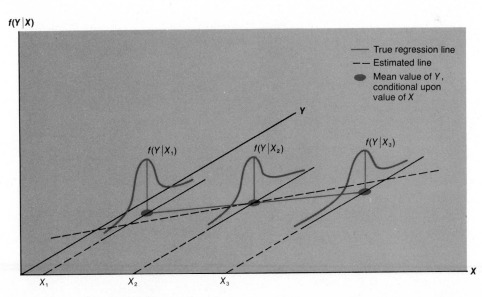

Figure 14.2

The regression line and the conditional probability distribution, $f(Y \mid X_i)$

Although the conditional distributions $f(Y \mid X_i)$ vary with the value of X, the following assumptions are made for the regression analysis:

1. The observations Y_i are statistically independent of one another. For example, if for some value X_1 we observed a corresponding value Y_1 below the regression line, it does not imply that the value Y_2 that corresponds to X_2 will also be below the line. Rather, the value Y_2 is independent of the value Y_1 and of the value of any other observation of the dependent variable.

2. The variance of the conditional probability distribution, $f(Y \mid X_i)$, which we denote by σ^2, *is the same for all values of X.* In the income and consumption example (Chapter 13), this assumption implies that the variance in consumption among households is the same for all levels of income. Formally we write $\mathrm{var}\,(Y \mid X_i) \equiv \mathrm{var}\,(Y_i) = \sigma^2$, where $\mathrm{var}\,(Y_i)$ should be understood to stand for the variance of Y when X is fixed at the level X_i.

3. The mean of the conditional probabilities of Y (denoted by μ_i) lies on the true regression line, meaning that the following equation holds:

$$E(Y \mid X_i) \equiv E(Y_i) \equiv \mu_i = \alpha + \beta X_i$$

where $E(Y_i)$ stands for the expected value of Y given that X is fixed at X_i.

14.2 Analysis of the Error Term, u

We saw earlier that the estimated regression line differs from the true line. Even if we knew what the true regression line was, the specific value of Y for any specified value of X would not be known because of the random return u. This has been illustrated numerically in Table 14.1, where the true regression line is assumed to have the parameters $\alpha = 10$ and $\beta = 2$. The random term u equals the deviation of the observation from the line. Figure 14.3 complements Table 14.1 in showing the relationship between Y_i, u_i, and the true regression line. The expected consumption level at income level X_i is equal to

$$\mu_i = \alpha + \beta X_i$$

Figure 14.3

The "true" regression line and the random term, u

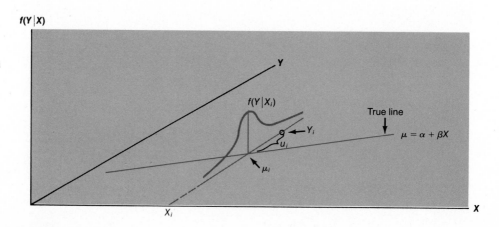

$f(Y \mid X)$

A particular observation at income level X_i is drawn from the conditional distribution shown in Figure 14.3, and the deviation of the value Y_i from the line is u_i. When u_i is positive, Y_i is above the line, and when u_i is negative, Y_i is below the line. The deviation u_i from the true regression line is attributable in principle to two factors:

1. Errors in measuring the dependent variable (nonsampling errors). Few people can state precisely their consumption expenditures in the past month or year; consequently, a sample of income and consumption is bound to include some degree of measurement error.
2. Even if measurement errors could be avoided, deviations from the line would still exist due to sampling errors. Households of equal income have different consumption levels for a variety of reasons, such as differences in their wealth (assets), their habits, the number of household members, and so on.

THE EXPECTED VALUE AND VARIANCE OF *u*

The above assumptions imply several things about the disturbance term, u. Starting with the model

$$Y = \alpha + \beta X + u$$

we can express u as follows:

$$u = Y - (\alpha + \beta X) \tag{14.3}$$

At any given level of X, X_i, Equation 14.3 becomes

$$u_i = Y_i - (\alpha + \beta X_i) \tag{14.4}$$

and since X_i is a single value of X—a constant—we get

$$E(u_i) = E(Y_i) - (\alpha + \beta X_i) \tag{14.5}$$

where $E(u_i)$ and $E(Y_i)$ should be interpreted again to represent the expected values of u and Y, respectively, when X is fixed at the level X_i. Recalling that $\alpha + \beta X_i = \mu_i$ (see assumption 3), we get the following equation:

$$E(u_i) = E(Y_i) - \mu_i = \mu_i - \mu_i = 0 \tag{14.6}$$

Taking the variance of Y at a given level of X, X_i, and recalling again that X_i is treated as a constant, we derive the following:

$$Y_i = \alpha + \beta X_i + u_i$$

This can be expressed in the following way:

$$\mathrm{var}\,(Y_i) = \mathrm{var}\,(\alpha + \beta X_i + u_i) = \mathrm{var}\,(u_i) \tag{14.7}$$

Thus

$$\mathrm{var}\,(Y_i) \equiv \sigma^2 = \sigma_u^2 \equiv \mathrm{var}\,(u_i) \tag{14.8}$$

An alternative formulation of the assumptions is as follows:

1. The u_i's are independent of one another.
2. $E(u_i) = 0$.
3. $\text{var}(u_i) = \sigma_u^2$ and is the same for all levels of X.

Note that at this stage we do not assume a specific probability distribution for Y_i (or u_i). Only at a later stage will we introduce the assumption that Y_i and u_i are normally distributed. Many of the derivations in this chapter are not based on the assumption of normality.

14.3 The Relationship between the Estimated Regression Line and the Distribution of the Error Term, *u*

The true regression line, as we saw, is given by Equation 14.9:

$$Y_i = \alpha + \beta X_i + u_i \qquad \textbf{(14.9)}$$

The application of the OLS method to sample data results in the following fitted line:

$$\hat{Y}_i = a + b X_i \qquad \textbf{(14.10)}$$

In Chapter 13 we saw that the scattered sample points generally deviate from the fitted line. We denoted the deviation of the ith observation from the fitted line by e_i (where $e_i = Y_i - \hat{Y}_i$), and substituting $\hat{Y}_i$ from Equation 14.10, we find that

$$e_i = Y_i - (a + b X_i)$$

or

$$Y_i = a + b X_i + e_i \qquad \textbf{(14.11)}$$

Equations 14.9 and 14.11 differ in two respects:

1. Equation 14.9 is formulated in terms of the true but unobserved parameters, α and β; Equation 14.11 is formulated in terms of their respective estimators, a and b.
2. The deviation u_i in Equation 14.9 is a deviation from the true line, while the deviation e_i in Equation 14.11 is a deviation from the estimated line.

Obviously, when we fit the estimated line through the scattered data we aim to get as close as possible to the true line. But how close we come depends, among other things, on the distribution of the term u. The more condensed the values of u are around their mean, the closer the fitted line will come to the true line (other things being equal, of course). This becomes clear with a glance at Figure 14.4, which shows estimated versus true regression lines under two alternative shapes of the conditional distributions:[1] in Figure

[1] The conditional distributions in Figure 14.4 are drawn as normal distributions, but this need not necessarily be the case.

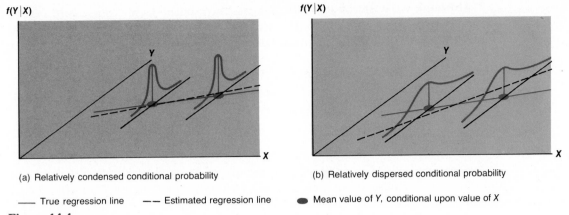

(a) Relatively condensed conditional probability   (b) Relatively dispersed conditional probability

—— True regression line   — — Estimated regression line   ● Mean value of Y, conditional upon value of X

Figure 14.4

The true and the estimated regression lines with different dispersions of the conditional probability distribution, $f(Y \mid X)$

14.4*a* the distributions are relatively condensed and the estimated regression line is relatively close to the true line. In Figure 14.4*b*, which illustrates more dispersed conditional distributions (but leaves all other factors the same), the estimated regression line is subject to greater sampling errors and may not be as close to the true line as in 14.4*a*. In the extreme case, where we assume the conditional distributions of u_i have no variance (that is, $\sigma_u^2 = 0$), $u_i = 0$, the model is deterministic and is given by $Y_i = \alpha + \beta X_i$. In this case all the sample points (X_i, Y_i) lie precisely on the straight line and a sample of size $n = 2$ or greater always yields an estimated line that coincides with the true line. As we shall see later, the dispersion of the conditional distribution of Y is not the only factor that determines the sampling errors in the constants a and b. The sample size used for the estimation of the coefficients, as well as the dispersion in the observed values of X, are other important factors.

14.4 Properties of the Ordinary Least-Squares Estimators

We saw in Chapter 13 that the ordinary least-squares method of fitting a line to scattered data yields the following estimators of the slope and the intercept:

$$b = \frac{\Sigma (X_i - \overline{X})(Y_i - \overline{Y})}{\Sigma (X_i - \overline{X})^2} = \frac{\Sigma X_i Y_i - n\overline{X}\overline{Y}}{\Sigma X_i^2 - n\overline{X}^2} \tag{14.12}$$

$$a = \overline{Y} - b\overline{X} \tag{14.13}$$

These two estimators have the following important properties:

1. Each is a linear combination of the observations Y_i; thus they are said to be linear estimators.
2. They are unbiased, meaning that the expected values of a and b are equal, respectively, to α and β.
3. A theorem called the **Gauss-Markov theorem** shows that among all the unbiased linear estimators, the OLS estimators a and b have the lowest variance.

Although we will not prove the Gauss-Markov theorem, you should realize that it provides a major justification for use of the ordinary least-squares method of fitting a line to sample data. Obviously, the smaller the variance of the estimator, the better, since the smaller the variance, the higher the chance of fitting the line close to the true line.

The variance of b is given by Equation 14.14:

$$\text{var}\,(b) \equiv \sigma_b^2 = \frac{\sigma_u^2}{\Sigma\,(X_i - \overline{X})^2} \tag{14.14}$$

According to the Gauss-Markov theorem, there is no other linear unbiased estimator for b with variance smaller than that given by Equation 14.14. Therefore, the OLS estimator b is called **BLUE**: **"best linear unbiased estimator."**

14.5 The Estimated Standard Deviation of the Dependent Variable Y and of the Error Term, e

In order to assess analytically the sampling errors in our regression analysis, we should first clarify the difference between two important estimators: the estimator of the standard deviation of Y, which we denote by S_Y, and the estimator of the *conditional* standard deviation of Y, given any X value, which we denote by S_e and which is known as the **standard error of estimate.** Obviously, both S_Y and S_e are the sample counterparts of the true measures σ_Y and σ_u, respectively.

THE ESTIMATED STANDARD DEVIATION OF Y

The standard deviation of Y, σ_Y, measures the total variability of the variable Y, regardless of the value of the variable X. Thus, if data are collected relating household consumption (Y) to household income (X), the measure S_Y relates to the diversity of consumption in the total sample, regardless of the differences in income among households in the sample. Obviously, some differences in consumption still exist among households, even after we account for differences in income. This gives rise to the error term u in Equation 14.2, whose σ_u is estimated by the standard deviation of the sample error terms e_i and denoted by S_e. In other words, the standard error of estimate, S_e, measures the degree of scatter of the data about the estimated regression line.

In Figure 14.5, a representative observation (Y_i) deviates from the sample average $(\overline{Y})$ by the amount identified as "total deviation," which equals $Y_i - \overline{Y}$. This "total deviation" is composed of two parts:

1. The deviation $\hat{Y}_i - \overline{Y}$, which is known as the "explained deviation" or "deviation due to the regression."
2. The "unexplained deviation" or "deviation around the regression line," which is given by $Y_i - \hat{Y}_i$.

The relationship between these deviations is as follows:

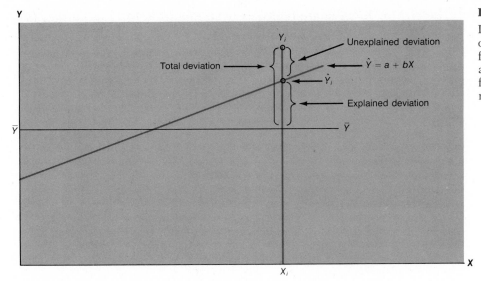

Figure 14.5
Deviation of an observation from the sample average and from the regression line

Total deviation = explained deviation + unexplained deviation.

$$Y_i - \overline{Y} = (\hat{Y}_i - \overline{Y}) + (Y_i - \hat{Y}_i) \qquad \text{(14.15)}$$

THE ESTIMATED STANDARD DEVIATION OF THE ERROR TERM, e

When the sample data become available, we can use the total deviation to *estimate* the total standard deviation of Y (σ_Y) and the unexplained deviation to *estimate* the standard deviation of Y around the regression line (σ_u). The estimators of σ_Y and σ_u are given by Equations 14.16 and 14.17:[2]

$$S_Y = \sqrt{\frac{\Sigma(Y_i - \overline{Y})^2}{n - 1}} \qquad \text{(14.16)}$$

$$S_e = \sqrt{\frac{\Sigma(Y_i - \hat{Y}_i)^2}{n - 2}} = \sqrt{\frac{\Sigma e_i^2}{n - 2}} \qquad \text{(14.17)}$$

Both equations may be simplified to give

$$S_Y = \sqrt{\frac{\Sigma Y_i^2 - n\overline{Y}^2}{n - 1}} \qquad \text{(14.18)}$$

$$S_e = \sqrt{\frac{\Sigma Y_i^2 - a\,\Sigma Y_i - b\,\Sigma X_i Y_i}{n - 2}} \qquad \text{(14.19)}$$

[2] Note that to obtain the deviation $Y_i - \overline{Y}$ we must first estimate $\overline{Y}$, and thereby lose one degree of freedom. To obtain $\hat{Y}_i$ we must first derive the coefficients a and b, so that two degrees of freedom are lost. This explains why the sum of squares in S_Y is divided by $n - 1$ and the sum of squares in S_e is divided by $n - 2$.

Note that S_e can be calculated directly from Equation 14.17. Equation 14.19, however, provides a computational formula. Another computational formula for S_e may be derived by use of the following relationship:[3]

$$\Sigma(Y_i - \overline{Y})^2 = b^2 \Sigma(X_i - \overline{X})^2 + \Sigma e_i^2$$

Having the sum of squares of Y and X as well as the regression slope, b, one can calculate Σe_i^2. Dividing all terms by $n - 2$, we obtain either

$$S_e^2 = \frac{\Sigma e_i^2}{n-2} = \frac{\Sigma(Y_i - \overline{Y})^2}{n-2} - b^2 \frac{\Sigma(X_i - \overline{X})^2}{n-2}$$

or

$$S_e = \sqrt{\frac{\Sigma(Y_i - \overline{Y})^2}{n-2} - b^2 \frac{\Sigma(X_i - \overline{X})^2}{n-2}} \qquad (14.19')$$

It seems that Equation 14.19' is the simplest way to calculate S_e.[4] (In the problems at the end of this chapter we use various ways to obtain S_e.)

If there are no deviations about the line, $e_i = 0$ for every observation, and hence $S_e = 0$. In this case we have a perfect fit in the sample and all the variance of Y_i is explained by the relationship between Y_i and X_i. In the consumption and income example, this means that all the variability that one finds in consumption (Y_i) is due solely to differences in income (X_i) in the sample observations. In general, the better the fit of the regression line to the data, the smaller the ratio of S_e to S_Y.

COMPUTING THE STATISTICS S_Y AND S_e

Before proceeding, let us compute the statistics S_Y and S_e for the income and consumption example of Chapter 13. In Table 13.2 the following magnitudes were computed:

$$\Sigma Y_i = \$165.00$$

$$\Sigma Y_i^2 = \$2,993.50$$

$$\Sigma X_i Y_i = \$2,879.50$$

$$\overline{Y} = \$16.50$$

$$n = 10$$

On the basis of the sample data we found

$$a = 5.3$$

$$b = 0.705$$

[3] It is easy to prove this formula, but the proof involves some tedious algebra. See J. Johnston, *Econometric Methods* (New York: McGraw-Hill, 1960), p. 31.
[4] With the definition of a and b and some algebra it can be shown that Equations 14.19 and 14.19' are equivalent.

Using these figures, we calculate

$$S_Y = \sqrt{\frac{2,993.5 - (10)(16.5)^2}{10 - 1}} = \sqrt{\frac{271}{9}} = \sqrt{30.11} = 5.49$$

and

$$S_e = \sqrt{\frac{2,993.5 - (5.3)(165.0) - (0.705)(2,879.5)}{10 - 2}} = \sqrt{\frac{88.95}{8}} = \sqrt{11.12} = 3.33$$

Thus, while the total variability of consumption (S_Y) is estimated to equal $5,490 (that is, 5.49 thousands of dollars), a given portion of it is related to differences in income. When we account for these differences, we find that the remaining unexplained standard deviation of Y is equal to $3,330.

14.6 The Probability Distribution of b and the Testing of Hypotheses

While β, the slope of the true regression line, is a constant underlying the relationship of the variables X and Y, its estimator, b, is a random variable whose value depends on the particular set of observations in the sample. As such, b has a probability distribution that can be employed in testing hypotheses.

THE PROBABILITY DISTRIBUTION OF b

The particular probability distribution of the statistic b depends on the probability distributions of the Y_i's, or more precisely on the probability distributions of the disturbance terms, the u_i's. In quite a few cases, the assumption that u follows the normal distribution is plausible, and if u is indeed normally distributed, the distribution of the estimated coefficient b is also normal. If the sample used is large enough, the distribution of b will be approximately normal (by the central limit theorem) even if u is not normally distributed.

If b is normally distributed, we may standardize it and get

$$Z = \frac{b - \beta}{\sigma_b} \tag{14.20}$$

Since σ_u is unknown and since from Equation 14.19 we know that

$$\sigma_b = \frac{\sigma_u}{\sqrt{\sum(X_i - \overline{X})^2}} = \frac{\sigma_u}{\sqrt{\sum X_i^2 - n\overline{X}^2}}$$

it follows that σ_b is also unknown. With σ_b unknown, it is impossible to construct confidence intervals for β or to test hypotheses concerning it by means of the statistic Z of Equation 14.20. But the sample counterpart of σ_b is measurable, and may be called in to the rescue. To obtain the estimator of σ_b, we refer back to Equation 14.19 and substitute S_e for σ_u, which yields

$$S_b = \frac{S_e}{\sqrt{\Sigma X_i^2 - n\overline{X}^2}} \qquad (14.21)$$

where S_b is the estimator of σ_b.

Under the assumption of normality of u, we get

$$\frac{b - \beta}{S_b} \sim t^{(n-2)} \qquad (14.22)$$

When we use the probability distribution in Equation 14.22, the construction of a confidence interval for β is a simple matter. As shown in Figure 14.6, the probability that $\dfrac{b - \beta}{S_b}$ will take on a value in the interval from

Figure 14.6

Probability distribution of the test statistic, $\dfrac{b - \beta}{S_b}$

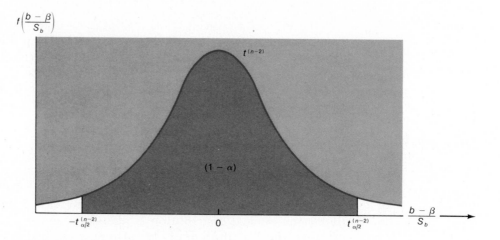

$-t_{\alpha/2}^{(n-2)}$ to $t_{\alpha/2}^{(n-2)}$ is equal to $1 - \alpha$. This probability can also be expressed by an equation:

$$P\left[-t_{\alpha/2}^{(n-2)} \leq \frac{b - \beta}{S_b} \leq t_{\alpha/2}^{(n-2)} \right] = 1 - \alpha \qquad (14.23)$$

To state it another way:

$$P[b - S_b \cdot t_{\alpha/2}^{(n-2)} \leq \beta \leq b + S_b \cdot t_{\alpha/2}^{(n-2)}] = 1 - \alpha \qquad (14.24)$$

Equation 14.24 provides a confidence interval for β. It states that there is a probability of $1 - \alpha$ that the interval from $b - S_b \cdot t_{\alpha/2}^{(n-2)}$ to $b + S_b \cdot t_{\alpha/2}^{(n-2)}$ includes within it the slope β of the true regression line.

TESTING HYPOTHESES CONCERNING β

With Equation 14.22, it is also possible to test hypotheses concerning β. In testing the null hypothesis

$$H_0: \quad \beta = \beta_0$$

against

$$H_1: \quad \beta \neq \beta_0$$

or against a one-sided alternative ($H_1: \beta > \beta_0$ or $\beta < \beta_0$) we simply use $\dfrac{b - \beta_0}{S_b}$ as our test statistic and employ the regular procedure for testing hypotheses of the mean when the standard deviation is unknown. If σ is known, we use the normal rather than the t distribution.

To illustrate, let us once again consider the income and consumption example of Chapter 13 to construct a 95 percent confidence interval for the slope and then test hypotheses concerning β. Given that $S_e = 3.33$, we use Equation 14.21 to estimate the standard deviation of the slope:

$$S_b = \frac{S_e}{\sqrt{\Sigma X_i^2 - n\overline{X}^2}} = \frac{3.33}{\sqrt{2{,}891 - (10)(15.9)^2}} = \frac{3.33}{19.05} = 0.1748$$

Recalling that $b = 0.705$ and noticing that $t_{0.025}^{(8)} = 2.306$, we may construct the 95 percent confidence interval for β by using Equation 14.24. Thus

$$b - S_b \cdot t_{\alpha/2}^{(n-2)} = 0.705 - (0.1748)(2.306) = 0.3019$$

and

$$b + S_b \cdot t_{\alpha/2}^{(n-2)} = 0.705 + (0.1748)(2.306) = 1.1081$$

Our confidence interval, then, is between 0.3019 and 1.1081.

Now suppose we want to test the hypothesis (using a significance level of 1 percent) that β is equal to zero against the one-sided alternative that β is greater than zero:

$$H_0: \quad \beta = 0$$
$$H_1: \quad \beta > 0$$

With $\dfrac{b - \beta_0}{S_b}$ as our test statistic, the critical value is $t_{0.01}^{(8)} = 2.896$ and our decision rule is to reject H_0 if $\dfrac{b - \beta_0}{S_b}$ is greater than 2.896 and to accept H_0 otherwise. Since $\dfrac{b - \beta_0}{S_b} = \dfrac{0.705}{0.1748} = 4.033$, we reject the null hypothesis and favor the alternative, which implies that consumption really tends to rise with income.

We can also test hypotheses where H_0 does not set β at zero. For example, when testing $H_0: \beta = 0.40$ versus $H_1: \beta > 0.40$, we find

$$\frac{b - \beta_0}{S_b} = \frac{0.705 - 0.40}{0.1748} = \frac{0.305}{0.1748} = 1.745$$

Using $\alpha = 0.01$, we cannot reject H_0, since $1.745 < 2.896$.

14.7 Confidence Intervals for the Mean μ_0 and the Individual Observation Y_0

We have emphasized several times that as a rule the estimated regression line deviates from the true line. It does so again in Figure 14.7, where a true regression line is shown alongside a series of estimated lines marked 1, 2, 3, 4, and 5. Let us consider this in more depth.

Figure 14.7

A true regression line and five estimated lines

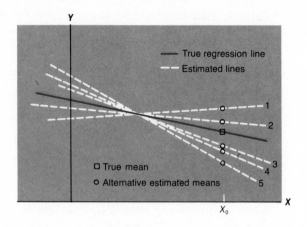

USING $\hat{Y}_0$ AS AN ESTIMATOR OF μ_0

Our estimate for $\mu_i = \alpha + \beta X_i$ (a point on the true line equal to the expected value of Y_i, given that the independent variable X takes on the value X_i) is the value $\hat{Y}_i$ (which is a point on the estimated line given the same level of the dependent variable, X_i). For the sake of clarity we will be using the notations X_0 and Y_0 to designate specific values of X_i and Y_i. Thus the expected value of $\hat{Y}_0$ is μ_0:

$$\mu_0 = \alpha + \beta X_0$$

To estimate μ_0, we use the corresponding value on the estimated regression line, that is, the value $\hat{Y}_0$:

$$\hat{Y}_0 = a + bX_0$$

Since a and b are unbiased—that is, $E(a) = \alpha$ and $E(b) = \beta$—it follows that $\hat{Y}_0$ is an unbiased estimator of μ_0:

$$E(\hat{Y}_0) = E(a + bX_0) = E(a) + E(b) \cdot X_0 = \alpha + \beta X_0 = \mu_0 \qquad \textbf{(14.25)}$$

DERIVING A CONFIDENCE INTERVAL FOR μ_0

It can be shown that the variance of $\hat{Y}_0$ is given by Equation 14.26:

$$\sigma_{\hat{Y}_0}^2 = \sigma_u^2 \left[\frac{1}{n} + \frac{(X_0 - \overline{X})^2}{\Sigma(X_i - \overline{X})^2} \right] \qquad \textbf{(14.26)}$$

The quantity given within brackets is available from the sample data. If σ_u is known, we can construct a confidence interval around μ_0 by using the standard normal distribution. A 95 percent confidence interval is given by

$$\mu_0 = \hat{Y}_0 \pm Z_{0.025} \cdot \sigma_u \sqrt{\frac{1}{n} + \frac{(X_0 - \overline{X})^2}{\Sigma(X_i - \overline{X})^2}} \qquad (14.27)$$

where both limits of the interval are written on the right-hand side: the minus sign applies to the lower limit and the plus sign to the upper limit. If σ_u is unknown, we substitute its estimator, S_e, switch to the $t^{(n-2)}$ distribution, and construct the interval as follows:

$$\mu_0 = \hat{Y}_0 \pm t_{0.025}^{(n-2)} \cdot S_e \sqrt{\frac{1}{n} + \frac{(X_0 - \overline{X})^2}{\Sigma(X_i - \overline{X})^2}} \qquad (14.28)$$

Holding all other factors unchanged, we obtain the narrowest confidence interval for μ_0 at $X_0 = \overline{X}$, since at that point the second term under the square root is equal to zero. As X_0 is shifted away from $\overline{X}$, to either left or right, the value $(X_0 - \overline{X})^2$ increases and so does the confidence interval. This causes the confidence interval in Equations 14.27 and 14.28 to vary with the distance of X_0 from $\overline{X}$. The closer X_0 to $\overline{X}$, the narrower the confidence interval; the farther X_0 from $\overline{X}$, the wider the confidence interval.

DERIVING A CONFIDENCE INTERVAL FOR AN INDIVIDUAL OBSERVATION, Y_0

So far we have derived the confidence interval for the mean μ_0 when $\hat{Y}_0$ served as its estimator. It is also of interest to derive a confidence interval for an individual observation Y_0 (which deviates from $\hat{Y}_0$ by the random error e_0):

$$Y_0 = \hat{Y}_0 + e_0 \qquad (14.29)$$

Consider again the variance of $\hat{Y}_0$ when σ_u is unknown (when σ_u is known, we merely substitute σ_u for S_e):

$$S_{\hat{Y}_0}^2 = S_e^2 \left[\frac{1}{n} + \frac{(X_0 - \overline{X})^2}{\Sigma(X_i - \overline{X})^2} \right]$$

The variance of an individual observation Y_0 includes the variance of the observation about the regression line, S_e^2, in addition to $S_{\hat{Y}_0}^2$. Because $\hat{Y}_0$ and e_0 are independent, $S_{Y_0}^2 = S_{\hat{Y}_0}^2 + S_e^2$. More explicitly we write

$$S_{Y_0}^2 = S_e^2 \left[\frac{1}{n} + \frac{(X_0 - \overline{X})^2}{\Sigma(X_i - \overline{X})^2} \right] + S_e^2$$

or

$$S_{Y_0}^2 = S_e^2 \left[\frac{1}{n} + \frac{(X_0 - \overline{X})^2}{\Sigma(X_i - \overline{X})} + 1 \right] \qquad (14.30)$$

Consequently, the 95 percent confidence interval for a single observation is

$$Y_0 = \hat{Y}_0 \pm t_{0.025}^{(n-2)} \cdot S_e \sqrt{\frac{1}{n} + \frac{(X_0 - \bar{X})^2}{\Sigma(X_i - \bar{X})^2} + 1} \qquad \text{(14.31)}$$

Of course, if σ_u is known, we can substitute it for S_e. The confidence interval for Y_0 is similar to that of μ_0, except that it is wider. Figure 14.8 shows the **confidence bands** for μ_0 and Y_0. They are simply bands that show the confidence limits for all values of X in a given range.

Continuing with our income and consumption example, we found that $\bar{X} = 15.9, S_e = 3.33, \Sigma(X - \bar{X})^2 = 362.9, a = 5.3, b = 0.705$, and $n = 10$. Suppose we want to construct a 95 percent confidence interval for the consumption, at an income level of 25 thousand dollars. The estimated consumption for 25 thousand dollars income is:

$$\hat{Y}_0 = 5.3 + 0.705 \cdot 25 = 22.925$$

Calculating $\sqrt{\dfrac{1}{n} + \dfrac{(X_0 - \bar{X})^2}{\Sigma(X_i - \bar{X})^2}}$ we obtain $\sqrt{\dfrac{1}{10} + \dfrac{(25 - 15.9)^2}{362.9}} = 0.573$, and since $t_{\alpha/2}^{(n-2)} = t_{0.025}^{(8)} = 2.306$, our confidence interval is given by:

$$\mu_0 = 22.925 \pm 2.306 \cdot 3.33 \cdot 0.573$$

or

$$\mu_0 = 22.925 \pm 4.400$$

In other words, we found a confidence interval from 19.525 to 27.325 thousand dollars for the *average* consumption of households with 25 thousand dollars income. The 95 percent confidence interval for the consumption of one (yet unknown) household with this level of income is found by calculating

$$\sqrt{\frac{1}{n} + \frac{(X_0 + \bar{X})^2}{\Sigma(X - \bar{X})^2} + 1} = \sqrt{\frac{1}{10} + \frac{(25 - 15.9)^2}{362.9} + 1} = 1.152$$

Figure 14.8

An estimated regression line and confidence bands for μ_0 and Y_0

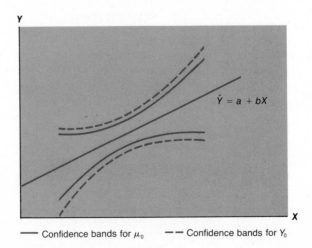

— Confidence bands for μ_0 – – Confidence bands for Y_0

Substituting in Equation 14.31 we obtain:

$$Y_0 = 22.925 \pm 2.306 \cdot 3.33 \cdot 1.152$$

or

$$Y_0 = 22.925 \pm 8.846$$

The confidence interval for an individual household is thus considerably wider and ranges from 14.079 to 31.771 thousand dollars.

14.8 APPLICATION 1:
COST ACCOUNTING: ALLOCATION OF INDIRECT COST

Firms with more than one production line face some difficulties in measuring the separate production costs of their various products. One must know the separate production costs in order to make decisions with regard to the optimal output of each product. The difficulty is due to the fact that apart from direct costs, there are indirect costs—management salary, rent, and the like—which should be allocated to the various production lines by some sort of accounting technique. Although indirect costs cannot be allocated precisely, one can find some key variables (number of workers, number of machines, machine hours, labor hours, and the like) which represent activity level in the various production lines, and allocate the indirect costs in relation to activity level. (The selected variables by which the firm allocates its indirect costs to the various production lines are known in accounting as the *allocation base*.) It is common to use regression techniques in allocating the indirect costs to the various production lines or departments. The following is an example of the way the indirect-cost allocation can be handled.

Typing Equipment, Inc., is a large manufacturing firm organized in four departments. Department *A* handles all administrative matters, while Departments *B*, *C*, and *D* handle various production processes. Among the various types of costs of Departments *B*, *C*, and *D* is the indirect cost, which includes some items that vary directly with direct labor hours (to be denoted by DLH). Among such cost items are electricity, water, and a portion of the maintenance cost. These costs are known as **indirect variable costs.** Among the cost items that *do not* vary directly with DLH are depreciation and rent. These costs are known as **fixed costs.** The total of fixed costs and indirect variable costs makes up the **indirect cost,** which will be denoted by IC.

Data of recent months show the following combinations of IC (combined for Departments *B*, *C*, and *D*) and DLH:

DLH (thousands of hours)	IC (thousands of dollars)
100	$200
170	230
150	225
212	270
308	310
345	330
310	325
340	330

Since the IC tends to vary directly with DLH, we choose DLH as the application base.

TABLE 14.2
Work Sheet for Calculation of Regression Line

Y (IC)	X (DLH)	Y^2	X^2	XY
200	100	40,000	10,000	20,000
230	170	52,900	28,900	39,100
225	150	50,625	22,500	33,750
270	212	72,900	44,944	57,240
310	308	96,100	94,864	95,480
330	345	108,900	119,025	113,850
325	310	105,625	96,100	100,750
330	340	108,900	115,600	112,200
$\Sigma Y = 2,220$	$\Sigma X = 1,935$	$\Sigma Y^2 = 635,950$	$\Sigma X^2 = 531,933$	$\Sigma XY = 572,370$
$\overline{Y} = 277.50$	$\overline{X} = 241.875$			

What is the estimated IC for each of the three departments for the next month, if DLH will be 40,000 in Department *B*, 150,000 in Department *C*, and 160,000 in Department *D?* We begin by running a regression line, using IC as a dependent variable and DLH as an explanatory variable. To do this we first develop the work sheet shown in Table 14.2. The regression's estimated parameters follow directly:

$$b = \frac{\Sigma XY - n\overline{X}\,\overline{Y}}{\Sigma X^2 - n\overline{X}^2} = \frac{572,370 - (8)(241.875)(277.50)}{531,933 - (8)(241.875)^2}$$

$$= \frac{572,370.0 - 536,962.5}{531,933.0 - 468,028.0} = \frac{35,407.5}{63,905.0} = 0.554$$

$$a = \overline{Y} - b\overline{X} = 277.5 - (0.554)(241.875) = 143.50$$

The estimated equation is then

$$\widehat{IC} = 143.50 + 0.554\text{DLH}$$

Figure 14.9 presents the scatter diagram and estimated line. Given the estimated line, let us proceed with the cost allocation. Total DLH anticipated for the next month is 40 + 150 + 160 = 350 thousands of hours. Total estimated *IC* is

$$\widehat{IC} = 143.50 + 0.554 \cdot 350 = 143.50 + 193.90 = 337.40$$

Out of the $337,400 in total indirect cost, a sum of $143,500 is estimated to exist even if DLH = 0; thus the $143,500 should be identified as the estimated fixed cost. The balance, $193,900, is the estimated indirect variable cost. The allocation of the fixed cost is proportional to the department's DLH. Table 14.3 shows how the IC of $337,400 is allocated.

The allocation described in Table 14.3 is the popular allocation procedure in accounting, known as the flexible budget.[5] The procedure is useful and important for

[5] See, for example, Charles T. Horngren, *Cost Accounting: A Managerial Emphasis*, 3d ed. (Englewood Cliffs, N.J.: Prentice-Hall, 1972).

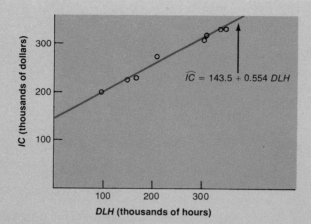

many reasons. In some cases, for example, the price of a product is determined by its cost of production plus some percentage of profit (this pricing method is known as "cost-plus"). This sort of pricing is often done when the firm is under regulation or working under a special contract. The determination of the indirect cost for each product becomes important in such cases. Suppose the firm in our example needs to decide whether Department *D* should be closed down. Let the revenue of that department be $500,000 per year and its direct cost be $400,000. Should we close the department? Looking at the flexible budget, we see that the total (direct and indirect) cost of the department is

$$\text{Total cost}_D = \$400,000 + \$154,230 = \$554,230$$

Hence the total cost of the department is greater than its revenue, and it seems that it should be closed. A closer look at the figures, however, shows that this is not the case. Out of the $154,230 indirect cost of Department *D*, $65,600 are fixed costs, which the firm would have to expend even if Department *D* were to be closed. In the short run, then, only $88,630 indirect variable cost of Department *D* should be considered. The total costs are thus

$$\text{Total costs}_D = \$400,000 + \$88,630 = \$488,630$$

Closing the department will save $488,630 in costs but will decrease revenues by $500,000. Hence the department should stay open in the short run. Later in this chapter

TABLE 14.3
Allocation of Indirect Cost, by Department

(1) *Department*	*(2)* *DLH* *(thousands of hours)*	*(3)* *Percent DLH/100*	*(4)* *Fixed costs ($143,500 · col. 3)*	*(5)* *Variable indirect cost ($193,900 · col. 3)**	*(6)* *Total IC (col. 4 + col. 5)*
B	40	0.1143	16,400	22,160	38,560
C	150	0.4286	61,500	83,110	144,610
D	160	0.4571	65,600	88,630	154,230
Total	350	1.0000	143,500	193,900	337,400

Note: Figures have been rounded.
* Also equal to 0.554·col.2·1,000.

we shall discuss multiple regression analysis, in which more than one independent variable is used to explain variations of one dependent variable. The application base of indirect cost allocation often includes more than just one variable, and multiple regression analysis is then used to derive the flexible budget.

14.9 APPLICATION 2:
FINANCE: MEASURING SECURITY RISK

The desirability of any financial investment depends heavily (though not exclusively) on its riskiness. Many stockbrokers, financial consultants, and academicians use a model known as the beta model to identify the riskiness of securities. For example, the large Merrill Lynch, Pierce, Fenner & Smith brokerage firm provides the well-known beta (or β) risk index as a regular service to its customers. The security's beta coefficient is measured by the simple regression technique presented in this chapter. The dependent variable is the security's rate of return (which will be denoted by Y, or alternatively by SRR), and the independent variable is the rate of return on a portfolio of all the securities traded in the market.[6] This portfolio is known as the "market portfolio," and as its proxy, consultants and brokers usually use the rate of return on some known and readily available market index, such as the Dow Jones average or Standard & Poor's (S&P) average. The rate of return on such an index serves as our independent variable (we denote the independent variable by X, or alternatively by MRR, for "market rate of return"). Using past rates of return, we may run the following simple regression:

$$SRR_i = a + bMRR_i + e_i$$

The estimated regression's slope, b, is used as a risk measure for the particular security whose rates of return are used as the dependent variable. The reason will become clear later. For now, let us illustrate the numerical computations. Suppose data on annual rates of return have been obtained for eight years, as shown in Table 14.4. The broker needs to estimate the regression line and compute S_Y and S_e. To estimate the regression line, we first develop Table 14.5, which serves as a work sheet, and then continue by computing the constants a and b:

$$b = \frac{\Sigma XY - n\overline{X}\overline{Y}}{\Sigma X^2 - n\overline{X}^2} = \frac{473 - (8)(2)(2.5)}{284 - (8)(2)^2} = \frac{473 - 40}{284 - 32} = \frac{433}{252} = 1.72$$

and

$$a = \overline{Y} - b\overline{X} = 2.50 - (1.72)(2) = 2.50 - 3.44 = -0.94$$

Given these two estimates, we may write the relationship between SRR and MRR as follows:

$$\widehat{SRR} = -0.94 + 1.72MRR$$

where $\widehat{SRR}$ is the estimated SRR value. The results should be interpreted to mean that a one-unit change (in this case the unit is one percentage point) in the average market

[6] The rate of return is the percentage profit or loss obtained in an investment.

TABLE 14.4
Rates of Return on the Market Portfolio and on One Stock
(percent)

Independent variable (MRR): average market return X	Dependent variable (SRR): return on stock Y
1.0	2.0
−5.0	−10.0
12.0	18.0
5.0	9.0
7.0	12.0
0.0	−1.0
−6.0	−12.0
2.0	2.0

TABLE 14.5
Work Sheet for Calculation of Regression Estimates

(1) Observation i	(2) MRR X_i	(3) SRR Y_i	(4) X_i^2	(5) Y_i^2	(6) X_iY_i (col. 2 · col. 3)
1	1	2	1	4	2
2	−5	−10	25	100	50
3	12	18	144	324	216
4	5	9	25	81	45
5	7	12	49	144	84
6	0	−1	0	1	0
7	−6	−12	36	144	72
8	2	2	4	4	4
	$\Sigma X_i = 16$ $\overline{X} = 2$	$\Sigma Y_i = 20.0$ $\overline{Y} = 2.5$	$\Sigma X_i^2 = 284$	$\Sigma Y_i^2 = 802$	$\Sigma X_iY_i = 473$

rate of return is estimated to be accompanied on the average by 1.72 units of change in the rate of return on the particular stock examined. (See Figure 14.10.) The slope, 1.72, is the estimated beta coefficient of the particular stock considered; it reflects the sensitivity of the stock's rate of return to changes in the rates of return on the market

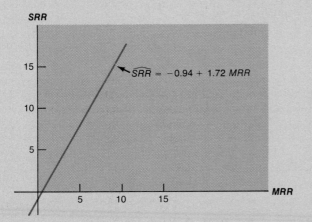

Figure 14.10

Estimated regression line relating security rate of return to market rate of return as a whole. Thus, if the rate of return on the market as a whole goes up 10 percentage points, the stock on hand is estimated to go up 17.2 percentage points; similarly, a decline of 10 percentage points in the rate of return on the market as a whole is estimated to be associated with a decline of 17.2 percentage points in this stock's rate of return. This is why the slope of this particular regression line is so significant to securities analysts. It is customary to identify all securities with beta coefficients greater than 1.0 as *aggressive* securities, since their rates of return change upward and downward more sharply than the average rates of return of the market. Securities with beta coefficients greater than 1.0 are considered risky for the same reason. Similarly, securities whose beta coefficients are less than 1.0 are identified as *defensive* securities, since their rates of return move upward and downward more moderately than the average rate of return for the market as a whole. Their rates of return are thus more stable and they are considered to be less risky securities. Note that for a zero rate of return on the market, the estimated rate of return on the stock is -0.94 percent.

We now continue by computing S_Y and S_e. Using Equation 14.18, we obtain

$$S_Y = \sqrt{\frac{\Sigma Y^2 - n\overline{Y}^2}{n-1}} = \sqrt{\frac{802 - (8)(2.5)^2}{8-1}} = \sqrt{107.4} = 10.36$$

and Equation 14.19 yields

$$S_e = \sqrt{\frac{\Sigma Y^2 - a\Sigma Y - b\Sigma XY}{n-2}} = \sqrt{\frac{802 - (-0.94)(20) - (1.72)(473)}{8-2}}$$

$$= \sqrt{1.21} = 1.10$$

While the total variability of the rate of return on the stock, as measured by S_Y, is 10.36, a substantial portion of this variability is related to fluctuations in the average rate of return on the market index. When we account for the fluctuations in X, we find that the remaining standard deviation of Y, S_e, is 1.10.

Consider now an investor who is interested in buying a given common stock, provided that the stock is not too risky. Specifically, he is willing to buy the stock if its beta coefficient is not greater than 1.50. What advice should we give him if the stock considered is the one for which we just computed $b = 1.72$? One thing we know is that our point estimate ($b = 1.72$) is greater than the investor's maximum acceptable beta coefficient. On the other hand, we know that b is a random variable and the true parameter could be (and probably is) somewhat different from 1.72. In fact, the standard deviation of b is estimated to be as follows:

$$S_b = \frac{S_e}{\sqrt{\Sigma X^2 - n\overline{X}^2}} = \frac{1.10}{\sqrt{252}} = \frac{1.10}{15.87} = 0.0693$$

A sound procedure here is to test the hypotheses

$$H_0: \quad \beta \leq 1.50$$
$$H_1: \quad \beta > 1.50$$

For illustration, let us use a 1 percent level of significance. The test statistic is $T = \dfrac{b - \beta_0}{S_b}$ and its distribution is $t^{(n-2)}$. Since $n = 8$, the critical value is $t_{0.01}^{(6)} = 3.143$.

We then accept H_0 if $T = \dfrac{b - \beta_0}{S_b}$ is less than or equal to 3.143 and reject H_0 if it is greater than 3.143 (see Figure 14.11). The value of T in this example is

$$T = \frac{b - \beta_0}{S_b} = \frac{1.72 - 1.50}{0.0693} = \frac{0.22}{0.0693} = 3.175$$

and since $T = 3.175$ is greater than 3.143, we reject H_0 and conclude that with a level of significance of 99 percent we should consider β to be greater than 1.50. Our recommendation to the investor, then, is to search for a less risky stock.

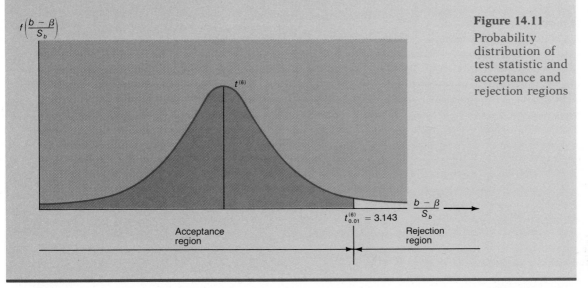

Figure 14.11

Probability distribution of test statistic and acceptance and rejection regions

14.10 APPLICATION 3:
MARKETING: INVESTMENT IN ADVERTISING

A company's sales of a given product are a function of many factors. Among them one may list variables relating to the general economic conditions in the area in which the product is to be marketed, the effectiveness of competition, and the aggressiveness of the firm marketing the product. Obviously, analyzing the effect on sales of one particular factor while ignoring the simultaneous effects of all other factors leads to misleading results. The simultaneous consideration of many explanatory variables in regression analysis will be presented later in this chapter; for the time being we shall deal with a simplified example in which the effect of advertising on sales is to be estimated, while all other factors are ignored.

Consider a company whose products are sold nationwide. The marketing network of the company is organized in 20 regional distribution centers generating roughly similar revenues each year. A new product is now undergoing marketing experimentation by the firm. It has been introduced in 10 regional areas, with different advertising expenditures in each. The advertising expenditures and sales of the product are shown in Table 14.6.

TABLE 14.6
Advertising Expenditures and Sales of New Product, by Region
(thousands of dollars)

Regional area	Advertising expenditures	Sales
1	$20	$160
2	10	120
3	30	220
4	16	120
5	40	235
6	35	225
7	20	160
8	25	200
9	33	220
10	19	120

To determine the effect of advertising expenditures on sales, a simple regression analysis is performed, using sales as the dependent variable (to be denoted by Y, or alternatively by *SALES*) and advertising expenditure as the independent variable (to be denoted by X, or alternatively by *AD*). The computation work sheet is presented in Table 14.7. Substituting into the slope formula, we get

$$b = \frac{\Sigma XY - n\overline{X}\overline{Y}}{\Sigma X^2 - n\overline{X}^2} = \frac{47{,}935 - (10)(24.8)(178.0)}{6{,}956.0 - (10)(24.8)^2}$$

$$= \frac{3{,}791.0}{805.6} = 4.706$$

The intercept is

$$a = \overline{Y} - b\overline{X} = 178.0 - (4.706)(24.8) = 61.291$$

so that the estimated regression equation is

$$\widehat{SALES} = 61.291 + 4.706AD$$

We may use the equation above to predict sales at a given advertising budget. If, for example, $22,000 is to be used for advertising in one of the 20 regional areas, what is the predicted sales level? Substituting 22 for X, we get

$$\widehat{SALES} = 61.291 + (4.706)(22.0) = 164.823$$

or $164,823. What if $80,000 is to be used for advertising in a given regional area? If we again use the estimated equation we get $437,771, since

$$\widehat{SALES} = 61.291 + (4.706)(80.0) = 437.771$$

This prediction, however, is very unreliable and constitutes an improper use of regression analysis. Why? Because in no region has a budget in the neighborhood of $80,000 and its resulting sales been observed. Extrapolation of the estimated regression line to an

TABLE 14.7
Work Sheet for Calculation of Regression Estimates

i	AD X	SALES Y	X^2	XY
1	20.0	160	400	3,200
2	10.0	120	100	1,200
3	30.0	220	900	6,600
4	16.0	120	256	1,920
5	40.0	235	1,600	9,400
6	35.0	225	1,225	7,875
7	20.0	160	400	3,200
8	25.0	200	625	5,000
9	33.0	220	1,089	7,260
10	19.0	120	361	2,280
	$\Sigma X = 248.0$ $\overline{X} = 24.8$	$\Sigma Y = 1,780$ $\overline{Y} = 178$	$\Sigma X^2 = 6,956$	$\Sigma XY = 47,935$

advertising budget of $80,000 is justified only if we have good reason to believe that the linear relationship we have observed holds true for budgets significantly greater than the ones we have observed. Marketing research studies have indicated the existence of a saturation level—a sort of upper ceiling beyond which sales cannot be made (for a given product price, of course).[7] The relationship between sales and advertising expenditures may thus be described as in Figure 14.12. The figure clearly shows that the global relationship between sales and advertising expenditure is nonlinear but may be approximated by a linear relationship for budgets less than X^* (see Figure 14.12). If $80,000 is beyond the value X^*, our prediction could be in great error, as Figure 14.13 shows.

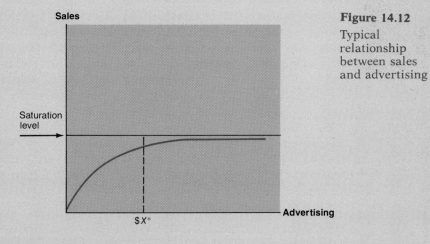

Sales

Saturation
level →

X^*

Advertising

Figure 14.12

Typical
relationship
between sales
and advertising

[7] See, for example, Julian L. Simon, *Issues in the Economics of Advertising* (Urbana: University of Illinois Press, 1970).

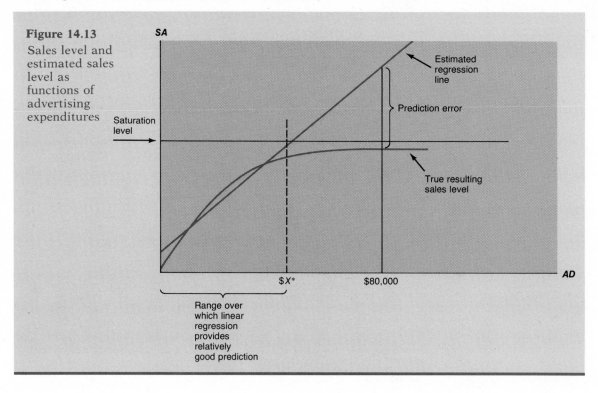

Figure 14.13
Sales level and estimated sales level as functions of advertising expenditures

14.11 Multiple Regression Analysis

Since real-world situations in business and economics are normally quite complex, accuracy in estimation and prediction of a dependent variable can often be achieved only if more than one explanatory variable is brought into the analysis. For example, the demand for a new cleanser depends on the amount of money spent to promote it, on the price of the new cleanser in relation to the prices of competitive cleansers, on whether it is a powder or a liquid, and, of course, on some measure of quality. Similarly, sales of a given car model depend on its price, the prices of similar models, the price of gasoline, the reputation of the manufacturer, and many other variables. A student's grade on a forthcoming test may be related to his grade-point average, the number of hours devoted to preparation, the number of other tests he will take in the same week, and so on.

We shall now deal with regression analysis in which a dependent variable is *linearly* related to a set of K explanatory variables, where K is greater than or equal to 2. This type of regression analysis is known as **multiple regression analysis.** The simplest case of multiple regression is that in which $K = 2$, and we shall deal with this case first.

MULTIPLE REGRESSION WITH TWO EXPLANATORY VARIABLES

Let a dependent variable Y be a linear function of two explanatory variables, X_1 and X_2, and a random variable u. We can then write the equation

$$Y = \beta_0 + \beta_1 X_1 + \beta_2 X_2 + u \tag{14.32}$$

where β_0, β_1, and β_2 are parameters (constants).

Graphically, Equation 14.32 may be described in a three-dimensional space as illustrated in Figure 14.14, where hypothetical sample observations are presented. Each observation is represented by a point in the space and is

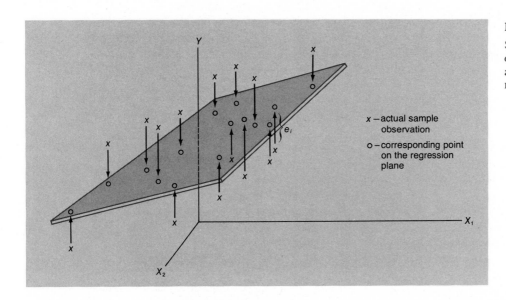

Figure 14.14

Scatter diagram of sample points about the regression plane

determined by the combination of the values of Y, X_1, and X_2. The estimated regression equation may be stated as

$$\hat{Y} = b_0 + b_1 X_1 + b_2 X_2 \tag{14.33}$$

This is an equation of a plane in a three-dimensional space, as depicted in Figure 14.14. It is clear that for each observation, Y, we can identify a sample error term, e; just as we did for simple regression, we drop or raise a vertical line from each point representing an observation (Y) to the regression plane ($\hat{Y}$). This vertical distance is the "error term," e, of the observation (see Figure 14.14):

$$e = Y - \hat{Y} \tag{14.34}$$

The ordinary least-squares principle may now be applied directly. We first substitute the value of $\hat{Y}$ from Equation 14.33 in Equation 14.34 and get

$$e = Y - b_0 - b_1 X_1 - b_2 X_2 \tag{14.35}$$

Now we square and sum all the sample observations and get

$$\Sigma e^2 = \Sigma (Y - b_0 - b_1 X_1 - b_2 X_2)^2 \tag{14.36}$$

If we take the derivatives of Σe^2 with respect to b_0, b_1, and b_2, equate to zero (to find the minimum of Σe^2), and solve, we obtain the following equations:

$$\Sigma Y = nb_0 + b_1 \Sigma X_1 + b_2 \Sigma X_2$$

$$\Sigma X_1 Y = b_0 \Sigma X_1 + b_1 \Sigma X_1^2 + b_2 \Sigma X_1 X_2 \qquad \textbf{(14.37)}$$

$$\Sigma X_2 Y = b_0 \Sigma X_2 + b_1 \Sigma X_1 X_2 + b_2 \Sigma X_2^2$$

These are the well-known **normal equations.**

Once the sample is observed and the data become available, all the quantities in Equation 14.37 aside from b_0, b_1, and b_2 may be directly computed. This leaves three equations with three unknowns (b_0, b_1, and b_2)— a solvable system of equations.

14.12 Interpretation of the Regression Coefficients

The true regression relationship with two explanatory variables is given by

$$Y = \beta_0 + \beta_1 X_1 + \beta_2 X_2 + u \qquad \textbf{(14.38)}$$

where β_0, β_1, β_2, and u are unknown coefficients.

Let us examine the meaning of the coefficient β_1. We add 1 to X_1 in Equation 14.38 and write

$$Y' = \beta_0 + \beta_1(X_1 + 1) + \beta_2 X_2 + u$$

or

$$Y' = (\beta_0 + \beta_1 X_1 + \beta_2 X_2 + u) + \beta_1 = Y + \beta_1$$

Similarly, if we subtract 1 from X_1 we get

$$Y'' = \beta_0 + \beta_1(X_1 - 1) + \beta_2 X_2 + u$$

$$= (\beta_0 + \beta_1 X_1 + \beta_2 X_2 + u) - \beta_1 = Y - \beta_1$$

This shows that Y increases (decreases) by β_1 units for every one-unit increase (decrease) in X_1, when X_2 is held constant. Similarly, it can be easily shown that Y increases (decreases) by β_2 units for every one-unit increase (decrease) in X_2, when X_1 is held constant. When more than two independent variables are used in the regression, the meaning of the coefficient of any one variable is the units of increase (decrease) in Y for each one-unit increase (decrease) in the independent variable, while all the other independent variables remain unchanged.

We interpret the coefficients of the estimated regression line, $\hat{Y} = b_0 + b_1 X_1 + b_2 X_2$, similarly. The coefficient b_1 is the estimated number of units of change in Y for a one-unit change in X_1 *when the variable X_2 is held unchanged.* The coefficient b_2 is the estimated number of units of change in Y for a one-unit change in X_2 *when the variable X_1 is held unchanged.* The coefficient b_0 is the estimated value of Y when both X_1 and X_2 are equal to zero.

As the case of simple linear regression, in order to conclude that the regression estimators b_0, b_1, and b_2 are Best Linear Unbiased Estimators (BLUE) of the true model parameters β_0, β_1, and β_2, we merely have to assume that the random terms u of the various observations are independent of one another, their mean is equal to zero, and their variance, σ_u^2, is the same for all possible values of the independent variables. If we want to determine confidence intervals and test hypotheses concerning a regression coefficient β, we must add the assumption that the random terms u are normally distributed.

14.13 Multiple Regression with Two, Three, or More Explanatory Variables

The generalization of regression analysis from two explanatory variables to more than two is conceptually simple: a dependent variable Y is assumed to be related to K independent variables, $X_1, \ldots, X_K$, in a statistical manner, as follows:

$$Y = \beta_0 + \beta_1 X_1 + \beta_2 X_2 + \cdots + \beta_K X_K + u \qquad (14.39)$$

We estimate the relationship by choosing that set of constants b_0, b_1, $\ldots$, b_K (the estimators of β_0, β_1, $\ldots$, β_K, respectively) which bring the quantity $\Sigma e^2 = \Sigma(Y - \hat{Y})^2$ to a minimum, where $\hat{Y}$ is given by:

$$\hat{Y} = b_0 + b_1 X_1 + \cdots + b_K X_K \qquad (14.40)$$

Generally speaking, it is impractical to estimate the constants $b_0, \ldots, b_K$ by hand calculations. If calculation must be done by hand, it is likely to be handled by linear algebra. We shall assume that a computer is available to do the calculations, and therefore focus on interpretation of the results.

As in the case of two explanatory variables, the constant b_0 is the predicted Y value when X_1 through X_K are all equal to zero. Each of the constants b_1 through b_K is the estimated change in Y as a result of a one-unit change in the respective explanatory variable, *when all the rest are held unchanged*.

As in simple regression analysis, the standard error of estimate (S_e) in multiple regression is an estimate of the standard deviation of the residuals around the regression plane, and its equation is

$$S_e = \sqrt{\frac{\Sigma e_2}{n - K - 1}} = \sqrt{\frac{\Sigma(Y - \hat{Y})^2}{n - K - 1}} \qquad (14.41)$$

The denominator under the square root is $n - K - 1$ because $\hat{Y}$ (which is a prerequisite for computing Σe^2) involves the estimation of $K + 1$ variables ($\beta_0, \beta_1, \beta_2, \ldots, \beta_K$), and as usual, one degree of freedom is lost for every parameter that is estimated.

We turn now to an example through which we shall discuss the standard deviation of the estimated regression coefficients, and the standard error of estimate.

EXAMPLE 14.1

Professional Athletics, a sporting-goods manufacturer, is interested in improving the process it uses to select new salesmen. Instead of using subjective evaluation, the company has decided to use objective criteria. Professional Athletics' management is looking into the relationship between a performance index (*PI*) and the following variables: IQ score (*IQ*); experience (*EXP*), measured in number of years on the job; score on a test related to personality traits (*PER*); and age (*AGE*). Data concerning currently employed salesmen are presented in Table 14.8. Given this sample, we want to estimate the multiple regression coefficients, using *PI* as the dependent variable and *IQ, EXP, PER* and *AGE* as the independent variables.

When the data of Table 14.8 are used in the equation

$$\widehat{PI} = b_0 + b_1 IQ + b_2 EXP + b_3 PER + b_4 AGE$$

the following are the resulting estimated coefficients provided by the computer:

$$b_0 = 75.4357$$

$$b_1 = 0.1730$$

$$b_2 = 0.0684$$

$$b_3 = 0.2124$$

$$b_4 = -0.3532$$

The performance index that we predict for a 50-year-old salesman with

TABLE 14.8

Sample Data on Personal Characteristics

i	(PI) Y	(IQ) X_1	(EXP) X_2	(PER) X_3	(AGE) X_4
1	96.8	80.0	9.0	90.3	40.0
2	92.4	100.0	12.0	90.7	40.0
3	104.2	120.0	6.0	110.5	50.0
4	107.8	105.0	1.0	101.0	19.0
5	94.5	118.0	2.0	85.1	40.0
6	96.6	80.0	11.0	103.2	55.0
7	94.7	89.0	22.0	107.9	52.0
8	99.5	93.0	3.0	90.0	39.0
9	97.3	109.0	30.0	85.2	60.0
10	102.8	94.0	9.0	100.3	26.0
11	109.5	114.0	16.0	101.8	36.0
12	109.2	101.0	7.0	88.2	26.0
13	101.1	98.0	18.0	113.3	40.0
14	90.8	82.0	6.0	96.7	42.0
15	104.4	88.0	5.0	109.6	23.0
16	112.6	115.0	4.0	116.8	30.0
17	95.2	107.0	25.0	79.4	43.0
18	107.0	92.0	2.0	119.0	35.0
19	94.3	92.0	11.0	120.6	61.0
20	104.4	87.0	7.0	99.0	27.0
21	106.5	114.0	13.0	117.1	30.0
22	110.1	108.0	12.0	117.5	30.0

an IQ of 100.0, eight years' experience, and a score of 110.0 on the personality trait test is

$$\widehat{PI} = 75.4357 + (0.1730)(100.0) + (0.0684)(8)$$

$$+ (0.2124)(110.0) - (0.3532)(50)$$

$$= 75.4357 + 17.3000 + 0.5472$$

$$+ 23.3640 - 17.6600 = 98.9869$$

The computer results are shown in Figure 14.15.[8] To produce these results you do *not* need to be a computer programmer. All you need to do is supply the data, identify the dependent and independent variables, and set up your deck of cards with the help of your university computer center personnel, if necessary.

The explanation of Figure 14.15, line by line, is as follows:

Line 1   A statement identifying the problem title as "Personal Characteristics Regression."

Line 2   Identifies the sample as one consisting of 22 observations.

Line 3   Titles for the next five lines.

Line 4   Statistics of the variable Y (variable number 1). Its mean and standard deviations are 101.44091 and 6.52137 respectively; the ratio of the standard deviation to the mean is 0.06429; the minimum and maximum sample values of Y are 90.80000 and 112.60000 respectively.

Line 5   Statistics of the variable X_1 (variable number 2).

Line 6   Statistics of the variable X_2 (variable number 3).

Line 7   Statistics of the variable X_3 (variable number 4).

Line 8   Statistics of the variable X_4 (variable number 5).

Line 9   Regression title.

Line 10 Identifies the variable Y as the dependent variable.

Line 11 The tolerance level is 0.0100. An explanatory variable that does not contribute at least 0.0100 to the regression's coefficient of determination (R^2) will not be included in the regression. We discuss the coefficient of determination in Chapter 15.

Line 12 The statement indicates that all the sample data will be included in the regression. An alternative situation would be the inclusion of the data in parts, creating a piecewise regression analysis.

Line 13 The multiple correlation coefficient (see Chapter 15) is equal to 0.8448, and the standard error of estimate S_e is 3.8781.

Line 14 R^2 (see Chapter 15) is equal to 0.7137.

Line 15 A title for the next three lines, in which an analysis of variance is presented.

Line 16 Titles for the next two lines.

Line 17 The sum of squares $\Sigma(\hat{Y} - \bar{Y})^2$ is equal to 637.415. The degrees of freedom of this sum of squares is 4, and the mean squares is $637.415/4 = 159.354$. The F ratio, 10.595, shows that the

---

[8] The computer printouts shown in the chapter were generated using the Biomedical Computer Programs, P series, by Health Sciences Computing Facility, University of California, Los Angeles.

explanatory variables, as a group, provide a significant explanation of the variable Y. The significance is at all values of α that are greater than 0.00017.

Line 18 The sum of squares of the residuals, $\Sigma(Y - \hat{Y})^2$, is equal to 255.679. It has 17 degrees of freedom and the mean squares is $255.679/17 = 15.040$.

Line 19 Titles for the next five lines.

Line 20 The regression intercept is equal to 75.436.

Line 21 The regression coefficient of X_1 (variable number 2) is 0.173, and its standard error is 0.069. The standardized regression coefficient (0.334) was not discussed in this book; it is generally of minor importance. The t value is 2.516 ($0.173/0.069 = 2.516$), where the t distribution here has $n - K - 1 = 22 - 4 - 1 = 17$ degrees of freedom. If we use $\alpha = 0.01$ so that our critical value is $t_{0.01}^{(17)} = 2.567$, a hypothesis such as $H_0: \beta_1 = 0$ cannot be rejected. Here, however, is where we must be very careful in interpreting the regression results. The sample shows that the coefficient b_1 is *different* from zero. It equals 0.173, and so our best inference about β_1 must be that it is positive and equals 0.173. Only if we have some reason to believe that β_1 is in truth equal to zero–perhaps because of a theory or because of some other empirical study from another place or time–will we question our estimate and test a hypothesis such as $H_0: \beta_1 = 0$. Accepting H_0 would merely mean that our sample data do not show strong enough evidence to warrant its rejection.

Line 22 Statistics of the coefficient of X_2.

Line 23 Statistics of the coefficient of X_3.

Line 24 Statistics of the coefficient of X_4.

```
 1. PROBLEM TITLE . . . . . . .  PERSONAL CHARACTERISTICS REGRESSION
 2. NUMBER OF CASES READ . . . . . . . . . . . . . . .          22
 3. VARIABLE        MEAN  STANDARD DEVIATION  ST.DEV/MEAN    MINIMUM      MAXIMUM
 4.  1  Y         101.44091      6.52137         .06429      90.80000    112.60000
 5.  2  X1         99.36364     12.58701         .12668      80.00000    120.00000
 6.  3  X2         10.50000      7.72596         .73581       1.00000     30.00000
 7.  4  X3        101.96364     12.60165         .12359      79.40000    120.00000
 8.  5  X4         38.36364     11.74550         .30616      19.00000     61.00000
 9.   REGRESSION TITLE . . . . . . . . .  PERSONAL CHARACTERISTICS REGRESSION
10.   DEPENDENT VARIABLE . . . . . . . . . . . . . . . . .   1  Y
11.   TOLERANCE . . . . . . . . . . . . . . . . . . .  .0100
12. ALL DATA CONSIDERED AS A SINGLE GROUP
13. MULTIPLE R           .8448        STD. ERROR OF EST.      3.8781
14. MULTIPLE R-SQUARE    .7137
15. ANALYSIS OF VARIANCE
16.                   SUM OF SQUARES   DF   MEAN SQUARE      F RATIO    P(TAIL)
17.   REGRESSION         637.415        4     159.354        10.595     .00017
18.   RESIDUAL           255.679       17      15.040
                                             STD. REG
19.  VARIABLE      COEFFICIENT STD.ERROR       COEFF        T       P(2 TAIL)
20. INTERCEPT       75.436
21.  X1      2        .173       .069         .334       2.516       .022
22.  X2      3        .068       .133         .081        .513       .615
23.  X3      4        .212       .069         .410       3.086       .007
24.  X4      5       -.353       .086        -.636      -4.107       .001
```

Figure 14.15

Results from computer regression printout

14.14 Association between Explanatory Variables

One important question concerning the estimated slope coefficients in a multiple regression is their significance. We saw earlier that the statistic $\dfrac{b - \beta}{S_b}$ in a simple regression has a t distribution with $(n - 2)$ degrees of freedom, where n is the number of observations. Similarly, the significance of any estimated slope coefficient b_i, in a multiple regression analysis with K independent variables, is determined by the statistic $\dfrac{b_i - \beta_i}{S_{b_i}}$ which has a t distribution with $(n - K - 1)$ degrees of freedom. Note that in a simple regression analysis K (i.e., the number of independent variables) is equal to 1, and $n - K - 1$ is equal to $n - 2$, so that it is nothing but a special case of multiple regression analysis. Hand calculation of S_{b_i} is somewhat complex, but as in simple regression it increases with S_e. Virtually all computer multiple regression programs provide the value of S_{b_i} in their standard output. In those outputs we typically also obtain the value of $\dfrac{b_i - \beta_i}{S_{b_i}}$ under the assumption that the null hypothesis is H_0: $\beta_i = 0$. In other words, in addition to b_i, the computer typically provides the values of S_{b_i} and $\dfrac{b_i}{S_{b_i}}$ (see Figure 14.15).

While we do not present the calculation of S_{b_i} here, we should note that the higher the association between the explanatory variables, the higher the standard errors of the slope coefficients. As a result, for a given estimate b_i, the statistic $\dfrac{b_i}{S_{b_i}}$ is lower as the association between the eplanatory variables increases. This means also that when the association between the explanatory variables increases, the confidence we have in the estimated slope coefficients decreases, since their values depend more heavily on the particular observations that happen to be in the sample.

Problems

14.1. "In a simple linear regression model, an increase in the variance of the random term u_i implies an increase in the variance of Y_i." Do you agree? Why?

14.2. Suppose that in the regression equation

$$Y_i = \alpha + \beta X_i + u_i$$

the coefficients are $\alpha = 1$ and $\beta = 2$. Suppose also that the probability distribution of u_i is discrete, as follows:

$$u_i = -5 \quad \text{with a probability of } \tfrac{1}{4}$$

$$u_i = 0 \quad \text{with a probability of } \tfrac{1}{2}$$

$$u_i = +5 \quad \text{with a probability of } \tfrac{1}{4}$$

Calculate the expected value and the variance of Y_i for $X_i = 4$. How does the variance of Y_i compare with the variance of u_i?

14.3. The independent variable X explains the behavior of the dependent variable Y. Why is it, then, that for a given level of X, say X_0, we expect to find a given variability of Y_0? Can you explain the reasons for such variability? Give an example.

14.4. "If the variance of u, σ_u^2, is equal to zero (that is, $\sigma_u^2 = 0$), the estimated regression line coincides with the true regression line." Prove this statement

   (a) Mathematically.
   (b) By a diagram (see Figure 14.4).

14.5. "If in a sample all the error terms are equal to zero (that is, all $e_i = 0$), the implication is that a perfect fit exists between the dependent and the independent variables." Do you agree? Explain.

14.6. Use the consumption and income example of Table 13.1 in the text to determine the confidence interval for μ_0 and for Y_0 at the following levels of X: 5.0, 15.0, 20.0, 25.0, 35.0. Draw a chart showing the confidence bands for μ_0 and Y_0.

14.7. Is it true that for very large samples the 95 percent confidence bands for μ_0 at the point $X_0 = \bar{X}$ are almost tangent to the estimated regression line? Explain. Would you change your answer for a 99 percent confidence interval?

14.8. Would you change your answer to Problem 14.7 if the confidence bands were for Y_0 rather than μ_0? Explain.

14.9. Explain the difference between the estimated and the "true" regression lines. Is it possible to get more than one estimated line for one "true" line?

14.10. Is it possible that a "true" regression line has a positive slope while an estimated line has a negative slope? Explain.

14.11. Suppose a sample consists of 10 observations and the estimated regression line is $\hat{Y} = a + bX$.

   (a) Is it possible that all the points in the scatter diagram will be below the estimated line?
   (b) Is it necessary that exactly 5 points are above the line and 5 points below?

14.12. It is well known that the ordinary least-squares regression estimators are best linear unbiased estimators (BLUE).

   (a) Explain the properties of BLUE. What is their importance?
   (b) Suppose you can sample as many observations as you want from an infinite population with no difficulty. Which is the one property of BLUE that will be most important then?

14.13. When testing hypotheses with respect to the slope coefficient β, we use the statistic $Z = \dfrac{b - \beta}{\sigma_b}$ if σ_u is known and $T = \dfrac{b - \beta}{S_b}$ if σ_u is unknown.

   (a) Suppose $\sigma_b = 0.2$, but its value is unknown. Suppose also the estimated value, S_b, happens to equal exactly 0.2. Is it possible to reject the null hypothesis H_0: $\beta = \beta_0$ using the t statistic, given that H_0 would have been accepted had σ_b been known and had the Z statistic been used with the same significance level? Explain.
   (b) Under conditions similar to those, except that S_b does not necessarily equal 0.2, of part a, is it possible that H_0 would be accepted when the statistic Z is used and rejected when t is used?

14.14. Let Y and X be income (measured in dollars) and education (measured in school years). Suppose we find that in the regression $\hat{Y} = a + bX$, $a = \$5,000$ and $b = \$100$.

   (a) If we measure X in school years, but measure income (Y) in pounds (£) using an exchange rate of £1 = \$2, how would the change affect the coefficients of a and b?
   (b) Suppose we test the null hypothesis H_0: $\beta = \beta_0$ once using dollars for income and once using pounds. What is the relationship between the value of the statistic computed for the first test (using dollars) and that of the second test (using pounds)? Prove your answer.

14.15. Underwriters are large institutional investors who undertake the preparation and marketing of new stocks and bonds. Table P14.15 presents the total volume of securities sold and the number of security issues underwritten during 1978 by the top 25 underwriters.

TABLE P14.15

Underwriters	Volume (millions of dollars)	Number of issues
Goldman Sachs	$1,967.4	98
Blyth Eastman Dillon	1,211.1	64
First Boston	1,167.5	45
Kidder Peabody	1,138.2	68
Merrill Lynch White Weld	1,125.0	74
E. F. Hutton	1,096.9	51
Smith Barney, Harris Upham	1,014.2	44
Salomon Brothers	1,013.3	45
Bache Halsey Stuart Shields	582.4	31
Paine Webber	549.2	38
Lehman Brothers Kuhn Loeb	425.3	19
Rothschild, Unterberg, Towbin	395.1	38
Alex. Brown & Sons	372.2	14
Wm. R. Hough	357.9	21
Dean Witter Reynolds	345.2	31
Butcher & Singer	318.7	41
Loeb Rhoades Hornblower	312.7	20
John Nuveen	267.7	23
Wertheim	258.8	5
Dain, Kalman & Quail	243.0	20
Matthews & Wright	233.0	15
First Kentucky Securities	223.1	4
Lazard Frères	214.6	10
Piper, Jaffray & Hopwood	193.3	18
Baker Watts	192.0	8

Source: *Institutional Investor*, March 1979.

(*a*) Estimate the regression equation $\hat{Y} = a + bX$ where Y and X are the volume (in millions of dollars) and number of issues, respectively.

(*b*) Test the null hypothesis

$$H_0: \beta = 20$$

versus

$$H_1: \beta \neq 20$$

(*c*) For an underwriter who will underwrite 110 issues, what are your point and your interval estimate for volume at the 95 percent confidence level?

14.16. Earnings per share are a measure of a company's profitability obtained by dividing the company's total net earnings by the number of shares outstanding.

Suppose the relationship between the share price and earnings per share across many firms is

$$\hat{Y} = 15 + 10X$$

where Y is price per share and X is earnings per share.

(*a*) What is the estimated price per share if earnings per share are $4?

(*b*) If you needed to assess the accuracy of your estimate, how would you go about it? What data would you need?

14.17. Consumer awareness of the existence of new products on the market may be measured by the percentage of customers who have heard about the product a certain length of time (say, six months) after it has been introduced.

A study is supposed to examine the relationship between the amount spent on advertising a new product and consumer awareness of that product. Suppose a sample shows the following data:

Consumer awareness (percent) (Y)	Advertising expenditure (thousands of dollars) (X)
52%	$200
21	180
10	100
90	800
64	610
64	450
56	370
40	190

(a) Estimate the regression line $\hat{Y} = a + bX$.

(b) Would a regression line based on consumer awareness as the independent variable and advertising expenditure as the dependent variable have any meaning? Explain.

(c) Using a 95 percent confidence level, construct an interval estimate of consumer awareness of a single product, assuming that $700,000 is spent for its promotion.

(d) Using a 5 percent significance level, test the hypothesis that advertising expenditure has no impact on consumer awareness.

14.18. A sample of 66 families reveals the following relationship between annual family income in thousands of dollars (X) and *percentage* of income spent for food (Y):

$$\hat{Y} = 80 - 1.5X$$

The regression line is appropriate only for the range of X between $10,000 and $40,000.

(a) Explain the meaning of the regression coefficients.

(b) On the basis of this regression estimate, does an average family with $40,000 income spend more on food than an average family with income of $10,000? Explain.

14.19. Some economists hold the view that the inflation rate and the unemployment rate tend to be inversely correlated over time. They claim that monetary expansion may cause inflation but simultaneously it tends to increase the demand for goods and services, an increase that tends to decrease the rate of unemployment. Table P14.19 presents the annual inflation and unemployment rates in the United States in the years 1956 through 1980. Regard the unemployment rate as the dependent variable and the inflation rate as the independent variable.

(a) Draw a scatter diagram and calculate the regression coefficients for the period 1956–80. Is the slope coefficient significantly different from zero? Use a 5 percent significance level.

(b) Repeat part a, this time using the data for the period 1956–67 only.

(c) Repeat part a once again, this time using the data for the period 1968–80 only.

(d) What is the meaning of the slope coefficient of the regression obtained in part b?

(e) Do the data justify breaking the period up into two subperiods, or should the analysis of the relationship between the inflation and unemployment rates be done on the data of the whole period, as in part a?

(f) What is the lesson we should learn from this analysis?

14.20. Maximization of sales or market share is frequently considered to be the goal of the business

TABLE P14.19

Year	Unemployment rate	Inflation rate
1956	4.1%	3.1%
1957	4.3	3.4
1958	6.8	1.6
1959	5.5	2.2
1960	5.5	3.1
1961	6.7	1.1
1962	5.5	1.2
1963	5.7	1.2
1964	5.2	1.3
1965	4.5	1.7
1966	3.8	2.9
1967	3.8	2.8
1968	3.6	4.2
1969	3.5	5.4
1970	4.9	5.9
1971	5.9	4.3
1972	5.6	3.3
1973	4.9	6.2
1974	5.6	11.0
1975	8.5	9.1
1976	7.7	5.8
1977	7.0	6.5
1978	6.0	7.6
1979	5.8	11.5
1980	7.1	13.5

Source: *Survey of Current Business*, various issues.

firm. Many, however, claim that business firms consider profit maximization to be their goal. Some argue that the two goals coincide. Table P14.20 provides data on sales and return on invested capital in the auto industry for the first nine months of 1978.

(a) Draw a scatter diagram in which returns on invested capital are measured along the vertical axis and sales are measured along the horizontal axis. Can you determine from the diagram whether sales and return on invested capital are closely related?
(b) Calculate the regression coefficient of return on invested capital (dependent) on sales (independent), and test the significance of the slope coefficient.
(c) In light of your results, do you agree that the two goals coincide? If so, to what degree?

TABLE P14.20

Company	Sales (millions of dollars)	Return on invested capital (percent)
American Motors	$ 2,026	7.84%
Arvin Industries	352	8.46
Bendix	2,802	10.50
Cummins Engine	1,100	9.75
Dana	1,707	12.04
Eagle-Picher	385	9.87
Eaton	1,949	9.18

TABLE 14.20 *(continued)*

Company	Sales (millions of dollars)	Return on invested capital (percent)
Federal Mogul	421	12.40
Ford Motor Co.	31,538	13.41
Fruehauf	1,623	9.85
General Motors	45,477	15.11
International Harvester	5,472	7.94
Questor	313	3.36
Sheller-Globe	459	7.18
Smith (A. O.)	601	9.51
TRW	2,755	10.32
Timken	808	9.93
Total	$99,788	166.65%
Average	5,869.8824	9.8029

Data from *Business Week*, January 8, 1979, p. 34.

14.21. In Section 14.9 we introduced beta as a measure of a security's risk and defined the concept of aggressive and defensive securities (those with $\beta > 1$ and $\beta < 1$, respectively). Table P14.21 gives the annual rates of return (SRR) (in percent) on three securities—General Motors (GM), International Business Machines (IBM), and American Motors (AMC)—for the period 1970–80. The last column in the table gives proxy data for the market rate of return (MRR) as measured by the annual rates of return of the Fisher stock market index for the same period.

    (*a*) Determine the beta estimates of the three securities by regressing each SRR on MRR. On the basis of these point estimates, which of the stocks is aggressive and which is defensive?

    (*b*) Test whether each of the above three beta estimates is significantly different from 1. (Use a 5 percent significance level.)

    (*c*) Summarize your results in *a* and *b* above.

TABLE P14.21

Year	GM	IBM	AMC	Stock market index
1970	22.3	−11.4	−33.3	1.4
1971	4.3	7.5	21.6	15.9
1972	6.5	21.2	17.8	17.8
1973	−37.8	−22.1	7.5	−16.9
1974	−27.6	−30.0	−62.3	−26.8
1975	97.1	37.9	65.4	37.7
1976	47.4	28.3	−27.9	26.3
1977	−11.4	1.8	−6.5	−4.8
1978	−5.4	13.9	31.0	7.4
1979	2.3	−9.5	47.9	21.8
1980	−4.3	11.3	−42.9	32.8

Source: Center for Research in Security Prices Tape, University of Chicago.

14.22. Economists argue that the interest rate tends to increase with inflation. The reason is, of course, that *with high inflation* the nominal interest paid should cover some real interest plus compensation for the depreciation of the value of money. Table P14.22 presents the prime interest rate and the annual inflation rate for the years 1968–80. The prime interest rate is the rate banks charge for loans given to their preferred customers.

TABLE P14.22

Year	Prime interest rate	Inflation rate
1968	5.90%	4.20%
1969	7.83	5.37
1970	7.72	5.92
1971	5.11	4.30
1972	4.69	3.30
1973	8.15	6.23
1974	9.87	10.97
1975	6.29	9.14
1976	5.19	5.77
1977	5.59	6.45
1978	8.11	7.60
1979	11.04	11.47
1980	12.78	13.46

Source: *Statistical Abstract of the United States*, 1975; *Survey of Current Business*, various issues; Morgan Guaranty Trust, *World Financial Markets*, various issues.

(a) Draw a scatter diagram for the data, measuring the prime rate along the vertical axis and the inflation rate along the horizontal axis.
(b) Calculate the regression coefficient when the prime rate is taken as the dependent variable and the inflation rate as the independent variable. What is your interpretation of the results?
(c) Test the significance of the regression slope at the 1 percent significance level.
(d) What is your interpretation of the regression slope in this regression? What is your interpretation of the slope?

14.23. The following is an estimated equation of a regression plan:

$$\hat{Y} = 2 + 3X_1 + 2X_2$$

Calculate the value $\sum_{i=1}^{3} e_i^2$ for the following three sample points:

$$Y = 8; \quad X_1 = 1; \quad X_2 = 1$$
$$Y = 10; \quad X_1 = 2; \quad X_2 = 4$$
$$Y = 10; \quad X_1 = 3; \quad X_2 = 4$$

14.24. Suppose we obtain the following estimated regression equation:

$$\hat{Y} = 5 + 2X_1 - 4X_2$$

Is it true that an increase of X_1 by two units is equivalent to a decrease of X_2 by one unit, in terms of the impact on $\hat{Y}$? Explain.
14.25. Using your own numerical example, demonstrate that the coefficient b_i of an independent variable X_i in a multiple regression equation represents the impact of a one-unit increase in X_i when all of the other explanatory variables are held *unchanged*.
14.26. Let Y be annual consumption (in thousands of dollars), X_1 annual income (in thousands of dollars), and X_2 formal education (in number of years of schooling). A study shows the following relationship between the variables:

$$\hat{Y} = 3.000 + 0.500X_1 + 0.125X_2$$

Explain the meaning of each of the three coefficients.
14.27. An insurance company pays its sales representatives a fixed weekly wage (X_1) plus a percentage

of net sales made directly by the representative (X_2). The firm, which operates nationwide, is interested in paying the representatives a combination of fixed wage and a percentage of sales that will maximize the sales of insurance policies. Currently the company estimates that the net sales made by a representative are related to his income components in the following way:

$$\hat{Y} = 40X_1 + 9X_2$$

where $\hat{Y}$ and X_1 are measured in hundreds of dollars and X_2 is measured in percentage points (e.g., if $X_2 = 1$, the representative's commission is 1 percent of his net sales). Currently, $X_1 = 2$ and $X_2 = 1$.

(a) What is the estimated net sales of a representative?

(b) The firm considers adding $100 to the fixed wage component (X_1) or alternatively increasing X_2 from 1 to 1.5. Which of these two alternatives results in a higher estimated average net sales?

14.28. The prices of stocks are related to the average level of earnings per share (EPS) as well as to the variance of EPS. Below is a sample of stock prices, average EPS level in recent years, and standard deviation of the respective EPS.

Price	Average EPS	Standard deviation of EPS
50	6	2
46	5	3
12	1	1
27	6	7
30	3	3
19	5	6

Estimate the coefficients of the regression

$$\hat{Y} = b_0 + b_1X_1 + b_2X_2$$

where Y is the price, X_1 is average EPS, and X_2 is the standard deviation of EPS. Explain the coefficients.

14.29. To test how coupons help advertisers increase sales, a regression analysis was applied to the use of coupons to promote the consumption of processed orange juice. The variables considered were:

1. Total number of coupons redeemed (CR).
2. Total coupon drop (the number of coupons issued) (C).
3. Coupon value (V)
4. Time elapsed since initial coupon drop (T)

The reason the relationship of these variables should be studied is that it provides answers to questions that must be addressed during the planning stage of a coupon drop. What, for example, is the size of coupon drop needed to achieve a particular level of consumer redemption? What is the lagged response to an initial coupon drop? How does one plan the budget in view of the anticipated sales activities following a coupon drop?

The data are shown in Table P14.29.

TABLE P14.29

Observation	CR (thousands of coupons)	C (thousands of coupons)	V (cents)	T (months)
1	6,149	40,000	10	5
2	5,566	70,000	15	1
3	7,060	30,000	15	10
4	5,542	50,000	15	1
5	2,171	20,000	10	6
6	4,064	50,000	7	5
7	9,007	40,000	10	12
8	7,569	30,000	20	4
9	5,368	50,000	10	2
10	3,200	60,000	5	3
11	8,863	36,000	20	10
12	1,540	15,000	4	8

(a) Run a simple regression of *CR* (dependent) on *C* (independent). Explain the meaning of the regression equation, and predict the coupon redemption for a 60-million coupon drop.

(b) Run a multiple regression analysis on the computer, using *CR* as the dependent variable and *C*, *V*, and *T* as the independent variables. Write down the regression equation and explain it.

(c) Compare the regression coefficients of the variable *C* in part *a* and in part *b*. What is the importance of including *all* of the relevant variables in a regression analysis?

(d) On the basis of the results of part *b*, what is the best prediction of coupon redemption in an advertising campaign for processed orange juice if each coupon is good for 5 cents and the coupon drop is 65 million coupons? Give your prediction for coupon redemption at the end of each of the first twelve months after the coupon drop. Repeat your calculations for 10-cent and 15-cent coupon values and draw a chart showing three graphs of the predicted coupon redemption (vertical axis) as a function of time (horizontal axis). What is the predicted *value* of the total coupons redeemed after six months for coupons worth 8 cents?

(e) Again on the basis of the multiple regression equation, predict the coupon redemption after six months if each coupon is good for 10 cents. Give your prediction for the following alternative coupon drops (in millions of coupons): 10, 20, 30, 40, 50, 60, and 70.

   Draw a chart showing the relationship between coupon redemption and coupon drop six months after the drop for a coupon value of 10 cents.

(f) Do you think the time variable (*T*) is indeed linearly related to *CR*, as we assumed in this problem? Explain.

CHAPTER FIFTEEN OUTLINE

15.1 The Simple Correlation Coefficient

The Sample Coefficient of Determination
The Sample Correlation Coefficient
The Coefficient of Determination as a Measure of
the Strength of the Relationship between X and Y
Inference Concerning ρ: When X Is Not a Random
Variable
Inference Concerning ρ: When X and Y Are Both
Random Variables

15.2 The Multiple Correlation Coefficient

Key Terms

coefficient of determination
simple correlation coefficient
multiple correlation coefficient

15

CORRELATION ANALYSIS

The coefficient of correlation measures the relationship between variables. In the simple regression context, it measures the direction as well as the strength of the relationship between the dependent variable Y and the explanatory variable X. The coefficient of correlation is denoted by the Greek letter ρ (rho). The squared value of ρ, ρ^2, is the **coefficient of determination** in the population. The estimated value of ρ is denoted by R, and the estimated coefficient of determination is the squared R: R^2.

A correlation coefficient that pertains to only two variables is called a **simple correlation coefficient.** If three or more variables are analyzed simultaneously, the correlation is **multiple.** In this chapter, we first relate the simple correlation coefficient to a simple regression analysis, then discuss the multiple correlation coefficient in connection with multiple regression analysis.

15.1 The Simple Correlation Coefficient

Consider Figure 15.1, in which a scatter diagram is shown along with the OLS regression line. A given observation Y_i is shown, and the distance from the sample average $\overline{Y}$ is identified as the "total deviation." For the same observation, the distance from the regression line is identified as the "unexplained deviation," and finally the distance from the regression line ($\hat{Y}$) to the average Y value ($\overline{Y}$) is labeled as the "explained deviation." A similar diagram was presented in Chapter 14 (see Figure 14.5), when we introduced the standard error of estimate S_e; the scatter diagram is reproduced here for convenience.

Figure 15.1

The total deviation, explained deviation, and unexplained deviation of an observation from the regression line

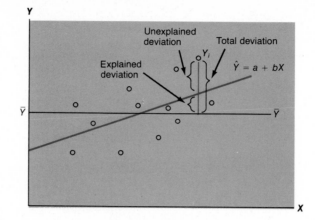

It is clear that the sum of the unexplained deviation and the explained deviation is equal to the total deviation:

Total deviation = unexplained deviation + explained deviation

Using our notation, we may write the equation as

$$Y - \overline{Y} = (Y - \hat{Y}) + (\hat{Y} - \overline{Y}) \tag{15.1}$$

As one can easily verify, Equation 15.1 holds for observations below the regression line as well as for those above the line.

THE SAMPLE COEFFICIENT OF DETERMINATION

It can be shown that the sum of squares $\Sigma(Y - \overline{Y})^2$, which is the sum of the squared total deviations across all the sample observations, is equal to the sum of squared explained deviations plus the sum of squared unexplained deviations:

$$\Sigma(Y - \overline{Y})^2 = \Sigma(Y - \hat{Y})^2 + \Sigma(\hat{Y} - \overline{Y})^2 \tag{15.2}$$

With Equation 15.2 we can express the sum of squares of the explained deviations as follows:

$$\Sigma(\hat{Y} - \overline{Y})^2 = \Sigma(Y - \overline{Y})^2 - \Sigma(Y - \hat{Y})^2 \tag{15.3}$$

Noting that $(Y - \hat{Y}) = e$, so that $\Sigma(Y - \hat{Y})^2 = \Sigma e^2$, and dividing all three terms in Equation 15.3 by $\Sigma(Y - \overline{Y})^2$, we get

$$\frac{\Sigma(\hat{Y} - \overline{Y})^2}{\Sigma(Y - \overline{Y})^2} = \frac{\Sigma(Y - \overline{Y})^2}{\Sigma(Y - \overline{Y})^2} - \frac{\Sigma e^2}{\Sigma(Y - \overline{Y})^2} \tag{15.4}$$

or

$$\frac{\Sigma(\hat{Y} - \overline{Y})^2}{\Sigma(Y - \overline{Y})^2} = 1 - \frac{\Sigma e^2}{\Sigma(Y - \overline{Y})^2} \tag{15.5}$$

The term on the left-hand side of Equation 15.5 is a measure of the explained variability of Y in the sample, expressed as a proportion of the total variability of Y. This measure is a sample coefficient of determination, R^2.

Using the notation R^2 for the estimated coefficient of determination, we rewrite Equation 15.5 as follows:

$$R^2 = 1 - \frac{\Sigma e^2}{\Sigma (Y - \overline{Y})^2} \qquad \cdot \tag{15.6}$$

The coefficient of determination, R^2, measures the fraction of total variability of Y "explained" by the regression. If $R^2 = 0.80$, for example, it means that 80 percent of the total sample variation in Y is explained by the variable X, while 20 percent of the sample variation remains unexplained. Thus the coefficient of determination has a straightforward interpretation as a measure of the strength of the relationship between X and Y.

Recall the definition of the standard error of estimate, S_e:

$$S_e = \sqrt{\frac{\Sigma e^2}{n - 2}} \tag{15.7}$$

After squaring and cross-multiplying we get

$$\Sigma e^2 = S_e^2 \cdot (n - 2) \tag{15.8}$$

Similarly, S_Y is given by

$$S_Y = \sqrt{\frac{\Sigma (Y - \overline{Y})^2}{n - 1}} \tag{15.9}$$

so that after squaring and cross-multiplying we get

$$\Sigma (Y - \overline{Y})^2 = S_Y^2 \cdot (n - 1) \tag{15.10}$$

After dividing Equation 15.8 by Equation 15.10, we get

$$\frac{\Sigma e^2}{\Sigma (Y - \overline{Y})^2} = \frac{S_e^2}{S_Y^2} \cdot \frac{n - 2}{n - 1}$$

and by substituting in Equation 15.6 we obtain the following equation:

$$R^2 = 1 - \frac{S_e^2}{S_Y^2} \cdot \frac{n - 2}{n - 1} \qquad \cdot \tag{15.11}$$

When the sample is large, the term $\dfrac{n-2}{n-1}$ is close to 1, and R^2 may be approximated by the following:

$$R^2 = 1 - \frac{S_e^2}{S_Y^2} \qquad (15.12)$$

THE SAMPLE CORRELATION COEFFICIENT

From Equation 15.11 it follows that the estimated correlation coefficient, R, is given by

$$R = \pm\sqrt{1 - \frac{S_e^2}{S_Y^2} \cdot \frac{n-2}{n-1}} \qquad (15.13)$$

where the sign of R is the same as the sign of the regression slope: if the relationship between the variables is direct, R is positive; if it is an inverse relationship, R is negative. A computational formula for R is

$$R = \frac{\Sigma(X - \overline{X})(Y - \overline{Y})}{\sqrt{\Sigma(X - \overline{X})^2}\,\sqrt{\Sigma(Y - \overline{Y})^2}} \qquad (15.14)$$

Equation 15.14 may be further simplified to give us the following:

$$R = \frac{\Sigma XY - n\overline{X}\,\overline{Y}}{\sqrt{\Sigma X^2 - n\overline{X}^2}\,\sqrt{\Sigma Y^2 - n\overline{Y}^2}} \qquad (15.15)$$

THE COEFFICIENT OF DETERMINATION AS A MEASURE OF THE STRENGTH OF THE RELATIONSHIP BETWEEN X AND Y

The correlation coefficient takes on values in the range between -1 and $+1$: $-1 \le R \le +1$. It follows that R^2 ranges between 0 and $+1$:

$$0 \le R^2 \le 1 \qquad (15.16)$$

Indeed, this is only logical, and can be easily verified from Equation 15.12, since R^2 measures the fraction of the total variation explained by the regression, and as such it must range between 0 and 1. The stronger the relationship between the variables X and Y, the closer R^2 is to 1, and conversely, the weaker the relationship, the closer R^2 is to 0.

That R^2 ranges between 0 and 1 can also be seen in another way. Figure 15.2 shows three alternative scatter diagrams. In Figure 15.2a all the sample

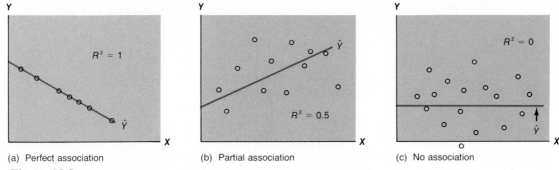

Figure 15.2

Alternative scatter diagrams and coefficients of determination

observations lie on the regression line. This is the case when all the e^2s are equal to zero. As a consequence, $\Sigma e^2 = 0$, and from Equation 15.6 it is clear that in this case $R^2 = 1$. In Figure 15.2c no relationship seems to exist between the two variables. As a result, the regression line has a zero slope and all the $\hat{Y}_i$'s are equal to $\overline{Y}$ (that is, the regression line of Y on X is horizontal and coincides with $\overline{Y}$). In this case the quantity $(\hat{Y} - \overline{Y})$ is equal to zero for all the observations and so $\Sigma(\hat{Y}_i - Y)^2 = 0$ and the left-hand side of Equation 15.3 is clearly zero, indicating that $R^2 = 0$. Naturally these are the most extreme cases. Figure 15.2b shows a scatter diagram for the case in which $R^2 = 0.5$. Here 50 percent of the variation in Y is explained by the regression line. In general, the closer the observations to the regression line (other things held the same), the greater R^2.

INFERENCE CONCERNING P: WHEN X IS NOT A RANDOM VARIABLE

In making inferences about the population correlation coefficient ρ, based on the sample coefficient R, we must distinguish between two situations that may arise. The first is that in which the dependent variable is a random variable while the independent variable is not. Basically, this is the way we treated the variables X and Y in regression analysis: we assumed that Y relates to X and to a random variable, u; thus Y itself was a random variable, but X was not.

The most relevant question about ρ in this case is whether it is equal to zero or not. We will show now that testing $H_0: \rho = 0$ is equivalent to testing $H_0: \beta = 0$ where β is the regression slope coefficient.

Denoting

$$S_X^2 = \frac{\Sigma(X - \overline{X})^2}{n - 1} \qquad \textbf{(15.17)}$$

we can show that the regression slope coefficient b and the correlation coefficent R are related in the following way:

$$b = R \cdot \frac{S_Y}{S_X} \qquad \textbf{(15.18)}$$

This means that when R equals zero, b also equals zero, and vice versa. The same relationship can be similarly shown to exist for the population:

$$\beta = \rho \cdot \frac{\sigma_Y}{\sigma_X} \tag{15.19}$$

Thus testing $H_0: \beta = 0$ is equivalent to testing $H_0: \rho = 0$.

INFERENCE CONCERNING ρ: WHEN X AND Y ARE BOTH RANDOM VARIABLES

The second situation that may arise when we are making inferences about ρ is that in which both X and Y are treated as *random* variables. We shall assume here that the population is infinite. If we can assume that the *joint* distribution of X and Y is *bivariate normal*—that is, that for any X value the distribution of Y is normal, and also that for any Y value the distribution of X is normal—then in testing the hypothesis that $\rho = 0$, we can use a statistic that has a t distribution. Specifically, if we test $H_0: \rho = 0$, then (even for small samples) we can use the following test statistic:

$$T = R \sqrt{\frac{n-2}{1-R^2}} \tag{15.20}$$

where the statistic T has a t distribution with $n - 2$ degrees of freedom.

EXAMPLE 15.1

There are many financial institutions—commercial banks, savings and loan associations, mutual savings banks, pension funds, insurance companies, and the like, known as "financial intermediaries"—which invest tens of billions of dollars in the U.S. capital markets. Economists are interested in studying the behavior of those institutions in the capital markets. As part of a study of this kind, suppose we want to determine the way interest rates in the bond market affect the amount of federal government bonds demanded by corporate and state and local government pension funds. Thus our dependent variable will be the total value of United States bonds held by those pension funds. We denote this variable by GB (for "government bonds"). Since we are dealing with simple correlation, we shall assume only one explanatory variable: the rate of interest on government bonds, which will be denoted by IGB. To avoid unnecessary calculations, we shall assume only 10 observations. Each observation represents a three-month average level. The data are given in Table 15.1. A work sheet for calculation of the correlation coefficient is provided in Table 15.2. Using Equation 15.15, we may now find the coefficient of correlation:

$$R = \frac{135,700.4 - (10)(6.68)(1,981.6)}{(\sqrt{464.5 - (10)(6.68)^2})(\sqrt{40,245,976 - (10)(1,981.6)^2})}$$

$$= \frac{3,329.52}{(\sqrt{18.276})(\sqrt{978,591})} = \frac{3,329.52}{(4.275)(989.237)} = 0.7873$$

From here $R^2 = 0.7873^2 = 0.6198$, meaning that 61.98 percent of the total variance of GB is explained by the interest rate. The positive sign of R indicates a direct relationship between GB and IGB, indicating that pension funds tend to increase their purchases of government bonds when the interest on those bonds rises. To test the hypothesis $H_0: \rho = 0$, we may calculate the regression slope coefficient, b, and then test the equivalent hypothesis: $H_0: \beta = 0$. If $H_0: \beta = 0$ is accepted, we should also accept $H_0: \rho = 0$, and vice versa. If, however, the variable may be assumed to be a *random* variable having a bivariate normal distribution with the variable Y, and we want to test $H_0: \rho = 0$ against $H_1: \rho \neq 0$, then we may use the test statistic T:

$$T = R \sqrt{\frac{n-2}{1-R^2}}$$

In this case we get

$$T = 0.7873 \cdot \sqrt{\frac{10-2}{1-0.7873^2}} = 3.612$$

Since T has a t distribution with $n - 2 = 8$ degrees of freedom, we must compare the value of 3.612 with the critical value and then make our decision accordingly.

TABLE 15.1

Government Bonds Held by Pension Funds and Rate of Interest Paid during Ten Quarters

Quarter	Bonds (GB) (millions of dollars)	Interest (IGB) (percent)
1	$2,030.0	7.1%
2	1,710.0	6.0
3	1,790.0	5.3
4	1,970.0	5.2
5	2,440.0	7.0
6	2,326.0	8.4
7	2,440.0	9.2
8	1,900.0	7.6
9	1,740.0	6.0
10	1,470.0	5.0

TABLE 15.2

Work Sheet for Calculation of Correlation Coefficient

Quarter	GB Y	IGB X	Y^2	X^2	XY
1	2,030	7.1	4,120,900	50.41	14,413.0
2	1,710	6.0	2,924,100	36.00	10,260.0
3	1,790	5.3	3,204,100	28.09	9,487.0
4	1,970	5.2	3,880,900	27.04	10,244.0
5	2,440	7.0	5,953,600	49.00	17,080.0
6	2,326	8.4	5,410,276	70.56	19,538.4
7	2,440	9.2	5,953,600	84.64	22,448.0
8	1,900	7.6	3,610,000	57.76	14,440.0
9	1,740	6.0	3,027,600	36.00	10,440.0
10	1,470	5.0	2,160,900	25.00	7,350.0
Total	19,816	66.8	40,245,976	464.50	135,700.4
Average	1,981.6	6.68			

Assuming $\alpha = 0.05$, we get $t_{0.025}^{(8)} = 2.306$. Since $3.612 > 2.306$, we reject H_0 and conclude that ρ is indeed different from zero.

15.2 The Multiple Correlation Coefficient

Multiple correlation in multiple regression analysis is the counterpart of simple correlation in simple regression analysis. As we saw in Chapter 14, a model of multiple regression such as

$$Y = \beta_0 + \beta_1 X_1 + \beta_2 X_2 + \cdots + \beta_K X_K + u$$

may be estimated to yield

$$\hat{Y} = b_0 + b_1 X_1 + b_2 X_2 + \cdots + b_K X_K$$

As in simple regression, the estimated coefficient of determination is given by the following equation:

$$R^2_{Y.12\ldots K} = 1 - \frac{\Sigma e^2}{\Sigma (Y - \overline{Y})^2} = 1 - \frac{S_e^2}{S_Y^2} \cdot \frac{n - K - 1}{n - 1} \qquad \textbf{(15.21)}$$

This is the analog of Equation 15.11 for simple correlation. The subscript of R^2 indicates that Y is the dependent variable and X_1 through X_K are the independent variables. For a large enough sample, R^2 may be approximated by the following:

$$R^2_{Y.12\ldots K} = 1 - \frac{S_e^2}{S_Y^2} \qquad \textbf{(15.22)}$$

This is equivalent to Equation 15.12. The estimated **multiple correlation coefficient** is simply the square root of R^2. Note, though, that in multiple regression, the correlation coefficient is always assigned a positive sign. Therefore we write the following:

$$R_{Y.12\ldots K} = +\sqrt{R^2_{Y.12\ldots K}} \qquad \textbf{(15.23)}$$

While the sign of R is determined by the sign of the slope coefficient, b, in the simple regression case, the sign of $R_{Y.12\ldots K}$ in the multiple regression case is positive, since different slope coefficients can have different signs.

EXAMPLE 15.2

Let us now extend Example 15.1 to a multiple correlation analysis. To avoid undesirable complications, we shall introduce only one more explanatory variable, the interest rate on corporate bonds, which will be denoted by *ICB*. The estimated equation will be

$$\widehat{GB} = b_0 + b_1 IGB + b_2 ICB$$

As one can expect, the coefficient b_1 will be positive, since an increase in the interest rate on government bonds (*IGB*) is expected to induce pension funds to buy more bonds (*GB*). On the other hand, when all other factors are held constant, we can expect to find that when the interest rate on corporate bonds (*ICB*) goes up, the demand for government bonds (*GB*) goes down, and thus we expect to see that b_2 is negative. Suppose the data are as in Table 15.3. Figure 15.3 shows results from a computer printout of the multiple regression

$$\widehat{GB} = b_0 + b_1 IGB + b_2 ICB$$

The coefficient of determination, $R^2_{Y.12}$, is equal in this case to 0.9884, meaning that 98.84 percent of the variance of Y is explained by the two explanatory variables *IGB* and *ICB*. According to the computer's cal-

TABLE 15.3
Government Bonds Held by Pension Funds and Interest Rate on Government and Corporate Bonds during Ten Quarters

Quarter	GB (millions of dollars)	IGB (percent)	ICB (percent)
1	$2,030.0	7.1%	9.2%
2	1,710.0	6.0	9.2
3	1,790.0	5.3	8.1
4	1,970.0	5.2	7.0
5	2,440.0	7.0	7.5
6	2,326.0	8.4	9.5
7	2,440.0	9.2	10.0
8	1,900.0	7.6	9.9
9	1,740.0	6.0	9.0
10	1,470.0	5.0	9.0

culations, the variance of *GB* is equal to 108,732 (the standard deviation is shown in Figure 15.3). Using this number and others provided in Figure 15.3, we see that $R^2_{Y.12}$ can also be obtained from Equation 15.21:

$$R^2_{Y.12} = 1 - \frac{S^2_e}{S^2_Y} \cdot \frac{n - K - 1}{n - 1} = 1 - \frac{40.201^2}{108,732.000} \cdot \frac{10 - 2 - 1}{10 - 1}$$

$$= 0.9884$$

The multiple correlation coefficient is equal to

$$R_{Y.12} = +\sqrt{R^2_{Y.12}} = +\sqrt{0.9884} = +0.9942$$

The positive sign of $R_{Y.12}$ does not mean that the regression slope coefficients are positive. In fact, although the slope coefficient with respect to the first variable is positive, the slope with respect to the second variable is negative.

```
   PROBLEM  TITLE . . . . . .  GOVERNMENT  BOND  HOLDING
 NUMBER  OF  CASES  READ . . . . . . . . . . . . . 10
 VARIABLE            MEAN  STANDARD  DEVIATION  ST.DEV/MEAN      MINIMUM       MAXIMUM
   1  GB        1981.60000          329.74576        .16640   1470.00000    2440.00000
   2  IGB          6.68000            1.42501        .21333      5.00000       9.20000
   3  ICB          8.84000             .99688        .11277      7.00000      10.00000
   REGRESSION  TITLE . . . . . . . . . . . .  GOVERNMENT  BOND  HOLDING
   DEPENDENT  VARIABLE . . . . . . . . . . .        1  GB
   TOLERANCE . . . . . . . . . . . . . . . .     .0100
 ALL  DATA  CONSIDERED  AS  A  SINGLE  GROUP
 MULTIPLE  R              .9942         STD.  ERROR  OF  EST.      40.2010
 MULTIPLE  R-SQUARE       .9884
 ANALYSIS  OF  VARIANCE
                    SUM OF SQUARES    DF    MEAN  SQUARE       F  RATIO    P(TAIL)
   REGRESSION         967277.559       2     483638.780       299.259      .00000
   RESIDUAL            11312.841       7       1616.120
                                            STD. REG
   VARIABLE      COEFFICIENT  STD.  ERROR    COEFF       T    P(2  TAIL)
 INTERCEPT         2284.446
 IGB        2       292.090      11.940      1.262    24.464    0.000
 ICB        3      -254.978      17.067      -.771   -14.940    0.000
```

Figure 15.3

Computer printout reproduction showing results of regression of *GB* (denoted Y) on *IGB* (denoted X_1) and *ICB* (denoted X_2)

Problems

15.1. When the coefficient of determination between Y and X is calculated, it is found that

$$\Sigma X^2 = 100 \qquad \Sigma Y^2 = 100$$

$$n\overline{X}^2 = 90 \qquad n\overline{Y}^2 = 90$$

(a) Is it possible that $\Sigma XY = 100$?

(b) What is the maximum value that ΣXY is conceivably equal to, given the above data?

15.2. What is the coefficient of determination between the variable X and itself? Use Equation 15.15 to prove your answer.

15.3. What is the coefficient of determination between the variable X and $-X$? Use Equation 15.15 to prove your answer.

15.4. Calculate R^2 and R for X and Y, once using the data of Table P15.4a and once using the data of Table P15.4b.

<table>
<tr><td colspan="2">**TABLE P15.4a**</td><td colspan="2">**TABLE P15.4b**</td></tr>
<tr><td>X</td><td>Y</td><td>X</td><td>Y</td></tr>
<tr><td>10</td><td>5</td><td>10</td><td>-5</td></tr>
<tr><td>20</td><td>10</td><td>20</td><td>-10</td></tr>
<tr><td>30</td><td>15</td><td>30</td><td>-15</td></tr>
</table>

15.5. "If $\Sigma(X_i - \overline{X})(Y_i - \overline{Y}) = \Sigma(X_i - \overline{X})^2$, then the coefficient of determination between X and Y, R^2, equals the slope of the following estimated regression line: $\hat{X} = a + bY$." Do you agree? Explain.

15.6. "If $\Sigma(X_i - \overline{X})^2 = \Sigma(Y_i - \overline{Y})^2$, then the coefficient of determination between X and Y, R^2, is equal to b_1^2 *and* to b_2^2 where b_1 and b_2 are the slopes of the following regression lines:

$$\hat{Y} = a_1 + b_1 X$$

and

$$\hat{X} = a_2 + b_2 Y."$$

Evaluate this statement and prove your answer.

15.7. The long-distance telephone bills and the respective number of long-distance calls of a small sample of telephone customers are as follows:

Bill	Number of calls
$25	10
67	15
2	1
30	3

Using the data of the above sample, show numerically that Equations 15.2, 15.5, and 15.6 indeed hold.

15.8. Using the data of Problem 15.7, show numerically the equivalence of Equations 15.6 and 15.14.

15.9. Suppose the coefficient of determination between X and Y is equal to 0.50. What is the correlation coefficient between $X - 10$ and $Y - 20$? What is the correlation coefficient between $\frac{1}{2}X$ and $\frac{3}{4}Y$?

15.10. A firm's personnel department hired employees for a given job, primarily on the basis of the results of an aptitude test administered to job applicants. The test grades are 0 through 10 (10 being the best possible grade). The performance of those hired was rated on the same scale by their supervisor a year after they were hired. A sample of the test grades and the supervisor's assigned grades is as follows:

Test grade	Supervisor's grade
5	8
6	8
10	10
2	5
9	7
5	4
7	10
7	6

(a) Draw a scatter diagram showing the test grade on the horizontal axis and the supervisor's grade on the vertical axis.

(b) Can you guess the sign and magnitude of the correlation coefficient before doing any calculations?

(c) Calculate the correlation coefficient.

(d) On the basis of your results, do you think the test is a good indicator of employees' performance on the job? Is it a very good indicator? Explain.

15.11. Table P15.11 presents the average daily prices of 30 industrial stocks, 20 transportation stocks, and 20 bonds at the close of each trading day during the month of November 1978.

Financial analysts commonly think that stocks tend to have their upswings and downswings together as a result of some common economywide factors (high inflation, high interest rates, and the like), while a similar relationship between prices of stocks and bonds does not exist. In order to examine the validity of these views, use Table P15.11 to do the following:

(a) Calculate the value of R^2 between the closing prices of industrial stocks and those of transportation stocks. Show your calculations in detail.

(b) Calculate the value of R^2 between the closing prices of industrial stocks and those of bonds.

(c) Calculate the value of R^2 between the closing prices of transportation stocks and those of bonds.

In view of your results, do you agree with the common view stated above?

Note: Carry all your calculations to four decimal places; otherwise substantial errors may result.

TABLE P15.11

Day	30 industrials	20 transport companies	20 bonds
1	827.79	219.03	86.67
2	816.96	215.04	86.62
3	823.11	216.84	86.33
6	814.88	215.04	86.44
7	800.07	211.14	86.36
8	807.61	211.53	86.41
9	803.97	210.90	86.34
10	807.09	213.62	86.24
13	792.01	207.64	86.12
14	785.26	205.49	86.26
15	785.60	206.76	86.54
16	794.18	209.49	86.65
17	797.73	210.41	86.91
20	805.61	211.63	86.93
21	804.05	211.04	87.02
22	807.00	212.36	86.96
23	————————HOLIDAY————————		
24	810.12	214.60	86.98
27	813.84	215.04	86.71
28	804.14	211.87	86.50
29	790.11	208.71	86.44
30	799.03	212.36	86.41

Source: *Barron's*, December 4, 1978.

15.12. Table P15.12 shows six data series for the years 1968–80. The first is a series of average weekly earnings in manufacturing and the second is the nominal prime interest rates (the interest rates banks charge on loans made to their preferred customers). Next, the Consumer Price Index (CPI) is given, followed by the annual inflation rate. The next variable, average earnings in 1967 dollars, is obtained by dividing average weekly earnings in manufacturing by the CPI and then multiplying the result by 100. For example, the average weekly earnings in manufacturing in 1977 was $227.50 and the CPI was 181.50, or 81.50 percent higher than in

TABLE P15.12

Year	Nominal average weekly earnings in manufacturing	Nominal prime interest rate	CPI	Inflation rate	Real average weekly earnings in manufacturing	Real prime interest rate
1968	$122.50	5.75%	104.2	4.20%	$117.56	1.48%
1969	129.50	7.61	109.8	5.37	117.94	2.14
1970	133.70	7.31	116.3	5.92	114.96	1.31
1971	142.40	4.85	121.3	4.30	117.39	0.53
1972	154.70	4.47	125.3	3.30	123.46	1.13
1973	165.70	8.08	133.1	6.23	124.49	1.74
1974	176.00	9.89	147.7	10.97	119.16	−0.97
1975	189.50	6.29	161.2	9.14	117.56	−2.61
1976	207.60	5.19	170.5	5.77	121.76	−0.55
1977	227.50	5.59	181.5	6.45	125.34	−0.88
1978	249.27	8.11	195.3	7.60	127.63	0.47
1979	269.34	11.04	217.7	11.47	123.72	−0.39
1980	288.62	12.78	247.0	13.46	116.85	−0.60

Source: *Statistical Abstract of the United States*, 1975; *Survey of Current Business*, various issues; Morgan Guaranty Trust, *World Financial Markets*, various issues.

1967. Thus the 1977 average weekly earnings figure, expressed in 1967 dollars, is $\frac{\$227.50}{181.50} \cdot 100 =$ $125.34. Finally the real prime interest rate is shown; it is obtained by dividing the nominal rate plus one by the inflation rate plus one, then subtracting one from the result. For example, we compute the real rate in 1978 to be $\frac{1 + 0.0811}{1 + 0.076} - 1 = 0.0047$, or 0.47 percent. By expressing the variables in *real* terms, we remove the impact of inflation on those variables.

  (a) Draw a scatter diagram of the *nominal* average weekly earnings in manufacturing and the *nominal* interest rate. Calculate the correlation coefficient between the two variables. Can you think of an economic explanation for the association between these two variables? Explain.

  (b) Draw a scatter diagram of the *real* average weekly earnings in manufacturing and the *real* interest rate. Calculate the correlation coefficient between these two variables.

  (c) Compare the results of parts *a* and *b*. How do you explain the difference?

15.13. Table P15.13 presents the classifications of loans made by commercial banks in June 1977 by purpose or group of borrowers. The data are given both in dollar amounts and in percentages. Calculate the correlation coefficient between the dollar amount of loans made for each purpose or to each group of borrowers and the percent of total. Is your result surprising? Explain.

TABLE P15.13

Purpose or group of borrowers	Amount (millions of dollars)	Percent of total
Federal and state government	$ 294.8	3.5%
Statutory authorities	110.4	1.3
Agriculture	546.1	6.4
Manufacturing	1,491.0	17.5
Housing	771.9	9.0
Building and construction	543.1	6.4
Real estate	346.1	4.1
Mining and quarrying	98.0	1.1
General commerce	1,801.8	21.1
Business purposes	439.7	5.2
Transport, storage, and communication	122.5	1.4
Hotels, restaurants, and boardinghouses	121.0	1.4
Financial institutions	461.4	5.4
Foreign trade bills	495.8	5.8
All others	890.8	10.4
Total	$8,534.4	100.0%

Source: *Euromoney,* March 1978.

15.14. Table P15.14 shows the value of the U.S. dollar expressed in terms of other currencies on selected dates.

  (a) Calculate the coefficient of correlation between the values of the dollar in terms of the Swiss and German currencies.

  (b) Calculate the coefficient of correlation between the values of the dollar in terms of the Japanese and Swedish currencies.

TABLE P15.14

Country	12/31/76	12/28/77	1/4/78	1/11/78	1/18/78	1/25/78	2/1/78	2/8/78
United States	1.00	1.00	1.00	1.00	1.00	1.00	1.00	1.00
Japan	293.23	240.63	237.76	240.07	242.66	241.48	241.98	241.78
Germany	2.35	2.12	2.06	2.11	2.13	2.10	2.11	2.11
Sweden	4.11	4.69	4.60	4.64	4.69	4.64	4.66	4.65
Switzerland	2.44	2.03	1.93	1.97	2.01	1.96	1.98	1.96

Source: *Euromoney,* March 1978.

PART 5

OTHER SELECTED TOPICS

Part 5 of this book is devoted to selected topics that are not directly discussed earlier in the book. Chapter 16 presents the analysis of time series; Chapter 17 deals with index numbers—a subject that has become increasingly important as a result of rising inflation rates in recent years; Chapter 18 covers the very important topic of nonparametric statistics; Chapter 19 is devoted to decision making under uncertainty and to Bayesian statistics.

The applications and examples in this part of the book concern the trend of newsprint production, the random walk theory, investment in mutual funds, and many other topics.

CHAPTER SIXTEEN OUTLINE

16.1 Introduction

16.2 The Secular Trend
Types of Secular Trends
Secular Trends in Actual Economic Data

16.3 Seasonal Variations
Isolating the *SI* Factor
Isolating the Seasonal Index
Deseasonalized Data

16.4 Cyclical Variations
Identifying the Current State of the Cycle
Forecasting Cyclical Movement: The Use of
Business Indicators

Key Terms

time series
secular trend
seasonal variations
cyclical variations
irregular variations
multiplicative model

sensitivity analysis
ratio to moving average
leading indicators
roughly coincidental
   indicators
lagging indicators

16
TIME SERIES ANALYSIS

16.1 Introduction

This chapter is devoted to the description and analysis of time-series movements. A **time series** consists of *data concerning a variable and collected over time*, such as monthly sales of a given firm, quarterly inventory levels of an industry, and annual domestic air traffic. Forecasting future values of time series is of great importance to management and government, as errors in prediction may lead to substandard performance and policy blunders or even insolvency. In order to provide quality forecasts, we must familiarize ourselves with time-series patterns and learn how to apply our knowledge of past patterns to short-term and long-term time-series forecasts.

At first glance one might think that time-series trends can be studied fully by means of regression analysis. Most time series, however, are subject to time-related movements that violate the basic assumptions of regression analysis. The data points of time series are often *not* scattered around the regression line in a random manner. Rather, they tend to *cluster* above and below the line, contrary to what regression analysis presumes. While the regression line is used in time-series analysis to project the long-term trend line, time-series analysis recognizes that short- and intermediate-term forecasts must consider the movement of the series *around* the regression line.

This leads us to consider four components of time-series data:

1. Secular trend (denoted by T).
2. Seasonal variations (denoted by S).
3. Cyclical variations (denoted by C).
4. Irregular variations (denoted by I).

The **secular trend** is the *long-term pattern* of the data. **Seasonal variations** are *variations in the activity level of a variable relative to the average annual level of activity;* they are therefore variations of the activity within the course of a year. **Cyclical variations** are *variations in the level of activity about the long-term trend.* While seasonal variations are occurring within the course of a year, the cyclical movements are those of annual activity levels around the long-term trend. **Irregular variations** are, as the name indicates, *erratic and irregular activity variations,* specific to the specific points in time in which they occur, and are not explained by the other three components of time series.

The time-series model we present here is known as the **multiplicative model,** since it expresses Y, the variable whose activity level is being studied, as the product of the time-series components. Thus

$$Y = TCSI \tag{16.1}$$

In Equation 16.1, the value of T is measured in the same units as Y (usually dollars), and the value of each of the other three components is measured as an index number divided by 100. For example, if S is equal to 1.10, the activity level in the season is 10 percent above the average, and if S is equal to 0.90, the activity level is 10 percent below the average. The multiplicative model, then, identifies factors associated with each component in such a way that when the components are multiplied, the value of Y is obtained. Let us deal with and explain each of the four components in detail, starting with the secular trend.

16.2 The Secular Trend

The secular trend, denoted by T, is the long-term trend of a dependent variable Y, the variable whose behavior is being studied—sales, expenditures, consumption, production, or whatever.

TYPES OF SECULAR TRENDS

Although each time series exhibits its own specific trend, such trends may be classified into several types most frequently observed. These types of trends are depicted in Figure 16.1. In Figure 16.1*a*, an increasing linear trend is presented. Along this trend line the variable Y grows by an equal absolute amount each period. The relationship between Y and t (time) is given by

$$\hat{Y} = a + bt \tag{16.2}$$

Although a linear growth pattern is not very typical when the activity level Y is observed over a very long time (such as several decades), fitting a linear trend is sometimes useful in approximating the pattern of a shorter period, such as 10 to 20 years.

Figure 16.1*b* is an example of an exponentially increasing trend. Along this line, the *rate of increase* in Y is constant. Thus, the absolute increase in Y

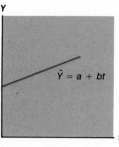

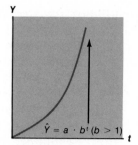

Figure 16.1
Alternative
secular-trend
patterns

(a) Increasing linear trend

(b) Exponentially increasing trend

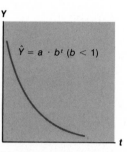

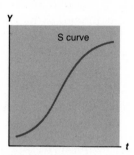

(c) Exponentially decreasing trend

(d) Elongated S trend

grows as time passes.[1] The estimated relationship between Y and t in this case is

$$\hat{Y} = a \cdot b^t \qquad (b > 1) \qquad \text{(16.3)}$$

By taking the logarithm, we may transform Equation 16.3 into a linear equation, as follows:

$$\log \hat{Y} = \log a + t \log b \qquad \text{(16.4)}$$

and if we denote $Y^* = \log Y$, $a^* = \log a$, and $b^* = \log b$, we may simplify and get

$$Y^* = a^* + b^* t \qquad \text{(16.5)}$$

which we can estimate by a linear regression procedure using Y^* as the dependent variable and t as an independent variable.

In Figure 16.1c we present a declining trend with a constant rate of

---

[1] Suppose $a = 1,000$ and $b = 1.10$. Let t be arbitrarily equal to 10, 11, and 12 and compute the respective $\hat{Y}$ values:

$$\hat{Y}_{10} = 1,000 \cdot 1.10^{10} = 2,593.74$$
$$\hat{Y}_{11} = 1,000 \cdot 1.10^{11} = 2,853.12$$
$$\hat{Y}_{12} = 1,000 \cdot 1.10^{12} = 3,138.43$$

The *absolute change* between t_{11} and t_{12} is $3,138.43 - 2,853.12 = 285.31$, and the *absolute change* between t_{10} and t_{11} is $2,853.12 - 2,593.74 = 259.38$. Thus, the change in the second period is greater when measured in the absolute. In both periods, however, there is a 10 percent increase over the previous period's Y level: $\dfrac{285.31}{2,853.12} = \dfrac{259.38}{2,593.74} = 0.10$ (or 10 percent).

decrease. This function is also exponential, except that the constant b is less than 1:

$$\hat{Y} = a \cdot b^t \qquad (b < 1) \qquad \textbf{(16.6)}$$

This kind of trend is typical of the consumption or production of a product or line of products or services with a declining demand—often because of the introduction of a preferred substitute. We may estimate this trend line by linear regression after taking the logarithms of Equation 16.6.

Finally, Figure 16.1*d* describes the most commonly observed growth curve. Its shape is like an elongated *S* and is indeed often called an *S* curve. In fact, the growth patterns described in Figure 16.1*a* and 16.1*b* could sometimes serve as approximations to some segments of the *S* curve. Thus, while linear and potential curves are often used to project conditions 10 to 15 years in the future, longer-term projections often should use an *S*-curve trend pattern.

SECULAR TRENDS IN ACTUAL ECONOMIC DATA

Let us now examine some examples of secular trends of actual and typical economic time series. First, we present the per capita consumption of cotton and synthetic fibers in the United States from 1948 to 1979. The data are given in Table 16.1 and presented graphically in Figure 16.2.

Several important observations can be made concerning per capita fiber consumption. First, the decline of cotton consumption is related to the increase in synthetic fiber consumption. Thus, long-run projections of these two items are likely to be affected by mutual interaction. Second, although the trend of consumption of both types of fiber is nonlinear, a linear trend can provide a fairly good approximation for time intervals within the 1948–79 period, such as the periods 1948–60 and 1968–79. This implies that linear trends can be used for short-term projection if we are careful to avoid using linear trend lines when trend inflection points are approached. Third, it is far easier to fit a trend line to past data than to predict how the trend will continue. The fiber-consumption example clearly shows that long-term projections take more than just plain past statistics. Take synthetic fiber consumption. The only clear indication that we can obtain from past consumption is that the trend is generally upward and follows an exponential pattern. It is basically our knowledge of other economic activity trends that tells us that the growth rate must soon start to slow. How soon this slowdown will come and how sharp it will be are indeed hard to predict, and require more knowledge of fiber consumption than that provided by technical trend-line fitting. To narrow our prediction interval we may, for example, decide to try to find out how much of the increase in synthetic fiber consumption came at the expense of a reduction in the consumption of cotton. It may be easier to predict the trend line of per capita cotton consumption, since experience with other fibers that have lost some of their popularity may be studied (wool, for example). Then, given projections of future cotton consumption, it may be easier to predict the derived synthetic fiber consumption trend.

Regardless of the technique used, however, it will always be difficult to predict future events, and the farther into the future we try to predict, the less accurate our predictions are likely to become. To help us to deal with

TABLE 16.1

Per Capita Consumption of Cotton and Synthetic Fibers,
United States, 1948–79

(pounds)

Year	Cotton fibers	Synthetic fibers
1948	27.5	7.8
1949	23.3	6.7
1950	29.4	9.5
1951	29.2	9.0
1952	26.5	8.9
1953	26.3	9.0
1954	23.8	8.7
1955	25.3	11.0
1956	25.0	9.7
1957	22.6	9.9
1958	21.3	9.6
1959	24.0	11.3
1960	23.3	9.9
1961	21.9	10.7
1962	22.9	12.4
1963	21.8	14.3
1964	22.8	16.2
1965	23.9	18.3
1966	25.1	20.2
1967	23.5	21.4
1968	21.9	26.7
1969	20.9	27.9
1970	19.7	27.8
1971	20.5	33.1
1972	20.0	37.8
1973	18.5	42.1
1974	16.2	36.3
1975	15.1	35.2
1976	17.2	38.1
1977	16.0	41.8
1978	16.2	43.2
1979	15.1	42.7

Source: *Commodity Yearbook,* various issues.

possible errors in prediction, we often conduct a **sensitivity analysis,** in which we change the underlying assumptions of the trend projections. This sensitivity analysis will provide "optimistic," "average," and "pessimistic" projections, which can be translated into relevant quantities of sales, costs, profits, and so on.

Let us consider one more example of an actual time series. Table 16.2 presents newsprint production in the United States and Canada in the period 1919–80. The data are presented also in Figure 16.3.

A glance at the data shows that newsprint production has increased significantly over the years, following an exponentially increasing trend. We note, too, that when a shorter period is considered, such as 1950–80, a linear trend fits the data well. Also notable in Figure 16.3 is the fact that the data points are not randomly scattered around the trend line. Rather, they tend to cluster together above or below the trend line, a fact that indicates that further analysis of the data is required. For example, all the

Figure 16.2

Per capita consumption of cotton and synthetic fibers, United States, 1948–79

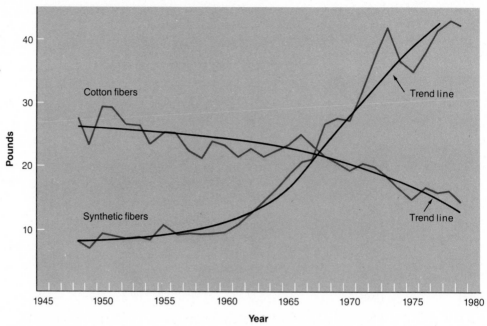

TABLE 16.2

Total Newsprint Production, United States and Canada, 1919–80

(thousands of short tons)

Year	Production	Year	Production	Year	Production
1919	2,184	1940	4,783	1961	8,829
1920	2,387	1941	4,786	1962	8,845
1921	2,033	1942	4,407	1963	8,848
1922	2,528	1943	4,024	1964	9,562
1923	2,748	1944	3,985	1965	9,900
1924	2,824	1945	4,316	1966	10,827
1925	3,052	1946	5,277	1967	10,672
1926	3,566	1947	5,646	1968	10,966
1927	3,572	1948	5,850	1969	11,990
1928	3,799	1949	6,076	1970	11,917
1929	4,121	1950	6,293	1971	11,593
1930	3,786	1951	6,641	1972	12,242
1931	3,379	1952	6,834	1973	12,571
1932	2,923	1953	6,805	1974	13,029
1933	2,963	1954	7,195	1975	11,370
1934	3,560	1955	7,743	1976	12,651
1935	3,665	1956	8,186	1977	12,859
1936	4,130	1957	8,222	1978	13,481
1937	4,594	1958	7,853	1979	13,714
1938	3,713	1959	8,358	1980	14,179
1939	4,114	1960	8,777		

Source: Standard & Poor's *Trade and Securities Statistics, Current Statistics,* 1981, pp. 24 and 284.

points for the period 1964–74 are above the trend line, while all the points for the period 1975–80 are below the line. This fact indicates that the exponential trend line does not in itself capture all of the nonrandom behavior of the data. Furthermore, the secular trend of the data may show what the production of newsprint will be in the coming years *if* future

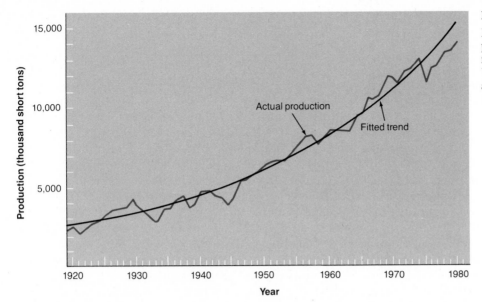

Figure 16.3

Newsprint production and fitted trend line, United States and Canada, 1919–80

conditions resemble past conditions. Significant changes in prices, taxes, and the competitiveness of substitute products could cause a change in the secular trend. All of these factors demonstrate that long-term projections require a good statistical approach and also knowledge of the specific circumstances of the industry or product involved.

16.3 Seasonal Variations

Analysis of short-term movements and forecasts of time series require the development of a seasonal index, *S*, to measure seasonal variations. A seasonal index is a measure of the relative activity of a season in comparison with the average *annual* activity level. It most commonly takes the form of a quarterly or monthly index. To illustrate, suppose the November seasonal index is 105 for a given activity. This should be interpreted to mean that the November activity level is typically 5 percent above the average for the year. Similarly, if a seasonal index is 88 for a given month, it expresses a level typically 12 percent (100 − 88 = 12) below average.

ISOLATING THE *SI* FACTOR

The seasonal component is the most predictable component of a time series in terms of both timing and strength. Yet because the data are subject to irregular variations too, a special statistical technique is required to separate out the seasonal index. Several techniques are used, ranging from simple and somewhat crude methods to sophisticated ones. The **ratio to moving average** is a superior method from both theoretical and practical standpoints.

Recall that the multiplicative model described in Section 16.1 assumes that $Y = TCSI$. The moving average of the time series is a measure of the product of T and C. So, by dividing the original data, Y, by the moving average, we obtain

$$\frac{Y}{\text{Moving average}} = \frac{TCSI}{TC} = SI \qquad \textbf{(16.7)}$$

and the isolation of the seasonal index itself, S, is accomplished by averaging the factor SI over several years.

Let us now illustrate the isolation of S in detail. Consider Table 16.3, which provides hypothetical sales figures for Seasonal Salad Dressing. In Column 2 the original quarterly sales figures are provided. Column 3 shows the moving total for four quarters. The first four-quarter total ($1,252) is the total for the first four-quarter sales data (that is, winter, spring, summer, and fall of 1982). To obtain the second four-quarter total ($1,272), we move up and sum the last three quarters of 1982 and the first quarter of 1983. Note that the first four-quarter total, $1,252, is printed in Column 3 halfway between the figures for the spring and summer of 1982. The reason is that the midpoint of the year happens to lie between the *end* of the second and the *beginning* of the third quarter. The second four-quarter total, $1,272, is printed in Column 3 halfway between the figures for the summer and fall of 1982. The reason is again that the midpoint of the period lies between the *end* of the summer quarter and the *beginning* of the fall quarter of 1982.

TABLE 16.3
Isolating the *SI* Factor: Seasonal Salad Dressing Sales

(1) Quarter	(2) Sales	(3) Four-quarter moving total	(4) Eight-quarter moving total	(5) Eight-quarter moving average (Col. 4 ÷ 8)	(6) Original data as percentage of moving average [SI = (Col. 2 ÷ Col. 5) · 100]
1982 winter	$302				
spring	350				
		$1,252			
summer	280		$2,524	$315.5	88.7%
		1,272			
fall	320				
1983 winter	322				

In order to obtain a quarterly average sales volume, we add up the two four-quarter totals and obtain an eight-quarter total. This total appears in Column 4 and is positioned exactly *between* the two four-quarter totals. As Table 16.3 shows, the eight-quarter total falls in the summer of 1982.

We now simply divide the eight-quarter total by 8 to get a quarterly moving average that corresponds to the summer of 1982. The quarterly moving average is our measure for the product TC for the quarter. Being an average for a period of more than a year, the moving average is expected to "average out" the seasonal and irregular movements within the year considered. Thus, the SI factor is now easily isolated. Following Equation 16.7, we divide the original data by the moving average and derive the factor SI. In Table 16.3 we compute for the summer quarter of 1982

$$SI = \frac{Y}{TC} = \frac{280.0}{315.5} = 0.887$$

meaning that the sales in the summer of 1982 were 88.7 percent of the average for the period winter 1982 through winter 1983.

ISOLATING THE SEASONAL INDEX

Let us turn to another example, using actual rather than hypothetical data, through which we will show how the isolation of the seasonal index *S* from the irregular index *I* is accomplished. In Table 16.4 we present the quarterly data of family clothing store sales in the United States in the period 1976 through 1980. Along with the original data, we present the derivation of the ratio to moving average (that is, the *SI* factor), as was just described.

TABLE 16.4

Ratio to Moving Average of Family Clothing Store Sales, United States, 1976–80 (millions of dollars)

(1) Quarter		(2) Sales	(3) Four-quarter moving total	(4) Eight-quarter moving total	(5) Eight-quarter moving average (Col. 4 ÷ 8)	(6) Ratio to moving average [$SI = $ (Col 2 ÷ Col. 5) · 100] (percent)
1976	Winter	$1,373				
	Spring	1,676				
			$7,224			
	Summer	1,714		$14,620	$1,827.50	93.79%
			7,396			
	Fall	2,461		15,018	1,877.25	131.10
			7,622			
1977	Winter	1,545		15,465	1,933.12	79.92
			7,843			
	Spring	1,902		15,898	1,987.25	95.71
			8,055			
	Summer	1,935		16,134	2,016.75	95.95
			8,079			
	Fall	2,673		16,144	2,018.00	132.46
			8,065			
1978	Winter	1,569		16,240	2,030.00	77.29
			8,175			
	Spring	1,888		16,428	2,053.50	91.94
			8,253			
	Summer	2,045		16,611	2,076.38	98.49
			8,358			
	Fall	2,751		16,767	2,095.88	131.26
			8,409			
1979	Winter	1,674		16,961	2,120.13	78.96
			8,552			
	Spring	1,939		17,217	2,152.13	90.10
			8,665			
	Summer	2,188		17,436	2,179.50	100.39
			8,771			
	Fall	2,864		17,664	2,208.00	129.71
			8,893			
1980	Winter	1,780		17,803	2,225.38	79.99
			8,910			
	Spring	2,061		18,037	2,254.63	91.41
			9,127			
	Summer	2,205				
	Fall	3,081				

Source: U.S. Department of Commerce, *Survey of Current Business*, various issues.

We isolate the seasonal index, *S*, by applying a procedure for averaging out the ratio to moving average (*SI*) over time. In Table 16.5 we organize the ratio to moving average of Table 16.4 by quarters.

A glance at Table 16.5 reveals how the seasonal index is calculated. For each season (quarter in our case) we first calculate the modified mean, which is the mean of the two or three central values of ratio to moving average (after they have been arranged in ascending order) for the season. For example, the modified mean for the fall is the arithmetic average of the two central values for the fall (131.10 and 131.26). The reason that the extreme values for each season are left out is that the extremes are likely to

TABLE 16.5
Derivation of the Seasonal Index

| Year | Ratio to moving average | | | | Total |
	Winter	Spring	Summer	Fall	
1976			93.79	131.10	
1977	79.92	95.71	95.95	132.46	
1978	77.29	91.94	98.49	131.26	
1979	78.96	90.10	100.39	129.71	
1980	79.99	91.41			
Modified Mean	79.44	91.68	97.22	131.18	399.52
Adjusted Modified Mean					
= Modified Mean $\cdot \dfrac{400}{399.52}$	79.54	91.79	97.34	131.34	400.00

be influenced by strong irregular factors that will not be averaged out. Therefore, when the seasonal index is computed, these extreme values are eliminated. Finally, since the sum of the four modified means is not exactly 400, we multiply each modified mean by $\dfrac{400}{399.52}$. Table 16.5 shows that family clothing sales have strong seasonal components. Sales are particularly high in the fall (back to school, Christmas season) and particularly low in the winter.

Once derived, the seasonal index serves two main purposes: to deseasonalize data and to help make short-term forecasts.

DESEASONALIZED DATA

Because activity levels vary from one season to another, we may be interested to known the extent to which a change in a given activity level is attributable to the seasonal factor as opposed to other factors, particularly trend and cyclical movements.

For example, if we look back at Table 16.4, we find that in the fall of 1979 family clothing sales reached $2,864 million, an increase over the $2,188 million in the summer of 1979. To what extent, if any, is the season per se the cause of that increase? To answer that question we refer to the seasonal index. We find that the seasonal index for the summer is 97.34 (see Table 16.5) and the index for the fall is 131.34. The greater volume of clothing sales in the fall compared to the summer quarter is therefore a typical phenomenon. Still, we may want to find out whether all of the increase between the summer and fall quarters of 1979 is accounted for by the seasonal factor, or whether some of it may perhaps indicate some longer-term change. To find out, we deseasonalize the original data by simply dividing each original *Y* value by the respective seasonal index and then multiplying by 100. This is done in Table 16.6.

The deseasonalized data show a *decline* rather than an increase between the summer and fall quarters of 1979. This means that although the original

TABLE 16.6
Deseasonalizing the Family Clothing Store Sales Data

(1) Quarter		(2) Family clothing sales (millions of dollars)	(3) Seasonal index	(4) Deseasonalized data [(Col. 2 ÷ Col. 3) · 100]
1976	Winter	$1,373	79.54	1,726.18
	Spring	1,676	91.79	1,825.91
	Summer	1,714	97.34	1,760.84
	Fall	2,461	131.34	1,873.76
1977	Winter	1,545	79.54	1,942.42
	Spring	1,902	91.79	2,072.12
	Summer	1,935	97.34	1,987.88
	Fall	2,673	131.34	2,035.18
1978	Winter	1,569	79.54	1,972.59
	Spring	1,888	91.79	2,056.87
	Summer	2,045	97.34	2,100.88
	Fall	2,751	131.34	2,094.56
1979	Winter	1,674	79.54	2,104.60
	Spring	1,939	91.79	2,112.43
	Summer	2,188	97.34	2,247.79
	Fall	2,864	131.34	2,180.60
1980	Winter	1,780	79.54	2,237.87
	Spring	2,061	91.79	2,245.34
	Summer	2,205	97.34	2,265.26
	Fall	3,081	131.34	2,345.82

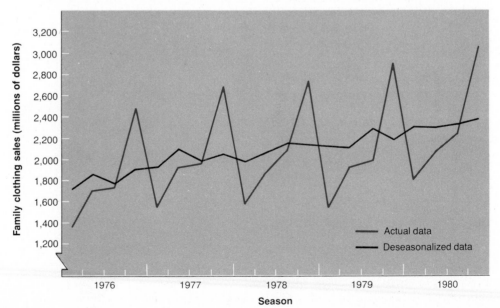

Figure 16.4

Family clothing store sales, actual and deseasonalized, 1976–80

figures show an increase, that increase was *less than* we would have expected to see if only the seasonal factor were affecting the activity level. The decline of the deseasonalized data could (but does not necessarily) indicate a slowdown in the secular trend and/or a downturn of a cyclical movement. One should not jump to conclusions, however, since the deseasonalized data are not free of the irregular factor.

Figure 16.4 presents both the original and the deseasonalized time series of family clothing store sales for the period 1976–80. It is evident from the figure that the deseasonalized data are smoother and do not have seasonal peaks and troughs; thus they are more reflective of the long-term movement of the time series.

The seasonal index can also be used to provide short-term forecasts of an activity level. After the trend forecast has been adjusted for cyclical movements (see Section 16.4), the forecast data are multiplied by the seasonal index and divided by 100. The result is a forecast that takes the seasonal variations into consideration.

16.4 Cyclical Variations

The seasonal index does not reflect the extent to which the activity level of the year as a whole varies from the long-term trend. Those movements of annual economic activity levels about the long-term trend are known as *cyclical variations*. The forecasting of those cycles is pivotal to short-term and intermediate-term planning of successful business operations.

In Figure 16.5 we present the deseasonalized quarterly data of retail sales of general merchandising stores in the period 1952–80. The diagram is based on the data of Table 16.7. As one can see, the deseasonalized data, although free of seasonal variations, move in cycles, above and below the long-term trend line.

Unlike seasonal variations, which are predictable with reasonable accuracy with respect to both timing and strength, cyclical fluctuations are troublesome to the forecaster. They are less predictable in their timing, duration, and intensity. Particularly difficult is the forecasting of turning points (peaks and troughs) and the speed at which a forthcoming upswing or downswing

Figure 16.5

Deseasonalized sales of general merchandising stores and fitted trend, 1952–80

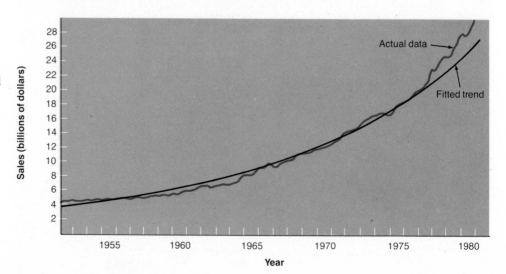

TABLE 16.7
Retail Sales of General Merchandising Stores, Deseasonalized, 1952–80
(billions of dollars)

Year	Winter	Spring	Summer	Fall
1952	$ 4.47	$ 4.59	$ 4.64	$ 4.88
1953	4.68	4.79	4.79	4.73
1954	4.53	4.73	4.72	4.81
1955	4.85	4.99	5.10	4.96
1956	5.06	5.18	5.31	5.17
1957	5.18	5.25	5.52	5.21
1958	5.24	5.32	5.48	5.52
1959	5.69	5.84	5.85	5.95
1960	5.97	5.99	5.95	6.01
1961	6.03	6.13	6.27	6.49
1962	6.58	6.77	6.89	7.03
1963	7.06	7.12	7.31	7.29
1964	7.62	7.89	8.11	8.40
1965	8.56	8.67	8.99	9.33
1966	9.72	9.76	10.04	10.14
1967	9.83	10.11	10.29	10.50
1968	10.79	11.05	11.36	11.47
1969	11.55	11.76	11.98	12.10
1970	12.17	12.22	12.33	12.97
1971	13.17	13.58	13.85	14.09
1972	14.47	14.79	15.37	15.78
1973	16.12	16.47	16.52	16.88
1974	17.29	17.55	17.59	17.30
1975	17.45	18.30	18.60	18.89
1976	19.20	19.13	19.59	20.45
1977	20.61	21.22	22.21	23.82
1978	23.46	24.56	25.03	25.65
1979	25.39	26.19	27.23	28.09
1980	28.42	28.02	28.84	30.28

Source: Standard & Poor's *Trade and Securities Statistics, Current Statistics,* various issues.

will take place. Failure to predict a business upswing could result in loss of opportunities to make sales, while a failure to predict a business downswing could lead to overinvestment and accumulation of excess inventories, resulting in substantial loss. Continuous failure to make predictions could ultimately lead to business failure.

In studying cyclical fluctuation we need first to learn how to determine the current state of the activity, and then how to go about predicting future cyclical movements. Let us consider these two facets of the study of cyclical movements.

IDENTIFYING THE CURRENT STATE OF THE CYCLE

Technically, the isolation of the cyclical components of a time series is a simple matter. After isolating the trend (T) and the seasonal index (S), we may derive the product, CI, in the following way:

$$\text{Cyclical and irregular components} = \frac{Y}{TS} = \frac{TCSI}{TS} = CI \qquad \textbf{(16.8)}$$

The product, CI, is of course a mixture of a cyclical factor and an irregular movement. In order to separate the cyclical component, C, from the irregular factor, I, we average out a few successive periods. Let us take a close look at the isolation of the cyclical component in the example of general merchandise sales. We note that in order to apply Equation 16.8, we need to know the secular trend line. Figure 16.5 shows an exponential trend line for the data. The trend line is given by

$$\hat{Y} = a \cdot b^t$$

and after transforming this equation to logarithmic form we rewrite it as

$$\hat{Y}^* = a^* + b^* t$$

where $\hat{Y}^* = \log \hat{Y}$, $a^* = \log a$, and $b^* = \log b$. Using the quarterly data for the period 1952–80 (Table 16.7), we find that

$$\hat{Y}^* = 1.308124 + 0.0171940t$$

where $t = 0$ for the first quarter of 1952 and increases by increments of 1 per quarter. On the basis of this trend line, we can isolate the cyclical factor by the process described in Table 16.8.

The table deserves a few comments. First, the trend line was derived from data for the period 1952–80. The first quarter of 1976 happens to be the

TABLE 16.8

Isolating the Cyclical Component for General Merchandising Sales, 1976–80

(1) Quarter		(2) t	(3) Deseasonalized data	(4) Trend	(5) CI [(Col. 3 ÷ Col. 4) · 100]	(6) Weighted three-quarter moving total	(7) C (Col. 6 ÷ 4)
1976	Winter	96	$19.20	$19.27	99.64%		
	Spring	97	19.13	19.61	97.55	$392.94	$ 98.24
	Summer	98	19.59	19.95	98.20	394.74	98.69
	Fall	99	20.45	20.29	100.79	399.59	99.90
1977	Winter	100	20.61	20.65	99.81	401.46	100.37
	Spring	101	21.22	21.00	101.05	405.84	101.46
	Summer	102	22.21	21.37	103.93	418.48	104.62
	Fall	103	23.82	21.74	109.57	429.13	107.28
1978	Winter	104	23.46	22.12	106.06	430.85	107.71
	Spring	105	24.56	22.50	109.16	433.73	108.46
	Summer	106	25.03	22.89	109.35	437.99	109.50
	Fall	107	25.65	23.29	110.13	436.79	109.20
1979	Winter	108	25.39	23.69	107.18	433.16	108.29
	Spring	109	26.19	24.10	108.67	435.57	108.89
	Summer	110	27.23	24.52	111.05	443.36	110.84
	Fall	111	28.09	24.95	112.59	448.21	112.05
1980	Winter	112	28.42	25.38	111.98	445.07	111.27
	Spring	113	28.02	25.82	108.52	438.80	109.70
	Summer	114	28.84	26.27	109.78	441.40	110.35
	Fall	115	30.28	26.72	113.32		

ninety-sixth quarter from the winter of 1952. The trend for the first quarter of 1976 is given by

$$\hat{Y} = 3.69923 \cdot 1.017343^{96} = 19.27$$

where 3.69923 is the antilogarithm of 1.308124 and 1.017343 is the antilogarithm of 0.0171940. The trend is determined for the other quarters in a similar way. The deseasonalized data were computed by use of a seasonal index derived from the data for the period 1952–80. Column 6 of Table 16.8 is the total of the previous quarter's value in Column 5, twice the current value of Column 5, and the next quarter's value in Column 5. For example, the first number in Column 6 was computed as follows:

$$392.94 = 99.64 + 2 \cdot 97.55 + 98.20$$

Last, the cyclical component itself is determined by dividing Column 6 by 4.

FORECASTING CYCLICAL MOVEMENT: THE USE OF BUSINESS INDICATORS

Experience shows that only by studying the simultaneous behavior of many economic variables can we hope to make high-quality predictions of cyclical movements. Key variables such as the gross national product, personal income, industrial production, retail sales, wholesale prices, stock market prices, and unemployment figures must be considered. In particular, it is desirable to learn the relationships between these variables to establish a network of relevant data that might give early signals of future cyclical trends.

The National Bureau of Economic Research has developed a sophisticated method for such predictions: the use of business indicators. These indicators are used by the U.S. Department of Commerce and are published in *Business Conditions Digest*. Business indicators are time series of economic variables that tend to indicate how the economy is progressing along its cyclical trend, and were developed by careful study of the interrelationships among economic and business time series. The indicators come in three major categories: **leading indicators, roughly coincident indicators,** and **lagging indicators.** The full list of indicators includes 88 variables, of which 36 are leading, 25 are coincident, 11 are lagging, and 16 are unclassified by timing but nevertheless represent important factors of business cycles.[2] In Table 16.9 we present what is known as the *short list* of 25 business indicators, 12 leading, 7 roughly coincident, and 6 lagging.

The "quality" of a time series as a business indicator is measured by the following criteria:

1. Economic significance.
2. Statistical adequacy.
3. Historical conformity to business cycles.
4. Consistency during business cycles.
5. Smoothness.
6. Promptness of publication.

These criteria are used to score the indicators on a scale from 0 to 100.

[2] See Geoffrey H. Moore and Julius Shiskin, *Indicators of Business Expansions and Contractions,* Occasional Paper no. 103 (New York: National Bureau of Economic Research, 1967).

TABLE 16.9

Cyclical Indicators: Short List of National Bureau of Economic Research

Leading indicators
  Average hourly workweek, production workers, manufacturing
  Average weekly initial claims, state unemployment insurance
  Index of net business formation
  New orders, durable goods industries
  Contracts and orders, plant and equipment
  Index of new building permits, private housing units
  Change in book value, manufacturing and trade inventories
  Index of industrial materials prices
  Index of stock prices, 500 common stocks
  Corporate profits after taxes (quarterly)
  Index: ratio, price to unit labor cost, manufacturing
  Change in consumer installment debt

Roughly coincident indicators
  GNP in current dollars
  GNP in 1958 dollars
  Index of industrial production
  Personal income
  Manufacturing and trade sales
  Sales of retail stores
  Employees on nonagricultural payrolls
  Unemployment rate, total

Lagging indicators
  Unemployment rate, persons unemployed 15 weeks or over
  Business expenditures, new plant and equipment
  Book value, manufacturing and trade inventories
  Index of labor cost per unit of output in manufacturing
  Commercial and industrial loans outstanding in large commercial banks
  Bank rates on short-term business loans

Source: U.S. Department of Commerce.

The leading indicators typically reach their peaks and troughs ahead of the coincident and lagging indicators. Similarly, the coincident indicators typically reach their peak and trough levels ahead of the lagging indicators. A typical though simplified set of movements of these indicators in relation to each other is shown in Figure 16.6. Consider the economy at time t_0. Here the leading indicators are starting a downturn while both the coincident and lagging indicators continue their expansion. If we indeed know that the

Figure 16.6

Typical but simplified relationship among leading, roughly coincident, and lagging indicators

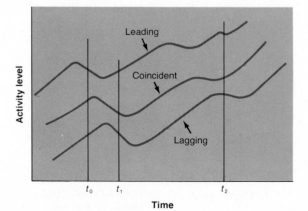

leading indicators precede the others at their peaks and troughs, we predict a downturn of the coincident and lagging indicators soon. The coincident indicators precede the lagging. Thus, when the coincident indicators also turn downward, their movement reinforces our earlier prediction and leads to a conclusion that the lagging indicators will soon turn downward. At time t_1, the coincident and lagging indicators are still on their downward trend, but the leading indicators are turning upward, indicating an upcoming trend in the other indicators as well.

Of course we would like the business indicators to be an early warning system as dependable as the one we have been describing. Unfortunately, this is not exactly the case, for several reasons. First, the indicators vary in their reliability. As we have mentioned, their "quality" is scored on a scale from 0 to 100. A low score for an indicator could mean that the signals it gives may turn out either to come too late or to be false. Second, since there are many indicators in each category (leading, coincident, and lagging), what if some indicators move in the opposite direction from others? Should we interpret such a movement as a signal of expansion or contraction? To deal with this problem, a composite index in each category is compiled that reflects the overall direction of the indicators in the category. Third, the action of the indicators is subject not only to cyclical fluctuations but also to irregular movements due to specific circumstances. To illustrate the problem, consider time t_2 in Figure 16.6. At this time the economy is in the midst of an upswing, but a short-term slowdown of leading indicators has occurred for some reason. At time t_2 it is hard to know whether the economy is on the verge of a new slowdown or simply at a temporary disturbance during a longer upswing movement. (Such a situation arose after the deep recession of 1975–76, when the economy, on its way to recovery, experienced a two-month decline in the composite index of leading indicators.) On such occasions we must sometimes rely on speculation and guesswork as to the direction the economy will go in future months.

Clearly, the study of business indicators contributes a great deal to our competence in predicting future cyclical trends. We must remember, however, that future events will always differ in some way from what we anticipate. Thus, studying the business indicators must be taken as a way to *reduce* rather than *eliminate* uncertainty about the future. The more knowledgeable we become, the more we can reduce uncertainties, but we must accept the fact that some degree of uncertainty is unavoidable.

Problems

16.1. Why do we need a special technique to study and analyze time series? Why is it not recommended that regular regression techniques be used?

16.2. Identify each of the four components of time series: secular trend, seasonal variations, cyclical variations, and irregular variations.

16.3. What are some common types of secular trends?

16.4. Why do businesspeople need to have forecasts of the secular trends of sales of their products?

16.5. Why do businesspeople need to forecast the cyclical fluctuations of business?

16.6. "Seasonal fluctuations are not very important for the business executive, since his or her investment decisions are based on long-run business prospects, and for those the executive needs to forecast the secular trend and the cyclical variations, not the seasonal variations." Discuss.

16.7. What are business indicators? Distinguish among leading, coincident, and lagging indicators.

16.8. Explain what a diffusion index is and how we can use it to forecast the cyclical trend.

16.9. The sales of the City Variety Company have been projected to grow 2.5 percent per quarter. The sales level for fall 1982 has been projected to be $1.5 million.

(*a*) What are the projected sales for each of the four quarters of 1983?
(*b*) Is the trend line linear in this case? Explain.

16.10. The sales of the Outdoor Furniture Company have been projected to grow $50,000 every quarter. If sales for fall 1985 have been projected to be $2.3 million, what are the projected sales for each of the four quarters in 1986? Does the trend indicate increasing sales at a decreasing rate or increasing sales at an increasing rate? Explain.

16.11. Table P16.11 gives the annual sales volumes of Firm *A* and Firm *B* for the years 1972–81.

TABLE P16.11

Year	Firm A	Firm B
1972	$108	$ 87
1973	106	97
1974	120	108
1975	118	114
1976	131	123
1977	133	138
1978	137	153
1979	148	171
1980	139	182
1981	151	198

(*a*) Using the ordinary least-squares method, fit a straight line to both series.
(*b*) Plot the sales series of Firm *A* and the computed trend values on one diagram. Do the same for Firm *B*.
(*c*) Does a straight line seem to fit well the sales series of Firm *A*? Does it seem to fit well the sales series of Firm *B*?

16.12. Table P16.12 gives a time series of the total book value of manufacturing and trade inventories by quarters for the years 1975–80.

TABLE P16.12

Year	Quarter	Inventories (millions of dollars)	Year	Quarter	Inventories (millions of dollars)
1975	1	$ 811,379	1978	1	$1,028,576
	2	800,856		2	1,070,154
	3	785,352		3	1,087,404
	4	798,499		4	1,133,448
1976	1	804,405	1979	1	1,167,658
	2	836,306		2	1,211,412
	3	869,790		3	1,237,577
	4	900,837		4	1,279,235
1977	1	940,573	1980	1	1,309,776
	2	961,939		2	1,342,852
	3	968,326		3	1,344,001
	4	1,002,102		4	1,375,136

Source: U.S. Department of Commerce, Bureau of Economic Analysis, *Survey of Current Business*, various issues.

(*a*) Derive the seasonal index using the ratio to moving average.
(*b*) Deseasonalize the data.

16.13. Table P16.13 gives data on United Kingdom imports of crude oil, natural gas liquids, and feedstocks by quarters, 1977–80.

TABLE P16.13

Year	Quarter	U.K. imports of crude oil, natural gas liquids, and feedstocks (thousands of metric tons)
1977	1	19,300
	2	18,405
	3	16,372
	4	16,621
1978	1	17,519
	2	16,423
	3	15,978
	4	18,223
1979	1	15,267
	2	14,797
	3	14,945
	4	15,373
1980	1	13,381
	2	12,333
	3	10,100
	4	10,903

Source: Organization for Economic Cooperation and Development, *Quarterly Oil Statistics*, various issues.

(*a*) Estimate the (linear) trend of the data.
(*b*) Derive the seasonal index using the ratio to moving average.
(*c*) Deseasonalize the data.
(*d*) Using your seasonal index, forecast quarterly imports of the materials in 1981, assuming that the past linear trend will continue and that the cyclical trend will bring a 1 percent decrease per quarter.

16.14. Table P16.14 gives a time series of monthly retail sales of passenger cars in the United States, 1977–80.

(*a*) Derive the seasonal index using the ratio to moving average.
(*b*) Deseasonalize the data.

TABLE P16.14

Year	Month	Sales (thousands of cars)	Year	Month	Sales (thousands of cars)
1977	1	725	1979	1	784
	2	811		2	841
	3	1,084		3	1,116
	4	1,029		4	988
	5	1,054		5	1,053
	6	1,117		6	905
	7	913		7	886
	8	931		8	916
	9	829		9	775
	10	1,014		10	899
	11	881		11	775
	12	795		12	733

TABLE P16.14—Continued

Year	Month	Sales (thousands of cars)	Year	Month	Sales (thousands of cars)
1978	1	687	1980	1	806
	2	777		2	812
	3	1,078		3	895
	4	1,043		4	743
	5	1,160		5	697
	6	1,138		6	702
	7	930		7	772
	8	958		8	686
	9	828		9	672
	10	1,034		10	847
	11	909		11	698
	12	769		12	650

Source: U.S. Department of Commerce, Bureau of Economic Analysis, *Survey of Current Business*, various issues.

16.15. A power company had the following demand in the period 1972–81:

Year	Demand (millions of kilowatt hours)
1972	100
1973	110
1974	120
1975	136
1976	147
1977	160
1978	173
1979	190
1980	206
1981	223

(a) Plot the above time series on a graph.

(b) Estimate the demand trend line, assuming that the trend is linear and using the OLS estimators.

(c) Draw the estimated trend line. Do you think a linear trend line is appropriate for this series? Why?

(d) What is your projection for 1984 sales, using the linear trend?

16.16. For the data of Problem 16.15, assume that the trend line is as follows:

$$\hat{Y} + a \cdot b^t$$

After taking logarithms on both sides, we get

$$\log \hat{Y} = \log a + t \ \log b$$

(a) Using the last equation, estimate the trend line $\hat{Y} = s \cdot b^t$.

(b) What is the projected demand for 1984? Compare your answer here with your answer to part *d* of Problem 16.15. How do you explain the difference?

(c) What is the annual percentage growth in demand implied by your answer to part *a* of this problem?

16.17. Use the data of Problem 16.13 as well as the trend and deseasonalized data of that problem. For each quarter of the years 1977–80, identify the state of the cycle.

CHAPTER SEVENTEEN OUTLINE

Key Terms

index numbers
Consumer Price Index (CPI)
unweighted index
average of price relatives
weighted (arithmetic) average
  index
Laspeyres price index
Paasche price index
fixed-weights index
quantity indexes

Laspeyres quantity index
Paasche quantity index
value index
deflation
technical shifting
splicing
Standard & Poor's stock
  price indexes
leading indicators

17

INDEX NUMBERS

17.1 The Nature and Meaning of Index Numbers

Index numbers *enable us easily to express the level of an activity or phenomenon in relation to its level at another time or place.* In this chapter we deal primarily with relationships over time. Suppose the price of a certain car was $6,000 in 1981 and the price of the same car in 1982 was $6,300. One may construct an index, I, which indicates the *relative level* of the 1982 price compared to the 1981 price. The index is calculated as follows:

$$I = \frac{\text{price 1982}}{\text{price 1981}} \cdot 100 = \frac{\$6,300}{\$6,000} \cdot 100 = 105$$

The index in this example is obtained by dividing the 1982 price by the 1981 price and multiplying the result by 100. The value 105 indicates that the price in 1982 is 5 percent higher *relative to the base-period price*. The base period is 1981, and the 1981 price is assigned the index 100.

Note that the index may be either greater than, equal to, or less than 100. For example, the price of a stock traded on the New York Stock Exchange was $10 on March 1 and $8 on the following April 1. If March 1 is taken as the base, the $10 price is given the index value 100 and the $8 price is given the index value 80:

$$I = \frac{8}{10} \cdot 100 = 80$$

Whenever the index is above 100, the interpretation is that there has been an increase in the measured quantity compared to the base period. Likewise, whenever the index is below 100, a decrease has occurred in the measured activity.

Business-related indexes may be divided into three general types:

1. Price indexes of the sort discussed above.
2. Quantity indexes, which measure relative economic change in physical units. (An example is the Index of Industrial Production.)
3. Value indexes, which reflect the combined impact of price and quantity changes.

We shall continue now to discuss price indexes alone. They constitute perhaps the most important indexes in business; also, the principal features of index numbers are common to all types of indexes, so that a detailed discussion of one type of index will suffice to clarify the problems involved in constructing, interpreting, and using others as well.

Economists, businessmen, and financial analysts usually examine indexes over more than just two periods. Here is a hypothetical example based on data on beef prices spanning several years:

Year	Beef price (cents per pound)	Price index
1977	80	100.0
1978	85	106.3
1979	100	125.0
1980	90	112.5
1981	110	137.5
1982	120	150.0

Each year's index indicates the price level of beef relative to the price in the base year, 1977. For example, the index value 137.5 for 1981 is calculated as follows:

$$I_{1981} = \frac{\text{price } 1981}{\text{price } 1977} \cdot 100 = \frac{110}{80} \cdot 100 = 137.5$$

The 1981 price was 37.5 percent higher than the price in 1977. Note that a price index that never falls below 100 does not mean that there are no periods of price decline, and an index that never rises above 100 does not mean that there are no periods of price increase. For example, the index dropped from a high of 125.0 in 1979 to a low of 112.5 in 1980. Although both values are above 100, it is clear that beef prices declined in 1980. Indeed, the price dropped from $1.00 per pound in 1979 to $0.90 in 1980. (Incidentally, the percentage change in the price may be calculated either from the original data or from the index. The percentage change in price between 1979 and 1980 was $\frac{90}{100} - 1 = \frac{112.5}{125} - 1 = -0.1$, or a drop of 10 percent.)

17.2 Choosing the Base Period

One of the most critical points in constructing an index is the choice of the appropriate base period and the time interval to be covered. Since an index is designed to facilitate the analysis of changes in activities over time, great care must be taken to avoid a situation in which misleading conclusions could be drawn because the base period has not been carefully selected. Consider a good whose price decreased steadily from 1963 through 1973 a total of 40 percent, and then increased steadily through 1982 back to its 1963 level. If 1963 is selected as the base year, the index declines from 100 to a low of 60 and then rebounds to 100 in 1982. This price index is illustrated in Figure 17.1*a*. Looking at the figure, we conclude that the price was the same in 1982 as it was in 1962. This conclusion is based on factual data and is in fact correct. If we choose 1973 as the base year, however (see Figure 17.1*b*), we conclude that the price has increased in the period 1973–82. This conclusion is also correct, but it may lead to a different interpretation.

The appropriate base period depends on relevant economic factors. For example, if the period following the year 1973 represents a stage of a new economic situation for the specific item under consideration, then 1973 should be preferred as a base period and the years 1962 through 1972 may be regarded as insignificant—perhaps even irrelevant, since that period's economic conditions are no longer pertinent. The disparity between the two indexes in our example demonstrates that the base period must be carefully selected so that the index will reflect relevant variations in activity level.

As another example, suppose the profits of a certain corporation have declined over a given period and then started to increase continuously. Consider a financial analyst who needs to examine the attractiveness of investment in this corporation's stock. By considering the entire period, the analyst may reach the conclusion that the firm's profitability has changed little, if at all, since the current profitability is similar to its level at the beginning of the period. On the other hand, if the trough in profitability led to a replacement of old management by new management, the analyst would be better advised to choose the trough as a base. In this way, the index will show a steady increase in profitability, reflecting the performance of the new management. Profitability growth projections will be more meaningful if the old management's performance is not taken into consideration.

In most real situations, and particularly when the index represents more than one item, it is hard to find a time in which economic conditions turn around in so clear-cut a fashion. The real world is, for better or for worse, a bit more complex. Index-number series normally show some fluctuations

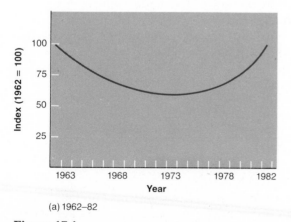

(a) 1962–82

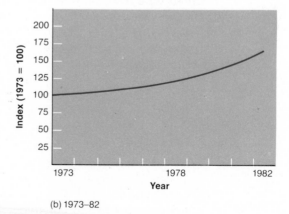

(b) 1973–82

Figure 17.1

Hypothetical price index pattern

over time. As a result, additional factors must be taken into consideration when the base period is selected.

First, the base period should be a "normal" period with regard to the relevant index. Obviously, "normality" is a rather elusive concept, but unless there are clear indications to the contrary, a period of relative price stability should be preferred to a period of strong fluctuations. Most indexes put out by the United States government have a three-year period for their base; that is, the average price over the three-year period is given the index 100.

Another consideration in choosing the base period is that it should not be too far in the past. Because of the dynamics of economic development, it is more meaningful to compare current conditions with those of a period from the recent past than with those of a bygone era.

Most indexes, as we shall see in Section 17.3, are designed to measure the relative levels of activity of a *group* of items. Because new products are constantly coming on the market and consumers are continuously shifting their demand from old to new products, and also because old products keep improving in quality, a comparison with a period too far in the past will not be as interesting as a comparison with a period closer to the present. You can easily think of a lot of products that are widely used today which did not even exist a few years ago. To cope with this problem, it is advisable to shift the base period forward every few years. The United States government indexes have their base periods shifted forward approximately every decade.

17.3 Aggregative Index: An Index of More Than One Item

The usefulness of indexes of one item only is limited and their contribution is marginal, since one can use the original data and obtain the same information without much extra difficulty. Index numbers contribute significantly more when they reflect the aggregate activity level of more than one item. For example, the well-known **Consumer Price Index (CPI)** *measures price changes of a representative "basket" of items over time.* It is extremely difficult to reach a conclusion regarding changes in the prices of various items without combining them into an index, and the contribution of the index here is very significant. Another famous index is the Dow-Jones (D-J), about which we all hear every day in the news. This index informs us about the aggregate price level of the stocks of a group of large corporations.

Let us illustrate the way aggregate indexes are constructed. Suppose the prices of beef and matches in 1981 (the base year) and 1982 (the nonbase year) were as follows:

	Price (cents)	
Product	*1981 (base year)*	*1982 (nonbase year)*
Beef (pound)	110	120
Matches (box)	5	10

For the sake of simplicity, suppose the consumer buys only beef and matches, and that we want to construct an index of the "consumption basket." The index may be constructed in several ways, each of which leads to a different result.

UNWEIGHTED INDEX NUMBERS

The **unweighted index** is simply the *sum of all the prices in the nonbase period divided by the sum of all the prices in the base period.* In our example this is simply

$$I_{1982} = \frac{120 + 10}{110 + 5} \cdot 100 = \frac{130}{115} \cdot 100 = 113.0$$

In a more general form, the index is obtained by summing the prices in the nonbase period, p_n, and dividing the total by the corresponding sum of all the prices in the base year, p_0:

UNWEIGHTED INDEX

$$I = \frac{\Sigma p_n}{\Sigma p_0} \cdot 100 \qquad\qquad (17.1)$$

The advantage of this index is its simplicity. It is easy to calculate and easy to understand. The disadvantage is that the index is a function of the units that we choose to use. Suppose that instead of looking at the price of one box of matches, we take the price of a big package containing a dozen small boxes. In this case we get the following numbers:

	Price (cents)	
Product	*1981 (base year)*	*1982 (nonbase year)*
Beef (pound)	110	120
Matches (dozen boxes)	60 (= 12 · 5)	120 (= 12 · 10)

The simple unweighted index based on these prices is

$$I = \frac{120 + 120}{110 + 60} \cdot 100 = \frac{240}{170} \cdot 100 = 141.2$$

compared with the 113.0 that we obtained earlier. Thus, despite its simplicity, the unweighted index is not adequate because its numerical value depends heavily on the units we select for the participating items. Since the units can be chosen arbitrarily, we obtain an arbitrary result. In order to overcome this problem, one can use the average of price relatives, which we discuss next.

THE AVERAGE OF PRICE RELATIVES

The idea behind the **average of price relatives** is to calculate a separate index for each item, and then compute the average of all the separate indexes. This technique overcomes the problem of arbitrary choice of units. Again, take the first example above:

$$\text{Index for matches} = \frac{10}{5} \cdot 100 = 200$$

$$\text{Index for beef} = \frac{120}{110} \cdot 100 = 109$$

and the average index is $\dfrac{200 + 109}{2} = 154.5$.

Now, for the second case, when the price of matches is given per dozen boxes,

$$\text{Index for matches} = \frac{120}{60} \cdot 100 = 200$$

$$\text{Index for beef} = \frac{120}{110} \cdot 100 = 109$$

and the arithmetic average of these two separate indexes remains 154.5.

The average of price relatives is also simple to calculate, and it has the advantage of not being vulnerable to disparities in the selected units (pounds, tons, or whatever).

In a more general form, when we have m items in the basket, we use the following formula:

AVERAGE OF PRICE RELATIVES

$$I = \frac{\sum \dfrac{p_n}{p_0} \cdot 100}{m} \tag{17.2}$$

The summation is over the m items included in the basket, and the ratio $\dfrac{p_n}{p_0}$ is calculated for each item separately.

As we have seen, this index has the advantage of not being dependent on the selected units. On the other hand, its disadvantage is that the separate price indexes are unweighted. To see why the weights are important, consider the cost-of-living index.

In several countries (Israel, Finland, Brazil), wages are linked to the cost-of-living index. When the index goes up by a certain percentage, employees get a proportionate wage increase. Since the United States experienced double-digit inflation in the mid-1970s, some economists (including the Nobel Prize-winner Milton Friedman) have suggested the establishment of some mode of indexing so that key economic variables (wages, interest rates, and so on) could be linked to the cost-of-living index. Even without a comprehensive indexing system, however, the wages of millions of American workers change with the cost-of-living index. In almost all collective bargaining contracts, both parties watch very carefully for changes in the Consumer Price Index and use them to bolster their arguments in their negotiations.

Suppose that we want to establish an index that will provide an adequate measure for employee compensation. Would you recommend using the average of price relatives?

Using this index in the beef-and-matches example, one would come to the conclusion that prices had increased 54.5 percent between 1981 and 1982 and hence wages should be increased by the same percentage in order to keep up with price increases. Is such an increase in wages indeed justified? The employees would be happy with such a cost-of-living adjustment, but obviously 54.5 percent will more than compensate for changes in the price level. The reason is that the change in the price of matches, though large in itself, has only a small impact on consumers, since they spend a very small fraction of their income on matches, while a substantially larger proportion is spent on beef. Thus, since the purpose of the cost-of-living adjustment is to keep the wage earner's welfare unchanged despite changes in prices, the wage increase should be based on the weighted average of price changes, the weights being a function of the relative amount of money spent on each item.

WEIGHTED INDEX NUMBERS

Let us continue with the beef-and-matches example, assuming that consumers spend 90 percent of their income on beef and only 10 percent on matches (recall that we have assumed for simplicity that only these two products are consumed). Thus, to enable consumers to purchase the same basket of goods at the new prices, we should calculate an index that takes into account the weight of each product in the consumption basket. The price of matches increased 100 percent and hence its index is 200 (see earlier calculations); the index of beef is 109, so the **weighted average index** is

$$I = \frac{200 \cdot 0.1 + 109 \cdot 0.9}{0.1 + 0.9} = \frac{20 + 98}{1} = 118$$

Thus, a wage increase of only 18 percent should enable employees to buy the same basket of goods they used to buy before the price increase. Take, for example, an employee who used to earn $200 a week, and bought $180 worth of beef ($200 · 0.9 = $180) and $20 worth of matches ($200 · 0.1 = $20). Assuming the old prices prevail, his earnings are sufficient for $\frac{180}{1.10} = 163.64$ pounds of beef and $\frac{20}{0.05} = 400$ boxes of matches. Suppose now the new prices prevail and the employee gets $236 per week—18 percent over the old earnings. To buy the previous quantities, the employee needs 163.64 · $1.20 = $196.37 for beef and 400 · $10 = $40 for matches. He therefore needs a total budget of $196.37 + $40 = $236.37. (The difference between this figure and the $236 obtained when we used the weighted index is only the result of rounding.)

From this example it is obvious that each product included in the basket should have an impact on the index proportional to its importance to consumers. Thus we calculate the weighted average index by the following formula:

WEIGHTED ARITHMETIC AVERAGE INDEX

$$I_w = \frac{\sum \left(\frac{p_n}{p_0} \cdot 100 \right) w}{\sum w} \tag{17.3}$$

Here w is the weight of the item in total spending during the base period.

In the specific case in which we have m items all with equal weights—that is, $w = \dfrac{1}{m}$—the weighted index reduces to the simple average of price relatives.

Obviously, with weighted averages, the question is what weights to employ. Here we have several options.

Laspeyres Price Index

The **Laspeyres price index**, named after Etienne Laspeyres, who first introduced it in the eighteenth century, uses the quantities consumed in the base period as weights. The following formula for it applies:

LASPEYRES PRICE INDEX

$$I_L = \frac{\Sigma p_n q_0}{\Sigma p_0 q_0} \cdot 100 \tag{17.4}$$

Here p_n is the price of a given item in the basket in the current period, p_0 is the price of the same item in the base period, and q_0 is the quantity of that item purchased in the base period.

Returning to the cost-of-living example, if the index currently stands at 110, it means that a 10 percent wage increase is required to enable our employees to purchase *the same quantities they purchased in the base period* (q_0).

Note that $I_L = I_w$ when we define the weights as $w = p_0 q_0$. To see this, simply replace w in Equation 17.3 by $p_0 q_0$ to obtain

$$I_w = \frac{\Sigma\left(\dfrac{p_n}{p_0} \cdot 100\right) p_0 q_0}{\Sigma p_0 q_0} = \frac{\Sigma p_n q_0}{\Sigma p_0 q_0} \cdot 100 = I_L$$

The p_0's that appear in the numerator cancel out.

Paasche Price Index

The **Paasche price index** is also a weighted index, but it uses the current quantities, q_n, as weights, rather than those of the base period, q_0. In the cost-of-living example, the Paasche index indicates what the wage level should be if our objective is to compensate for price increases in such a way that wage earners can buy the *current basket of items* (i.e., a different relative quantity of items) rather than the basket they consumed in the base period. The following formula applies:

PAASCHE PRICE INDEX

$$I_P = \frac{\Sigma p_n q_n}{\Sigma p_0 q_n} \cdot 100 \tag{17.6}$$

Note that I_P is obtained from I_w by use of the weights $w = p_0 q_n$. To show this, simply substitute $p_0 q_n$ for w in I_w to obtain

$$I_w = \frac{\sum \left(\frac{p_n}{p_0} \cdot 100 \right) w}{\sum w} = \frac{\sum \left(\frac{p_n}{p_0} \cdot 100 \right) p_0 q_n}{\sum p_0 q_n} = \frac{\sum p_n q_n}{\sum p_0 q_n} \cdot 100 = I_P \qquad \textbf{(17.7)}$$

Fixed-Weights Price Index

The weighted index numbers are theoretically preferred to the simple unweighted indexes, but they are also more difficult to calculate. The Laspeyres index is used more widely than the Paasche index. It requires information about the weights q_0—about consumer spending on each item in the base period. To determine the weights, then, it is necessary to carry out a survey only once, in the base period. With the Paasche index, one has to determine the quantities q_n every period, which is impractical in most cases, since a separate survey is required every period.

In practice, a modified Laspeyres index is often used; it is known as the **fixed-weights index,** I_F. The quantities, though fixed, do not really stay unchanged forever; they are changed from time to time, as new survey data provide updated information of quantities.

In principle, the fixed-weights index is very similar to the Laspeyres index, since both use quantities for weights. The difference is that in the fixed-weights index the weights are not necessarily representative of the base period, and every now and then they can be changed and updated to reflect current consumer behavior. For example, the Consumer Price Index, published by the Bureau of Labor Statistics, uses weights derived from a consumer survey for the years 1972–73, although the base period is 1967. These fixed weights will remain in use until a new consumer survey is completed and more updated weights are obtained.

17.4 Quantity Indexes

In Section 17.3 we explored several types of price indexes in some detail. All of them are designed to measure *value* changes when *quantities are held constant.* **Quantity indexes,** on the other hand, are designed to *measure value changes when prices are held constant.* The purpose of such indexes is to provide a measure of change in physical quantity, such as that of industrial production.

Technically, an unweighted average of relative quantities can be computed by averaging all of the separate $\frac{q_n}{q_0} \cdot 100$ values. This average, however, is meaningless, since the various items are measured in different units. Therefore, we make use of the **Laspeyres and Paasche quantity indexes,** given by Equations 17.8 and 17.9:

LASPEYRES QUANTITY INDEX

$$I_L = \frac{\sum p_0 q_n}{\sum p_0 q_0} \cdot 100 \qquad \textbf{(17.8)}$$

PAASCHE QUANTITY INDEX

$$I_P = \frac{\Sigma p_n q_n}{\Sigma p_n q_0} \cdot 100 \tag{17.9}$$

The Laspeyres quantity index measures the change in value that would have occurred as a result of quantity changes had the prices been kept at their *initial* (base-period) level. The Paasche quantity index, in contrast, measures the change in value that would have occurred as a result of quantity changes had the prices been equal to the nonbase period prices in both the base and nonbase periods.

The most widely used quantity index in the United States is the Federal Reserve Board's Index of Industrial Production.

17.5 Value Index

Suppose that the United States currently exports to Japan quantities q_n of various products at unit prices p_n. The respective quantities and prices in the base period are q_0 and p_0. The ratio of $\Sigma p_n q_n$ to $\Sigma p_0 q_0$ represents the change in the value of exports, a change that has two components: changes in export prices and changes in the quantity of exported products. When the ratio is multiplied by 100, it gives an index called a **value index.** It does not measure the separate changes in prices or in quantities; rather, it measures the combined change in the value of our exports—the total amount of dollars we receive from our trade partner. Formally

VALUE INDEX

$$I_V = \frac{\Sigma p_n q_n}{\Sigma p_0 q_0} \cdot 100 \tag{17.10}$$

Let us now look more closely at the Laspeyres and Paasche price and quantity indexes, as well as the value index, using a numerical example.

EXAMPLE 17.1

Suppose a typical consumption basket consists of three products only: beef, bread, and milk. The following were the quantities consumed in the base period, 1981, and in the latest available nonbase year, 1982:

Product	1981 (base year)	1982 (nonbase year)
Beef (pounds)	200	210
Bread (loaves)	400	410
Milk (quarts)	200	600

The corresponding prices were (in cents):

Product	1981 (base year)	1982 (nonbase year)
Beef (pound)	100	110
Bread (loaf)	40	50
Milk (quart)	30	40

Let us first calculate the Laspeyres and Paasche *price* indexes:

$$I_L = \frac{\Sigma p_n q_0}{\Sigma p_0 q_0} \cdot 100 = \frac{110 \cdot 200 + 50 \cdot 400 + 40 \cdot 200}{100 \cdot 200 + 40 \cdot 400 + 30 \cdot 200} \cdot 100 = \frac{50{,}000}{42{,}000} \cdot 100 = 119.0$$

$$I_P = \frac{\Sigma p_n q_n}{\Sigma p_0 q_n} \cdot 100 = \frac{110 \cdot 210 + 50 \cdot 410 + 40 \cdot 600}{100 \cdot 210 + 40 \cdot 410 + 30 \cdot 600} \cdot 100 = \frac{67{,}600}{55{,}400} \cdot 100 = 122.0$$

Next, let us calculate the Laspeyres and Paasche *quantity* indexes:

$$I_L = \frac{\Sigma p_0 q_n}{\Sigma p_0 q_0} \cdot 100 = \frac{100 \cdot 210 + 40 \cdot 410 + 30 \cdot 600}{100 \cdot 200 + 40 \cdot 400 + 30 \cdot 200} \cdot 100 = \frac{55{,}400}{42{,}000} \cdot 100 = 131.9$$

$$I_P = \frac{\Sigma p_n q_n}{\Sigma p_n q_0} \cdot 100 = \frac{110 \cdot 210 + 50 \cdot 410 + 40 \cdot 600}{110 \cdot 200 + 50 \cdot 400 + 40 \cdot 200} \cdot 100 = \frac{67{,}600}{50{,}000} \cdot 100 = 135.2$$

Finally, the value index, I_V, is given by

$$I_V = \frac{\Sigma p_n q_n}{\Sigma p_0 q_0} \cdot 100 = \frac{110 \cdot 210 + 50 \cdot 410 + 40 \cdot 600}{100 \cdot 200 + 40 \cdot 400 + 30 \cdot 200} \cdot 100 = \frac{67{,}600}{42{,}000} \cdot 100 = 161.0$$

What is the interpretation of these five indexes?

The Laspeyres price index, I_L, indicates how much more the consumer needed to spend in 1982 to purchase the same basket of items as in 1981. Thus it represents the effect of a change in prices on a *given base-period* basket of items. If wages are linked to prices, I_L indicates that a 19 percent increase is in order.

The Paasche price index measures the change in consumer spending resulting from a change in prices on a *given nonbase-period* basket of items. Thus, if a change in consumption patterns has occurred between the base and the current period, the Paasche price index uses the after-the-change quantities. The most notable change in quantities in our example occurs in milk consumption. The Paasche index uses the 1982 quantity, 600 quarts, whereas the Laspeyres index uses the 1981 quantity, 200 quarts. Since the price of milk rose between 1981 and 1982, the Paasche index simply gives more weight to this price increase, so that I_P turns out higher than I_L. (Of course, the index represents the changes in prices of the other two items as well.)

Similarly, the Paasche quantity index turns out higher than the Laspeyres index. Since prices are used here as weights, this is not surprising. Milk prices rose proportionately higher than other prices, so that Paasche gives milk a greater weight than Laspeyres does. Since the rise in consumption was greatest in milk, the Paasche quantity index is higher than the Laspeyres. (Again, it is clear that the index considers other items besides milk as well.)

The value index, I_V, simply measures the amount of dollars spent in period n relative to the amount of dollars spent in the base period. As we indicated earlier, it answers the following question: How much more (or less) money does the consumer need in a given period (relative to the base period) in order to buy the new basket of items, considering the simultaneous change in prices and quantities?

Note that the product of the Laspeyres price ratio and the Paasche quantity ratio is equal to the product of the Laspeyres quantity ratio and the Paasche price ratio. This product gives the value ratio:

$$\underbrace{\frac{\Sigma p_n q_0}{\Sigma p_0 q_0}}_{\substack{\text{Laspeyres}\\\text{price}}} \cdot \underbrace{\frac{\Sigma p_n q_n}{\Sigma p_n q_0}}_{\substack{\text{Paasche}\\\text{quantity}}} = \underbrace{\frac{\Sigma p_0 q_n}{\Sigma p_0 q_0}}_{\substack{\text{Laspeyres}\\\text{quantity}}} \cdot \underbrace{\frac{\Sigma p_n q_n}{\Sigma p_0 q_n}}_{\substack{\text{Paasche}\\\text{price}}} = \underbrace{\frac{\Sigma p_n q_n}{\Sigma p_0 q_0}}_{\text{value}}$$

In our example

$$1.19 \cdot 1.352 = 1.22 \cdot 1.319 = 1.61$$

17.6 How to Use Index-Number Series

Some methods for using index-number series are presented in this section. We shall discuss *deflation* of value series, *technical shifting* of the base period, and *splicing*.

DEFLATION OF VALUE SERIES BY PRICE INDEXES

A common use of price indexes is the adjustment of value series for changes in the purchasing power of the dollar. This procedure is known as **deflation,** and it results in a restatement of the original value series in terms of "constant dollars." The reason such deflation is important is that values stated in "current" or "nominal" figures could change significantly over time or from place to place, solely as a result of inflation. Thus, in order to eliminate the impact of inflation and obtain a picture of the "real" change, not the "nominal" change, we need to deflate the series.

Consider the following hypothetical average annual salary of professors in 1970 and 1982:

Year	Average annual salary	Consumer Price Index	"Real" average annual salary
1970	$14,000	100	$14,000
1982	$27,300	210	$13,000

The nominal average salary indeed increased during this period by 95 percent $\left(\dfrac{27,300}{14,000} = 1.95\right)$; but were the professors better off in 1982? The answer, of course, is no. The $27,300 salary in 1982 is not sufficient to buy the same basket of items that $14,000 could buy in 1970. This fact is easily revealed by a comparison of 1970 and 1982 real salaries. To obtain the real salary we deflate the nominal 1982 salary by first dividing the nominal salary by the price index (210) and then multiplying by 100:

$$\frac{27,300}{210} \cdot 100 = \$13,000$$

(Not all of the professors are to be pitied; some found consulting work to supplement their incomes, and some even began writing textbooks.)

The following is a more comprehensive example of value-series deflation based on actual data for the United States and West Germany.

EXAMPLE 17.2

The Gross National Products (GNPs) of the United States and West Germany for the years 1974–80, along with their Consumer Price Indexes, are given in Table 17.1. During this period, the German GNP rose from 986.9 billion to 1,491.9 billion Deutsche marks, so that in 1980 it was 51.2 percent higher than in 1974. During the same period, the U.S. GNP rose from $1,413.3. to $2,626.1 billion, or by 85.6 percent. The differences between the nominal GNPs are also shown in Figure 17.2*a*. The question is whether the U.S. standard of living did indeed grow that much more rapidly during the period. To answer the question, we must first switch to real currency figures by deflating each GNP series by its respective CPI. When we do that, a different picture emerges (see Table 17.2).

TABLE 17.1
Gross National Products and Consumer Price Indexes, West Germany and United States, 1974–80

| | West Germany | | United States | |
| | GNP (billions of Deutsche marks) | CPI (1975 = 100) | GNP (billions of dollars) | CPI (1975 = 100) |
Year				
1974	DM 986.9	94.4	$1,413.3	91.6
1975	1,034.9	100.0	1,549.2	100.0
1976	1,125.0	104.3	1,718.0	105.8
1977	1,200.6	108.1	1,918.0	112.7
1978	1,290.7	111.1	2,156.1	121.2
1979	1,398.2	115.6	2,413.9	134.9
1980	1,491.9	122.0	2,626.1	153.1

Source: International Monetary Fund, *International Financial Statistics*, September 1981.

TABLE 17.2
GNP in Real Terms, West Germany and United States, 1974–80

Year	West Germany (billions of 1975 Deutsche marks)	United States (billions of 1975 dollars)
1974	DM 1,045.4	$1,542.9
1975	1,034.9	1,549.2
1976	1,078.6	1,623.8
1977	1,110.6	1,701.9
1978	1,161.7	1,779.0
1979	1,209.5	1,789.4
1980	1,222.9	1,715.3

As the numbers as well as Figure 17.2*b* show, the real growth in the GNP of both countries was substantially more moderate than the nominal growth because of the high inflation rates during the period. Furthermore, the growth in real terms was substantially smaller in the

Figure 17.2

GNP of the
United States
and West
Germany,
current and
"real"
currencies,
1974–80

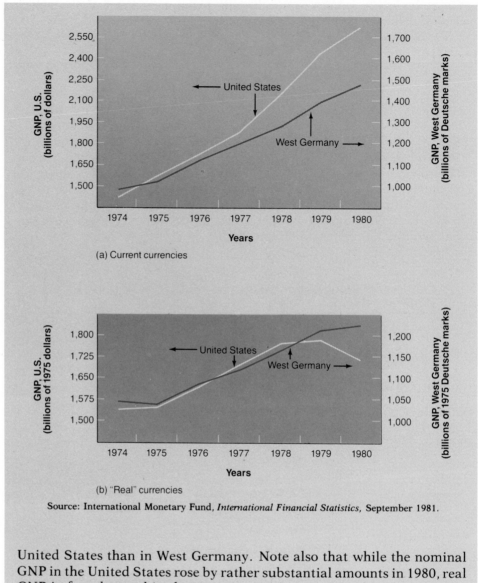

(a) Current currencies

(b) "Real" currencies

Source: International Monetary Fund, *International Financial Statistics*, September 1981.

United States than in West Germany. Note also that while the nominal
GNP in the United States rose by rather substantial amounts in 1980, real
GNP in fact dropped in that year.

SHIFTING THE BASE YEAR

Sometimes it is of interest to compare two indexes either within the same
country (such as Standard and Poor's stock index and the Consumer Price
Index) or between two countries. Since the indexes may have different base
years, they should be adjusted to a common base period so that a meaningful
comparison may be made. The procedure, known as **technical shifting**, is
quite simple and is best illustrated by a numerical example. Table 17.3
presents two indexes with different base years. In Column 1 we have a
hypothetical Consumer Price Index whose base year is 1973. Suppose we
want to shift the base year to 1978. All we need do is *deflate* the series by the
original 1978 index value. If we divide each number in Column 1 by 122 and
then multiply the result by 100, we will obtain the same index with 1978 as

its new base year, as shown in Column 2. For example, for 1973 we get $\frac{100}{122} \cdot 100 = 82.0$, for 1974 we get $\frac{110}{122} \cdot 100 = 90.2$, and so on. Note that the percentage change in the index numbers from one year to the next is the same in both columns, and shifting the base year is only a technical adjustment used for presentation purposes.

TABLE 17.3
Hypothetical Consumer Price Indexes Based on 1973 and 1978, Respectively

Year	(1) Consumer Price Index (1973 = 100)	(2) [(Col. 1) ÷ 122] · 100 Consumer Price Index (1978 = 100)
1973	100	82.0
1974	110	90.2
1975	112	91.8
1976	114	93.4
1977	118	96.7
1978	122	100.0
1979	130	106.6
1980	135	110.7
1981	140	114.8
1982	150	122.9

To calculate the percentage change in an index from, say, 1981 to 1982, we divide the 1982 price by the 1981 price, subtract 1.0, and multiply the result by 100. Here are the calculations of the percentage change in the index numbers of both columns (deviations are due to rounding):

$$\text{Column 1} \qquad \text{Column 2}$$
$$\left(\frac{150}{140} - 1\right) \cdot 100 = 7.14\% \qquad \left(\frac{122.9}{114.8} - 1\right) \cdot 100 = 7.06\%$$

SPLICING

Occasionally a published index will be discontinued or significantly revised. If a revision occurs, the next index and the old index *do not* measure exactly the same thing and they therefore must be distinguished. At times, however, we need to use the index over a period that requires data from the old series as well as the new. Despite all the theoretical limitations of using two different indexes as if they were one and the same, this is sometimes the best alternative available. *Combining the two indexes* is known as **splicing,** and it can be done only if the old and new series have at least one overlapping period, which is normally the case when a revision occurs. Technically, splicing is a very easy procedure and is illustrated in Table 17.4. For the period 1978–82, the spliced index is identical to the revised index. For the years 1973–77 the spliced index is the same as the old index after it has been deflated by its 1978 level. That is to say, to obtain the spliced index for the 1973–77 period we simply divide the old index figures by 108.7 and then multiply by 100. For 1973, for example, we get $\frac{90.1}{108.7} \cdot 100 = 82.9$, and so on.

TABLE 17.4
Old, Revised, and Spliced Indexes

Year	Old index (1975 = 100)	Revised index (1978 = 100)	Spliced index (1978 = 100)
1973	90.1		82.9
1974	95.0		87.4
1975	100.0		92.0
1976	103.3		95.0
1977	104.6		96.2
1978	108.7	100.0	100.0
1979		110.1	110.1
1980		108.3	108.3
1981		114.2	114.2
1982		121.8	121.8

While the technical work involved in splicing is simple, the index must be used with great caution, since the figures for the "old" years can at best be regarded as rough approximations.

17.7 Some Important U.S. Indexes

THE CONSUMER PRICE INDEX (CPI)

One of the most important indexes compiled by the U.S. government is the Consumer Price Index (CPI). It is a widely accepted measure of inflation, and, as mentioned earlier, is frequently used in wage contract bargaining. The CPI often serves also as a deflator of economic series. It is a major indicator of the general health of the U.S. economy. In some countries major economic variables, such as wages and interest rates, are linked to those countries' Consumer Price Indexes.

The name Consumer Price Index was adopted by the Bureau of Labor Statistics (BLS) and the National Industrial Conference Board during World War I. Initially it measured changes in the retail prices of the goods and services bought by city wage earners and clerical workers. In 1978, a second index, the CPI-U, was also introduced to expand coverage to all urban consumers. A revised version of the old CPI, called the CPI-W, still measures changes in expenditures by city wage earners and clerical workers only. Both indexes are continuous with the old CPI.

The CPI became important at the end of World War I, when data were in demand for use in wage negotiations in shipbuilding cities. Changes in the cost of living were first published in the BLS *Monthly Labor Review* in October 1919, and regular publication began in February 1921. Since 1978, prices of food, fuels, and a few other items have been collected monthly in all cities, and prices of most commodities and services have been collected monthly in the five largest cities, and bimonthly in the remaining cities the index surveys.

Weights used in calculating the index are based on studies of actual expenditures. Quantities and qualities of items in the "market basket" remain essentially the same between consecutive pricing periods, so that the index measures the effect of *price changes alone* on the cost of living. It does not measure changes in the total amount families spend. A study conducted during 1917–19 provided the weights used until 1935. Since then, the index has undergone five major revisions, each of which involved bringing the "market basket" of goods and services up to date, revising the weights, and improving the sample and methodology. The most recent revision, instituted in 1978, adopted 1967 as the reference year (1967 = 100), and introduced new expenditure weights based on a 1972–73 Consumer Expenditure Survey of 216 areas.

The list of items currently priced for the index includes approximately 400 goods and services. For some items several different qualities are priced. Characteristics of each item are no longer specified in great detail, but every effort is made to ensure that differences in reported prices are measures of price change only. Researchers attempt to obtain the prices actually paid by consumers, not list prices from which discounts normally are given. All taxes directly associated with the purchase or use of the items are incorporated in the index.

The CIP is now based on prices collected in 85 areas, including cities, suburbs, and urbanized places; until 1978, only 56 areas were sampled. Area definitions used in the current indexes are those established by the Office of Management and Budget in 1973, based on data collected in the 1970 census. Between scheduled survey dates, prices are held at the level of their last pricing for all goods and services except new automobiles. Price data for the 85 areas are combined for the United States with weights based on 1970 population density in the various sample areas. Regional indexes are also published for 28 separate metropolitan areas.

Unlike the old CPI, the new indexes are not divided into specific commodity groups. General categories like "household expenses" and "transportation" are now used to distinguish between types of expenses, to reflect the new item selection process, called "disaggregation," which the Bureau of Labor Statistics believes allows its agents to purchase a more representative range of items than they were able to obtain under older selection processes.

STANDARD & POOR'S (S&P) STOCK PRICE INDEXES

Many indexes measure changes in stock prices. Among them are the Dow Jones Index, the New York Stock Exchange Index, and the S&P indexes. All of these indexes are used as indicators for forecasts of changes in the economy, because pessimistic or optimistic attitudes of investors with regard to future corporate earnings are reflected in stock prices.

The **S&P stock price indexes** have been steadily expanding their coverage over the years to supply a dependable measure of the composite price pattern of the majority of stocks. Back in 1923, S&P pioneered with the issuance of a scientifically constructed stock price index organized according to leading industrial groups. At that time, 26 subgroup indexes, based on 233 stock issues, were compiled. Five hundred stocks are now covered, broken down into 88 groups that make up the four main categories: industrials, railroads, utilities, and the 500 composites. There are also four supplementary group series: capital goods companies, consumer goods, high-grade common stocks, and low-priced common stocks. In addition there are 11 group indexes that are not included in the S&P Series of 500 Stocks. These include indexes of bank stocks (New York City and outside New York City), investment companies, property, liability and life insurance companies, trucking companies, and discount stores, and four other groups that were added in 1970 in order to trace the price movement of some of the more recently developed industries: air freight, atomic energy, conglomerates, and multiline insurance companies. Also added in 1970 and included in the Series of 500 Stocks are indexes for forest products, hotels/motels, offshore drilling, real estate, and restaurant operators.

These S&P stock price indexes, which are based on the aggregate market value of the common stocks of all of the companies in the sample, express the observed market value as a percentage of the average market value during the base period. Originally the base period was 1926; it was subsequently shifted to the average of the 1935–39 period, and then finally to the currently used base period, 1941–43, with the average stock value of 1941–43 being set at 10. This base results in a price index level that is more realistic than that of most popular composite stock price measures, in that it is not absurdly distant from the average price level of all stocks listed on the New York Stock Exchange. The group indexes added since 1957 are based on various recent periods.

The 1957 revision of the S&P stock price indexes marked a giant step in another direction. Thanks to electronic computer and input feeders, the four main group indexes are now computed at five-minute intervals. S&P does not publish these frequent readings, but does maintain a record of them. Hourly indexes are published in the Daily News section of *S&P Corporation Records*. Daily high, low, and closing indexes are published in the weekly *Outlook* and the monthly *Current Statistics*.

The formula adopted by S&P after much testing is a value index generally defined as a "base-weighted aggregative" expressed in relatives with the average value for the base period (1941–43) equal to 10. This method of computation has two distinct advantages over most index number series: (1) it has the flexibility needed to adjust for arbitrary price changes caused by the issuance of rights, stock dividends, splits, and the like, and (2) the resultant index numbers are accurate and have a relatively high degree of continuity, which is especially important when long-term comparisons are to be made. Certain modifications of the basic formula have been introduced to make it possible to maintain the best possible representation over the years. The character of the stock market is subject to gradual but continuous change, and it is only by periodic checks of coverage that true representation can be maintained.

Each component stock is weighted, so that it will influence the index in proportion to its importance in its respective market. The most suitable weighting factor for this purpose is the number of shares outstanding. The price of any share multiplied by the number of shares outstanding gives the current market value of that particular stock. The market value determines the relative importance of the security.

The base value of a group of stocks is the average of the weekly group values for the period 1941–43. The current group value is expressed as a relative number by dividing it by its base-period value and multiplying the result by 10. In this relative form an index number attains its maximum usefulness for statistical purposes.

The formula for the base-weighted aggregative index is

$$\text{Index} = \frac{\Sigma p_1 q_1}{\Sigma p_0 q_0} \cdot 10$$

where p_1 represents the current market price, p_0 the market price in the base period, q_1 the number of shares currently outstanding, and q_0 the number of shares outstanding in the base period. The denominator of the index is adjusted to reflect new issues of stocks.

Note that instead of choosing a base of 100, S&P chose 10 as the base index number, but this arbitrary choice does not change the conception inherent in the index.

Figure 17.3 gives the CPI and the S&P index for the period 1918–80. Note that while the CPI increased about fivefold during this period, the S&P index increased twelvefold. Roughly, this means that if an investor held one dollar in cash (not invested) since 1918, it would be worth only about 20 cents in 1980. If the investor invested the dollar in 1918 in a portfolio of stocks corresponding to the S&P index, it would be worth $12 in nominal terms or about $2.40 in real terms in 1980 (since CPI increased almost fivefold during this period: 12 ÷ 5 = 2.4).

Note that despite the general upward trend, the year-by year fluctuations in the S&P index are quite large; this implies a substantial risk to investors. For example, between 1972 and mid-1974 the index dropped from about 120 to 80, and hence investors lost about one-third of the 1972 value of their investments. Note also that we shifted the CPI base year so that the two indexes (CPI and S&P) have the same base of 10 in 1941–43.

Among other important indexes, which we shall not discuss in detail, are the Wholesale Price Index, the Industrial Production Index, the Corporate Profits Index, the Gross National Product (GNP), and the Population Index.

LEADING ECONOMIC INDICATORS

The National Bureau of Economic Research (NBER) established twelve economic indicators, including some important indexes, which attempt to

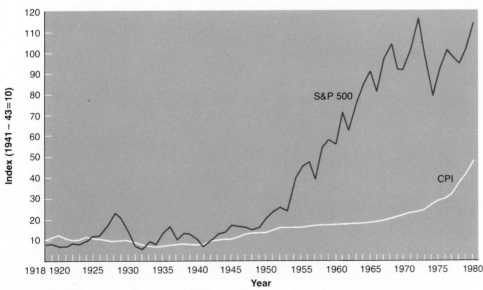

Figure 17.3

Consumer Price Index and Standard & Poor's 500, 1918–80 (1941–43 = 10 for both indexes)

Sources: *Standard & Poor's Trade and Securities Statistics, Security Price Index Record* (New York, 1980), p. 4; U.S. Department of Commerce, Bureau of the Census, *Historical Statistics of the United States, Colonial Times to 1970* (Washington, D.C., 1975), pt. 1, Series E 135–166, pp. 210–11; U.S. Department of Commerce, Bureau of Economic Analysis, *Business Conditions Digest*, September 1981, p. 7.

approximate changes in the economy and may thus be used to indicate future business trends or economic cycles. The average of these **leading indicators** is important in policy making. Charts of the twelve leading economic indicators selected by the NBER are regularly published in *Business Conditions Digest*. Some of the indexes included among these indicators are the net business formations (business starts minus failures), new orders by firms, housing permits, inventory accumulation, prices of raw materials, corporate profits, and stock price level (the S&P index).

17.8 International Comparison of Indexes

We often wish to compare one country's activity with another's. International comparison of indexes serves to explain many economic phenomena and helps in the formulation of customs policy, export-import policy, and the

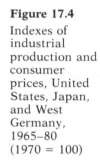

Figure 17.4

Indexes of industrial production and consumer prices, United States, Japan, and West Germany, 1965–80 (1970 = 100)

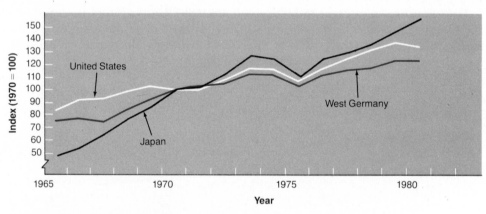

(a) Total industrial production

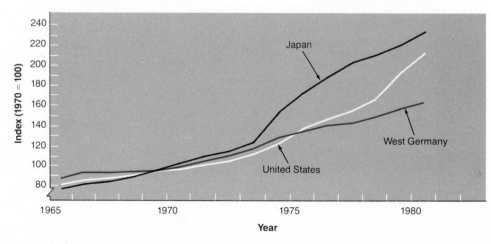

(b) Consumer prices

Source: International Monetary Fund, *International Financial Statistics*, various issues.

like. Obviously, in order to get a meaningful picture one has to shift the indexes of all countries included in the study to a common base year.

Figure 17.4*a* gives the indexes of total industrial production in the United States, West Germany, and Japan. As we can see, industrial production grew considerably faster in Japan than in the United States and West Germany during the 1965–80 period. In all three countries, the index declined around 1973, in the wake of the oil crisis that hit the industrial world. Japan, which is totally dependent on imported oil, suffered the most from the crisis. After 1975, however, all three countries overcame the oil crisis and industrial production resumed its growth, returning in 1977 to the level it had reached before the oil crisis.

Figure 17.4*b* compares consumer prices in the same three countries over the same period of time. The figure shows a greater diversity in consumer price behavior in the years following the oil crisis (1973) than in earlier years. Prices in West Germany rose the least during the period. The United States experienced similar price behavior before 1973, but more rapidly increasing prices afterward. Japan was affected the most by the oil crisis: its prices increased significantly faster than in the other two countries. However, consumer prices also rose more rapidly in Japan before the oil crisis, that is, during the 1965–73 period.

Figure 17.5 compares the indexes of hourly earnings in the same three countries and over the same period of time, 1965–80. While earnings in West Germany increased somewhat faster than in the United States, the most striking fact seen in the figure is that earnings have increased in Japan much more rapidly than in either West Germany or the United States. The higher rate of inflation in Japan (see Figure 17.4*b*) accounts only in part for the steeper increase in wages: the balance represents a genuine increase in real earnings and improvement in the Japanese standard of living. The general trend appears to operate in the direction of a diminishing wage differential among the industrial countries, with the result that traditional cheap-labor countries are losing their advantage.

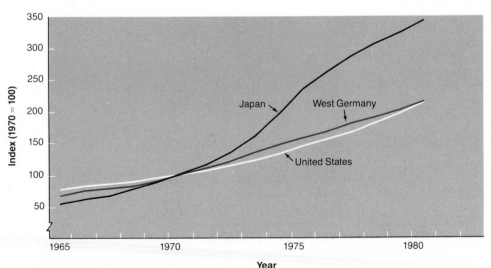

Figure 17.5

Index of hourly earnings in the United States, West Germany, and Japan, 1965–80

Source: International Monetary Fund, *International Financial Statistics*, various issues.

17.9 A Word of Caution

Index numbers can serve only as a proxy for a description of a given phenomenon, and generalizations may be very dangerous. We shall list only some of the major difficulties involved in index numbers.

1. Choosing the base period. As we show in Figure 17.1, one can "play around" with the base period and give various impressions of the real trend of the variable. An index based on a given year could lead one to accept the notion that a company's profitability was in good shape, while one based on a different year could suggest the contrary. When we interpret such an index we must always consider the fact that the index measures the firm's profitability only in *relation* to the base period.

2. Items included in the index. A change of 10 percent in the Consumer Price Index may be completely irrelevant to a consumer whose consumption pattern is unlike that of the "average consumer." If I consume only bread and milk and the price of these two items has not changed, there is a zero increase in the price level as far as I am concerned. Perhaps somewhat more interesting is the case of the Dow Jones Industrial Index, which consists of 30 stocks of some of the largest corporations in the United States. In 1939 IBM was removed from the Dow Jones index and AT&T was substituted. Had IBM remained, the index would have reached 1,017.39 in December 1961 instead of only 734.91.

3. Quality changes. Virtually all goods and services in the economy change qualitatively over the years, and sometimes it is hard to reflect this qualitative change in price comparisons between two periods. The more often we conduct a survey to update the weights, the better is the index. Such surveys are conducted only every decade or so, however, since they are quite expensive and complicated.

Problems

17.1. Why is the selection of a base period important for proper presentation of data by an index? What are the guidelines for choosing a base period?

17.2. What is the disadvantage of an unweighted index?

17.3. The Consumer Price Index (CPI) in a certain country was 100.0 in 1970 and 200.0 in 1980. What was the average annual price increase over the period 1970–80?

17.4. The Consumer Price Index (CPI) in a certain country in 1960, 1970, and 1980 was as follows:

Year	CPI
1960	100.0
1970	100.0
1980	300.0

(a) What was the average annual price increase in the period 1960–70?
(b) What was the average annual price increase in the period 1970–80?
(c) What was the average annual price increase in the period 1960–80?
(d) Was the average annual price increase in the period 1960–80 as calculated in part c equal to the average of your answers to parts a and b? Explain.

17.5. The sales of DAB Corporation in billions of dollars for the years 1972–82 were as follows:

Year	Sales
1972	$5.0
1973	4.5
1974	4.0
1975	3.5
1976	3.0
1977	3.5
1978	4.0
1979	4.5
1980	4.5
1981	4.5
1982	5.0

(*a*) Construct an index for the dollar value of sales, using 1972 as the base year.

(*b*) Do you think that 1972 is a good year to serve as a base period? What information do you need in order to select the base period?

17.6. The following are the prices of four products and services bought by consumers during the years 1979, 1980, and 1981:

Product	1979	1980	1981
Bread (cents per loaf)	60.0	65.0	70.0
Transportation (dollars per 10 miles)	1.0	1.5	2.0
Rent (dollars per month)	250.0	260.0	270.0
Recreation (dollars per activity)	8.0	8.0	9.0

Suppose the average consumer spends 35 percent of his income on bread, 15 percent on transportation, 30 percent on rent, and 20 percent on recreation.

(*a*) Compute the unweighted index for the above items, using 1979 as the base.

(*b*) Compute the average of price relatives on the 1979 base.

(*c*) Compute the weighted index on the 1979 base.

17.7. The following are the prices of five commodities during the years 1978–81:

Commodity	1978	1979	1980	1981
A	$ 45	$ 45	$ 45	$ 45
B	60	58	57	51
C	100	120	140	160
D	1,000	1,200	1,400	1,800
E	38	43	61	54

(*a*) Compute the unweighted index for the five commodities, using 1980 as the base.

(*b*) Compute the average of price relatives on the 1980 base.

17.8. The prices and quantities consumed for commodities *A* and *B* during the years 1970 and 1980 are as follows:

Commodity	Unit price		Quantities consumed	
	1970	1980	1970	1980
A	$1.00	$1.60	30	70
B	2.30	1.80	10	72

(a) Compute the 1980 index on the base of 1970 using the average of price relatives.

(b) What are the Paasche and Laspeyres price indexes for the above data when the base year is 1970?

(c) What are the Paasche and Laspeyres quantity indexes for the above data when 1970 is the base year?

(d) Calculate the value index for the data on the 1970 base.

(e) Verify the following relationship using the above data:

$$\left(\begin{array}{c}\text{Laspeyres}\\\text{price}\end{array}\right) \cdot \left(\begin{array}{c}\text{Paasche}\\\text{quantity}\end{array}\right) = \left(\begin{array}{c}\text{Laspeyres}\\\text{quantity}\end{array}\right) \cdot \left(\begin{array}{c}\text{Paasche}\\\text{price}\end{array}\right) = \left(\begin{array}{c}\text{Value}\\\text{index}\end{array}\right)$$

17.9. Figure P17.9 shows the annual percentage gains in telephones served by the United Telephone System and by the American telephone industry for the years 1966–78.

Figure P17.9

Percent gains in telephones served, by United Telephone and by the telephone industry, 1966–78

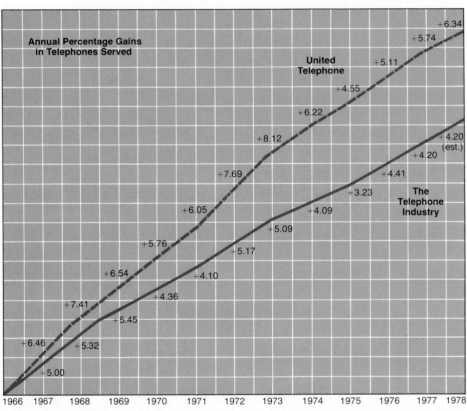

Source: Advertisement for United Telecommunications, Inc., in *Institutional Investor*, March 1979. Reprinted by permission of United Telecom.

(a) From Figure P17.9 construct an index for the number of telephones served by the United Telephone System and by the telephone industry in the United States. Use 1966 as the base year for both indexes.

(b) What is the percentage gain of the United Telephone System and of the telephone industry in the period 1971–78? What is the percentage change in the period 1976–78?

17.10. Canada's index of hourly earnings in the manufacturing field for the years 1973–80 and the Consumer Price Index for those years are shown in Table P17.10. Derive the wage index in real terms (note that 1975 = 100.0).

TABLE P17.10

Year	Wage index	CPI
1973	81.4	76.3
1974	90.3	86.6
1975	100.0	100.0
1976	107.5	113.8
1977	116.1	126.1
1978	126.5	135.2
1979	138.1	147.0
1980	152.1	161.9

Source: International Monetary Fund, *International Financial Statistics,* October 1980 and July 1981.

17.11. The 1969–78 revenues, net earnings, and earnings per share of Community Psychiatric Centers are shown in Table P17.11. Derive three indexes, one for each of the variables, using 1975 as the base year. Draw the three indexes on one diagram.

TABLE P17.11

Fiscal year ended November 30	Revenues	Net earnings	Earnings per share
1969	$ 2,797,000	$ 250,000	$0.15
1970	4,513,000	446,000	0.20
1971	6,363,000	684,000	0.24
1972	8,495,000	1,050,000	0.32
1973	12,192,000	1,360,000	0.42
1974	15,656,000	1,514,000	0.51
1975	17,227,000	1,800,000	0.63
1976	20,693,000	2,136,000	0.76
1977	27,680,000	3,352,000	1.17
1978	33,585,000	4,881,000	1.65

Source: Advertisement for Community Psychiatric Centers, San Francisco, in *Institutional Investor,* April 1979.

17.12. An index of industrial production in Japan for the years 1973–80 is given in Table P17.12.

TABLE P17.12

Year	Index of industrial production (1975 = 100)
1973	102.8
1974	106.4
1975	100.0
1976	105.5
1977	107.8
1978	112.3
1979	117.5
1980	115.7

Source: International Monetary Fund, *International Financial Statistics,* October 1980 and July 1981.

(*a*) Calculate the percentage of each year's change in industrial production.
(*b*) Change the base period of the index from 1975 to 1973.

17.13. The prices and quantities consumed of commodies *A*, *B*, and *C* for the years 1980–82 were as follows:

Commodity	Unit price			Quantities consumed		
	1980	1981	1982	1980	1981	1982
A	$ 10	$ 12	$ 14	50	52	56
B	8	7	8	3	3	3
C	108	118	132	20	21	24

(*a*) Compute the price index for the commodities using the average of price relatives, with 1980 as the base year.
(*b*) What are the Paasche and Laspeyres price indexes for the above data when 1980 is the base year?
(*c*) What are the Paasche and Laspeyres quantity indexes for the above data when 1980 is the base year?

17.14. A wholesale price index for a given country is based on 1978 prices:

Year	Index
1978	100.0
1979	104.0
1980	111.5
1981	114.7
1982	121.6

The index for earlier years is based on 1973 prices:

Year	Index
1973	100.0
1974	102.1
1975	102.9
1976	103.6
1977	104.7
1978	109.9

(*a*) Splice the above index series and provide one index series on the base of 1973.
(*b*) Splice the index series and provide one index series on the base of 1978.
(*c*) Splice the index series and provide one index series on the base of 1982.

17.15. The annual salary of an employee in a large manufacturing firm for the years 1977–82 was as follows:

Year	Annual salary
1977	$10,000
1978	11,000
1979	12,000
1980	13,000
1981	14,000
1982	15,000

(*a*) Suppose consumer prices have gone up 2 percent per year during the period. Calculate the employee's "real" salary in each of the years 1977–82.
(*b*) Rework part *a*, assuming that prices went up 5 percent per year.
(*c*) Rework part *a* again, assuming that prices went up 10 percent per year.

17.16. Table P17.16 shows indexes of average hourly earnings and the Consumer Price Index of the United States, West Germany, and Switzerland in the years 1975–80.

TABLE P17.16

Year	Average hourly earnings			CPI		
	West Germany	Switzerland	U.S.A.	West Germany	Switzerland	U.S.A.
1975	100.0	100.0	100.0	100.0	100.0	100.0
1976	106.4	101.6	108.1	104.3	101.7	105.8
1977	113.9	103.6	117.6	108.1	103.3	112.7
1978	120.0	106.7	127.7	111.1	104.1	121.2
1979	126.9	109.9	138.7	115.6	107.9	134.9
1980	135.3	116.2	150.5	122.0	112.2	153.1

Source: International Monetary Fund, *International Financial Statistics*, July 1981.

(*a*) What was the average annual percentage increase of hourly earnings in each of the three countries?
(*b*) What was the average annual percentage increase of consumer prices in each of the three countries?
(*c*) Derive an index of "real" average hourly earnings for the three countries.
(*d*) What was the average annual percentage increase of "real" hourly earnings in each of the countries?

17.17. Table P17.17 shows the net imports of crude oil, natural gas liquids, and feedstocks in selected countries. For each country derive an index of the imports with 1975 as the base year.

TABLE P17.17

Year	U.S.A.	Japan (thousand metric tons)	U.K.
1974	180,810	237,839	112,815
1975	207,806	223,302	91,360
1976	301,071	228,699	90,466
1977	374,256	236,508	70,698
1978	349,054	230,181	68,143
1979	355,600	239,154	60,382
1980	288,635	216,840	46,717

Source: Organization for Economic Cooperation and Development, *Quarterly Oil Statistics*, 1981.

17.18. Table P17.18 lists annual percentage changes of selected indicators of labor markets in several countries, 1976–80. The percentage changes reflect changes from the previous to the current year. For example, the 13.8 percent change in Canada's hourly earnings in 1976 means that the average hourly earnings in 1976 were 13.8 percent larger than they had been in 1975.

(*a*) Use 1975 as the base period and derive indexes for hourly earnings, employment, and productivity for 1975–80 in the countries listed.
(*b*) Draw three diagrams, one for each indicator, showing the indexes for the four countries. Summarize the changes verbally.

TABLE P17.18

Indicators	1976	1977	1978	1979	1980
Canada					
Hourly earnings	13.8	10.1	7.1	8.7	10.1
Employment	1.4	−1.4	1.0	3.0	−1.8
Productivity	5.5	2.2	4.2	4.6	−1.5
United States					
Hourly earnings	8.1	8.8	8.6	8.6	8.5
Employment	3.2	3.9	5.1	5.1	3.6
Productivity	10.7	5.9	5.8	4.2	−3.4
Japan					
Hourly earnings	12.5	9.2	7.1	5.9	6.4
Employment	−2.0	−0.9	−2.3	−0.7	0.6
Productivity	11.1	4.1	6.2	8.3	7.0
Italy					
Hourly earnings	20.9	27.6	16.2	19.3	21.9
Employment	−1.4	1.0	−1.0	0.3	0.6
Productivity	12.4	1.1	2.0	6.6	5.6

Source: International Monetary Fund, *International Financial Statistics*, July 1981.

CHAPTER EIGHTEEN OUTLINE

Key Terms

nonparametric statistical
  methods
binomial test
chi-square test
runs test
run
random walk theory
sign test

Mann-Whitney test
Kruskal-Wallis test
Kruskal-Wallis statistic
Spearman rank correlation
  coefficient
Kendall rank correlation coef-
  ficient

18

NONPARAMETRIC METHODS

In most of the previous chapters on hypothesis testing, we assumed some knowledge of the distribution of observations (normal distribution, t distribution, and so on) and often made additional assumptions, such as independence of the observations or homogeneity of the population variances.

If one can be sure that the specific assumptions needed for the particular test indeed hold, the parametric tests described in earlier chapters present no difficulties and may be appropriately used. In reality, however, we seldom have complete knowledge of the relevant distributions, and in many studies, particularly those relying on small samples, the normality assumption is questionable. For example, we may say that it is *reasonable to assume* that the height distribution of human beings is very close to normal. It is not precisely normal, but we make the *assumption* of normality and proceed with hypothesis testing. The fact that the distribution is not exactly normal introduces an error into our test whose magnitude is usually unknown. Very rarely is any attention paid to the error caused by the fact that the distribution is not exactly normal.

18.1 The Scope of Nonparametric Methods

Suppose that we would like to test the hypothesis, at a significance level of 5 percent, that the average monthly income of a certain population is $1,000, against the alternative hypothesis that the average income is greater than $1,000. We sample 16 observations and calculate the sample mean, which happens to be $\overline{X} = \$1,200$. For simplicity, assume that the standard deviation in the population is known and is $400. Assuming that the sample is drawn from a normal distribution, we have $\dfrac{\overline{X} - \mu}{\sigma/\sqrt{n}} \sim N(0, 1)$, where μ is the population mean, σ is the standard deviation, and n is the number of observations. If the null hypothesis holds, we should have $\dfrac{\overline{X} - 1,000}{400/\sqrt{16}}$

$= \dfrac{\overline{X} - 1,000}{100} \sim N(0, 1)$. Since $\overline{X} = \$1,200$, we get $\dfrac{\overline{X} - 1,000}{100} = \dfrac{1,200 - 1,000}{100}$

$= 2.0$.

Should we accept or reject the null hypothesis, which says $\mu = \$1,000$? Since we have a one-tailed test at the 5 percent significance level, we reject H_0 if the statistic $\dfrac{\overline{X} - \mu}{\sigma/\sqrt{n}}$ is greater than 1.645, or if $\overline{X} > \mu + 1.645 \cdot \dfrac{\sigma}{\sqrt{n}}$

$= 1,000 + 1.645 \cdot 100 = 1,164.50$. In our case $\overline{X} = \$1,200$ and hence we reject the null hypothesis (see curve A in Figure 18.1).

Figure 18.1

Rejection region under true distribution (curve B) and under assumed distribution (curve A)

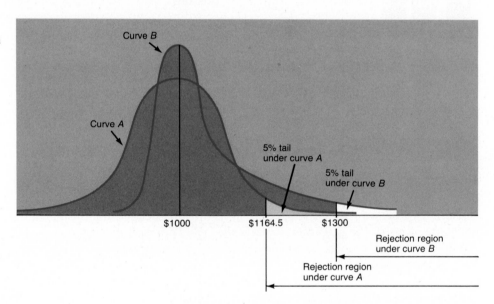

Now suppose that the normality assumption does not hold. Indeed, the mean income in the population is $\mu = \$1,000$, but the distribution is not normal and is given by curve B rather than curve A (see Figure 18.1). Since the distribution of the sample average is given by curve B, we should reject the null hypothesis only if $\overline{X} > \$1,300$: there is a 5 percent probability that $\overline{X} > \$1,300$ with curve B. Thus, the erroneous normality assumption (that is, assuming that the distribution follows the normal curve A rather than curve B) has led us to reject the null hypothesis, H_0: $\mu = \$1,000$. In other words, $\overline{X} = \$1,200$ falls in the rejection region if curve A is the true distribution and in the acceptance region if the true distribution is given by curve B. Hence, by assuming that the distribution is normal when it is not, we calculate the wrong probabilities and reach a wrong decision.

Nonparametric statistical methods of hypothesis testing are available for use in situations in which the underlying distributions are unknown. These methods make no assumptions about the distributions.

As the name "nonparametric" implies, these methods assume nothing about the population parameters. They are "distribution-free" in the sense that they enable us to test a hypothesis and reach conclusions regardless of the shape of the population distribution. To be more specific, the two major assumptions that are made in most statistical tests—homogeneity of variance (of the two or more populations under consideration) and normality—can be relaxed when we use nonparametric tests.

The disadvantage of nonparametric tests is that they are less powerful than parametric tests. In other words, for a given significance level (that is, a given probability of committing a Type I error), the probability of committing a Type II error is greater with nonparametric tests than with parametric tests, which implies that the power of the parametric test is greater. In many cases, however, when the number of observations is sufficiently large, the power of the nonparametric test approaches the power of the corresponding parametric test.

There are many nonparametric tests, and many textbooks in statistics devoted solely to this subject. In this chapter we describe what we believe to be the most useful test methods for business and economics. One can classify the various nonparametric tests according to the measurement scale of the data and according to the type of test, that is, its assumptions and its purpose.

Nonparametric tests can be used for the following types of data:

1. *Categories:* This is when qualitative data are coded. For example, Catholic = 0, Protestant = 1, Jewish = 2, Moslem = 3. The order of the codes has no meaning and we could decide on alternative codes that would do an equally good job; for example, Protestant = 10, Catholic = 9, Jewish = 8, Moslem = 7.
2. *Ranks:* The observations are ordered from lowest to highest (or vice versa) and assigned ranks by their order. For example, suppose three toothpaste brands are ranked by a certain quality. The ranks 1, 2, and 3 indicate an order which should not be changed. However, the ranks do not tell us by how much one toothpaste is better than another.
3. *Cardinal Scale:* With this type of data one can not only rank the order of the observations, but also determine by how much a given observation is greater than another. Income, for example, is measured on a cardinal scale. Given the observations $500, $1,000, and $12,000, we can determine that the second observation is twice as much as the first, the third observation is 20 percent higher than the second, and so on. This is not possible to do with ranked data.

Some of the nonparametric tests deal with one sample and some deal with two or more samples. Some two-sample tests are designed for independent samples, yet others are dealing with related samples. A description and details of some of these tests are given in this chapter.

18.2 The Binomial Test

The **binomial test** is one that *tests the value of a proportion*. It may be considered a nonparametric test, since the probability of a given event can be calculated simply by counting the number of "success" trials and dividing the sum by the total number of trials in the binomial experiment; no distribution parameters are needed. For example, suppose we want to test the hypothesis that there is a probability $p = \frac{1}{2}$ that the price of a certain stock will go up, and a probability $q = \frac{1}{2}$ that it will either go down or stay unchanged. One can collect a sample of 20 daily price changes and count the number of pluses (price increases) and the number of minuses (price decreases or no change), and then see whether the sample results fall in the rejection or acceptance region by simply calculating the probability of the observed event under the null hypothesis ($p = \frac{1}{2}$). Since we have devoted considerable space to the binomial distribution elsewhere (the testing procedure is described in detail in Chapter 10), we shall not extend the discussion of this test here.

Note that though the *test* is indeed nonparametric, the binomial distribution is parametric; its parameters are n and p. In order to calculate a given probability, we must know these parameters. In testing hypotheses, however, the sample size, n, is given, and p is provided by the null hypothesis; hence it is considered a nonparametric test.

18.3 The Chi-Square Test

Under certain conditions having to do with the minimum number of observations in each cell, the **chi-square test** can also be considered a nonparametric test, since nothing is assumed about the distribution of the observations. Goodness-of-fit tests can be carried out by use of the χ^2 distribution. Again, since we cover this distribution in detail in Chapter 11, we shall not repeat it here. We would like to emphasize, however, that some applications of the chi-square test are parametric (e.g., testing the null hypothesis that $\sigma^2 = 100$), while others are nonparametric (e.g., testing for goodness of fit). In the first case we have to assume normality, while in the second case such an assumption is not necessary.

18.4 Test for Randomness: Runs Test

The one-sample binomial test and the chi-square test use the *frequency* of events in the sample. They ignore the order in which the events are generated. But in certain situations the order of appearance of the events is no less important than the frequency.

WHEN IS THE RUNS TEST APPROPRIATE?

Suppose we observe ten successive daily price changes of a certain stock and get five plus readings (price up) and five minus readings (price unchanged or down). The binomial test based on this frequency information does not enable us to reject the hypothesis that the number of pluses is equal to the number of minuses; that is, the hypothesis H_0: $p = 0.5$ is not rejected, where p stands for the probability of "price up" on a given day.

Indeed, take $\alpha = 0.1$. In this case the 10 percent two-tailed critical region of the binomial distribution under the null hypothesis for a sample of ten observations is roughly $n_1 \leq 2$ and $n_1 \geq 8$, where n_1 is the number of plus readings.[1] In other words, there is a 0.90 probability of getting between 2 and 8 plus readings under the null hypothesis H_0: $p = 0.5$. Since the sample gave $n_1 = 5$, the null hypothesis cannot be rejected at this significance level.

The chi-square test (see Chapter 11) gives similar results. Consider the data in Table 18.1. Here the chi-square statistic is zero:

$$\chi^2 = \sum_{i=1}^{2} \frac{(o_i - e_i)^2}{e_i} = 0$$

Obviously, the null hypothesis H_0: $p = 0.5$ cannot be rejected at the $\alpha = 0.1$ level.

TABLE 18.1
Calculation of the Chi-Square Test Statistic for Testing the Hypothesis H_0: $p = 0.5$

Category	Number of pluses	Number of minuses	Total
Observed frequency o_i	5	5	10
Expected frequency e_i	5	5	10
$o_i - e_i$	0	0	0
$\Sigma(o_i - e_i)^2/e_i$	0	0	$\chi^2 = 0$

These two tests, based on the frequency of outcomes, apparently support the hypothesis that there is an equal probability of getting plus or minus. And yet there are several ways we might have obtained 5 pluses and 5 minuses from ten observations. For example, we could have registered samples with the following sequences of pluses and minuses over the 10 days of observations of stock price changes:

1. + + + + + − − − − −
2. − − − − − + + + + +
3. + − + − + − + − + −
4. + + − − + + − − + −
5. + + − + − − + − + −

Each of these samples contains 5 pluses and 5 minuses. The frequencies are the same, but the order of appearance of the two events changes. It is hard to imagine that the first two series, in which five successive outcomes of one kind are followed by five successive outcomes of the other kind, do not differ fundamentally from a random sequence of pluses and minuses. Similarly, the third series, with its systematic alternating order of plus and minus, does not look like a random sample generated by a process with equal probabilities of obtaining plus and minus. Most observers will intuitively classify the fifth sample as random whereas opinions will probably vary with regard to the fourth sample.

In short, the order of the observations in the sample reveals information about the process generating the observations: it indicates whether the

[1] The critical values are taken from the binomial probabilities table in the Appendix (Table A.1).

process is random or not. The example shows that tests that use the frequency of events while ignoring their order may lose a lot of important information. Thus, when the order of appearance (and not only the proportion of pluses and minuses) is important, the binomial test and the chi-square test are not appropriate.

IDENTIFYING THE RUNS PATTERN

The **runs test** is designed to use the information contained in the *order* of events. A **run** is defined as an *unbroken succession of outcomes of the same kind* in a sample consisting of *outcomes in two categories* (a dichotomous or dichotomizable variable).

Thus the last of our series of stock price ups and downs (sample 5 above) begins with a run of two pluses, followed by a run of one minus, then a run of one plus, a run of two minuses, and so on. The number of events of the same kind in a run is the *length* of the run. Identifying the runs in the above series, we write the corresponding run lengths:

$$+ \ + \ - \ + \ - \ - \ + \ - \ + \ -$$
$$2 \quad 1 \ \ 1 \quad 2 \ \ 1 \ \ 1 \ \ 1 \ \ 1$$

There is a total of 8 runs in this sample: 2 runs of length 2 and 6 runs of length 1.

The first series of daily stock price changes has 2 runs, each of a length of 5 observations:

$$+ \ + \ + \ + \ + \ - \ - \ - \ - \ -$$
$$5 \qquad \qquad \quad 5$$

The third series has 10 runs, each of length 1:

$$+ \ - \ + \ - \ + \ - \ + \ - \ + \ -$$
$$1 \ \ 1 \ \ 1 \ \ 1 \ \ 1 \ \ 1 \ \ 1 \ \ 1 \ \ 1 \ \ 1$$

WHAT THE RUNS PATTERN INDICATES

If the number of runs in a given sample is too small, as in the first two series, a certain grouping of outcomes is indicated: outcomes of the same kind tend to follow one another and the sample is not random. If such bunching actually occurred in stock price changes, it would tend to contradict the hypothesis that plus and minus appear at random, and investors could take advantage of this price change pattern.

At the other extreme, if the number of runs is too large, as in the third series (10 runs in a sample of 5 pluses and 5 minuses), a certain type of systematic dependence among the events is indicated: each event is always followed by an event of the opposite type, which also can be used by quick investors in the stock market to their advantage.

If the sample is random, then there is neither a tendency for bunching of identical outcomes nor the opposite type of dependence, in which outcomes of one category are generally followed by outcomes of the other. This, in effect, is the null hypothesis of the runs test, which tests for sample randomness against the alternative of a nonrandom sample.

Note that the symbol we attach to the various outcome categories is not important. In our example, we are dealing with a variable naturally classified into two categories: the increase of the price of a stock and the decrease or lack of change of the price. We may label the category "price up" by "plus" or assign a numerical label, 1. The category "price unchanged or down" may be labeled "minus" or, numerically, 0. We could have reversed the labels, representing plus as 0 and minus as 1, or assigned different numerical labels altogether, such as plus = 15, minus = 107. The price behavior in this setting is adequately described as long as we use two distinct numerical labels for "up" and "unchanged or down."

APPLYING THE RUNS TEST

To apply the runs test, we have to determine the number of runs, r; the number of outcomes in the first category, n_1; and the number of outcomes in the second category, n_2 ($N = n_1 + n_2$ is the sample size). The runs test thus uses both order information (as represented by r) and frequency information (as represented by n_1 and n_2).

The critical values of r for the runs test are usually listed in two tables, one pair for each value of the significance level, α. The two tables correspond to the two tails of the critical region. Appendix Tables A.6a and A.6b at the end of the book present, respectively, the lower and the upper tails of the runs distribution for $\alpha = 0.05$ under the null hypothesis of randomness for samples with various combinations on n_1 and n_2. Table A.6a defines, in effect, the number of runs that is "too small" for randomness: the values of r in Table A.6a are so small that the probability of their occurrence under the null hypothesis is $p = \frac{\alpha}{2} = 0.025$ or less. Table A.6b defines the number of runs that is "too large" for randomness: the values of r in Table A.6b are so large that the probability of their occurrence under the null hypothesis of randomness is $p = \frac{\alpha}{2} = 0.025$ or less.

In a two-tailed test, if the number of runs, r, corresponding to the observed n_1 and n_2 is less than or equal to the critical value in Table A.6a or greater than or equal to the critical value in Table A.6b, we reject the null hypothesis of randomness at the $\alpha = 0.05$ level of significance. If the direction of deviation from randomness is suspected from the start ("too many" or "too few" runs), a one-tailed test is appropriate and only one of the tables, A.6a or A.6b, should be used. The corresponding one-tailed significance level in this case, of course, is $\alpha = 0.025$.

Note that Tables A.6a and A.6b extend only up to $n_1 = 20$ and $n_2 = 20$. This is the *small-sample case*. For *large-samples*, when either n_1 or n_2 is greater than 20, the standard normal approximation applies (see the application below).

Let us now apply the runs test to the five series of daily stock price changes described at the beginning of this section. In each case we shall employ a two-tailed test, and the null hypothesis is that daily price changes over time are random.

Series 1: $+ + + + + - - - - -$

In this series we have 5 pluses and 5 minuses, so $n_1 = n_2 = 5$. The pluses and minuses form 2 runs, so $r = 2$.

We enter Table A.6*a* at the row with $n_1 = 5$ and move to the column with $n_2 = 5$. The critical value in the table is $r_a = 2$. Since $r = 2 \leq r_a$, we reject the null hypothesis at the $\alpha = 0.05$ level. The sample contains too few runs to be random. Thus, the stock price changes are dependent over time, and one can use this information for investment decision making.

Series 2: $- \; - \; - \; - \; - \; + \; + \; + \; + \; +$

Here $n_1 = 5, n_2 = 5, r = 2$, just as in Series 1. We reject the null hypothesis at the $\alpha = 0.05$ level.

Series 3: $+ \; - \; + \; - \; + \; - \; + \; - \; + \; -$

Here $n_1 = 5, n_2 = 5, r = 10$. Enter Table A.6*b* at the row with $n_1 = 5$ and move to the column with $n_2 = 5$. The critical value in the table is $r_b = 10$. Since $r = 10 \geq r_b$, we reject the null hypothesis at the $\alpha = 0.05$ level. The sample contains too many runs to be random.

Series 4: $+ \; + \; - \; - \; + \; + \; - \; - \; + \; -$

Here $n_1 = 5, n_2 = 5, r = 6$. From Table A.6*a* we have $r_a = 2$; from Table A.6*b* $r_b = 10$. Thus $r_a < r < r_b$, and the null hypothesis cannot be rejected at the $\alpha = 0.05$ level.

Series 5: $+ \; + \; - \; + \; - \; - \; + \; - \; + \; -$

Here $n_1 = 5, n_2 = 5, r = 8$. Again $r_a < r < r_b$, and the null hypothesis cannot be rejected at the $\alpha = 0.05$ level. The last two series thus appear to be consistent with the hypothesis that stock price changes are random.

In the application that follows we test again for randomness of stock price changes, this time assuming a large series of changes and employing the normal approximation. We illustrate this test by using an approximation.

18.5 APPLICATION:
THE RUNS TEST AND THE RANDOM WALK THEORY

An interesting subject in economics and finance is the analysis of time series of prices. The **random walk theory** asserts that the analysis of past data—in particular past market-price data—cannot be used to forecast future price changes. Thus, those who support the random walk hypothesis imply that the many investors and Wall Street financial analysts who analyze past price data for the purpose of forecasting future prices are wasting their time. In other words, price changes over time are independent of previous price changes.

The random walk theory is not limited to securities prices; it is applied as well to prices of other economic values, such as real estate and commodity prices. Most empirical studies that test the validity of the random walk theory, however, use either commodity prices or stock prices.

In order to test whether past information can be used to forecast future prices and to make decisions, it is natural to apply the runs test for independence over time. For small samples, we simply apply the above procedure and use Tables A.6*a* and A.6*b*. In most cases, however, more than 20 daily price changes are available, and we use the normal approximation in order to test the random walk theory. By the random walk model, price increases and decreases have no tendency to bunch together, nor should there be any noticeable tendency toward "reverses." Any given sequence of daily price changes should thus form a random series of plus and minus values.

A year-long series of prices for stock *A* during 1979 yielded 242 daily price changes, of which 109 were plus ($n_1 = 109$) and 133 were minus ($n_2 = 133$). The sample contained $r = 80$ runs.

Since both n_1 and n_2 are much greater than 20, Tables A.6*a* and A.6*b* are inapplicable. We are dealing with a large sample and the number of runs is approximately normally distributed, with mean

$$\mu_r = \frac{2n_1 n_2}{n_1 + n_2} + 1 \tag{18.1}$$

and standard deviation

$$\sigma_r = \sqrt{\frac{2n_1 n_2 (2n_1 n_2 - n_1 - n_2)}{(n_1 + n_2)^2 (n_1 + n_2 - 1)}} \tag{18.2}$$

Here μ_r and σ_r are the expected number of runs and the standard deviation of the number of runs under the null hypothesis of a random order of appearance (that is, when the random walk theory holds).

The null hypothesis in the case of a large sample may therefore be tested by use of the standard normal variable, Z_r:

$$Z_r = \frac{r - \mu_r}{\sigma_r} \tag{18.3}$$

In our example, for $n_1 = 109$ and $n_2 = 133$, we get

$$\mu_r = \frac{2 \cdot 109 \cdot 133}{242} + 1 = 120.8$$

$$\sigma_r = \sqrt{\frac{2 \cdot 109 \cdot 133 \cdot (2 \cdot 109 \cdot 133 - 109 - 133)}{(109 + 133)^2 \cdot (109 + 133 - 1)}} = \sqrt{59.06} = 7.7$$

Hence

$$Z_r = \frac{r - \mu_r}{\sigma_r} = \frac{80 - 120.8}{7.7} = -5.30$$

Since Z_r is negative, our sample contains fewer runs than a random sample. To determine whether the difference is significant at the $\alpha = 0.05$ level, we use the table of standard normal distribution (inside back cover) and define the two-tailed 5 percent critical region:

$$Z \leq -1.96$$

and

$$Z \geq 1.96$$

Since $Z_r = -5.30 < -1.96$, we reject the null hypothesis at the $\alpha = 0.05$ level. Our sample contains too few runs to be random and the daily price changes show a significant tendency to bunch: an increase in price is likely to lead to further price increases, and a decrease in price will be followed by further price decreases on subsequent days. Note that our conclusion from the empirical test is, as usual, subject to the possibility that we have committed a Type I error. While we have expected approximately 121 runs ($120.8 \cong 121$), we observed only 80, meaning that *bunching has actually occurred.* The deviation of the observed number of runs from those expected to occur, however, could be attributable only to chance variation.

As there is no parametric test for the randomness of a sequence of events, the runs test is unique and therefore particularly important.

18.6 The Sign Test: Two Related Samples

The **sign test** is probably the best known of the nonparametric tests. As its name indicates, the test uses plus and minus signs; pairs of matched observations from two related samples are represented by plus or minus signs, depending on which of the two observations is larger. The test is clearly nonparametric in that we assume nothing about the population from which the two related samples are drawn.

EXAMPLE 18.1

The production department of the Eillon Company has decided to compare the productivity of workers on the day shift with the productivity of workers on the night shift. The productivity (measured in units of output) of one group of workers was recorded for 15 consecutive day shifts and then for 15 consecutive night shifts. Table 18.2 summarizes the observed productivity data.

The sign test is applied to the matched observations in order to determine whether the number of plus signs is significantly greater than the number of minus signs (that is, whether day-shift productivity is significantly greater than night-shift productivity). The one observation with a tie, producing a reading of 0 in the last column, is omitted, and we take $n = 14$ in what follows.

The null hypothesis of the sign test is that the probability of getting plus is equal to the probability of getting minus, that is, the two signs are as if generated by the flipping of a balanced coin; hence the binomial tables can be used to calculate the probability of getting a given number of plus values in a sample of n observations.

Reference to the table of cumulative binomial probabilities (see Appendix Table A.1) shows that there is a probability of 0.212 (almost

22 percent) of getting 9 or more plus signs out of 14 trials under the null hypothesis of equal probability of plus and minus signs ($p = \frac{1}{2}$).

The results in favor of higher productivity on the day shift are thus not significant, since to establish superiority of the productivity of the day shift at the 0.10 level of significance, at least ten plus signs should have been observed in the last column of Table 18.2 (see Appendix Table A.1).

As with the binomial test, for large n one can use the normal approximation to calculate the probabilities of various events.

TABLE 18.2
Determining the Signs of the Matched Observations

Observations	Day-shift productivity X_D	Night-shift productivity X_N	Direction of difference	Sign
1	84	78	$X_D > X_N$	+
2	85	82	$X_D > X_N$	+
3	69	74	$X_D < X_N$	−
4	75	68	$X_D > X_N$	+
5	87	79	$X_D > X_N$	+
6	73	84	$X_D < X_N$	−
7	92	90	$X_D > X_N$	+
8	70	59	$X_D > X_N$	+
9	74	71	$X_D > X_N$	+
10	79	85	$X_D < X_N$	−
11	70	66	$X_D > X_N$	+
12	65	69	$X_D < X_N$	−
13	79	83	$X_D < X_N$	−
14	89	89	$X_D = X_N$	0
15	80	75	$X_D > X_N$	+

18.7 The Mann-Whitney Test: Two Independent Samples

Testing whether or not two independent samples come from the same population is usually carried out by the t test. When it is unreasonable to make the assumptions underlying the t test, one can use the **Mann-Whitney test,** which is one of the most powerful nonparametric tests relative to its parametric counterpart, the t test.

Suppose that we have two independent samples of the sizes n_A and n_B. Thus the total number of observations is $N = n_A + n_B$. We arrange all the N observations in ascending order and rank them. To identify the origin of the ranks, we attach the symbol A to each ranked observation from the first sample and B to each rank representing an observation from the second sample. Suppose that $n_A = 4$, $n_B = 6$; consider the following three series of ten observations each, in which they are ranked from lowest to highest:

1. *AAAABBBBBB*
2. *BBBBBBAAAA*
3. *AABABBBBAB*

In case 1, it is obvious that distribution B is located to the right of distribution A, that is, it has a higher mean. In case 2 it is obvious that the opposite situation prevails. Case 3 is more complicated, and one has to use a statistical test in order to determine whether the distribution of B is located significantly to the right of the distribution of A. The Mann-Whitney test is designed to analyze such a case. We will illustrate how to employ the test by means of an example.

EXAMPLE 18.2

A manager is considering the purchase of one of two competing computer models, which we label A and B. Apart from the other obvious factors (cost and so on), the manager is concerned about the reliability of the computers, measured by the amount of downtime as a percentage of computer capacity (or total use time). The percentage of monthly downtime out of total usage time figures supplied by several installations in which computers A and B operate is as follows:

Computer A	Computer B
2.54	2.81
2.84	2.14
1.26	2.56
2.12	2.94
	2.64

As you can see, the number of observations is $n_A = 4$ for computer A and $n_B = 5$ for computer B. The test does not require the constraint $n_A = n_B$, nor does it assume any other matching between the observations in the two samples.

Let us pool the nine observations, arrange them in ascending order, and rank them:

Computer	A	A	B	A	B	B	B	A	B
Ordered observations	1.26	2.12	2.14	2.54	2.56	2.64	2.81	2.84	2.94
Rank	1	2	3	4	5	6	7	8	9

The sum of the ranks of distribution A, which we denote by R_A, is $R_A = 1 + 2 + 4 + 8 = 15$, and similarly the sum of the ranks of distribution B is $R_B = 3 + 5 + 6 + 7 + 9 = 30$.

The statistics T_A and T_B are as follows:

$$T_A = R_A - \frac{n_A(n_A + 1)}{2} = 15 - \frac{4(4 + 1)}{2} = 5 \tag{18.4}$$

$$T_B = R_B - \frac{n_B(n_B + 1)}{2} = 30 - \frac{5(5 + 1)}{2} = 15 \tag{18.5}$$

The two statistics T_A and T_B are related by a simple equation: $T_A + T_B = n_A n_B$, so that if one value is known (T_A, for example), the other can be calculated from this relationship: $T_B = n_A n_B - T_A$. In our case, $n_A = 4$, $n_B = 5$, $T_A = 5$, so that $T_B = 4 \cdot 5 - 5 = 15$, as we have indeed obtained;

similarly, if T_B is known, we get $T_A = 4 \cdot 5 - 15 = 5$. Because of this relationship between T_A and T_B, the Mann-Whitney test tables are calculated and published for the smaller of the two statistics (T_A, in our example).

Suppose that we want to test the following null hypothesis at a significance level of $\alpha = 0.10$:

H_0: Both computers have equal downtime characteristics.
H_1: The two computers differ in their downtime characteristics.

This is clearly a two-tailed test. We enter Appendix Table A.7 under $n_A = 4$, $n_B = 5$ and for the appropriate value of p, which for the two-tailed test is simply half the significance level, $p = \alpha/2 = 0.05$. The corresponding critical value in the table is 3. Since T_A (the smaller of T_A and T_B) is greater than 3—that is, $T_A = 5 > 3$—we cannot reject the null hypothesis that the two computers have equal downtime characteristics.

ONE-TAILED TEST

Suppose that initially we believe computer A in Example 18.2 to be more reliable than computer B, so we test the following hypotheses:

H_0: Computer A is, at most, as reliable as computer B.
H_1: Computer A is more reliable than computer B.

Note that if A is indeed more reliable (less downtime), we expect to have low R_A and hence low T_A value. Obviously, we have a one-tailed test with $\alpha = 0.10$, as before. We enter Table A.7 for $n_A = 4$, $n_B = 5$, and $p = \alpha = 0.10$ (one-tailed test). The critical value is seen to be 5. Since $T_A = 5$, we cannot reject the one-tailed null hypothesis. Computer A is not confirmed to be more reliable at the 10 percent level of significance.

LARGE SAMPLES AND TIES

Let us continue with our two computers example and conduct a more extensive user survey, sending out questionnaires to a large sample of installations using one of the two computers. Thirty-five questionnaires are completed and returned. An average score is computed for each questionnaire, representing the user's overall evaluation of his computer. The higher the score, the better the computer.

We rank the scores in ascending order from 1 to 35; the results are summarized in Table 18.3. There are several ties in our observations: the score 67 appears twice (in both cases for computer B) and the scores 56 and 106 appear three times each (once for computer A and twice for computer B). Ties are usually assigned the same rank, a rank equal to the average of the consecutive ranks that would have been normally assigned to the tied observations in ascending order. Thus, the three observations producing score 56 are located in positions 7, 8, and 9 in our ranking. All three observations are assigned the same average rank, $\dfrac{7+8+9}{3} = \dfrac{24}{3} = 8$. Similarly the two observations with score 67 occupy positions 12 and 13; they are

TABLE 18.3
Scores of Two Competing Computers Calculated from User Evaluation Survey

Computer A		Computer B	
Score	Rank	Score	Rank
148	30.0	142	29.0
73	14.0	126	26.0
56	8.0	21	2.0
63	10.0	177	34.0
90	18.0	118	24.0
42	6.0	169	33.0
106	20.0	79	16.0
125	25.0	160	31.0
129	28.0	106	20.0
161	32.0	56	8.0
114	23.0	56	8.0
127	27.0	76	15.0
27	4.0	113	22.0
	$R_A = 245.0$	83	17.0
		64	11.0
		36	5.0
		24	3.0
		106	20.0
		67	12.5
		186	35.0
		67	12.5
		10	1.0
			$R_B = 385.0$

assigned the average rank $\dfrac{12 + 13}{2} = \dfrac{25}{2} = 12.5$. (Note that this is a general procedure for the ranking of ties in nonparametric tests.) We test the following hypotheses:

H_0:   Computer A is no better than computer B.
H_1:   Computer A is better than computer B.

A one-tailed test is obviously called for.

In our example we have $n_A = 13$, $n_B = 22$, $R_A = 245$, and $R_B = 385$. Since $n_B > 20$, we can use the normal approximation for the Mann-Whitney statistic. The expected value and the variance of the Mann-Whitney T where T is either T_A or T_B are given by the following:

$$\mu_T = \frac{n_A n_B}{2} \tag{18.6}$$

$$\sigma_T^2 = \frac{n_A n_B (n_A + n_B + 1)}{12} \tag{18.7}$$

In our case

$$\mu_T = \frac{13 \cdot 22}{2} = 143$$

$$\sigma_T^2 = \frac{13 \cdot 22 \cdot (13 + 22 + 1)}{12} = 858$$

and hence

$$\sigma_T = \sqrt{858} = 29.3$$

Now let us calculate T_A and T_B:

$$T_A = 245 - \frac{13 \cdot 14}{2} = 245 - 91 = 154$$

$$T_B = 385 - \frac{22 \cdot 23}{2} = 385 - 253 = 132$$

Note that if computer A is better than computer B, as H_1 claims, then we expect to have a large value of T_A: recall that

$$T_A = R_A - \frac{n_A(n_A + 1)}{2}$$

and if A is the better computer, it will accumulate higher scores and R_A will be large. This implies, for a given sample size n_A, that T_A is also large. Thus, we can hope to reject the null hypothesis in this one-tailed test for large values of T_A only.

To apply the normal approximation, calculate the standard normal value

$$Z_T = \frac{T_A - \mu_T}{\sigma_T} = \frac{154 - 143}{29.3} = 0.375$$

Since $Z_{0.90} = 1.28$ and $Z_T = 0.375$ falls in the acceptance region, we cannot reject the null hypothesis.

Suppose now that A on the whole got much higher scores, and T_A is equal to 203. In such a case, we have

$$Z_T = \frac{203 - 143}{29.3} = \frac{60}{29.3} = 2.05$$

Given these high scores for computer A, $Z_T = 2.05 > 1.28$ and we reject the null hypothesis at a significance level of $\alpha = 0.10$. The conclusion under these assumptions is that computer A is indeed better than computer B.

The Mann-Whitney test is very powerful. Its power is about 95.5 percent of the power of the t test for large N, and approximately 95 percent even for samples of modest size. Thus, one can use the Mann-Whitney instead of the t test with almost no loss of power. The benefit, of course, is that we do not have to adopt the various restrictive assumptions underlying the t test.

18.8 The Kruskal-Wallis Test: Analysis of Variance

The parametric t test is used to test the means of two populations for equality. The means of k populations, where $k > 2$, are tested for equality by the analysis of variance (distribution F). As we have seen, the Mann-Whitney test successfully replaces the t test for two independent samples when no strong evidence is available in favor of the assumptions underlying the t test. By the same argument, when there is no reason to assume normality and homogeneity of variances of the k populations under consideration, the nonparametric **Kruskal-Wallis test** may be used to replace the one-way analysis-of-variance F test.

The Kruskal-Wallis test is in fact a one-way analysis of variance in which ranks are used. It is designed to detect differences among k ($k > 2$) populations, of which k independent samples are drawn and ranked according to a certain property or attribute, and it tests the null hypothesis that all the samples are drawn from the same population against the hypothesis that the k populations vary significantly in the relevant attribute. As we are dealing with a *one-way* analysis of variance and the samples are assumed to be independent, no matching of the sample data is implied and different sample sizes are allowed.

EXAMPLE 18.3

A retail chain has decided to conduct a comparative credit-rating survey of customers at three stores in a metropolitan area: a downtown store, a suburban store in a predominantly young white-collar community, and a store in a shopping center frequented by a mixed cross-section of customers from several nearby communities. The percentage of accounts overdue by more than 30 days in each store is taken as the relevant measure and data have been collected for six months back:

Month	1	2	3	4	5	6
Store 1	5.5	6.2	5.8	4.9	6.0	5.9
Store 2	5.2	4.9	4.5	5.7	6.2	5.0
Store 3	6.8	6.7	5.9	6.1	6.3	5.9

Are there significant differences in the credit ratings of customers in the three stores? Since there is no evidence suggesting homogeneity of variance of the overdue accounts in the three stores and there is no reason to assume normal distribution of these accounts, we employ a nonparametric test.

To apply the Kruskal-Wallis test, we first rank all the 18 available observations as if they came from a single population, then calculate the rank sums, R_i, for each sample. These ranks are given in Table 18.4. (Note that we treat ties as in the Mann-Whitney test.) The rank sums R_i of the three samples are

$$R_1 = 2.5 + 6 + 8 + 10 + 12 + 14.5 = 53$$

$$R_2 = 1 + 2.5 + 4 + 5 + 7 + 14.5 = 34$$

$$R_3 = 10 + 10 + 13 + 16 + 17 + 18 = 84$$

Intuitively we know that the smaller the difference among R_1, R_2, and R_3—that is, the more similar the three populations—the smaller the chance that we will reject the null hypothesis of equal distributions.

For the Kruskal-Wallis test we use the **Kruskal-Wallis statistic**, denoted by H, which is given by the following formula:

$$H = \frac{12}{N(N+1)} \left(\sum_{i=1}^{k} \frac{R_i^2}{n_i} \right) - 3(N+1) \tag{18.8}$$

TABLE 18.4
Credit Rating and Ranks of 18 Observations
in Three Stores

Credit-rating measure	Rank	Store
4.5	1	2
4.9	2.5	2
4.9	2.5	1
5.0	4	2
5.2	5	2
5.5	6	1
5.7	7	2
5.8	8	1
5.9	10	1
5.9	10	3
5.9	10	3
6.0	12	1
6.1	13	3
6.2	14.5	2
6.2	14.5	1
6.3	16	3
6.7	17	3
6.8	18	3

The null hypothesis is rejected for large values of H, for given n_i's and given $N = \Sigma n_i$, the larger the differences among the R_i's, the larger is $\sum R_i^2$, and the larger $\sum \dfrac{R_i^2}{n_i}$. Hence the larger is the value of H.

In our case,

$$k = 3 \text{ (three independent samples)}$$

$$n_1 = n_2 = n_3 = 6 \text{ (six observations in each sample)}$$

$$N = n_1 + n_2 + n_3 = 18 \text{ (a total of 18 observations)}$$

Inserting these values and the three rank sums, we obtain

$$H = \frac{12}{18 \cdot 19} \left(\frac{53^2}{6} + \frac{34^2}{6} + \frac{84^2}{6} \right) - 3 \cdot 19$$

$$H = 0.0351 \cdot (468.1667 + 192.6667 + 1,176.0000) - 57.00$$

$$H = 0.0351 \cdot 1,836.8334 - 57.00$$

$$H = 64.473 - 57.000 = 7.473$$

The Kruskal-Wallis statistic, calculated for $k = 3$ samples, is approximately distributed χ^2 with $k - 1 = 2$ degrees of freedom. Using the chi-square table (Appendix Table A.4) we see that our result is significant at the $\alpha = 0.05$ significance level and the null hypothesis is rejected, since the sample statistic H is larger than the critical value $\chi^{2(2)}_{0.05} = 5.991$. Thus there appear to be significant differences in the credit standings of customers in the three stores. (The chi-square distribution provides an adequate approximation in our case because all three samples contain more than five observations.)

Now suppose that, because of technical difficulties, it is possible to cover only five months in Stores 1 and 2 and three months in Store 3. Thus, $k = 3$, as before, but the other values change:

$$n_1 = 5$$
$$n_2 = 5$$
$$n_3 = 3$$
$$N = n_1 + n_2 + n_3 = 13$$

A new ranking is now needed, based on the samples below:

Month	1	2	3	4	5
Store 1	6.2	5.8	4.9	6.0	5.9
Store 2	5.2	4.5	5.7	6.2	5.0
Store 3	6.8	6.7	6.1		

As in the parametric one-way analysis of variance, the nonparametric test may be carried out with unequal numbers of observations in the samples. The new ranking is given in Table 18.5. The new rank sums are given by

$$R_1 = 2 + 6 + 7 + 8 + 10.5 = 33.5$$
$$R_2 = 1 + 3 + 4 + 5 + 10.5 = 23.5$$
$$R_3 = 9 + 12 + 13 = 34$$

TABLE 18.5
Credit Rating and Ranking of 13 Observations in Three Stores

Credit-rating measure	Rank	Store
4.5	1	2
4.9	2	1
5.0	3	2
5.2	4	2
5.7	5	2
5.8	6	1
5.9	7	1
6.0	8	1
6.1	9	3
6.2	10.5	2
6.2	10.5	1
6.7	12	3
6.8	13	3

We can now calculate the Kruskal-Wallis statistic for the second set of data:

$$H = \frac{12}{13 \cdot 14} \left(\frac{33.5^2}{5} + \frac{23.5^2}{5} + \frac{34^2}{3} \right) - 3 \cdot 14$$

$$= 0.0659 \cdot (224.45 + 110.45 + 385.33) - 42.00$$

$$= 0.0659 \cdot 720.23 - 42.00 = 47.46 - 42.00 = 5.46$$

This result is significant at the $\alpha = 0.10$ level if we use the chi-square distribution with $k - 1 = 2$ degrees of freedom. The chi-square distribution is not a good approximation for H in the small-sample case, however, especially when none of the samples is larger than 5. In the small-sample case, we use a special table for the critical region of the Kruskal-Wallis statistic. Entering Appendix Table A.8 at $n_1 = 5$, $n_2 = 5$, $n_3 = 3$, we see that $H \geq 5.46$ is indeed significant at the $\alpha = 0.10$ level.[2] Note that the critical value for $\alpha = 0.10$ is given by 4.5451, and since the calculated value of H (5.46) is greater, we reject the null hypothesis.

[2] The order of the sample sizes makes no difference to the significance of H. Thus, the same critical value applies if, for example, $n_1 = 3$, $n_2 = 5$, and $n_3 = 5$.

To sum up the Kruskal-Wallis test, recall that when one of the k samples contains more than 5 observations ($n_1 > 5$ for some i), one can use the chi-square distribution to test the significance of H: if H is greater than $\chi^2_{\alpha(k-1)}$, where α is a significance level chosen in advance, we reject the null hypothesis, which asserts that all the k samples are drawn from the same population.

If all the n_i's are smaller than 5, we should consult special tables giving the exact distribution of the Kruskal-Wallis statistic in order to reach a decision. In the small-sample case, Table A.8 in the Appendix gives the critical values of H for all relevant combinations of n_1, n_2, n_3, and for various significance levels. After selecting the significance level α, we enter the table for appropriate combinations of n_1, n_2, and n_3 and obtain the critical value of H. If the observed sample statistic H is greater than the critical value that appears in the table, we reject the null hypothesis. Finally, note that tables for the exact distribution of the Kruskal-Wallis statistic are not readily available for more than three samples ($k \geq 4$), but one may use the approximate chi-square distribution with satisfactory results.

The Kruskal-Wallis test, like the Mann-Whitney test, is very powerful. For large samples its power is 95.5 percent of the power of the appropriate parametric test (the F test), and one does not have to make the restrictive assumptions required by the parametric F test.

18.9 Nonparametric Measures of Association

In Chapter 15 we discussed the correlation coefficient as a parametric measure of association. This correlation coefficient, however, is inapplicable to nominal or to ranked data. Moreover, to test the significance of the parametric correlation coefficient, one has to assume that the sample is drawn from a population characterized by the normal distribution, an assumption totally out of line where nominal or ranked data are concerned.

In this section we discuss nonparametric measures of association between two variables: the Spearman rank correlation (which we denote by r_s) and the Kendall rank correlation (which we denote by τ).

EXAMPLE 18.4

The buyer of women's fashions for a leading department store ranked the seven identically priced models of summer dresses she saw at a fashion preview according to anticipated sales in the forthcoming season. The model most likely to storm the market in the buyer's opinion was given the highest rank, 1, the candidate for the second-best-selling slot was given the rank 2, and so on. The buyer's ranking served as a basis for the store's purchasing policy for that season, and at the season's end the buyer compared her original ranking with the sales-volume performance of the various models stocked. The buyer's ranking and the actual sales are given in Table 18.6.

TABLE 18.6
Ranking of Previewed Models and Actual Sales for Entire Season

Rank of anticipated acceptance (x)	Actual sales volume (thousands of dollars) (y)
1	$196
2	167
3	152
4	161
5	164
6	121
7	132

The buyer is naturally interested in her record as reflected in her ability to rank correctly the sales of the various models before the season begins. To test the buyer's performance, we need to calculate the extent of association or correlation between the preview ranking (x) and the actual season's sales (y). The standard parametric correlation coefficient is inapplicable to this case (even if we can reasonably assume normality): one of the variables (x) is expressed in the form of ranks, and no numerical values are available for this variable. On the other hand, we can easily rank the observed sales figures, y, converting the numerical values y into ranks. The next step is naturally to calculate a *rank* correlation coefficient between the two rankings, x and y, in order to determine the degree of association between the buyer's prediction and the actual sales (both expressed in ranks).

We shall now calculate two standard rank correlation coefficients, using the data of Example 18.4.

SPEARMAN RANK CORRELATION COEFFICIENT

The **Spearman rank correlation coefficient** is given by the following equation:

$$r_s = 1 - \frac{6\Sigma d^2}{N(N^2 - 1)} \tag{18.9}$$

Here N is the number of pairs of ranks and d is the difference in the ranks of x and y.

Using this formula and the data of Table 18.7, we prepare the work sheet shown in that same table and obtain

$$r_s = 1 - \frac{6 \cdot 10}{7 \cdot 48} = 1 - 0.1786 = 0.8214$$

Note that the values of the Spearman correlation coefficient are bounded by -1 and $+1$. If all the paired observations have the same rank, then all the d values are equal to zero and hence $r_s = +1$. Similarly, if we have completely opposite rankings, it is possible to show that

$$\frac{6\Sigma d^2}{N(N^2 - 1)} = 2$$

and hence $r_s = 1 - 2 = -1$.

TABLE 18.7
Work Sheet for Calculation of Spearman Correlation Coefficient, r_s

Ranking of anticipated acceptance (x)	Ranking of observed sales (y)	$d = x - y$	d^2
1	1	0	0
2	2	0	0
3	5	−2	4
4	4	0	0
5	3	2	4
6	7	−1	1
7	6	1	1
			$\Sigma d^2 = 10$

No special assumptions are needed in order to test the significance of the sample correlation coefficient r_s. We have the following hypotheses:

H_0: The population correlation is zero.
H_1: The population correlation is greater than zero.

Obviously, this is a one-tailed test. Looking at Table A.9 in the Appendix, we find that for $N = 7$ and significance level of $\alpha = 0.05$, the critical value is 0.714. Since in our sample we obtained $r_s = 0.8214$, we reject the null hypothesis at a 5 percent significance level and conclude that the two variables are indeed positively associated. Note that in a case of suspected negative correlation in the sample, we would carry out a one-tailed test with the alternative hypothesis that r_s is smaller than zero. The decision rule for negative r_s remains as before, the only difference being that we ignore the sign of the sample r_s when entering Table A.9: if $|r_s|$ is greater than or equal to the critical value in Table A.9, we reject the null hypothesis.

Large Samples

If N is larger than 10, one can use the Student t distribution to test r_s for significance: the random variable

$$r_s \sqrt{\frac{N-2}{1-r_s^2}}$$

follows a t distribution with $N - 2$ degrees of freedom.

Spearman's rank correlation is very powerful, and it is free of the restrictions imposed by the parametric correlation. The power of the non-parametric measure of association r_s is 91 percent of the power of the parametric coefficient of correlation.

Let us now calculate the Kendall rank correlation coefficient by using the data given in Tables 18.6 and 18.7.

KENDALL RANK CORRELATION COEFFICIENT

The **Kendall rank correlation coefficient** is given by the following equation:

$$\tau = \frac{S}{\frac{1}{2}N(N-1)} \tag{18.10}$$

Here N is the number of pairs and the statistic S is calculated as follows:

1. Arrange one rank series in an increasing order (see first column in Table 18.7).
2. For each rank in the second column in Table 18.7, count the number of following ranks that are larger.
3. Subtract from this count the number of following ranks that are smaller.
4. Add up the differences obtained successively for each rank to get the statistic S.

In our case, the first rank in Column 2 of Table 18.7 is 1. It is followed by six ranks, all of which are larger. The first term in S is thus $(6 - 0)$. The second rank in Column 2 is 2; it is followed by five larger ranks and no smaller ranks; its contribution to S is thus $(5 - 0)$. The third rank in Column 2 is 5; it is followed by two larger ranks (7, 6) and two smaller ranks (4, 3); its contribution to S is thus $(2 - 2)$. Similarly, we find

$$S = (6 - 0) + (5 - 0) + (2 - 2) + (2 - 1) + (2 - 0) + (0 - 1) = 13$$

$$\tau = \frac{S}{\frac{1}{2}N(N-1)} = \frac{13}{\frac{1}{2} \cdot 7 \cdot 6} = 0.6190$$

A glance at Table A.10 in the Appendix shows that the probability of getting $S = 13$ and higher for $N = 7$, under the null hypothesis of no correlation, is 0.035. Thus we reject the null hypothesis at $\alpha = 0.05$ and conclude that there is a positive association between x and y.

Note that for given values of N and S, Table A.10 gives the *probability* of getting this or a larger value of S under the null hypothesis that $\tau = 0$. If we select a significance level of $\alpha = 0.05$ and get a probability of less than the chosen α from the table, the null hypothesis is rejected.

The table gives critical values for $N \leq 10$ only. For $N > 10$, we have a normal approximation where

$$\mu_\tau = 0 \qquad\qquad (18.11)$$

$$\sigma_\tau = \sqrt{\frac{2(2N + 5)}{9N(N - 1)}} \qquad\qquad (18.12)$$

and the statistic

$$Z_\tau = \frac{\tau - \mu_\tau}{\sigma_\tau} = \frac{\tau}{\sqrt{\dfrac{2(2N + 5)}{9N(N - 1)}}} \qquad\qquad (18.13)$$

is distributed according to the standard normal distribution.

Problems

18.1. Explain the difference between parametric and nonparametric methods.

18.2. Give an example in which an assumption with regard to the distribution of the random variable may cause an error in decision making. Illustrate your answer graphically.

18.3. Give an example of two assumptions commonly used in parametric tests.

18.4. In what respect are nonparametric tests inferior to parametric tests?

18.5. Give an example in which the binomial distribution is treated as a nonparametric distribution, and another example in which it is treated as a parametric distribution.

18.6. Give an example of the chi-square distribution in which it is treated as a nonparametric distribution and an example in which it is treated as a parametric distribution.

18.7. In his monthly report to the managing director, the production manager indicated that there had been a recent alarming increase in the number of rejects in Department C. He diagnosed the problem as being due to equipment obsolescence and recommended that a new machine be purchased. The cost accountant, summoned by the managing director for a second opinion, checked the previous two weeks' quality-control records and reported that the average number of rejects did not exceed the long-run daily average of 7.5 and she therefore saw no cause for alarm. The production manager countered that the average number of rejects did not mean much and that one should consider the bunching of rejects between successive maintenance sessions. To support his opinion he produced the quality-control records for the last ten working days between two successive maintenance sessions (see Table P18.7).

Propose a nonparametric test in order to check whether days with above-average reject counts do indeed bunch together abnormally. Apply the test to the production manager's data. What is your conclusion?

TABLE P18.7

Day after last maintenance	Number of rejects
1	1
2	1
3	1
4	2
5	5
6	12
7	12
8	12
9	14
10	15

18.8. Table P18.8 shows a composite monthly stock index for the period 1971–75 as a monthly average of Standard & Poor's stock price indexes (1941–43 = 100). Use a nonparametric method to test whether price increases and price decreases in the stock market, as reflected by this composite index, constitute a random series.

TABLE P18.8

Year	Jan.	Feb.	Mar.	Apr.	May	June	July	Aug.	Sept.	Oct.	Nov.	Dec.	Avg.
1975	72.56	80.10	83.78	84.72	90.10	92.40	92.49	85.71	84.67	88.57	90.07	88.70	86.16
1974	96.11	93.45	97.44	92.46	89.67	89.79	79.31	76.03	68.12	69.44	71.74	67.07	82.85
1973	118.4	114.2	112.4	110.3	107.2	104.8	105.8	103.8	105.6	109.8	102.0	94.78	107.4
1972	103.3	105.2	107.7	108.8	107.7	108.0	107.2	111.0	109.4	109.6	115.1	117.5	109.2
1971	93.49	97.11	99.60	103.0	101.6	99.72	99.00	97.24	99.40	97.29	92.78	99.17	98.29

Source: Standard & Poor's *Trade and Securities Statistics, Security Price Index Record*, 1976 edition, p. 121.

18.9. An electronic-instrument assembly plant receives weekly shipments of solid-state components from a semiconductor manufacturer. The components are packed for delivery in boxes of 1,000 units each and the standard quality-control procedure consists of selecting one box at random and inspecting all the units in it for possible defects. The following table provides a statistical summary of the past inspection records of all weekly shipments received:

Number of defects in inspected box (x)	Relative frequency of occurrence over entire period (f)
0	0.30
1	0.50
2	0.10
3	0.05
4	0.05
5 and over	0.00
	1.00

On the basis of the historical data, the staff statistician estimates that the average number of defects in weekly shipments has a Poisson distribution with $\lambda = 1.05$. The current policy is to accept the entire shipment if the average number of defects per box is less than 1.05; otherwise the entire shipment is rejected and returned to the component manufacturer. Since the accept/reject decision is based on a sample of one box from each shipment, an entire shipment may be rejected although the average number of defects per box is less than 1.05. If this happens, the shipment is erroneously rejected. The firm is prepared to agree to this error (Type I error) as long as its probability is not greater than 10 percent.

(a) Assuming that the distribution is indeed Poisson, as estimated by the statistician, what is the maximum number of defects, x, in the sampled box to justify acceptance of the shipment? For what values of x in the sample should the shipment be returned for Type I error of not larger than 10 percent? (Hint: On the basis of the historical data, $E(X) = \text{var}(X) = \lambda = 1.05$, which justifies the statistician's assumption of a Poisson distribution for the average number of defects. Verify this conclusion. Recall that the probability of getting precisely x defects in a Poisson distribution with parameter λ is given by $P(X = x) = \dfrac{e^{-\lambda}\lambda^x}{x!}$. You will need the probabilities $P(X \leq x)$ and $P(X > x)$ to solve the problem.)

(b) Suppose now that we are not willing to use the assumption of a Poisson distribution and accept the historical record in the table above as representing the "true" parameter-free distribution of the average number of defects per box in weekly shipments. What is the appropriate accept/reject decision in this case, assuming again a Type I error $\alpha = 10$ percent?

(*c*) What distribution causes more "acceptable" shipments to be rejected and returned? What is the percentage of returned shipments under the assumption of the Poisson distribution and what is the corresponding percentage under the parameter-free empirical distribution? (Hint: Refer to the table to estimate the percentage of shipments with the appropriate *x*.)

(*d*) Assume that the plant receives 100 shipments annually and the true distribution of the average number of defects per box is as given in the above table while the accept/reject decision is based on the Poisson distribution with λ = 1.05. The handling costs for each returned shipment are $1,000. What is the total *extra* cost incurred by the firm as a result of using the wrong distribution for its accept/reject decisions? (Hint: Calculate the *additional* cost above the cost that the plant is willing to bear in accepting a Type I error of 10 percent with the true distribution.)

(*e*) What did you learn about the "cost of making an assumption" regarding a distribution and the reason we need nonparametric methods?

18.10. Diversification of investment is a basic strategy intended to protect one's portfolio from extreme losses that could result from exposure to risk. By spreading the investment among various assets, one stabilizes one's return on investment—provided, of course, that one has not inadvertently invested in assets whose values increase or decrease all at the same time. For diversification to work as intended, while some assets depreciate in value, others must appreciate.

The latest fashion in portfolio management is international diversification—that is, spreading investment dollars among stocks of various countries, on the assumption that the stock markets in those countries do not rise or fall simultaneously. The last column in Table P18.10 shows the

TABLE P18.10

	High	Low	Close	Week's change
Australia	566.79	441.19	537.97	− 1.96
Austria	2,277.00	2,243.00	2,277.00	+ 12.00
Belgium	119.37	96.10	107.87	− 0.68
Canada	1,322.65	996.88	1,298.27	+ 14.42
France	127.35	78.10	121.64	+ 0.90
Italy	4,883.00	3,086.00	3,991.00	− 27.00
Japan	5,968.26	4,867.91	5,884.29	−146.07
Netherlands	100.20	85.90	89.80	− 1.00
Switzerland	323.70	279.00	N.A.	
United Kingdom	535.50	433.40	479.30	− 1.70
West Germany	863.80	759.40	819.60	− 1.60

Source: *Barron's*, December 25, 1978.

weekly changes in the stock market indexes of ten industrial countries during the week ending December 22, 1978. The changes are expressed in index points for each country, and the figures are thus not comparable numerically.

Use the sign test to test the hypothesis that the fluctuations among the ten stock indexes in the sample have equal probabilities. Can you justify the use of the sign test in this case? What other test could you suggest for this purpose?

18.11. Mutual funds provide the individual investor with a convenient medium for diversification. A mutual fund spreads its investment among numerous securities, so an individual who purchases a share in a mutual fund in fact invests in a highly diversified portfolio, something that he could hardly have achieved on his own with limited funds. Mutual funds tailor their diversification strategies to different investor groups. There are capital-growth funds, current-income funds, equity funds, bond funds, and so on. Table P18.11 lists the average annual total return for various categories of mutual funds for the period 1974–80. (The annual total return roughly represents the percentage of profit on invested dollars.)

(*a*) Use a nonparametric test to check whether there are any significant differences among the total returns that various fund strategies promise to the individual investor.

(*b*) What parametric test can be used for the same purpose? Why use a nonparametric test?

TABLE P18.11

	Average total returns (percent)						
Fund strategy	*1974*	*1975*	*1976*	*1977*	*1978*	*1979*	*1980*
Maximum capital gain	−27.6	39.3	29.9	8.2	15.7	40.0	41.7
Long-term growth	−26.7	32.7	23.4	0.4	12.5	28.9	35.1
Growth and current income	−21.4	33.7	25.5	−2.6	8.6	23.2	28.9
Balanced	−16.7	25.9	23.7	−1.0	4.2	13.6	19.6
Income funds:							
Common stock policy	−12.4	35.8	31.2	1.5	4.9	14.8	17.9

Source: Reprinted by permission from the *Wiesenberger Investment Companies Service*, 1981 edition, copyright © 1981, Warren, Gorham & Lamont, Inc., 210 South St., Boston, Mass. All Rights Reserved.

18.12. Three researchers used questionnaires to investigate some job-satisfaction variables of male and female salespersons in the pharmaceutical industry. Historically selling pharmaceuticals is a male-dominated occupation. Since earlier research had shown that women in traditionally male fields were less satisfied with their jobs than men who held the same types of jobs, the researchers tested the hypothesis

H_1:   female pharmaceutical salespersons will be less satisfied with their jobs than males

as an alternative to the null hypothesis

H_0:   female and male pharmaceutical salespersons will display the same job-satisfaction characteristics

TABLE P18.12

Variable	Sex	Distribution [a]			Chi-square [b]
		Low	*Med.*	*High*	
Satisfaction					
Pay	M	29%	38%	33%	0.797
	F	28	31	41	NS
Promotion	M	18	48	34	2.37
	F	28	38	35	NS
Supervision	M	21	18	62	6.36
	F	28	35	38	S
Work	M	18	70	13	6.95
	F	21	55	24	NS
Co-workers	M	27	37	36	18.43
Self-confidence	F	48	45	7	S**
Working with people	M	14	37	49	3.91
	F	14	45	41	NS
Product knowledge	M	9	72	19	7.07
	F	14	86	0	S
Calling on a specialist	M	10	67	23	32.08
	F	14	86	0	S**
Sales ability	M	12	67	21	11.94
	F	7	93	0	S*
Job security	M	13	59	29	8.35
	F	21	65	14	S
Salary	M	33	33	33	5.36
	F	48	37	15	NS

[a] Totals may not equal 100% due to rounding.
[b] S   = significant with probability less than or equal to 0.05.
   S* = probability less than or equal to 0.01.
   S** = probability less than or equal to 0.001.
   NS = not significant.

Source: J. E. Swann, C. M. Futrell, and J. T. Todd, "Same Job—Different Views: Women and Men in Industrial Sales," *Journal of Marketing*, 42 (January 1978, 95, published by the American Marketing Association). Reprinted with permission.

Table P18.12 gives the ranking (on a three-point scale of low, medium and high) assigned by 160 male and 29 female salespersons to several job-satisfaction variables.

Use the Mann-Whitney test to test the null hypothesis for the two variables *Pay* and *Co-workers* at the $\alpha = 5$ percent significance level. What are your conclusions from these two tests? (Hint: Convert the percentages in the table to absolute number of respondents and allow for ties in your combined ranking.)

18.13. Individuals often invest for the long run, so long-term returns are what really count. Investment decisions, however, are often made on impulse or on the basis of last week's performance. Table P18.13 gives the performance averages for eight categories of mutual funds employing different investment strategies. The figures include the *annual* performance for the year ending December 23, 1980 (long-term performance), and the *weekly* change for the week ending the same day (short-term results).

TABLE P18.13

Number and type of fund	Percent change, year to 12/23/80	Percent change, week ended 12/23/80
62 Capital appreciation	+36.68	+1.62
165 Growth	+34.23	+1.73
82 Growth and income	+26.66	+1.76
23 Balanced	+18.60	+2.27
30 Income	+10.38	+2.29
3 Insurance	+ 8.08	+3.20
7 Specialty	+15.81	+2.08
10 Option	+13.97	+1.93

Source: Reprinted by permission of *Barron's*, © Dow Jones & Company, Inc. (December 29, 1980). All Rights Reserved.

Suggest a nonparametric method to check whether investment decisions based on last week's figures will on the whole pick up the best performers on an annual basis.

18.14. In Problem 18.10 we investigated the conditions for international diversification. Now let us see if mutual funds that spread their investments over various countries manage to achieve significantly better results than the more traditional funds, which restrict their investment to U.S. assets.

Table P18.14 gives the 1977 total returns (with all distributions reinvested) of four mutual

TABLE P18.14

Fund	Total return (percent)
Internationally diversified funds	
Canadian Fund	−1.1
International Investors	32.9
Research Capital Fund	32.2
Scudder International Fund	−0.4
Templeton Growth Fund	20.3
Funds investing in U.S. securities only	
Maximum capital gains	7.3
Long-term growth	−0.4
Growth and current income	−3.2
Balanced	−0.9
Income funds: common stock policy	−0.4
Income funds: flexible policy	3.2
Income funds: senior securities policy	4.9
Insurance and bank stocks	5.7
Public utility stock	7.7
Tax-exempt bonds	6.8

Source: Reprinted by permission from the *Wiesenberger Investment Companies Service*, 1978 edition, copyright © 1978, Gorham and Lamont Inc., 210 South Street, Boston, Mass. All Rights Reserved.

funds specializing in international issues versus the average 1977 total returns of various categories of mutual funds investing in U.S. securities.

(a) Use a nonparametric test that will enable you to check whether the international funds did better in 1977 than mutual funds investing in U.S. securities only.

(b) Some will argue that Canadian Fund cannot properly be classified as an internationally diversified mutual fund. Exclude this fund from the sample and repeat the same test as in part a. Are the results any different? If so, can you explain why?

(c) What is the parametric counterpart of this test? Apply it to the same data.

(d) Discuss the results obtained by the two tests. Which is the more appropriate in our case?

18.15. Table P18.15 shows the after-tax return on average invested capital (in percent) for a sample of twenty-three computer companies in the first nine months of 1980 and the corresponding period in 1978.

(a) Calculate the Kendall rank correlation coefficient between the 1980 and 1979 returns. Is the result significant at the $\alpha = 5$ percent level? How do you interpret the results?

(b) Calculate the Spearman rank correlation coefficient and the ordinary Pearson (moment-product) correlation coefficient. Test these coefficients for significance.

(c) Compare the three coefficients. Which would you recommend for our problem?

TABLE P18.15

Company	1979	1980
AM International	0.6%	1.0%
Amdahl	11.8	7.5
Burroughs	16.0	3.8
Control Data	12.0	13.4
Data General	22.6	18.1
Datapoint	22.7	17.8
Diebold	16.8	17.2
Digital Equipment	19.7	18.2
Duplex Products	23.7	22.6
Four-Phase Systems	21.4	5.7
Honeywell	16.4	16.9
IBM	21.6	23.6
Lanier Business Products	24.9	27.4
Management Assistance	25.9	14.6
Memorex	19.3	−20.2
Mohawk Data Sciences	19.2	18.6
NCR	17.8	17.0
Nashua	21.3	15.0
Pitney Bowes	23.4	23.1
Prime Computer	57.4	50.9
Sperry Rand	15.9	15.8
Wang Laboratories	30.8	29.0
Xerox	19.6	18.7

Data from *Business Week*, March 16, 1981.

18.16. Suppose we ordered the income of males (M) and females (F). The following lists show ordered income by earner in five samples. Count the number of runs in each sample.

(a) M M M M M F F F

(b) F F F F F M M M

(c) M M M M M M M F

(d) M F M F M F M F

(e) F M F M F M F M

18.17. The stock price of GSB on eleven successive days in 1980 was as follows:

$$\$20, \$20\tfrac{1}{4}, \$20\tfrac{3}{4}, \$21, \$21\tfrac{1}{2}, \$22, \$21, \$21, \$21, \$19, \$18$$

Test the hypothesis that the daily price changes are random.

18.18. Repeat the test of Problem 18.17 with the following two alternative price series:

(*a*) $20, $19, $20, $18, 18\tfrac{1}{2}$, 18\tfrac{1}{4}$, 18\tfrac{1}{2}$, $18, $19, 18\tfrac{1}{2}$, $19

(*b*) $20, $21, $18, $19, $20, $19, $20, $21, $16, $15, $13

18.19. Suppose in the runs test we obtain $n_1 = 50$ (price up) and $n_2 = 50$ (price down). It is also given that the number of runs, r, is 51. The null hypothesis is that all of the observations were drawn at random from a given population. Would you accept or reject the null hypothesis? Draw your conclusion without the help of any statistical tables.

18.20. The average productivity of high school graduates and high school dropouts on a given production line is to be tested. Suppose a statistical test is needed to decide the equality or inequality of productivity. If the assumption of equal variances cannot be made and the sample is small, what test would you choose for this purpose? Explain your choice.

18.21. When the value of the Mann-Whitney statistic is calculated, the following values are computed:

$$T_A = 10$$

$$T_B = 50$$

$$n_A = 12$$

Find the value of n_B.

18.22. The monthly income of men and women in a sample of nine observations is as follows:

F	F	M	F	M	M	M	F	M
$600,	$650,	$700,	$700,	$750,	$800,	$850,	$900,	$1,000

Test the hypothesis that the income of men is equal to that of women, where H_1 is the hypothesis that the two groups have different averages.

18.23. Using the Kruskal-Wallis test with three groups where $n_1 = n_2 = n_3 = 3$, we found that $R_1 = R_2 = R_3$. What is the value of the statistic H? Use Equation 18.8. Would you reject the null hypothesis?

18.24. Suppose the monthly incomes of two samples (X and Y) are as follows:

X	Y
$1,000	$1,000
1,500	1,500
2,000	4,000

Write down the ranks of the observations for each of the samples. Do we lose information by switching from numerical data to ranks? Is it meaningful to calculate the mean and variance of ranks? Explain your answer.

18.25. One of the issues investigated by Leo Bogart in his article "Is All This Advertising Necessary?" in the *Journal of Advertising Research* for October 1978 is the following: is there a significant correlation between the amount of national advertising behind product brands in a given year and the amount of slogan identification by consumers in the next year? The data, taken from a series of national surveys conducted between September 1976 and February 1977, are presented in Table P18.25.

Calculate the Spearman and Kendall rank correlations and determine if they are significantly different from zero at $\alpha = 0.05$ significance level.

TABLE P18.25

Brand	1976 Advertising expenditure	Percent of consumers identifying slogan in early 1977
Charmin	$ 7,289,000	82%
Alka-Seltzer	11,730,000	79
Chiffon	1,166,000	58
Morton's (Salt)	1,016,000	57
Contact	8,952,000	55
Hertz	5,511,000	47
Ragú	5,357,000	45
Meow Mix	6,296,000	41
McDonald's	81,831,000	38
Dynamo	5,010,000	37
Aim	12,087,000	33
Schlitz	16,244,000	23
Coca-Cola	46,768,000	16

Source: R. H. Bruskin Associates, *Bruskin Report*, May 1977.

CHAPTER NINETEEN OUTLINE

Key Terms

payoff matrix
maximax criterion
maximin criterion
minimax criterion
minimin criterion
minimax regret criterion
regret table
maximum expected return
   rule
utility

maximum expected utility
   rule
diminishing marginal utility
risk premium
mean-variance rule
prior probabilities
posterior probabilities
Bayesian decision-making
decision tree

19
DECISION-MAKING UNDER UNCERTAINTY

Businesspeople and other individuals are often faced by the need to select one of several alternative activities. In a world of uncertainty, making a prudent choice is at times a complex matter. At the core of the problem is the obvious difficulty of anticipating the consequences of each activity. Quantitative approaches are available to deal with these situations. In some of these approaches, the various possible outcomes are evaluated without the use of probabilistic methods. Other approaches employ probability assessments derived either objectively or subjectively by the individual (or group of individuals) making the choice. In this chapter, we shall describe some of these methods.

19.1 The Payoff Matrix

Let us assume that a businessman faced by an investment decision has to choose among three possible activities. In his assessment of the future, he envisions three possible states of the economy: inflation, recession, and stagflation. (Stagflation, incidentally, is a relatively new phenomenon that began with the oil crisis of 1974. While under normal circumstances inflation is associated with an active economy, stagflation is a situation in which stagnation and inflation prevail simultaneously.)

We shall use A_1, A_2, and A_3 to denote the three activities among which our investor has to choose:

A_1 An investment project in the construction industry.
A_2 An investment project in the cigarette-manufacturing industry.
A_3 A project in the bakery-products industry.

The three possible states of the economy, which are frequently dubbed "states of nature," are denoted by S_1, S_2, and S_3:

S_1 Inflation.
S_2 Stagflation.
S_3 Recession.

To consider all these possibilities, the businessman sets up a **payoff matrix,** which is simply a table that lists the anticipated profit from each investment (in millions of dollars, say) under the various possible states of nature. Table 19.1 illustrates such a payoff matrix.

TABLE 19.1
Payoff Matrix

	State of nature		
Activity	S_1 *Inflation*	S_2 *Stagflation*	S_3 *Recession*
A_1 Construction	500	300	10
A_2 Cigarettes	40	150	100
A_3 Bakery products	80	60	70

Note that the consumption of bread, hence the profit from investment A_3, does not change much with the various states of nature: bread is a basic consumption good, which people don't give up easily. The construction industry, however, is highly vulnerable to economic conditions. People do not usually buy new or bigger houses in a recession. On the other hand, cigarettes sell better in bad times than in good: feeling nervous about job security, people may seek relief in increased smoking.

Given this payoff matrix, which project ("activity") would you recommend for the investor? Several possible decision rules are available.

19.2 Decisions Based on Extreme Values

Some decision rules are nonprobabilistic and are based on only one value per activity in the payoff matrix. Other decision criteria consider the entire distribution of payoffs. We shall start with the nonprobabilistic decision rules.

THE MAXIMAX CRITERION

Some individuals are born optimists. Whatever activity he chooses, the optimist expects the state of nature that is most favorable for that activity to prevail. Thus, if he invests in A_1, he expects inflation and a profit of 500. If he invests in A_2, he expects stagflation and a profit of 150. If he chooses A_3, he again expects inflation and a profit of 80. From each activity, he expects to get the maximum value of the appropriate row in the payoff matrix. Consequently, he naturally chooses the activity with the largest of these maximum values. His decision rule is correspondingly called **maximax**: the maximum value of the rows' maxima. In our specific example, we have

$$\text{Max } A_1 = 500$$

$$\text{Max } A_2 = 150$$

$$\text{Max } A_3 = 80$$

The maximum of these maxima is 500, and our investor who follows the maximax rule will choose to invest in construction (A_1).

THE MAXIMIN CRITERION

The **maximin criterion** is a decision rule for born pessimists. Whatever activity he chooses, the pessimist expects nature to work against him. If he is going to drop a buttered slice of bread, he believes that the buttered side will always hit the floor, no matter which side he decides to butter. Therefore, when choosing among activities, he considers the minimum value of each activity and chooses the one corresponding to the maximum of all the minima. In our specific example

$$\text{Min } A_1 = 10$$

$$\text{Min } A_2 = 40$$

$$\text{Min } A_3 = 60$$

Our pessimist now naturally picks the activity with the largest anticipated minimum payoff as the best of bad choices: the maximum of the anticipated minima is 60, and under the maximin criterion the investor will choose to invest in bakery products (A_3).

THE MINIMAX CRITERION

When the profit figures in a payoff table are replaced with costs (that is, money outlays required to achieve a given goal), the **minimax criterion**—the minimum of all maxima—naturally replaces the maximin criterion. For example, suppose that three possible activities may be undertaken to achieve a given volume of sales—specifically, three advertising methods, which we denote by A_1, A_2, and A_3. Table 19.2 describes the corresponding cost matrix.

TABLE 19.2
Cost Matrix

| | | State of nature | | |
Activity		S_1 Inflation	S_2 Stagflation	S_3 Recession
A_1	Advertise on TV	10	20	35
A_2	Advertise in newspapers	60	40	50
A_3	Advertise by mail	20	30	25

Our born pessimist always expects the worse, so whatever the promotion campaign, he expects the costs to reach the maximum figure. The maximum cost under each activity is given by

$$\text{Max } A_1 = 35$$

$$\text{Max } A_2 = 60$$

$$\text{Max } A_3 = 30$$

The minimum of all these maximum costs is 30; therefore the minimax criterion leads the investor to choose activity A_3, which ensures the lowest possible exposure in terms of maximum outlay.

Thus, the minimax criterion is conceptually related to the maximin criterion. The minimax is used when one deals with costs or outlays of cash, while the maximin is used when the payoff matrix represents profit or incoming cash flow.

THE MINIMIN CRITERION (WALD CRITERION)

An optimistic investor who applies the maximax criterion to a payoff matrix will use the **minimin criterion** when faced with a cost matrix. In Table 19.2 the anticipated minimum costs for each activity are

$$\text{Min } A_1 = 10$$

$$\text{Min } A_2 = 40$$

$$\text{Min } A_3 = 20$$

Thus the minimum of all minima is 10, and activity A_1 is chosen. This decision-maker is an optimist in the sense that whatever activity he chooses, he expects nature to act in his favor: for each activity he considers the lowest possible cost.

THE MINIMAX REGRET CRITERION

Leonard Savage suggests the **minimax regret criterion,** which advises us to calculate the maximum possible regret for each activity, and then to choose the activity that minimizes the maximum regret.[1] Suppose that the investor selected activity A_i and state of nature S_j actually occurred. If activity A_i gives the maximum profit (or minimum cost) in the given state of nature, there is no regret: the investor chose wisely. If, however, some activity other than A_i (A_k, say) promises the maximum profit in this state of nature, the investor can be said not to have chosen wisely; his regret is measured by the difference between the maximum profit in state of nature S_j (which he would have attained had he chosen A_k) and the profit he actually gets having chosen A_i.

To illustrate the minimax regret criterion, we turn again to the payoff matrix in Table 19.1. Suppose we choose A_1 and inflation occurs. Then the regret of choosing A_1 is clearly zero: it gives the maximum possible outcome in the column corresponding to inflation. If we had chosen A_2, the regret

[1] Leonard J. Savage, "The Theory of Statistical Decision," *Journal of the American Statistical Association* 46 (1951): 55–67.

would have been the difference between the maximum payoff attainable under inflation (500) and the payoff obtained with A_2 (40). So for A_2 we get Regret = 500 − 40 = 460, which is indeed the opportunity loss caused by choosing A_2 rather than the best activity under inflation, A_1. Similarly, if we had chosen A_3, the regret would have been 420: 500 − 80 = 420. Under stagflation we subtract the corresponding payoff from the column maximum of 300 and under recession we deduct the payoffs from the highest payoff attainable under this state of nature, which is 100. The resulting **regret table** is shown in Table 19.3.

TABLE 19.3
Regret Table

| | State of nature | | |
| | S_1 Inflation | S_2 Stagflation | S_3 Recession |
Activity			
A_1 Construction	0	0	90
A_2 Cigarettes	460	150	0
A_3 Bakery products	420	240	30

The maximum regret (or opportunity loss) for each activity is given by

$$\text{Max } A_1 = 90$$

$$\text{Max } A_2 = 460$$

$$\text{Max } A_3 = 420$$

Now we choose the minimum of these three values, 90. Activity A_1, the construction industry, should be selected by the minimax regret criterion.

19.3 The Maximum Expected Return Rule

All the decision rules mentioned so far in this chapter rely completely on one extreme value in each row. In other words, if a number in the table is not the extreme in its row, its specific value is not taken into account. If we go back to our original payoff table (Table 19.1) and use the maximax criterion, we select activity A_1. If we change the payoff strongly *against* A_1 and strongly *in favor* of A_2 and A_3, however, the alteration still may have no impact on the decision reached under the maximax rule. For example, let us introduce the changes as in Table 19.4. Still the maximax is 500 and A_1 is selected. Thus, dramatic changes in the payoff matrix had no impact on our decision.

TABLE 19.4
Revised Payoff Matrix

| | State of nature | | |
| | S_1 Inflation | S_2 Stagflation | S_3 Recession |
Activity			
A_1 Construction	500	0	0
A_2 Cigarettes	499	490	490
A_3 Bakery products	499	490	493

All the rules discussed so far rely on one extreme value per activity and therefore suffer from the same serious drawback. Rather than concentrating on one extreme value per activity, we now suggest looking at all the outcomes and their corresponding probabilities—and *then* selecting the option with the highest expected value.

In some cases objective probabilities are available for the various states of nature. (For example, one can calculate the expected value of a lottery prize, since the probability of each event is well defined.) The probabilities of other investments may be subjective, since they are a function of the investor's experience and judgment. In still other cases, when the probabilities are unknown and we do not wish to venture a guess, equal probabilities may be assigned to each state of nature.

Applying equal probabilities to the three states of nature in our original payoff matrix (Table 19.1) yields the following:

$$E(A_1) = 500 \cdot \tfrac{1}{3} + 300 \cdot \tfrac{1}{3} + 10 \cdot \tfrac{1}{3} = 270.0$$

$$E(A_2) = 40 \cdot \tfrac{1}{3} + 150 \cdot \tfrac{1}{3} + 100 \cdot \tfrac{1}{3} = 96.7$$

$$E(A_3) = 80 \cdot \tfrac{1}{3} + 60 \cdot \tfrac{1}{3} + 70 \cdot \tfrac{1}{3} \quad = 70.0$$

and activity A_1, with the highest expected payoff, is chosen. Note that our decision will be different if the revised payoff matrix (Table 19.4) is used. In this case we will get

$$E(A_1) = 500 \cdot \tfrac{1}{3} + 0 \cdot \tfrac{1}{3} + 0 \cdot \tfrac{1}{3} \quad = 166.7$$

$$E(A_2) = 499 \cdot \tfrac{1}{3} + 490 \cdot \tfrac{1}{3} + 490 \cdot \tfrac{1}{3} = 493.0$$

$$E(A_3) = 499 \cdot \tfrac{1}{3} + 490 \cdot \tfrac{1}{3} + 493 \cdot \tfrac{1}{3} = 494.0$$

Now our selection is A_3.

The rule we have used here, the **maximum expected return rule,** is based on the expected value of the outcomes. This rule overcomes the drawback of relying on a single extreme value, but it still is not wholly satisfactory. Most businesspeople and individual investors indicate by their revealed preferences that they are not satisfied with this rule. In other words, their everyday behavior is not consistent with it. Consider a numerical example. Suppose that by a flip of a coin you can either gain or lose $10,000. Heads, you pay $10,000; tails, you win $10,000. Are you willing to participate in it? Most surveys and experiments of this kind show that people are wary of such a game; they are not indifferent to its dangers. Although the expected outcome is zero, most people will not accept the challenge and will decline to play this "fair game." And yet the maximum expected return rule indicates that one should be indifferent to the dangers of such a game since the expected profit from it is exactly zero.

The reason we are reluctant to enter such a game is that if we lose $10,000, the hurt is much greater than the corresponding benefit we enjoy if we win $10,000. The words "hurt" and "enjoy" imply that we actually consider the **utility** derived from money rather than the amount of money as such.

This leads us to the next rule, the expected utility rule. But before we go into this in more detail, let us further demonstrate the fallacy of the maximum expected return rule by the well-known St. Petersburg paradox. This paradox is of historical interest: it occupied the minds of the best mathematicians of the eighteenth century and indeed paved the way for the concept of utility and the expected utility rule, which is now acknowledged by most academicians.

THE ST. PETERSBURG PARADOX

The classic problem known as the St. Petersburg paradox was first formulated by the Swiss mathematician Nicolas Bernoulli:[2] Peter tosses a coin and continues to do so until it lands "heads" when it comes to ground. He agrees to give Paul $1 if he gets heads on the very first throw, $2 if he gets it on the second, $4 if on the third, $8 if on the fourth, and so on, so that with each additional throw the number of dollars he must pay is doubled. Suppose we seek to determine the value of Paul's expectation.

In general, if heads first appears on the nth toss, the player is awarded a prize equal to 2^{n-1}. The size of the prize is uncertain, and depends on the results of each experiment, but when the coin lands heads for the first time, the game is over, so that only one prize is awarded per game. Naturally, the player would like heads to appear only after a long series of tails, since this would increase his prize. Note that the number of games played is a geometric random variable with the probability of "success" equal to $\frac{1}{2}$ (see Chapter 7).

What would be a fair price for Paul to pay for the opportunity to play such a game? If heads comes up on the first toss, the prize is $1; the probability of this outcome is $\frac{1}{2}$ since there is an equal probability of obtaining heads and tails when an unbiased coin is tossed. What is the probability of winning $2 on the second toss? The probability of getting tails on the first toss is $\frac{1}{2}$, and since the two tosses are independent events, the probability of getting heads on the second toss is also $\frac{1}{2}$. The joint probability of the event *TH* (that is, tails followed by heads) is given by the product of the two probabilities: $\frac{1}{2} \cdot \frac{1}{2} = \frac{1}{4}$.[3] Theoretically, the game can continue a long time before the coin lands heads for the first time, but the probability of the game's lasting for a great number of tosses is, of course, very small.

The maximum expected return rule suggests that the game's expected value constitutes the maximum price for Paul to pay for this fair gamble. To facilitate the calculation of the expected value, we have set out the possible results of the coin tossing and their probabilities in Table 19.5. If we denote the possible prize by X, the expected income of the St. Petersburg game can be calculated as follows:

$$E(X) = \tfrac{1}{2} \cdot 1 + \tfrac{1}{4} \cdot 2 + \tfrac{1}{8} \cdot 4 + \tfrac{1}{16} \cdot 8 + \cdots$$

$$= \tfrac{1}{2} + \tfrac{1}{2} + \tfrac{1}{2} + \tfrac{1}{2} + \cdots = \infty$$

Since there is no theoretical limit to the number of tosses, the mathematical expectation of the game is infinite; that is, invoking the maximum expected return rule, Paul should be prepared to pay any sum, however large, for the opportunity to play the game!

[2] The first published analysis of the problem was written by Daniel Bernoulli (a younger cousin) during his stay in St. Petersburg as a visiting scholar (1725–33), hence the name "St. Petersburg paradox." The problem as formulated by Nicolas Bernoulli is quoted by Daniel Bernoulli in "Specimen Theoriae Novae de Mensura Sortis," *Papers of the Imperial Academy of Sciences in Petersburg* 5 (1738). An English translation, "Exposition of a New Theory on the Measurement of Risk," appears in *Econometrica* 22, no. 1 (January 1954): 23–36.

[3] Since the outcome of each toss is independent of the outcomes of the other tosses, $P(H \cap T) = P(H) \cdot P(T) = \frac{1}{2} \cdot \frac{1}{2} = \frac{1}{4}$. See Chapter 5.

TABLE 19.5

St. Petersburg Game

Toss on which heads first appears	Result*	Probability of result	Prize
1	H	$\frac{1}{2}$	1
2	TH	$\frac{1}{4}$	2
3	TTH	$\frac{1}{8}$	4
4	TTTH	$\frac{1}{16}$	8
.	.	.	.
.	.	.	.
.	.	.	.
	$(n-1)$ times		
	$\overbrace{TT \cdots TH}$	$\frac{1}{2^n}$	
n			2^{n-1}

* *H* = heads; *T* = tails.

Now assume that you are offered the opportunity to play such a game. How much would you pay for the opportunity? An experiment conducted with a group of students revealed that most were prepared to pay only \$2 or \$3 for a chance to play. A few were willing to pay as much as \$8, but no one offered more than that. This contradiction between the amount that most people are willing to pay for an opportunity to play the game and its infinite mathematical expectation constitutes the so-called St. Petersburg paradox.

Special interest attaches to the solutions proposed independently by the mathematician Daniel Bernoulli and by his contemporary Gabriel Cramer, who sought to resolve the problem by rejecting the maximum expected return rule and substituting expected utility in its place. Their efforts constitute an important intellectual milestone leading to the modern theory of choice under conditions of uncertainty.

19.4 The Concept of Utility and the Maximum Expected Utility Rule

The St. Petersburg paradox, as well as common sense, indicates that each dollar has a certain utility and that in decision-making one should consider the utility derived from each activity and apply the **maximum expected utility rule** rather than the maximum expected return rule. As we shall see below, the expected utility rule resolves Peter's problems, and in addition it clarifies many problems that businesspeople are faced with every day.

Consider an individual who must choose between the two alternative investment projects outlined in Table 19.6. Note that the two projects under consideration have the same expected income, \$200. A glance at the two investments suffices to show that the uncertainty involved in investment *B* is greater than that of investment *A*. We shall show below that despite the equal expected returns, investment *A* is indeed preferred to investment *B*. The intuitive explanation for the preference of investment *A* can be clarified if we examine the difference in the monetary outcomes of two proposals. Suppose that the investor who tentatively chooses investment *A* considers shifting from *A* to *B*. What changes are induced by such a shift? The differences between the two projects can be summarized as follows: if the lower outcome occurs, the investor will realize \$100 less on investment *B* than on investment *A*, while if the higher outcome occurs, investment *B* yields $100 more. Thus, if the investor changes his mind and shifts from *A* to *B*, he has a 50 percent chance of gaining $100 but he also has a 50 percent chance of losing $100. Is it worthwhile to shift from investment *A* to investment *B*?

TABLE 19.6
Alternative Investments

	Investment A		Investment B	
Net income	Probability		Net income	Probability
$100	$\frac{1}{2}$		$ 0	$\frac{1}{2}$
300	$\frac{1}{2}$		400	$\frac{1}{2}$
Expected income $200			$200	

In general, most investors *will not* switch from *A* to *B*, since the subjective satisfaction (or utility gain) that they derive from the additional $100 is less than the dissatisfaction (or utility loss) that they face if they lose $100. To help to clarify this argument, consider an individual who uses the monetary return to buy consumer goods. Most individuals reveal **diminishing marginal utility,** which means that the satisfaction (utility gain) from consumption diminishes as consumption increases. That is to say, the consumer initially satisfies his more essential needs, and hence the utility that he derives from spending, say, the first $100 is relatively large. Once he has satisfied his more basic needs, we expect that the additional utility he derives from spending a second increment of $100 will be lower, and so on for each additional increment of income.

The concept of diminishing marginal utility is illustrated in Table 19.7. As we can see, total utility increases as income rises, so that the higher the income, the larger the satisfaction derived from it. The marginal (that is, incremental) utility is diminishing, however; the utility of the first $100 increment is 10 "utils," the additional utility derived from the second $100 increment is 8 utils, and the next two $100 increments add 7 and 5 utils, respectively. In Table 19.8 we present the expected utility calculations for investments *A* and *B*. The data of Table 19.8 indicate that while investments *A* and *B* are characterized by the same expected income, they differ with respect to expected utility. The expected utility derived from investment *A* is 17.5 utils, compared to investment *B*'s expected utility of only 15 utils. Thus, whereas the expected return rule cannot discriminate between investments *A* and *B*, the expected utility rule indicates a clear preference for investment *A*, which, as we have already noted, is considerably less uncertain with regard to possible future income. Moreover, the ranking of investment *A* over *B* holds for all utility functions, as long as the utility function has the

TABLE 19.7
The Utility Function

Income	Utility (utils)	Marginal utility (utils)
$ 0	0	
100	10	10
200	18	8
300	25	7
400	30	5

TABLE 19.8
Expected Utility Calculations

Investment A			Investment B		
Probability	*Income*	*Utility*	*Probability*	*Income*	*Utility*
$\frac{1}{2}$	$100	10	$\frac{1}{2}$	$ 0	0
$\frac{1}{2}$	300	25	$\frac{1}{2}$	400	30
Expected net income	$200			$200	
Expected utility*		17.5			15

* The expected utilities of investments *A* and *B* are given by $10 \cdot \frac{1}{2} + 25 \cdot \frac{1}{2} = 17.5$ and $0 \cdot \frac{1}{2} + 30 \cdot \frac{1}{2} = 15$, respectively.

property of diminishing marginal utility. This statement can be proved by an examination of individuals' attitudes toward risk.

Indeed, John von Neumann and Oskar Morgenstern have shown that when we have uncertain future income induced by a given activity, every rational investor should select his activity according to the expected utility rule rather than the expected return rule.[4] The mathematical proof of this claim and the way it solves the St. Petersburg paradox can be found in many books on finance and economics.[5] Here, we will take a look at the various ways individuals approach risk in their everyday dealings.

ALTERNATIVE ATTITUDES TOWARD UNCERTAINTY

In general, an investment or an activity that generates an uncertain income is said to be risky. The more uncertain the outcome, the riskier the investment. Thus, we use the words *uncertainty* and *risk* interchangeably.

It is convenient for the purposes of our analysis to distinguish among three classes of investors: those who dislike risk, whom we call "risk averters"; those who prefer more risky to less risky investments, whom we shall call "risk lovers"; and those who disregard risk altogether, whom we shall call "risk neutral." As the bulk of the theoretical and empirical evidence supports the view that the typical investor is a risk averter, we shall concentrate on that broad class of individuals.

Suppose an individual is offered the opportunity of investing $100 in the following project:

Income	Probability
$ 90	$\frac{1}{2}$
110	$\frac{1}{2}$

The expected income of such an investment is $90 \cdot \frac{1}{2} + 110 \cdot \frac{1}{2} = 100$. Thus, the expected value equals the initial investment. In other words, the expected monetary profit from the investment is zero. Can an individual be expected to invest in such a project? Since we have rejected the expected return rule

[4] John von Neumann and Oskar Morgenstern, *Theory of Games and Economic Behavior,* rev. ed. (Princeton, N.J.: Princeton University Press, 1953). A more popular version is given in Luce and Raiffa, *Games and Decisions.*

[5] See, e.g., Haim Levy and Marshall Sarnat, *Investment and Portfolio Analysis* (New York: Wiley, 1972).

and have tentatively replaced it with the expected utility rule, our answer depends on the individual's attitude toward risk: that is, on the degree to which he likes or dislikes to trade a safe prospect (the initial $100 investment) for an uncertain one (the income generated by the investment).

The preferences of risk averters are characterized by diminishing marginal utility, as in Table 19.7: the utility increases with income, but at an ever diminishing rate. It follows that every risk averter prefers a perfectly certain investment to an uncertain one that has the same expected return. A risk averter will not invest in the above project because in terms of utility, the possible loss of $10 more than offsets the possible gain of $10.

This conclusion can be confirmed by a graphic device. Figure 19.1 sets out the same investment problem: the investment of $100 in a project with equal probabilities of returning $90 and of returning $110. The possible dollar values of income are measured along the horizontal axis, and utility is measured along the vertical axis in utils. The concave form of the individual's utility function expresses the risk averter's characteristic of diminishing marginal utility.

Now remember that the individual's attention is ultimately focused on his level of utility. Money is important only insofar as it gives rise to utility. When the individual wants to decide whether or not to pay $100 for the investment, he approaches this problem in the following manner. The $90 income, if it occurs, will give rise to U_1 utils. If $110 turns up, his utility will be U_2 utils. On the average, then, he can expect $\bar{U} = U_1 \cdot \frac{1}{2} + U_2 \cdot \frac{1}{2}$ utils, which is the expected utility of the investment. For a risk averter, whose utility function is concave like the one in Figure 19.1, the utility of $100 (denoted by U_3) is greater than $\bar{U}$. Since the individual focuses on utility rather than on income per se, he will decide not to pay $100 for the uncertain income. Why should he give up U_3 utils only to get back a lesser (expected) number of utils, $\bar{U}$? He will obviously be better off keeping the $100 and refraining from investment. Graphically, the expected utility, $\bar{U}$, may be easily located when only two possible uncertain outcomes are involved. Raising a perpendicular line from the expected monetary value on the horizontal axis ($100 in our example), we find at its intersection with the

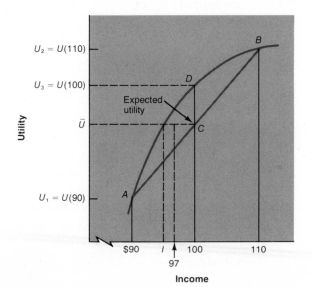

Figure 19.1

A utility function over income and expected utility

line segment connecting points *A* and *B* in Figure 19.1 the point *C*, whose height measured from the horizontal axis is equal to $\overline{U}$.

Now let us assume that the same individual is offered the same investment project but at a lower initial investment, say $97. The expected income remains $100, so the project has a positive expected net income ($100 − $97 = $3). Will the risk-averse individual invest in the project? Despite the lower price, Figure 19.1 clearly shows that he will not be willing to invest in this project because the utility of a perfectly certain sum of $97 still exceeds the expected utility of the risky project ($\overline{U}$). How far must the initial investment fall before our risk-averse investor will be willing to accept the project? Again the answer can be readily inferred from Figure 19.1. The maximum investment that he will be willing to make is represented by point *I* on the horizontal axis. At this point the utility of the initial investment, *U(I)*, is just equal to the expected utility of the risky investment. The distance between 100 and *I* measures the **risk premium** required to induce the risk-averse individual to invest in the project. At investments lower than *I* (points to the left of *I* on the horizontal axis) the project is attractive to our individual, since it represents a gain in expected utility. Conversely, at investments above *I*, as we have already seen, the risky project represents a loss of expected utility for the risk-averse investor.

Having explained the basic properties of risk-aversion concepts, let us make a graphic analysis of the original example in Table 19.6. Recall that we have asserted that all risk averters (that is, individuals characterized by diminishing marginal utility of money) will prefer investment *A* to investment *B*, because of the greater dispersion of the latter for an equal expected income. Figure 19.2 graphs the data of Table 19.6. Since we are assuming risk aversion, the utility function is drawn as a concave curve rising from the origin. The expected utility of project *A* is simply $U(100) \cdot \frac{1}{2} + U(300) \cdot \frac{1}{2} = 10 \cdot \frac{1}{2} + 25 \cdot \frac{1}{2} = 17.5$. This value corresponds to point *A* of Figure 19.2, the intersection of the vertical line rising from point 200 on the horizontal axis (which is the expected monetary return) and the appropriate line segment on the utility function. Similarly, point *B* indicates the expected utility from

Figure 19.2

Expected utility of investments *A* and *B* in Table 19.6

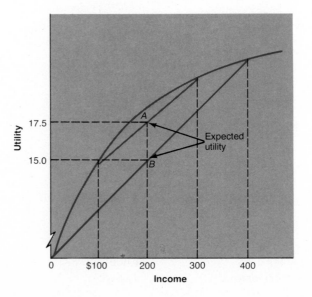

investment B. Since point A is higher than point B, it follows that the risk averter will prefer investment A to investment B.

The reason for this result can also be inferred from Figure 19.2. Other things being equal, risk averters do not like a wide dispersion of outcomes. From the graph it is clear that both projects have the same expected profit, but the range of outcomes of investment B is much greater than the range of outcomes of investment A. Hence all risk averters will prefer A over B.

Two additional classes of possible investors can be identified. First we have risk lovers, who have a preference for risk rather than an aversion to it. Such individuals are characterized by convex utility functions, implying that the marginal utility of each additional dollar increases. Upon reflection we see that this is not a very realistic assumption.

Then we have those investors who are risk-neutral, that is, are neither risk averters nor risk lovers. Such individuals have linear utility functions, and display constant marginal utility of money. Risk neutrality constitutes a middle ground between risk lovers and risk averters. Risk-neutral individuals choose their investments solely according to their respective expected incomes, and they completely ignore the dispersion of the various returns.

Figure 19.3 illustrates the utility functions of a risk lover and a risk-neutral person side by side.

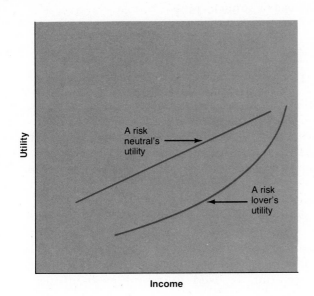

Figure 19.3
Utility functions of risk lover and risk-neutral

19.5 The Mean-Variance Rule

Although the expected utility rule is normatively the best criterion available, sometimes it is very hard to implement. We frequently do not know the investor's utility function; and furthermore, we often face a conflict-of-interest situation, as in the case of a manager who must act on behalf of many stockholders, each of whom has a different utility function from the others. Thus, despite its sound theoretical foundations, the expected utility rule is often replaced by the mean-variance rule, which provides a more practical decision criterion.

According to the **mean-variance rule,** the expected return (mean) measures an investment's profitability, while the variance (or dispersion) of returns measures its risks. Consider the following four projects, with the means and variances specified:

Investment project	Mean	Variance
A	$10	$100
B	9	80
C	8	90
D	7	95

A pairwise comparison of the investment projects shows that project B dominates projects C and D, since B has a higher profit and a lower risk than both C and D. There is no clear-cut decision with regard to project A, however, since $E(A) > E(B)$ but $\sigma^2(A) > \sigma^2(B)$. Here the investor, apart from and in addition to statistical tools, needs to use judgment when considering the trade-off between profit and risk. Some investors will find it worthwhile to undertake the higher risk in return for the higher expected return. Others will not wish to undertake project A. They will prefer the lower variance even at the expense of some expected return.

19.6 Bayesian Decision-Making: Prior and Posterior Probabilities

The more sophisticated decision-making methods—the maximum expected return rule, the maximum expected utility rule, the mean-variance rule, and others that have not been mentioned—are probabilistic in nature. Obviously, the methods can be only as good as the data they use. Generally, when additional information becomes available, a better decision is reached.

In this section, we discuss the way additional information can be incorporated into the decision-making process to revise the probabilities used in the probabilistic decision-making methods. The initial probabilities are referred to as **prior probabilities,** and the revised probabilities are called **posterior probabilities.** Probabilities may be revised by the collection of data or by the acquisition of relevant information in some other way. The way additional information is used to revise probabilities is illustrated by Example 19.1.

EXAMPLE 19.1

A company is considering the production of a luxury product on which it expects to make a $100 net profit per unit sold. The initial information available to the company is that 5 percent of American consumers will buy the product, most of them wealthier than the national average. Specifically, it is estimated that 26 percent of those with $30,000 or more in annual income will buy the product, while only 1 percent of those with lower annual incomes will buy it. Sixteen percent of American consumers have annual incomes of $30,000 or more, and 84 percent have annual incomes of less than $30,000. The company is considering the construction of a manufacturing plant at an investment of $4 million, and is planning to market the product in an area with a population of a million consumers. It will not make the investment unless $4 million or more is expected to be realized in net profit in the first year of operation.

The expected annual profit, not considering the $4 million investment, is

$$0.05 \cdot 1,000,000 \cdot \$100 = \$5,000,000$$

where 0.05 represents the percentage of consumers who are expected to buy the product, 1,000,000 is the number of potential consumers, and $100 is the profit per item sold. Since the $5 million is more than the minimum required by the firm ($4 million), the firm should decide to make the investment.

Before committing itself to a $4 million investment, however, the company may wish to find out the number of wealthier consumers ($30,000 or more) among the million potential consumers. Additional information is collected and reveals that only 10 percent of the potential consumers have annual incomes of $30,000 or more, compared to 16 percent in the U.S. population as a whole. With this additional information, the company's decision changes. The number of consumers expected to buy the product is equal to the sum of 26 percent of 100,000 "wealthy" consumers and 1 percent of the remaining 900,000. And since the profit per item is $100, we can expect the following annual profit: $(0.26 \cdot 100,000 + 0.01 \cdot 900,000) \cdot 100 = (26,000 + 9,000) \cdot 100 = \$3,500,000$. Since this amount will not cover the initial investment in the first year, the firm decides not to undertake the project.

Note that 5 percent of the U.S. population is expected to buy the product. So if we choose one person at random, the probability that he will be a buyer is 0.05. Denote the following events concerning a consumer chosen at random:

A_1 His annual income is $30,000 or more.
A_2 His annual income is less than $30,000.
B He is a buyer of the product.

It is given that in the total population we have

$$P(B \mid A_1) = 0.26$$

$$P(B \mid A_2) = 0.01$$

so that we can confirm the probability of B by calculating (see Chapter 5)

$$P(B) = P(B \mid A_1) \cdot P(A_1) + P(B \mid A_2) \cdot P(A_2)$$

$$= 0.26 \cdot 0.16 + 0.01 \cdot 0.84 = 0.05$$

Events A_1, A_2, and B are illustrated by a Venn diagram in Figure 19.4. Note that the part of event B in A_1 accounts for 26 percent of A_1 while its part in A_2 accounts for only 1 percent of A_2. On the whole, event B

Figure 19.4

Venn diagram:
U.S. consumers,
by income and
by willingness
to buy a specific
luxury product

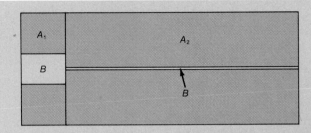

(shown in white) occupies 5 percent of the combined areas of A_1 and A_2.

The million potential consumers of the product, in the area where the firm plans to market the product, have an income distribution that is *different* from the overall U.S. distribution. Specifically, the percentage of wealthy people in that location is smaller. This means that the relevant diagram should show a smaller A_1 area than that in Figure 19.4. The area, of course, must be proportional to the percentage of wealthy people in the million potential consumers. The correct proportions are 10 percent for A_1 and 90 percent for A_2, which are shown in Figure 19.5.

Figure 19.5

Venn diagram:
consumers in
one specific
geographic
location, by
income and by
willingness to
buy the product

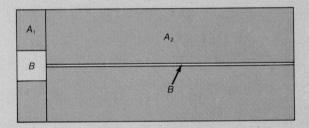

Comparing Figure 19.5 with Figure 19.4, we see that the relevant percentage of B in the total area is smaller. Indeed, the calculations show that

$$P(B) = P(B \mid A_1) \cdot P(A_1) + P(B \mid A_2) \cdot P(A_2)$$

$$= 0.26 \cdot 0.10 + 0.01 \cdot 0.90 = 0.035$$

meaning that we can expect only 3.5 percent of the potential customers actually to buy the product. The expected profit, then, is

$$0.035 \cdot 1,000,000 \cdot \$100 = \$3,500,000$$

as obtained earlier.

This process of decision-making, the revising of probabilities by means of Bayes' formula, is called **Bayesian decision-making.** One can also calculate additional relevant probabilities with Bayes' theorem. For example, if a person is selected at random among the potential customers, what is the probability that his income is \$30,000 or more? The probability is given by the percentage of people we called "wealthy" in the population: $P(A_1) = 0.10$, or 10 percent.

Suppose, however, we know that the individual is a buyer of the product. Then we calculate the probability this way:

$$P(A_1 \mid B) = \frac{P(A_1 \cap B)}{P(B)} = \frac{P(B \mid A_1) \cdot P(A_1)}{P(B \mid A_1) \cdot P(A_1) + P(B \mid A_2) \cdot P(A_2)}$$

$$= \frac{0.26 \cdot 0.10}{0.26 \cdot 0.10 + 0.01 \cdot 0.90} = \frac{0.026}{0.035} = 0.743$$

The additional information (that is, that the individual chosen is a buyer of the product) has raised the probability from 10 percent all the way up to 74.3 percent. This can also be described diagrammatically. Of the million potential customers, 35,000 (3.5 percent) buy the product. The individual chosen must then be one of the 35,000 buyers. There are 100,000 consumers with $30,000 or more in annual income, and 26,000 of them (26 percent) are buyers. Among the rest of the 900,000 consumers in the population, 1 percent are buyers, or 9,000 consumers. This means (see Figure 19.6) that the 35,000 buyers consist of 26,000 wealthy

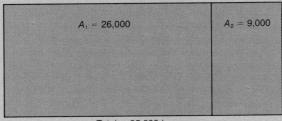

Total = 35,000 buyers

Figure 19.6

Venn diagram: potential purchasers of the product, grouped according to income

customers and 9,000 nonwealthy customers. The 26,000 wealthy ones indeed are 74.3 percent of the buyers: $\frac{26,000}{35,000} = 0.743$.

The probabilities used before the new information about the wealth of the potential customers was collected are called *prior probabilities*, while the revised (conditional) probabilities calculated on the basis of the additional information are called *posterior probabilities*.

In what follows we illustrate two aspects of the Bayesian approach: *a*) how sample information may be used in order to get a "better fix" on a parameter, namely how the prior distribution and sample information are combined to obtain a posterior distribution, and *b*) how the prior and posterior distributions can be used, after specifying an objective function, to make decisions under uncertainty. In particular, it highlights how the use of sample information may lead to an alternative decision to the one that would be reached without it.

Consider a company that has developed a new product and has to decide whether or not to introduce this product in a certain market area. The demand, that is, the proportion of the population that will buy the product at the fixed selling price of $5, is uncertain; but the manager estimates a prior probability distribution of the proportion, *p*. This distribution is shown in Table 19.9. Other data relevant to the decision are the following:

1. If the company enters the market, it will incur setup costs of $19,000.

TABLE 19.9
Prior Probability Distribution of the Proportion of the Population, p, Likely To Buy the Product

Value of p	Probability
0.10	0.50
0.20	0.25
0.30	0.15
0.50	0.10
	1.00

2. The profit per unit sold (disregarding the setup cost) is $1.
3. The total population of the market area is 100,000.

To get a "better" estimate of the true value of the parameter p, the manager takes a random sample of five people ($n = 5$) and determines the number of people indicating they would buy the product. (As a practical matter, a sample size of only five people would not provide reliable statistical evidence. However, there are also cost considerations involved in picking the optimal sample size. We will not consider these issues here.)

Suppose that only one person answers "yes (I will buy the product)," ($X = 1$). Under the appropriate assumptions, the probability distribution of observing x answers of "yes" in a sample of five, given a value of p, is given by the binomial distribution,

$$P(X = x \mid n, p_0) = \binom{n}{x} (p_0)^x (1 - p_0)^{n-x}$$

where

x stands for the number of successes ("yes") observed,
n stands for the sample size,
p_0 indicates that the probability is derived assuming that the true value of the random variable p is equal to p_0.

Consider the sample result $X = 1$. If indeed the true proportion is $p = 0.10$, then the likelihood of obtaining $X = 1$ in a sample of five observations is given by

$$P(X = 1 \mid n = 5, p = 0.10) = \binom{5}{1} (0.10)^1 (0.90)^4 = 0.3280$$

The probability of obtaining $X = 1$ if the true proportion is $p = 0.20$ is given by

$$P(X = 1 \mid n = 5, p = 0.20) = \binom{5}{1} (0.20)^1 (0.80)^4 = 0.4096$$

In a similar fashion we can derive the likelihood of the sample result ($X = 1$) for each of the alternative values which p can take on. These probabilities are shown in column 3 of Table 19.10.

The prior probability of $p = 0.10$ is equal to 0.50. Given the sample result $X = 1$, we may ask: "What is the conditional probability that $p = 0.10$ *given* that the sample shows $X = 1$?" This probability is the posterior probability of $p = 0.10$, and we can expect it to differ (in general) from the prior

TABLE 19.10

The Posterior Distribution for $n = 5$, $X = 1$

(1) p	*(2)* *Prior probability*	*(3)* *Likelihood*	*(4)* *(Prior probability) (Likelihood)*	*(5)* *Posterior probability*
0.10	0.50	0.32805	0.1640	0.4881
0.20	0.25	0.40960	0.1024	0.3047
0.30	0.15	0.36015	0.0540	0.1607
0.50	0.10	0.15625	0.0156	0.0465
Total	1.00	1.25405	0.3360	1.0000

NOTE: The likelihood (column 3) is calculated as follows:

$$P(X = 1 \mid n = 5, p = p_i) = \binom{5}{1} (p_i)^1 (1 - p_i)^4$$

For example, for $p = 0.10$ we get:

$$P(X = 1 \mid n = 5, p = 0.10) = \binom{5}{1} (0.10)^1 (1 - 0.10)^4 = 0.32805$$

probability of 0.50. The posterior probability is $P(p = 0.10 \mid X = 1)$ and we can calculate it using Bayes' theorem:

$$P(p = 0.10 \mid X = 1) = \frac{P(X = 1 \mid n = 5, p = 0.10)P(p = 0.10)}{\sum_i P(X = 1 \mid n = 5, p = p_i)P(p = p_i)}$$

$$= \frac{0.32805 \cdot 0.50}{0.3360} = 0.4881$$

where $P(X = 1 \mid n = 5, p = p_i)$ is the probability of obtaining the sample results assuming that the true value of p is p_i (which can be 0.10, 0.20, 0.30, or 0.50). This probability is called the likelihood (of the sample result), as we saw. The probability $P(p = p_i)$ is simply the prior probability of p_i. For example, $P(p = 0.10) = 0.50$. The denominator of the above formula is equal to 0.3360 (see Table 19.10). Table 19.10 gives the posterior probabilities for all the possible values of p.

At this point, it is instructive to compare the expected net profit (π), based on the prior and the posterior distributions.

Expected net profit based on the prior distribution:

$$\pi_0 = [0.50 \cdot 0.10 \cdot 100,000 \cdot \$1 + 0.25 \cdot 0.20 \cdot 100,000 \cdot \$1$$
$$+ 0.15 \cdot 0.30 \cdot 100,000 \cdot \$1 + 0.10 \cdot 0.50 \cdot 100,000 \cdot \$1] - \$10,000$$
$$= \$19,500 - \$19,000 = \$500$$

Expected net profit based on the posterior distribution:

$$\pi_1 = [0.4881 \cdot 1.10 \cdot 100,000 \cdot \$1 + 0.347 \cdot 0.20 \cdot 100,000 \cdot \$1$$
$$+ 0.1607 \cdot 0.30 \cdot 100.000 \cdot \$1 + 0.0465 \cdot 0.50 \cdot 100,000 \cdot \$1] - \$10,000$$
$$= \$18,121 - \$19,000 = -\$879$$

π_1 is less than π_0 because the sample suggests a lower probability for high values of p. Furthermore, π turns from positive to negative when posterior rather than prior probabilities are used.

Suppose the manager now takes another sample of $n = 5$ and finds that four people respond "yes," that is, $X = 4$. The manager is in a position to revise previously obtained posterior probabilities. The calculations are presented in Table 19.11. Note that the entries in column 2 of Table 19.11 are identical to the entries in column (5) of Table 19.10.

TABLE 19.11
Posterior Distribution Based on the Second Sample, $n = 5, X = 4$

(1) p	(2) *Prior probability*[a]	(3) *Likelihood*	(4) *(Prior probability) (Likelihood)*	(5) *Posterior probability*
0.10	0.4881	0.00045	0.0002196	0.0157
0.20	0.3047	0.00640	0.0019501	0.1394
0.30	0.1607	0.02835	0.0045558	0.3256
0.50	0.0465	0.15625	0.0072656	0.5193
Total	1.0000	0.19145	0.0139911	1.0000

NOTE: The likelihood (column 3) is calculated as follows:

$$P(X = 4 \mid n = 5, p = p_i) = \binom{5}{4} (p_i)^4 (1 - p_i)^1$$

For example, for $p = 0.10$ we get:

$$P(X = 4 \mid n = 5, p = 0.10) = \binom{5}{4} (0.10)^4 (0.9)^1 = 0.00045$$

[a] Values are from Table 19.10.

The expected net profit based on the second posterior distribution is as follows:

$$\begin{aligned}
\pi_2 = \ &[0.0157 \cdot 0.10 \cdot 100{,}000 \cdot \$1 + 0.1394 \cdot 0.20 \cdot 100{,}000 \cdot \$1 \\
&+ 0.3256 \cdot 0.30 \cdot 100{,}000 \cdot \$1 + 0.5193 \cdot 0.50 \cdot 100{,}000 \cdot \$1 - \$19{,}000 \\
= \ &\$38{,}678 - \$19{,}000 = \$19{,}678
\end{aligned}$$

The evidence from the second sample indicates that the value of p is likely to be higher than what is indicated by the first sample. Hence we get: $\pi_2 > \pi_0 > \pi_1$.

TABLE 19.12
Posterior Probabilities Calculated on the Basis of a Single Sample, $n = 10, x = 5$

(1) p	(2) *Prior probability*	(3) *Likelihood*	(4) *(Prior probability) (Likelihood)*	(5) *Posterior probability*
0.10	0.50	0.00149	0.000745	0.0157
0.20	0.25	0.02642	0.006605	0.1394
0.30	0.15	0.10292	0.015438	0.3257
0.50	0.10	0.24609	0.024609	0.5192
Total	1.00	0.37692	0.047397	1.0000

NOTE: The calculations are similar to those of Tables 19.10 and 19.11; column 2 entries are now given by

$$P(X = 5 \mid n = 10, p = p_i) = \binom{10}{5} (p_i)^5 (1 - p_i)^5$$

For example, with $p_i = 0.10$, we get:

$$P(X = 5 \mid n = 10, p = 0.10) = \binom{10}{5} (0.10)^5 (0.90)^5 = 0.00149$$

An important feature of the Bayesian approach is that when independent samples are taken, the same posterior distribution is obtained if we calculate the posteriors successively, or if we calculate the posterior after all the samples are taken. To see this, suppose we had taken a sample of ten people and observed that five of them responded "yes"; thus $n = 10$ and $X = 5$, which is the same as obtaining the results of the two samples all at once.

The differences between column 5 of Table 19.12 and column 5 of Table 19.11 are due solely to rounding errors.

19.7 Decision Trees

The decision flow diagram, or **decision tree,** facilitates decision-making when uncertainty prevails, especially when the problem involves a sequence of decisions.[6] In a sequential decision problem, in which the actions taken at one stage depend on actions taken earlier, the evaluation of alternatives can become very complicated. In such cases, the decision-tree technique facilitates project evaluation by enabling the firm to write down all the possible future decisions, as well as their monetary outcomes, in a systematic manner.

Perhaps the best way to explain the decision tree is to demonstrate its use through a specific example.

EXAMPLE 19.2

Suppose that an oil company owns drilling rights in the North Sea and that the company initially must decide whether or not to make a seismic test that will indicate the chances of finding oil in this area. Hence, undertaking the test, which is a costly venture, or proceeding without it is the first in our sequence of decisions (see Figure 19.7). In Stage 2, the firm again faces two alternatives: either to sell its drilling rights to another company or to drill with the hope of finding oil. As one can see in Figure 19.7, these two simple alternatives yield radically different monetary rewards, depending on the action taken in Stage 1, and (in the event that the firm decides to make the test) on the test's success or failure. Hence, the first stage is characterized in Figure 19.7 by two branches of possible action, denoted by A_1 and A_2. If the firm decides to make the seismic test (that is, to follow Branch A_2) it will, in the second stage, again be confronted by two possible decisions (Branches B_1 and B_2). Thus, each successive decision in the sequence has its own branches to represent further decisions: hence the name "decision tree."

If the firm decides not to carry out the seismic test, it can sell the drilling rights for $12 million (see Branch G_9); alternatively, the company can drill for oil without making the seismic test. In the latter event, the monetary outcome depends solely on whether or not oil is actually found. Suppose that the oil company estimates the probability of finding oil at 0.4 and the probability of finding a dry hole at 0.6. If we assume that the drilling cost is $15 million, there is a probability of 0.6 of losing this sum (see Branch G_8). On the other hand, there is a probability of 0.4

---

[6] Readers who wish to pursue this subject further may consult John F. Magee, "Decision Trees for Decision Making," *Harvard Business Review*, July–August 1964, and "How to Use Decision Trees in Capital Investment," *Harvard Business Review*, September–October 1964; or Howard Raiffa, *Decision Analysis: Introductory Lectures on Choices Under Uncertainty* (Reading, Mass.: Addison-Wesley, 1968).

Figure 19.7

Decision tree:
decisions
concerning
seismic testing,
drilling for oil,
and selling
drilling rights in
a specific
geographic area

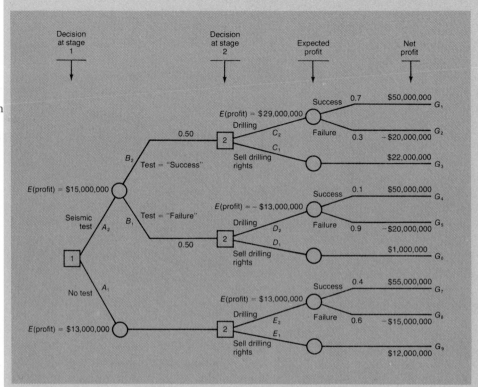

of striking oil, in which case the firm will earn a profit (after deduct-
ing the drilling and other costs) of $55 million (see Branch G_7). The ex-
pected profit, should the firm decide to drill *without* a seismic test, is
$13 million, since $E(\text{profit}) = 0.4 \cdot \$55,000,000 + 0.6 \cdot (-\$15,000,000)$
$= \$13,000,000$.

Let us turn now to the monetary consequences of following Branch
A_2: we now assume that the firm decides to make the seismic test in the
first stage. We further assume that this seismic test costs $5 million and
that there exists a probability of 0.5 that the test will yield good results
(denoted as "success" in Figure 19.7), and a probability of 0.5 that it
will fail. The decision at Stage 2 obviously will depend on the test
results. Should the company decide to sell the drilling rights after the
seismic test fails, it will realize a lower price than it could have obtained
without the seismic test. Clearly the poor results of the test (which we
assume are public knowledge) will lower the market value of the drilling
rights. Let us assume that if the test fails, the firm can sell its concession
for $6 million, which will net the firm only $1 million because we have
assumed that the seismic test costs $5 million (see Branch G_6). On the
other hand, the firm might still decide to drill despite the failure of the
seismic test. As a result of the failure of the test, however, the company
revises its estimates of the probability of finding oil (this is Bayesian
decision-making; see above). The probability of hitting oil is now
estimated to be only 0.1 (as compared with 0.4 without the additional
information provided by the seismic test). Should the firm fail to find
oil, the loss will be $20 million—the assumed drilling cost of $15 million
plus the $5 million cost of the seismic test (see Branch G_5). Should the

firm strike oil, the net profit will be $50 million, which reflects a profit of $55 million from the net oil revenues minus the cost of the seismic test (see Branch G_4). The expected profit should the firm decide to drill even if the seismic test fails turns into a loss of $13 million: $E(\text{profit}) = 0.10 \cdot \$50,000,000 + 0.90 \cdot -\$20,000,000 = -\$13,000,000$.

The probability of a successful seismic test is described by Branch B_2. Clearly a successful test will increase the value of the drilling rights, say to $27 million, and since the company spent $5 million on the seismic test, its net income from selling the rights would be $22 million (see Branch G_3). Obviously, a successful seismic test also increases the probability of finding oil, so we assume that the firm now estimates this probability at 0.7 (Branch G_1). Thus the expected profit, should the firm decide to drill following a successful seismic test, is $29 million: $E(\text{profit}) = 0.70 \cdot \$50,000,000 + 0.30 \cdot -\$20,000,000 = \$29,000,000$.

Now that we have obtained the monetary outcomes from all the possible branches of Figure 19.7, which decision sequence is optimal? Clearly, the decision depends on the utility that the firm attributes to each possible outcome. But in order to demonstrate the use of the decision-tree technique, let us assume for simplicity that the firm reaches its decision according to the criterion of maximum expected profit. Following this rule, we calculate the expected profit for each branch of Figure 19.7, and the course of action represented by the branch with the highest expected profit will be chosen. This is not as simple as it might seem, however. First we must examine the profit of Stage 2 in order to choose the optimal course of action (branch) for Stage 2; only then can we "fold back" the tree and choose the optimal decision for Stage 1.

Our first step is to compare the expected profit of the branches in Stage 2. Assuming that the firm makes the seismic test, the expected profit of Branch C_2 ($29 million) is higher than the profit from selling the contract, Branch C_1; hence the course of action denoted by Branch C_1 can be discarded and should be ignored in our further calculations. Similarly, Branch D_1 results in a higher expected profit ($1 million) than does Branch D_2, so D_2 can also be discarded. If, on the other hand, the firm decides not to make the seismic test, Branch E_2 has a higher expected profit ($13 million) than E_1, and therefore the latter can be discarded. Having first made these eliminations in the second-stage decisions, we can then evaluate the first-stage decision as follows: the expected profit of the seismic test becomes $15 million: $0.50 \cdot \$29,000,000 + 0.50 \cdot \$1,000,000 = \$15,000,000$.

The expected profit in the case of no seismic test is $13,000,000, as explained earlier. Note that this calculation of Stage 1 exploits the previous screening of the alternatives of Stage 2. If the test is successful, Branch C_2 is chosen, so the expected profit of that alternative is $29 million. If the test is unsuccessful, the best path to follow is Branch D_1, which results in a profit of $1 million. Similarly, if we fold back the other branches, the maximum expected profit when the seismic test is *not* made is the $13 million that results from the option of drilling without the test.

Examining our results, we find that the optimal decision at the first stage under the maximum expected return rule is to make the seismic test. If successful, the firm will go ahead at Stage 2 with the decision to drill for oil; should the test prove unsuccessful, the optimal second-stage decision will be to sell the drilling rights. In terms of Figure 19.7, the optimal path follows Branches A_2, B_2, and C_2 if the test is a success, and Branch A_2 followed by B_1 and D_1 should it fail. Note that in both cases we start with Branch A_2, the decision to make the seismic test; the *next decision in the sequence is taken only after the results of the test have been obtained.*

In summary, the decision-tree technique permits us to transport ourselves in conceptual time to the extremities of the tree, where expectations are calculated in terms of the alternative outcomes and their probabilities of occurrence. We then work our way back by folding back, so to speak, the branches of the tree, choosing only those paths that yield the maximum expected profit at each decision junction.

Recall that we selected the best sequence of decisions according to the maximum expected profit rule. The same technique can be employed when the objective is to maximize expected utility: for every monetary outcome, simply substitute the corresponding utility. Moreover, one can calculate the mean and the variance of each branch of the tree and use the mean-variance rule. No matter what decision rule is used, decision trees are useful when a sequence of decisions must be faced.

Problems

19.1. An entrepreneur is considering investing in one of the following three business lines: auto manufacturing, oil refining, or frozen-food manufacturing. His expected profit in each line as a function of future oil prices is given in the following payoff matrix:

	State of nature		
Activity	S_1 Price of oil is up	S_2 Price of oil is down	S_3 Price of oil is unchanged
A_1 Auto manufacturing	100	500	300
A_2 Oil refining	400	200	250
A_3 Frozen-food manufacturing	290	310	300

(a) What is the maximin strategy?
(b) What is the maximax strategy?
(c) If the entrepreneur decides to invest in the auto industry, is he a pessimist or an optimist? Explain.

19.2. Give an example of a payoff matrix in which you apply the minimin criterion.

19.3. Use the payoff table of Problem 19.1 to establish the regret table.

(a) Explain the notion of "regret."
(b) Which strategy would you choose according to the minimax regret criterion?

19.4. Reconsider Problem 19.1. Suppose that if the entrepreneur faces another oil crisis and oil prices go up, the government will give him a grant of $200, no matter what activity he has selected.

(a) Do you think that such a grant, given regardless of the activity selected, should change his decision?

(b) Reconstruct the payoff table of Problem 19.1 to include the effect of the $200 grant and find the selected strategy according to the maximax, maximin, and minimax regret rules.

(c) Compare your results with those you obtained in Problem 19.1. Which rules are independent of the grant of a constant amount of dollars in a given state of nature?

19.5. Repeat Problem 19.1 but add another alternative: investing directly in stocks of oil firms. Thus we have the following payoff table:

		State of nature		
		S_1 Price of oil is up	S_2 Price of oil is down	S_3 Price of oil is unchanged
	Activity			
A_1	Auto manufacturing	100	500	300
A_2	Oil refining	400	200	250
A_3	Frozen-food manufacturing	290	310	300
A_4	Purchase of oil stocks	600	150	400

(a) Which activity would you choose by the maximax rule? Which would you choose by the maximin rule? Which by the minimax regret criterion? Compare your results with those of Problem 19.1.

(b) According to the minimax regret criterion, purchase of oil stocks constitutes the "irrelevant alternative." Explain this concept.

19.6. Referring back to the payoff matrix of Problem 19.1, assume that the entrepreneur makes decisions according to the expected profit criterion.

(a) What activity will he choose if he believes that the following probabilities characterize the three states of nature: $P(S_1) = \frac{1}{3}$, $P(S_2) = \frac{1}{3}$, $P(S_3) = \frac{1}{3}$?

(b) What activity will he choose if it is given that $P(S_1) = 0.1$, $P(S_2) = 0.8$, $P(S_3) = 0.1$?

(c) In what respect does the maximum expected profit rule differ from the other rules (maximax, minimax, minimax regret)?

19.7. (a) Describe the St. Petersburg game. How much would you be willing to pay in order to participate in such a game? What is the expected value of the prize?

(b) Suppose that we change the St. Petersburg game as follows: if heads appears in one of the first 10 tosses (inclusive) you get $1,000, but if it first appears on the eleventh toss or later, you get 2^{n-1}, as in the original game. How much would you be willing to pay to participate in such a game? What is the expected value?

19.8. Suppose you have to choose one of the following two options:

Option A		Option B	
Probability	*Profit*	*Probability*	*Profit*
$\frac{1}{2}$	90	$\frac{1}{2}$	80
$\frac{1}{2}$	110	$\frac{1}{2}$	120

(a) Which option would you choose according to the expected profit rule?

(b) Which option would you choose according to the expected utility rule, knowing that the marginal utility of money diminishes? Illustrate your answer graphically.

19.9. Define risk averter, risk lover, and risk neutral.

19.10. Suppose there is a 50 percent chance that ABC stock will sell at $300 next week and a 50 percent chance that it will sell at $400. A risk averter is willing to pay a maximum of $320 for this stock. What is the risk premium? Show graphically at least three points on his utility function. (Draw two as you wish, and the third point must be determined by the data of this problem.)

19.11. Suppose that an individual's utility function is given by $U(x) = \sqrt{x}$. The individual has to select between the following two options:

Option A		Option B	
Probability	Net profit	Probability	Net profit
$\frac{1}{3}$	0	$\frac{3}{4}$	0
$\frac{1}{3}$	100	$\frac{1}{4}$	1,600
$\frac{1}{3}$	900		

(a) Which option has a higher expected profit?

(b) Which option would you select when $U(x) = \sqrt{x}$? Is the individual a risk averter or risk lover?

(c) Which option would you choose if your utility function is given by $U^*(x) = 10 + 2\sqrt{x}$?

(d) Compare and explain your answers to b and c.

19.12. Which of the two options in Problem 19.11 would you select as better if you used the mean-variance rule?

19.13. Suppose the means and variance of the profits of six investment options are as follows:

Option	Mean	Variance
A	5	14
B	7	10
C	8	8
D	9	3
E	10	12
F	4	8

Which of the options would you certainly reject if you used the mean-variance rule?

19.14. Explain the concept of Bayesian decision-making.

19.15. International Car, Inc., is considering investing $10 million in an African country in order to manufacture cars for the local market. Financial analysis has shown that International Car will make a net profit of $500 per car sold. The company will invest only if it estimates that all of the $100 million can be recovered in the first year (if 20,000 cars will be sold).

The population of the African country consists of 1 million families, 2 percent of whom are considered "rich" and 98 percent "poor." It is estimated that 90 percent of the rich and only 1 percent of the poor will buy cars from International Car.

(a) Should the firm invest in the country?

(b) Suppose there is a 10 percent chance of a revolution in the country (political risk). International Car will continue to operate, but 80 percent of the rich will leave the country. With this additional information, should International Car invest?

19.16. XYZ Company has just received an order for 1000 widgets. The production process has not been inspected in a while, and hence the proportion of defectives that are produced, θ, is uncertain. However, from experience the manager is able to estimate a prior probability distribution for θ (see Table P19.16). There are two corrective actions available: (a) Have the service company make a routine overhaul; this ensures that θ will be no higher than 0.10, that is, if the true θ was 0.15 or greater, then it is brought down to 0.10; if the true θ was 0.10 or less, the overhaul has no effect. This costs $25. (b) Have the service company make a thorough check and replace all worn-out parts. This ensures that θ will be 0.05. This costs $100.

First, the manager takes a sample of 10 items ($n = 10$) to study the state of the processes.

He observes 5 defectives ($r = 5$). Based on this sample, he obtains a revised or posterior distribution for θ.

Now the manager must choose one of the following actions:

(a) Leave the process as it is.
(b) Have the service company make a routine overhaul.
(c) Have the service company make a major overhaul.

He will choose the action that maximizes expected net profit. Other relevant data: net profit per item is $0.50, the cost of replacing defective items is $2.00 per item.

What action should he choose?

TABLE P19.16
Prior Probability Distribution of θ, the Proportion of Defectives

θ_i	$P(\theta = \theta_i)$
0.05	0.60
0.10	0.30
0.15	0.08
0.25	0.02
	1.00

APPENDIX A

Statistical Tables

Table A.1
The Cumulative Binomial Distribution

n	x	0.05	0.10	0.15	0.20	0.25	0.30	0.35	0.40	0.45	0.50
2	0	0.9025	0.8100	0.7225	0.6400	0.5625	0.4900	0.4225	0.3600	0.3025	0.2500
	1	0.9975	0.9900	0.9775	0.9600	0.9375	0.9100	0.8775	0.8400	0.7975	0.7500
	2	1.0000	1.0000	1.0000	1.0000	1.0000	1.0000	1.0000	1.0000	1.0000	1.0000
3	0	0.8574	0.7290	0.6141	0.5120	0.4219	0.3430	0.2746	0.2160	0.1664	0.1250
	1	0.9928	0.9720	0.9392	0.8960	0.8438	0.7840	0.7183	0.6480	0.5748	0.5000
	2	0.9999	0.9990	0.9966	0.9920	0.9844	0.9730	0.9571	0.9360	0.9089	0.8750
	3	1.0000	1.0000	1.0000	1.0000	1.0000	1.0000	1.0000	1.0000	1.0000	1.0000
4	0	0.8145	0.6561	0.5220	0.4096	0.3164	0.2401	0.1785	0.1296	0.0915	0.0625
	1	0.9860	0.9477	0.8905	0.8192	0.7383	0.6517	0.5630	0.4752	0.3910	0.3125
	2	0.9995	0.9963	0.9880	0.9728	0.9492	0.9163	0.8735	0.8208	0.7585	0.6875
	3	1.0000	0.9999	0.9995	0.9984	0.9961	0.9919	0.9850	0.9744	0.9590	0.9375
	4	1.0000	1.0000	1.0000	1.0000	1.0000	1.0000	1.0000	1.0000	1.0000	1.0000
5	0	0.7738	0.5905	0.4437	0.3277	0.2373	0.1681	0.1160	0.0778	0.0503	0.0313
	1	0.9774	0.9185	0.8352	0.7373	0.6328	0.5282	0.4284	0.3370	0.2562	0.1875
	2	0.9988	0.9914	0.9734	0.9421	0.8965	0.8369	0.7648	0.6826	0.5931	0.5000
	3	1.0000	0.9995	0.9978	0.9933	0.9844	0.9692	0.9460	0.9130	0.8688	0.8125
	4	1.0000	1.0000	0.9999	0.9997	0.9990	0.9976	0.9947	0.9898	0.9815	0.9688
	5	1.0000	1.0000	1.0000	1.0000	1.0000	1.0000	1.0000	1.0000	1.0000	1.0000
6	0	0.7351	0.5314	0.3771	0.2621	0.1780	0.1176	0.0754	0.0467	0.0277	0.0156
	1	0.9672	0.8857	0.7765	0.6554	0.5339	0.4202	0.3191	0.2333	0.1636	0.1094
	2	0.9978	0.9842	0.9527	0.9011	0.8306	0.7443	0.6471	0.5443	0.4415	0.3437
	3	0.9999	0.9987	0.9941	0.9830	0.9624	0.9295	0.8826	0.8208	0.7447	0.6563
	4	1.0000	0.9999	0.9996	0.9984	0.9954	0.9891	0.9777	0.9590	0.9308	0.8906
	5	1.0000	1.0000	1.0000	0.9999	0.9998	0.9993	0.9982	0.9959	0.9917	0.9844
	6	1.0000	1.0000	1.0000	1.0000	1.0000	1.0000	1.0000	1.0000	1.0000	1.0000

Table A.1
The Cumulative Binomial Distribution (Continued)

n	x	0.05	0.10	0.15	0.20	0.25	p 0.30	0.35	0.40	0.45	0.50
7	0	0.6983	0.4783	0.3206	0.2097	0.1335	0.0824	0.0490	0.0280	0.0152	0.0078
	1	0.9556	0.8503	0.7166	0.5767	0.4449	0.3294	0.2338	0.1586	0.1024	0.0625
	2	0.9962	0.9743	0.9262	0.8520	0.7564	0.6471	0.5323	0.4199	0.3164	0.2266
	3	0.9998	0.9973	0.9879	0.9667	0.9294	0.8740	0.8002	0.7102	0.6083	0.5000
	4	1.0000	0.9998	0.9988	0.9953	0.9871	0.9712	0.9444	0.9037	0.8471	0.7734
	5	1.0000	1.0000	0.9999	0.9996	0.9987	0.9962	0.9910	0.9812	0.9643	0.9375
	6	1.0000	1.0000	1.0000	1.0000	0.9999	0.9998	0.9994	0.9984	0.9963	0.9922
	7	1.0000	1.0000	1.0000	1.0000	1.0000	1.0000	1.0000	1.0000	1.0000	1.0000
8	0	0.6634	0.4305	0.2725	0.1678	0.1001	0.0576	0.0319	0.0168	0.0084	0.0039
	1	0.9428	0.8131	0.6572	0.5033	0.3671	0.2553	0.1691	0.1064	0.0632	0.0352
	2	0.9942	0.9619	0.8948	0.7969	0.6785	0.5518	0.4278	0.3154	0.2201	0.1445
	3	0.9996	0.9950	0.9786	0.9437	0.8862	0.8059	0.7064	0.5941	0.4770	0.3633
	4	1.0000	0.9996	0.9971	0.9896	0.9727	0.9420	0.8939	0.8263	0.7396	0.6367
	5	1.0000	1.0000	0.9998	0.9988	0.9958	0.9887	0.9747	0.9502	0.9115	0.8555
	6	1.0000	1.0000	1.0000	0.9999	0.9996	0.9987	0.9964	0.9915	0.9819	0.9648
	7	1.0000	1.0000	1.0000	1.0000	1.0000	0.9999	0.9998	0.9993	0.9983	0.9961
	8	1.0000	1.0000	1.0000	1.0000	1.0000	1.0000	1.0000	1.0000	1.0000	1.0000
9	0	0.6302	0.3874	0.2316	0.1342	0.0751	0.0404	0.0207	0.0101	0.0046	0.0020
	1	0.9288	0.7748	0.5995	0.4362	0.3003	0.1960	0.1211	0.0705	0.0385	0.0195
	2	0.9916	0.9470	0.8591	0.7382	0.6007	0.4628	0.3373	0.2318	0.1495	0.0898
	3	0.9994	0.9917	0.9661	0.9144	0.8343	0.7297	0.6089	0.4826	0.3614	0.2539
	4	1.0000	0.9991	0.9944	0.9804	0.9511	0.9012	0.8283	0.7334	0.6214	0.5000
	5	1.0000	0.9999	0.9994	0.9969	0.9900	0.9747	0.9464	0.9006	0.8342	0.7461
	6	1.0000	1.0000	1.0000	0.9997	0.9987	0.9957	0.9888	0.9750	0.9502	0.9102
	7	1.0000	1.0000	1.0000	1.0000	0.9999	0.9996	0.9986	0.9962	0.9909	0.9805
	8	1.0000	1.0000	1.0000	1.0000	1.0000	1.0000	0.9999	0.9997	0.9992	0.9980
	9	1.0000	1.0000	1.0000	1.0000	1.0000	1.0000	1.0000	1.0000	1.0000	1.0000

Table A.1
The Cumulative Binomial Distribution (Continued)

n	x						p				
		0.05	0.10	0.15	0.20	0.25	0.30	0.35	0.40	0.45	0.50
10	0	0.5987	0.3487	0.1969	0.1074	0.0563	0.0282	0.0135	0.0060	0.0025	0.0010
	1	0.9139	0.7361	0.5443	0.3758	0.2440	0.1493	0.0860	0.0464	0.0233	0.0107
	2	0.9885	0.9298	0.8202	0.6778	0.5256	0.3828	0.2616	0.1673	0.0996	0.0547
	3	0.9990	0.9872	0.9500	0.8791	0.7759	0.6496	0.5138	0.3823	0.2660	0.1719
	4	0.9999	0.9984	0.9901	0.9672	0.9219	0.8497	0.7515	0.6331	0.5044	0.3770
	5	1.0000	0.9999	0.9986	0.9936	0.9803	0.9527	0.9051	0.8338	0.7384	0.6230
	6	1.0000	1.0000	0.9999	0.9991	0.9965	0.9894	0.9740	0.9452	0.8980	0.8281
	7	1.0000	1.0000	1.0000	0.9999	0.9996	0.9984	0.9952	0.9877	0.9726	0.9453
	8	1.0000	1.0000	1.0000	1.0000	1.0000	0.9999	0.9995	0.9983	0.9955	0.9893
	9	1.0000	1.0000	1.0000	1.0000	1.0000	1.0000	1.0000	0.9999	0.9997	0.9990
	10	1.0000	1.0000	1.0000	1.0000	1.0000	1.0000	1.0000	1.0000	1.0000	1.0000
11	0	0.5688	0.3138	0.1673	0.0859	0.0422	0.0198	0.0088	0.0036	0.0014	0.0005
	1	0.8981	0.6974	0.4922	0.3221	0.1971	0.1130	0.0606	0.0302	0.0139	0.0059
	2	0.9848	0.9104	0.7788	0.6174	0.4552	0.3127	0.2001	0.1189	0.0652	0.0327
	3	0.9984	0.9815	0.9306	0.8389	0.7133	0.5696	0.4256	0.2963	0.1911	0.1133
	4	0.9999	0.9972	0.9841	0.9496	0.8854	0.7897	0.6683	0.5328	0.3971	0.2744
	5	1.0000	0.9997	0.9973	0.9883	0.9657	0.9218	0.8513	0.7535	0.6331	0.5000
	6	1.0000	1.0000	0.9997	0.9980	0.9924	0.9784	0.9499	0.9006	0.8262	0.7256
	7	1.0000	1.0000	1.0000	0.9998	0.9988	0.9957	0.9878	0.9707	0.9390	0.8867
	8	1.0000	1.0000	1.0000	1.0000	0.9999	0.9994	0.9980	0.9941	0.9852	0.9673
	9	1.0000	1.0000	1.0000	1.0000	1.0000	1.0000	0.9998	0.9993	0.9978	0.9941
	10	1.0000	1.0000	1.0000	1.0000	1.0000	1.0000	1.0000	1.0000	0.9998	0.9995
	11	1.0000	1.0000	1.0000	1.0000	1.0000	1.0000	1.0000	1.0000	1.0000	1.0000
12	0	0.5404	0.2824	0.1422	0.0687	0.0317	0.0138	0.0057	0.0022	0.0008	0.0002
	1	0.8816	0.6590	0.4435	0.2749	0.1584	0.0850	0.0424	0.0196	0.0083	0.0032
	2	0.9804	0.8891	0.7358	0.5583	0.3907	0.2528	0.1513	0.0834	0.0421	0.0193
	3	0.9978	0.9744	0.9078	0.7946	0.6488	0.4925	0.3467	0.2253	0.1345	0.0730
	4	0.9998	0.9957	0.9761	0.9274	0.8424	0.7237	0.5833	0.4382	0.3044	0.1938

Table A.1
The Cumulative Binomial Distribution (Continued)

n	x	0.05	0.10	0.15	0.20	0.25	0.30	0.35	0.40	0.45	0.50
	5	1.0000	0.9995	0.9954	0.9806	0.9456	0.8822	0.7873	0.6652	0.5269	0.3872
	6	1.0000	0.9999	0.9993	0.9961	0.9857	0.9614	0.9154	0.8418	0.7393	0.6128
	7	1.0000	1.0000	0.9999	0.9994	0.9972	0.9905	0.9745	0.9427	0.8883	0.8062
	8	1.0000	1.0000	1.0000	0.9999	0.9996	0.9983	0.9944	0.9847	0.9644	0.9270
	9	1.0000	1.0000	1.0000	1.0000	1.0000	0.9998	0.9992	0.9972	0.9921	0.9807
	10	1.0000	1.0000	1.0000	1.0000	1.0000	1.0000	0.9999	0.9997	0.9989	0.9968
	11	1.0000	1.0000	1.0000	1.0000	1.0000	1.0000	1.0000	1.0000	0.9999	0.9998
	12	1.0000	1.0000	1.0000	1.0000	1.0000	1.0000	1.0000	1.0000	1.0000	1.0000
13	0	0.5133	0.2542	0.1209	0.0550	0.0238	0.0097	0.0037	0.0013	0.0004	0.0001
	1	0.8646	0.6213	0.3983	0.2336	0.1267	0.0637	0.0296	0.0126	0.0049	0.0017
	2	0.9755	0.8661	0.6920	0.5017	0.3326	0.2025	0.1132	0.0579	0.0269	0.0112
	3	0.9969	0.9658	0.8820	0.7473	0.5843	0.4206	0.2783	0.1686	0.0929	0.0461
	4	0.9997	0.9935	0.9658	0.9009	0.7940	0.6543	0.5005	0.3530	0.2279	0.1334
	5	1.0000	0.9991	0.9925	0.9700	0.9198	0.8346	0.7159	0.5744	0.4268	0.2905
	6	1.0000	0.9999	0.9987	0.9930	0.9757	0.9376	0.8705	0.7712	0.6437	0.5000
	7	1.0000	1.0000	0.9998	0.9988	0.9944	0.9818	0.9538	0.9023	0.8212	0.7095
	8	1.0000	1.0000	1.0000	0.9998	0.9990	0.9960	0.9874	0.9679	0.9302	0.8666
	9	1.0000	1.0000	1.0000	1.0000	0.9999	0.9993	0.9975	0.9922	0.9797	0.9539
	10	1.0000	1.0000	1.0000	1.0000	1.0000	0.9999	0.9997	0.9987	0.9959	0.9888
	11	1.0000	1.0000	1.0000	1.0000	1.0000	1.0000	1.0000	0.9999	0.9995	0.9983
	12	1.0000	1.0000	1.0000	1.0000	1.0000	1.0000	1.0000	1.0000	1.0000	0.9999
	13	1.0000	1.0000	1.0000	1.0000	1.0000	1.0000	1.0000	1.0000	1.0000	1.0000
14	0	0.4877	0.2288	0.1028	0.0440	0.0178	0.0068	0.0024	0.0008	0.0002	0.0001
	1	0.8470	0.5846	0.3567	0.1979	0.1010	0.0475	0.0205	0.0081	0.0029	0.0009
	2	0.9699	0.8416	0.6479	0.4481	0.2811	0.1608	0.0839	0.0398	0.0170	0.0065
	3	0.9958	0.9559	0.8535	0.6982	0.5213	0.3552	0.2205	0.1243	0.0632	0.0287
	4	0.9996	0.9908	0.9533	0.8702	0.7415	0.5842	0.4227	0.2793	0.1672	0.0898
	5	1.0000	0.9985	0.9885	0.9561	0.8883	0.7805	0.6405	0.4859	0.3373	0.2120

Table A.1
The Cumulative Binomial Distribution (Continued)

n	x	p									
		0.05	0.10	0.15	0.20	0.25	0.30	0.35	0.40	0.45	0.50
	6	1.0000	0.9998	0.9978	0.9884	0.9617	0.9067	0.8164	0.6925	0.5461	0.3953
	7	1.0000	1.0000	0.9997	0.9976	0.9897	0.9685	0.9247	0.8499	0.7414	0.6047
	8	1.0000	1.0000	1.0000	0.9996	0.9978	0.9917	0.9757	0.9417	0.8811	0.7880
	9	1.0000	1.0000	1.0000	1.0000	0.9997	0.9983	0.9940	0.9825	0.9574	0.9102
	10	1.0000	1.0000	1.0000	1.0000	1.0000	0.9998	0.9989	0.9961	0.9886	0.9713
	11	1.0000	1.0000	1.0000	1.0000	1.0000	1.0000	0.9999	0.9994	0.9978	0.9935
	12	1.0000	1.0000	1.0000	1.0000	1.0000	1.0000	1.0000	0.9999	0.9997	0.9991
	13	1.0000	1.0000	1.0000	1.0000	1.0000	1.0000	1.0000	1.0000	1.0000	0.9999
	14	1.0000	1.0000	1.0000	1.0000	1.0000	1.0000	1.0000	1.0000	1.0000	1.0000
15	0	0.4633	0.2059	0.0874	0.0352	0.0134	0.0047	0.0016	0.0005	0.0001	0.0000
	1	0.8290	0.5490	0.3186	0.1671	0.0802	0.0353	0.0142	0.0052	0.0017	0.0005
	2	0.9638	0.8159	0.6042	0.3980	0.2361	0.1268	0.0617	0.0271	0.0107	0.0037
	3	0.9945	0.9444	0.8227	0.6482	0.4613	0.2969	0.1727	0.0905	0.0424	0.0176
	4	0.9994	0.9873	0.9383	0.8358	0.6865	0.5155	0.3519	0.2173	0.1204	0.0592
	5	0.9999	0.9978	0.9832	0.9389	0.8516	0.7216	0.5643	0.4032	0.2608	0.1509
	6	1.0000	0.9997	0.9964	0.9819	0.9434	0.8689	0.7548	0.6098	0.4522	0.3036
	7	1.0000	1.0000	0.9994	0.9958	0.9827	0.9500	0.8868	0.7869	0.6535	0.5000
	8	1.0000	1.0000	0.9999	0.9992	0.9958	0.9848	0.9578	0.9050	0.8182	0.6964
	9	1.0000	1.0000	1.0000	0.9999	0.9992	0.9963	0.9876	0.9662	0.9231	0.8491
	10	1.0000	1.0000	1.0000	1.0000	0.9999	0.9993	0.9972	0.9907	0.9745	0.9408
	11	1.0000	1.0000	1.0000	1.0000	1.0000	0.9999	0.9995	0.9981	0.9937	0.9824
	12	1.0000	1.0000	1.0000	1.0000	1.0000	1.0000	0.9999	0.9997	0.9989	0.9963
	13	1.0000	1.0000	1.0000	1.0000	1.0000	1.0000	1.0000	1.0000	0.9999	0.9995
	14	1.0000	1.0000	1.0000	1.0000	1.0000	1.0000	1.0000	1.0000	1.0000	1.0000
	15	1.0000	1.0000	1.0000	1.0000	1.0000	1.0000	1.0000	1.0000	1.0000	1.0000
20	0	0.3585	0.1216	0.0388	0.0115	0.0032	0.0008	0.0002	0.0000	0.0000	0.0000
	1	0.7358	0.3917	0.1756	0.0692	0.0243	0.0076	0.0021	0.0005	0.0001	0.0000
	2	0.9245	0.6769	0.4049	0.2061	0.0913	0.0355	0.0121	0.0036	0.0009	0.0002

Table A.1
The Cumulative Binomial Distribution (Continued)

n	x	\(p\) 0.05	0.10	0.15	0.20	0.25	0.30	0.35	0.40	0.45	0.50
	3	0.9841	0.8670	0.6477	0.4114	0.2252	0.1071	0.0444	0.0160	0.0049	0.0013
	4	0.9974	0.9568	0.8298	0.6296	0.4148	0.2375	0.1182	0.0510	0.0189	0.0059
	5	0.9997	0.9887	0.9327	0.8042	0.6172	0.4164	0.2454	0.1256	0.0553	0.0207
	6	1.0000	0.9976	0.9781	0.9133	0.7858	0.6080	0.4166	0.2500	0.1299	0.0577
	7	1.0000	0.9996	0.9941	0.9679	0.8982	0.7723	0.6010	0.4159	0.2520	0.1316
	8	1.0000	0.9999	0.9987	0.9900	0.9591	0.8867	0.7624	0.5956	0.4143	0.2517
	9	1.0000	1.0000	0.9998	0.9974	0.9861	0.9520	0.8782	0.7553	0.5914	0.4119
	10	1.0000	1.0000	1.0000	0.9994	0.9961	0.9829	0.9468	0.8725	0.7507	0.5881
	11	1.0000	1.0000	1.0000	0.9999	0.9991	0.9949	0.9804	0.9435	0.8692	0.7483
	12	1.0000	1.0000	1.0000	1.0000	0.9998	0.9987	0.9940	0.9790	0.9420	0.8684
	13	1.0000	1.0000	1.0000	1.0000	1.0000	0.9997	0.9985	0.9935	0.9786	0.9423
	14	1.0000	1.0000	1.0000	1.0000	1.0000	1.0000	0.9997	0.9984	0.9936	0.9793
	15	1.0000	1.0000	1.0000	1.0000	1.0000	1.0000	1.0000	0.9997	0.9985	0.9941
	16	1.0000	1.0000	1.0000	1.0000	1.0000	1.0000	1.0000	1.0000	0.9997	0.9987
	17	1.0000	1.0000	1.0000	1.0000	1.0000	1.0000	1.0000	1.0000	1.0000	0.9998
	18	1.0000	1.0000	1.0000	1.0000	1.0000	1.0000	1.0000	1.0000	1.0000	1.0000
	19	1.0000	1.0000	1.0000	1.0000	1.0000	1.0000	1.0000	1.0000	1.0000	1.0000
	20	1.0000	1.0000	1.0000	1.0000	1.0000	1.0000	1.0000	1.0000	1.0000	1.0000

Table A.2
The Cumulative Poisson Distribution

x	λ = 0.1	λ = 0.2	λ = 0.3	λ = 0.4	λ = 0.5
0	0.90484	0.81873	0.74082	0.67032	0.60653
1	0.99532	0.98248	0.96306	0.93845	0.90980
2	0.99985	0.99885	0.99640	0.99207	0.98561
3	1.00000	0.99994	0.99973	0.99922	0.99825
4		1.00000	0.99998	0.99994	0.99983
5			1.00000	1.00000	0.99999
6					1.00000

x	λ = 0.6	λ = 0.7	λ = 0.8	λ = 0.9	λ = 1.0
0	0.54881	0.49658	0.44933	0.40657	0.36788
1	0.87810	0.84419	0.80879	0.77248	0.73576
2	0.97688	0.96586	0.95258	0.93714	0.91970
3	0.99664	0.99425	0.99092	0.98654	0.98101
4	0.99961	0.99921	0.99859	0.99766	0.99634
5	0.99996	0.99991	0.99982	0.99966	0.99941
6	1.00000	0.99999	0.99998	0.99996	0.99992
7		1.00000	1.00000	1.00000	0.99999
8					1.00000

x	λ = 2	λ = 3	λ = 4	λ = 5	λ = 6
0	0.13534	0.04979	0.01832	0.00674	0.00248
1	0.40601	0.19915	0.09158	0.04043	0.01735
2	0.67668	0.42319	0.23810	0.12465	0.06197
3	0.85712	0.64723	0.43347	0.26503	0.15120
4	0.94735	0.81526	0.62884	0.44049	0.28506
5	0.98344	0.91608	0.78513	0.61596	0.44568
6	0.99547	0.96649	0.88933	0.76218	0.60630
7	0.99890	0.98810	0.94887	0.86663	0.74398
8	0.99976	0.99620	0.97864	0.93191	0.84724
9	0.99995	0.99890	0.99187	0.96817	0.91608
10	0.99999	0.99971	0.99716	0.98630	0.95738
11	1.00000	0.99993	0.99908	0.99455	0.97991
12		0.99998	0.99973	0.99798	0.99117
13		1.00000	0.99992	0.99930	0.99637
14			0.99998	0.99977	0.99860
15			1.00000	0.99993	0.99949
16				0.99998	0.99982
17				1.00000	0.99994
18					0.99998
19					1.00000

Table A.2
The Cumulative Poisson Distribution (Concluded)

x	$\lambda = 7$	$\lambda = 8$	$\lambda = 9$	$\lambda = 10$
0	0.00091	0.00033	0.00012	0.00004
1	0.00730	0.00302	0.00123	0.00050
2	0.02964	0.01375	0.00623	0.00277
3	0.08176	0.04238	0.02123	0.01034
4	0.17299	0.09963	0.05496	0.02925
5	0.30071	0.19124	0.11569	0.06709
6	0.44971	0.31337	0.20678	0.13014
7	0.59871	0.45296	0.32390	0.22022
8	0.72909	0.59255	0.45565	0.33282
9	0.83050	0.71662	0.58741	0.45793
10	0.90148	0.81589	0.70599	0.58304
11	0.94665	0.88808	0.80301	0.69678
12	0.97300	0.93620	0.87577	0.79156
13	0.98719	0.96582	0.92615	0.86446
14	0.99428	0.98274	0.95853	0.91654
15	0.99759	0.99177	0.97796	0.95126
16	0.99904	0.99628	0.98889	0.97296
17	0.99964	0.99841	0.99468	0.98572
18	0.99987	0.99935	0.99757	0.99281
19	0.99996	0.99975	0.99894	0.99655
20	0.99999	0.99991	0.99956	0.99841
21	1.00000	0.99997	0.99982	0.99930
22		0.99999	0.99993	0.99970
23		1.00000	0.99998	0.99988
24			0.99999	0.99995
25			1.00000	0.99998
26				0.99999
27				1.00000

Source: From *Poisson's Binomial Exponential Limit* by E. C. Molina © 1942. Reprinted by permission of Bell Laboratories.

Table A.3
Exponential Functions

x	e^x	e^{-x}	x	e^x	e^{-x}
0.00	1.000	1.000	3.00	20.086	0.050
0.10	1.105	0.905	3.10	22.198	0.045
0.20	1.221	0.819	3.20	24.533	0.041
0.30	1.350	0.741	3.30	27.113	0.037
0.40	1.492	0.670	3.40	29.964	0.033
0.50	1.649	0.607	3.50	33.115	0.030
0.60	1.822	0.549	3.60	36.598	0.027
0.70	2.014	0.497	3.70	40.447	0.025
0.80	2.226	0.449	3.80	44.701	0.022
0.90	2.460	0.407	3.90	49.402	0.020
1.00	2.718	0.368	4.00	54.598	0.018
1.10	3.004	0.333	4.10	60.340	0.017
1.20	3.320	0.301	4.20	66.686	0.015
1.30	3.669	0.273	4.30	73.700	0.014
1.40	4.055	0.247	4.40	81.451	0.012
1.50	4.482	0.223	4.50	90.017	0.011
1.60	4.953	0.202	4.60	99.484	0.010
1.70	5.474	0.183	4.70	109.947	0.009
1.80	6.050	0.165	4.80	121.510	0.008
1.90	6.686	0.150	4.90	134.290	0.007
2.00	7.389	0.135	5.00	148.413	0.007
2.10	8.166	0.122	5.10	164.022	0.006
2.20	9.025	0.111	5.20	181.272	0.006
2.30	9.974	0.100	5.30	200.337	0.005
2.40	11.023	0.091	5.40	221.406	0.005
2.50	12.182	0.082	5.50	244.692	0.004
2.60	13.464	0.074	5.60	270.426	0.004
2.70	14.880	0.067	5.70	298.867	0.003
2.80	16.445	0.061	5.80	330.300	0.003
2.90	18.174	0.055	5.90	365.037	0.003
3.00	20.086	0.050	6.00	403.429	0.002

Table A.4
The Chi-Square Distribution

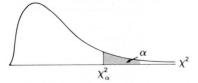

The following table provides the values of χ^2_α that correspond to a given right-tail area α and a specified number of degrees of freedom.

df	Right-tail area							
	0.99	0.975	0.95	0.90	0.10	0.05	0.025	0.01
1	0.00016	0.00098	0.0039	0.016	2.706	3.841	5.024	6.635
2	0.02001	0.05064	0.103	0.211	4.605	5.991	7.378	9.210
3	0.115	0.216	0.352	0.584	6.251	7.815	9.348	11.345
4	0.297	0.484	0.711	1.064	7.779	9.488	11.143	13.277
5	0.554	0.831	1.145	1.610	9.236	11.070	12.832	15.086
6	0.872	1.237	1.635	2.204	10.645	12.592	14.449	16.812
7	1.239	1.690	2.167	2.833	12.017	14.067	16.013	18.475
8	1.646	2.180	2.733	3.490	13.362	15.507	17.535	20.090
9	2.088	2.700	3.325	4.168	14.684	16.919	19.023	21.666
10	2.558	3.247	3.940	4.865	15.987	18.307	20.483	23.209
11	3.053	3.816	4.575	5.578	17.275	19.675	21.920	24.725
12	3.571	4.404	5.226	6.304	18.549	21.026	23.337	26.217
13	4.107	5.009	5.892	7.042	19.812	22.362	24.736	27.688
14	4.660	5.629	6.571	7.790	21.064	23.685	26.119	29.141
15	5.229	6.262	7.261	8.547	22.307	24.996	27.488	30.578
16	5.812	6.908	7.962	9.312	23.542	26.296	28.845	32.000
17	6.408	7.564	8.672	10.085	24.769	27.587	30.191	33.409
18	7.015	8.231	9.390	10.865	25.989	28.869	31.527	34.805
19	7.633	8.907	10.117	11.651	27.204	30.144	32.852	36.191
20	8.260	9.591	10.851	12.443	28.412	31.410	34.170	37.566
21	8.897	10.283	11.591	13.240	29.615	32.671	35.479	38.932
22	9.542	10.982	12.338	14.041	30.813	33.924	36.781	40.289
23	10.196	11.689	13.091	14.848	32.007	35.172	38.076	41.638
24	10.856	12.401	13.848	15.659	33.196	36.415	39.364	42.980
25	11.524	13.120	14.611	16.473	34.382	37.652	40.646	44.314
26	12.198	13.844	15.379	17.292	35.563	38.885	41.923	45.642
27	12.879	14.573	16.151	18.114	36.741	40.113	43.194	46.963
28	13.565	15.308	16.928	18.939	37.916	41.337	44.461	48.278
29	14.256	16.047	17.708	19.768	39.087	42.557	45.722	49.588
30	14.953	16.791	18.493	20.599	40.256	43.773	46.979	50.892

Source: Table A.4 is taken from Table IV of Fisher and Yates: *Statistical Tables for Biological, Agricultural and Medical Research*, published by Longman Group Ltd., London (previously published by Oliver & Boyd, Edinburgh), and by permission of the authors and publishers.

Table A.5 The F distribution

Table A.5a
Upper 5% Points (0.95 Fractiles)

$\nu_2 \backslash \nu_1$	1	2	3	4	5	6	7	8	9	10	12	15	20	24	30	40	60	120	∞
1	161.4	199.5	215.7	224.6	230.2	234.0	236.8	238.9	240.5	241.9	243.9	245.9	248.0	249.1	250.1	251.1	252.2	253.3	254.3
2	18.51	19.00	19.16	19.25	19.30	19.33	19.35	19.37	19.38	19.40	19.41	19.43	19.45	19.45	19.46	19.47	19.48	19.49	19.50
3	10.13	9.55	9.28	9.12	9.01	8.94	8.89	8.85	8.81	8.79	8.74	8.70	8.66	8.64	8.62	8.59	8.57	8.55	8.53
4	7.71	6.94	6.59	6.39	6.26	6.16	6.09	6.04	6.00	5.96	5.91	5.86	5.80	5.77	5.75	5.72	5.69	5.66	5.63
5	6.61	5.79	5.41	5.19	5.05	4.95	4.88	4.82	4.77	4.74	4.68	4.62	4.56	4.53	4.50	4.46	4.43	4.40	4.36
6	5.99	5.14	4.76	4.53	4.39	4.28	4.21	4.15	4.10	4.06	4.00	3.94	3.87	3.84	3.81	3.77	3.74	3.70	3.67
7	5.59	4.74	4.35	4.12	3.97	3.87	3.79	3.73	3.68	3.64	3.57	3.51	3.44	3.41	3.38	3.34	3.30	3.27	3.23
8	5.32	4.46	4.07	3.84	3.69	3.58	3.50	3.44	3.39	3.35	3.28	3.22	3.15	3.12	3.08	3.04	3.01	2.97	2.93
9	5.12	4.26	3.86	3.63	3.48	3.37	3.29	3.23	3.18	3.14	3.07	3.01	2.94	2.90	2.86	2.83	2.79	2.75	2.71
10	4.96	4.10	3.71	3.48	3.33	3.22	3.14	3.07	3.02	2.98	2.91	2.85	2.77	2.74	2.70	2.66	2.62	2.58	2.54
11	4.84	3.98	3.59	3.36	3.20	3.09	3.01	2.95	2.90	2.85	2.79	2.72	2.65	2.61	2.57	2.53	2.49	2.45	2.40
12	4.75	3.89	3.49	3.26	3.11	3.00	2.91	2.85	2.80	2.75	2.69	2.62	2.54	2.51	2.47	2.43	2.38	2.34	2.30
13	4.67	3.81	3.41	3.18	3.03	2.92	2.83	2.77	2.71	2.67	2.60	2.53	2.46	2.42	2.38	2.34	2.30	2.25	2.21
14	4.60	3.74	3.34	3.11	2.96	2.85	2.76	2.70	2.65	2.60	2.53	2.46	2.39	2.35	2.31	2.27	2.22	2.18	2.13
15	4.54	3.68	3.29	3.06	2.90	2.79	2.71	2.64	2.59	2.54	2.48	2.40	2.33	2.29	2.25	2.20	2.16	2.11	2.07
16	4.49	3.63	3.24	3.01	2.85	2.74	2.66	2.59	2.54	2.49	2.42	2.35	2.28	2.24	2.19	2.15	2.11	2.06	2.01
17	4.45	3.59	3.20	2.96	2.81	2.70	2.61	2.55	2.49	2.45	2.38	2.31	2.23	2.19	2.15	2.10	2.06	2.01	1.96
18	4.41	3.55	3.16	2.93	2.77	2.66	2.58	2.51	2.46	2.41	2.34	2.27	2.19	2.15	2.11	2.06	2.02	1.97	1.92
19	4.38	3.52	3.13	2.90	2.74	2.63	2.54	2.48	2.42	2.38	2.31	2.23	2.16	2.11	2.07	2.03	1.98	1.93	1.88
20	4.35	3.49	3.10	2.87	2.71	2.60	2.51	2.45	2.39	2.35	2.28	2.20	2.12	2.08	2.04	1.99	1.95	1.90	1.84
21	4.32	3.47	3.07	2.84	2.68	2.57	2.49	2.42	2.37	2.32	2.25	2.18	2.10	2.05	2.01	1.96	1.92	1.87	1.81
22	4.30	3.44	3.05	2.82	2.66	2.55	2.46	2.40	2.34	2.30	2.23	2.15	2.07	2.03	1.98	1.94	1.89	1.84	1.78
23	4.28	3.42	3.03	2.80	2.64	2.53	2.44	2.37	2.32	2.27	2.20	2.13	2.05	2.01	1.96	1.91	1.86	1.81	1.76
24	4.26	3.40	3.01	2.78	2.62	2.51	2.42	2.36	2.30	2.25	2.18	2.11	2.03	1.98	1.94	1.89	1.84	1.79	1.73
25	4.24	3.39	2.99	2.76	2.60	2.49	2.40	2.34	2.28	2.24	2.16	2.09	2.01	1.96	1.92	1.87	1.82	1.77	1.71
26	4.23	3.37	2.98	2.74	2.59	2.47	2.39	2.32	2.27	2.22	2.15	2.07	1.99	1.95	1.90	1.85	1.80	1.75	1.69
27	4.21	3.35	2.96	2.73	2.57	2.46	2.37	2.31	2.25	2.20	2.13	2.06	1.97	1.93	1.88	1.84	1.79	1.73	1.67
28	4.20	3.34	2.95	2.71	2.56	2.45	2.36	2.29	2.24	2.19	2.12	2.04	1.96	1.91	1.87	1.82	1.77	1.71	1.65
29	4.18	3.33	2.93	2.70	2.55	2.43	2.35	2.28	2.22	2.18	2.10	2.03	1.94	1.90	1.85	1.81	1.75	1.70	1.64
30	4.17	3.32	2.92	2.69	2.53	2.42	2.33	2.27	2.21	2.16	2.09	2.01	1.93	1.89	1.84	1.79	1.74	1.68	1.62
40	4.08	3.23	2.84	2.61	2.45	2.34	2.25	2.18	2.12	2.08	2.00	1.92	1.84	1.79	1.74	1.69	1.64	1.58	1.51
60	4.00	3.15	2.76	2.53	2.37	2.25	2.17	2.10	2.04	1.99	1.92	1.84	1.75	1.70	1.65	1.59	1.53	1.47	1.39
120	3.92	3.07	2.68	2.45	2.29	2.17	2.09	2.02	1.96	1.91	1.83	1.75	1.66	1.61	1.55	1.50	1.43	1.35	1.25
∞	3.84	3.00	2.60	2.37	2.21	2.10	2.01	1.94	1.88	1.83	1.75	1.67	1.57	1.52	1.46	1.39	1.32	1.22	1.00

Illustration: The value that bounds a 5% right-tail area under the $F_{(5,6)}$ distribution is 4.39.

Table A.5b
Upper 2.5% Points (0.975 Fractiles)

v_2 \ v_1	1	2	3	4	5	6	7	8	9	10	12	15	20	24	30	40	60	120	∞
1	647.8	799.5	864.2	899.6	921.8	937.1	948.2	956.7	963.3	968.6	976.7	984.9	993.1	997.2	1001	1006	1010	1014	1018
2	38.51	39.00	39.17	39.25	39.30	39.33	39.36	39.37	39.39	39.40	39.41	39.43	39.45	39.46	39.46	39.47	39.48	39.49	39.50
3	17.44	16.04	15.44	15.10	14.88	14.73	14.62	14.54	14.47	14.42	14.34	14.25	14.17	14.12	14.08	14.04	13.99	13.95	13.90
4	12.22	10.65	9.98	9.60	9.36	9.20	9.07	8.98	8.90	8.84	8.75	8.66	8.56	8.51	8.46	8.41	8.36	8.31	8.26
5	10.01	8.43	7.76	7.39	7.15	6.98	6.85	6.76	6.68	6.62	6.52	6.43	6.33	6.28	6.23	6.18	6.12	6.07	6.02
6	8.81	7.26	6.60	6.23	5.99	5.82	5.70	5.60	5.52	5.46	5.37	5.27	5.17	5.12	5.07	5.01	4.96	4.90	4.85
7	8.07	6.54	5.89	5.52	5.29	5.12	4.99	4.90	4.82	4.76	4.67	4.57	4.47	4.42	4.36	4.31	4.25	4.20	4.14
8	7.57	6.06	5.42	5.05	4.82	4.65	4.53	4.43	4.36	4.30	4.20	4.10	4.00	3.95	3.89	3.84	3.78	3.73	3.67
9	7.21	5.71	5.08	4.72	4.48	4.32	4.20	4.10	4.03	3.96	3.87	3.77	3.67	3.61	3.56	3.51	3.45	3.39	3.33
10	6.94	5.46	4.83	4.47	4.24	4.07	3.95	3.85	3.78	3.72	3.62	3.52	3.42	3.37	3.31	3.26	3.20	3.14	3.08
11	6.72	5.26	4.63	4.28	4.04	3.88	3.76	3.66	3.59	3.53	3.43	3.33	3.23	3.17	3.12	3.06	3.00	2.94	2.88
12	6.55	5.10	4.47	4.12	3.89	3.73	3.61	3.51	3.44	3.37	3.28	3.18	3.07	3.02	2.96	2.91	2.85	2.79	2.72
13	6.41	4.97	4.35	4.00	3.77	3.60	3.48	3.39	3.31	3.25	3.15	3.05	2.95	2.89	2.84	2.78	2.72	2.66	2.60
14	6.30	4.86	4.24	3.89	3.66	3.50	3.38	3.29	3.21	3.15	3.05	2.95	2.84	2.79	2.73	2.67	2.61	2.55	2.49
15	6.20	4.77	4.15	3.80	3.58	3.41	3.29	3.20	3.12	3.06	2.96	2.86	2.76	2.70	2.64	2.59	2.52	2.46	2.40
16	6.12	4.69	4.08	3.73	3.50	3.34	3.22	3.12	3.05	2.99	2.89	2.79	2.68	2.63	2.57	2.51	2.45	2.38	2.32
17	6.04	4.62	4.01	3.66	3.44	3.28	3.16	3.06	2.98	2.92	2.82	2.72	2.62	2.56	2.50	2.44	2.38	2.32	2.25
18	5.98	4.56	3.95	3.61	3.38	3.22	3.10	3.01	2.93	2.87	2.77	2.67	2.56	2.50	2.44	2.38	2.32	2.26	2.19
19	5.92	4.51	3.90	3.56	3.33	3.17	3.05	2.96	2.88	2.82	2.72	2.62	2.51	2.45	2.39	2.33	2.27	2.20	2.13
20	5.87	4.46	3.86	3.51	3.29	3.13	3.01	2.91	2.84	2.77	2.68	2.57	2.46	2.41	2.35	2.29	2.22	2.16	2.09
21	5.83	4.42	3.82	3.48	3.25	3.09	2.97	2.87	2.80	2.73	2.64	2.53	2.42	2.37	2.31	2.25	2.18	2.11	2.04
22	5.79	4.38	3.78	3.44	3.22	3.05	2.93	2.84	2.76	2.70	2.60	2.50	2.39	2.33	2.27	2.21	2.14	2.08	2.00
23	5.75	4.35	3.75	3.41	3.18	3.02	2.90	2.81	2.73	2.67	2.57	2.47	2.36	2.30	2.24	2.18	2.11	2.04	1.97
24	5.72	4.32	3.72	3.38	3.15	2.99	2.87	2.78	2.70	2.64	2.54	2.44	2.33	2.27	2.21	2.15	2.08	2.01	1.94
25	5.69	4.29	3.69	3.35	3.13	2.97	2.85	2.75	2.68	2.61	2.51	2.41	2.30	2.24	2.18	2.12	2.05	1.98	1.91
26	5.66	4.27	3.67	3.33	3.10	2.94	2.82	2.73	2.65	2.59	2.49	2.39	2.28	2.22	2.16	2.09	2.03	1.95	1.88
27	5.63	4.24	3.65	3.31	3.08	2.92	2.80	2.71	2.63	2.57	2.47	2.36	2.25	2.19	2.13	2.07	2.00	1.93	1.85
28	5.61	4.22	3.63	3.29	3.06	2.90	2.78	2.69	2.61	2.55	2.45	2.34	2.23	2.17	2.11	2.05	1.98	1.91	1.83
29	5.59	4.20	3.61	3.27	3.04	2.88	2.76	2.67	2.59	2.53	2.43	2.32	2.21	2.15	2.09	2.03	1.96	1.89	1.81
30	5.57	4.18	3.59	3.25	3.03	2.87	2.75	2.65	2.57	2.51	2.41	2.31	2.20	2.14	2.07	2.01	1.94	1.87	1.79
40	5.42	4.05	3.46	3.13	2.90	2.74	2.62	2.53	2.45	2.39	2.29	2.18	2.07	2.01	1.94	1.88	1.80	1.72	1.64
60	5.29	3.93	3.34	3.01	2.79	2.63	2.51	2.41	2.33	2.27	2.17	2.06	1.94	1.88	1.82	1.74	1.67	1.58	1.48
120	5.15	3.80	3.23	2.89	2.67	2.52	2.39	2.30	2.22	2.16	2.05	1.94	1.82	1.76	1.69	1.61	1.53	1.43	1.31
∞	5.02	3.69	3.12	2.79	2.57	2.41	2.29	2.19	2.11	2.05	1.94	1.83	1.71	1.64	1.57	1.48	1.39	1.27	1.00

Illustration: The value that bounds a 2.5% right-tail area under the $F_{(5,6)}$ distribution is 5.99.

Table A.5c
Upper 1% Points (0.99 Fractiles)

v_2 \\ v_1	1	2	3	4	5	6	7	8	9	10	12	15	20	24	30	40	60	120	∞
1	4052	4999.5	5403	5625	5764	5859	5928	5981	6022	6056	6106	6157	6209	6235	6261	6287	6313	6339	6366
2	98.50	99.00	99.17	99.25	99.30	99.33	99.36	99.37	99.39	99.40	99.42	99.43	99.45	99.46	99.47	99.47	99.48	99.49	99.50
3	34.12	30.82	29.46	28.71	28.24	27.91	27.67	27.49	27.35	27.23	27.05	26.87	26.69	26.60	26.50	26.41	26.32	26.22	26.13
4	21.20	18.00	16.69	15.98	15.52	15.21	14.98	14.80	14.66	14.55	14.37	14.20	14.02	13.93	13.84	13.75	13.65	13.56	13.46
5	16.26	13.27	12.06	11.39	10.97	10.67	10.46	10.29	10.16	10.05	9.89	9.72	9.55	9.47	9.38	9.29	9.20	9.11	9.02
6	13.75	10.92	9.78	9.15	8.75	8.47	8.26	8.10	7.98	7.87	7.72	7.56	7.40	7.31	7.23	7.14	7.06	6.97	6.88
7	12.25	9.55	8.45	7.85	7.46	7.19	6.99	6.84	6.72	6.62	6.47	6.31	6.16	6.07	5.99	5.91	5.82	5.74	5.65
8	11.26	8.65	7.59	7.01	6.63	6.37	6.18	6.03	5.91	5.81	5.67	5.52	5.36	5.28	5.20	5.12	5.03	4.95	4.86
9	10.56	8.02	6.99	6.42	6.06	5.80	5.61	5.47	5.35	5.26	5.11	4.96	4.81	4.73	4.65	4.57	4.48	4.40	4.31
10	10.04	7.56	6.55	5.99	5.64	5.39	5.20	5.06	4.94	4.85	4.71	4.56	4.41	4.33	4.25	4.17	4.08	4.00	3.91
11	9.65	7.21	6.22	5.67	5.32	5.07	4.89	4.74	4.63	4.54	4.40	4.25	4.10	4.02	3.94	3.86	3.78	3.69	3.60
12	9.33	6.93	5.95	5.41	5.06	4.82	4.64	4.50	4.39	4.30	4.16	4.01	3.86	3.78	3.70	3.62	3.54	3.45	3.36
13	9.07	6.70	5.74	5.21	4.86	4.62	4.44	4.30	4.19	4.10	3.96	3.82	3.66	3.59	3.51	3.43	3.34	3.25	3.17
14	8.86	6.51	5.56	5.04	4.69	4.46	4.28	4.14	4.03	3.94	3.80	3.66	3.51	3.43	3.35	3.27	3.18	3.09	3.00
15	8.68	6.36	5.42	4.89	4.56	4.32	4.14	4.00	3.89	3.80	3.67	3.52	3.37	3.29	3.21	3.13	3.05	2.96	2.87
16	8.53	6.23	5.29	4.77	4.44	4.20	4.03	3.89	3.78	3.69	3.55	3.41	3.26	3.18	3.10	3.02	2.93	2.84	2.75
17	8.40	6.11	5.18	4.67	4.34	4.10	3.93	3.79	3.68	3.59	3.46	3.31	3.16	3.08	3.00	2.92	2.83	2.75	2.65
18	8.29	6.01	5.09	4.58	4.25	4.01	3.84	3.71	3.60	3.51	3.37	3.23	3.08	3.00	2.92	2.84	2.75	2.66	2.57
19	8.18	5.93	5.01	4.50	4.17	3.94	3.77	3.63	3.52	3.43	3.30	3.15	3.00	2.92	2.84	2.76	2.67	2.58	2.49
20	8.10	5.85	4.94	4.43	4.10	3.87	3.70	3.56	3.46	3.37	3.23	3.09	2.94	2.86	2.78	2.69	2.61	2.52	2.42
21	8.02	5.78	4.87	4.37	4.04	3.81	3.64	3.51	3.40	3.31	3.17	3.03	2.88	2.80	2.72	2.64	2.55	2.46	2.36
22	7.95	5.72	4.82	4.31	3.99	3.76	3.59	3.45	3.35	3.26	3.12	2.98	2.83	2.75	2.67	2.58	2.50	2.40	2.31
23	7.88	5.66	4.76	4.26	3.94	3.71	3.54	3.41	3.30	3.21	3.07	2.93	2.78	2.70	2.62	2.54	2.45	2.35	2.26
24	7.82	5.61	4.72	4.22	3.90	3.67	3.50	3.36	3.26	3.17	3.03	2.89	2.74	2.66	2.58	2.49	2.40	2.31	2.21
25	7.77	5.57	4.68	4.18	3.85	3.63	3.46	3.32	3.22	3.13	2.99	2.85	2.70	2.62	2.54	2.45	2.36	2.27	2.17
26	7.72	5.53	4.64	4.14	3.82	3.59	3.42	3.29	3.18	3.09	2.96	2.81	2.66	2.58	2.50	2.42	2.33	2.23	2.13
27	7.68	5.49	4.60	4.11	3.78	3.56	3.39	3.26	3.15	3.06	2.93	2.78	2.63	2.55	2.47	2.38	2.29	2.20	2.10
28	7.64	5.45	4.57	4.07	3.75	3.53	3.36	3.23	3.12	3.03	2.90	2.75	2.60	2.52	2.44	2.35	2.26	2.17	2.06
29	7.60	5.42	4.54	4.04	3.73	3.50	3.33	3.20	3.09	3.00	2.87	2.73	2.57	2.49	2.41	2.33	2.23	2.14	2.03
30	7.56	5.39	4.51	4.02	3.70	3.47	3.30	3.17	3.07	2.98	2.84	2.70	2.55	2.47	2.39	2.30	2.21	2.11	2.01
40	7.31	5.18	4.31	3.83	3.51	3.29	3.12	2.99	2.89	2.80	2.66	2.52	2.37	2.29	2.20	2.11	2.02	1.92	1.80
60	7.08	4.98	4.13	3.65	3.34	3.12	2.95	2.82	2.72	2.63	2.50	2.35	2.20	2.12	2.03	1.94	1.84	1.73	1.60
120	6.85	4.79	3.95	3.48	3.17	2.96	2.79	2.66	2.56	2.47	2.34	2.19	2.03	1.95	1.86	1.76	1.66	1.53	1.38
∞	6.63	4.61	3.78	3.32	3.02	2.80	2.64	2.51	2.41	2.32	2.18	2.04	1.88	1.79	1.70	1.59	1.47	1.32	1.00

Illustration: The value that bounds a 1% right-tail area under the $F^{(5,6)}$ distribution is 8.75.

Source: Table 18 of E. S. Pearson and H. O. Hartley, *Biometrika Tables for Statisticians*, Vol. I, published for the Biometrika Trustees at the University Press, Cambridge. Reprinted by permission of the Biometrika Trustees.

Table A.6
Critical Values of *r* in the Runs Test

Given in the bodies of Table A.6a and Table A.6b are various critical values of *r* for various values of n_1 and n_2. For the one-sample runs test, any value of *r* which is equal to or smaller than that shown in Table A.6a or equal to or larger than that shown in Table A.6b is significant at the 0.05 level.

Table A.6a

n_1 \ n_2	2	3	4	5	6	7	8	9	10	11	12	13	14	15	16	17	18	19	20
2											2	2	2	2	2	2	2	2	2
3				2	2	2	2	2	2	2	2	2	2	3	3	3	3	3	3
4			2	2	2	3	3	3	3	3	3	3	3	3	4	4	4	4	4
5			2	2	3	3	3	3	3	4	4	4	4	4	4	4	5	5	5
6		2	2	3	3	3	3	4	4	4	4	5	5	5	5	5	5	6	6
7		2	2	3	3	3	4	4	5	5	5	5	5	6	6	6	6	6	6
8		2	3	3	3	4	4	5	5	5	6	6	6	6	6	7	7	7	7
9		2	3	3	4	4	5	5	5	6	6	6	7	7	7	7	8	8	8
10		2	3	3	4	5	5	5	6	6	7	7	7	7	8	8	8	8	9
11		2	3	4	4	5	5	6	6	7	7	7	8	8	8	9	9	9	9
12	2	2	3	4	4	5	6	6	7	7	7	8	8	8	9	9	9	10	10
13	2	2	3	4	5	5	6	6	7	7	8	8	9	9	9	10	10	10	10
14	2	2	3	4	5	5	6	7	7	8	8	9	9	9	10	10	10	11	11
15	2	3	3	4	5	6	6	7	7	8	8	9	9	10	10	11	11	11	12
16	2	3	4	4	5	6	6	7	8	8	9	9	10	10	11	11	11	12	12
17	2	3	4	4	5	6	7	7	8	9	9	10	10	11	11	11	12	12	13
18	2	3	4	5	5	6	7	8	8	9	9	10	10	11	11	12	12	13	13
19	2	3	4	5	6	6	7	8	8	9	10	10	11	11	12	12	13	13	13
20	2	3	4	5	6	6	7	8	9	9	10	10	11	12	12	13	13	13	14

Table A.6
Critical Values of r in the Runs Test (Continued)

Table A.6b

n_1 \ n_2	2	3	4	5	6	7	8	9	10	11	12	13	14	15	16	17	18	19	20
2																			
3																			
4				9	9														
5			9	10	10	11	11												
6			9	10	11	12	12	13	13	13	13								
7				11	12	13	13	14	14	14	14	15	15	15					
8				11	12	13	14	14	15	15	16	16	16	16	17	17	17	17	17
9					13	14	14	15	16	16	16	17	17	18	18	18	18	18	18
10					13	14	15	16	16	17	17	18	18	18	19	19	19	20	20
11					13	14	15	16	17	17	18	19	19	19	20	20	20	21	21
12					13	14	16	16	17	18	19	19	20	20	21	21	21	22	22
13						15	16	17	18	19	19	20	20	21	21	22	22	23	23
14						15	16	17	18	19	20	20	21	22	22	23	23	23	24
15						15	16	18	18	19	20	21	22	22	23	23	24	24	25
16							17	18	19	20	21	21	22	23	23	24	25	25	25
17							17	18	19	20	21	22	23	23	24	25	25	26	26
18							17	18	19	20	21	22	23	24	25	25	26	26	27
19							17	18	20	21	22	23	23	24	25	26	26	27	27
20							17	18	20	21	22	23	24	25	25	26	27	27	28

Source: Adapted from Swed, Frieda S., and Eisenhart, C. 1943. Tables for testing randomness of grouping in a sequence of alternatives. *Ann. Math. Statist.*, *14, 83–86*, with the kind permission of the authors and publishers.

Table A.7
Quantiles of the Mann-Whitney Test Statistic

n_A	p	$n_B=2$	3	4	5	6	7	8	9	10	11	12	13	14	15	16	17	18	19	20
2	0.001																			
	0.005																		1	1
	0.01												1	1	1	1	1	1	2	2
	0.025							1	1	1	1	2	2	2	2	2	3	3	3	3
	0.05				1	1	1	2	2	2	2	3	3	4	4	4	4	5	5	5
	0.10		1	1	2	2	2	3	3	4	4	5	5	6	6	6	7	7	8	8
3	0.001																1	1	1	1
	0.005								1	1	1	2	2	2	3	3	3	3	4	4
	0.01						1	1	2	2	2	3	3	3	4	4	5	5	5	6
	0.025				1	2	2	3	3	4	4	5	5	6	6	7	7	8	8	9
	0.05			1	2	3	3	4	5	5	6	6	7	8	8	9	10	10	11	12
	0.10	1	2	2	3	4	5	6	6	7	8	9	10	11	11	12	13	14	15	16
4	0.001									1	1	1	2	2	2	3	3	4	4	4
	0.005					1	1	2	2	3	3	4	4	5	6	6	7	7	8	9
	0.01				1	2	2	3	4	4	5	6	6	7	8	8	9	10	10	11
	0.025			1	2	3	4	5	5	6	7	8	9	10	11	12	12	13	14	15
	0.05		1	2	3	4	5	6	7	8	9	10	11	12	13	15	16	17	18	19
	0.10	1	2	4	5	6	7	8	10	11	12	13	14	16	17	18	19	21	22	23

NOTE: The entries in this table are the quantiles t_p of the Mann-Whitney test statistic T_A, for selected values of p. Note that $P(T_A < t_p) \le p$. Upper quantiles may be found from the equation

$$t_{1-p} = n_A n_B - t_p$$

Critical regions correspond to values less than (or greater than) but not including the appropriate quantile.

Source: Adapted by permission of the Biometrika Trustees from Table 1, in L. R. Verdooren "Extended Tables of Critical Values for Wilcoxon's Test Statistic," Biometrika, 50, pp. 177–86.

Table A.7
Quantiles of the Mann-Whitney Test Statistic (Continued)

n_A	p	$n_B=2$	3	4	5	6	7	8	9	10	11	12	13	14	15	16	17	18	19	20
5	0.001							1	2	2	3	3	4	4	5	6	6	7	8	8
	0.005					2	2	3	4	5	6	7	8	8	9	10	11	12	13	14
	0.01			1	2	3	4	5	6	7	8	9	10	11	12	13	14	15	16	17
	0.025		1	2	3	4	6	7	8	9	10	12	13	14	15	16	18	19	20	21
	0.05	1	2	3	5	6	7	9	10	12	13	14	16	17	19	20	21	23	24	26
	0.10	2	3	5	6	8	9	11	13	14	16	18	19	21	23	24	26	28	29	31
6	0.001							2	3	4	5	5	6	7	8	9	10	11	12	13
	0.005				2	3	4	5	6	7	8	10	11	12	13	14	16	17	18	19
	0.01			2	3	4	5	7	8	9	10	12	13	14	16	17	19	20	21	23
	0.025		2	3	4	6	7	9	11	12	14	15	17	18	20	22	23	25	26	28
	0.05		3	4	6	8	9	11	13	15	17	18	20	22	24	26	27	29	31	33
	0.10		4	6	8	10	12	14	16	18	20	22	24	26	28	30	32	35	37	39
7	0.001					1	2	3	4	6	7	8	9	10	11	12	14	15	16	17
	0.005			1	2	4	5	7	8	10	11	13	14	16	17	19	20	22	23	25
	0.01		1	2	4	5	7	8	10	12	13	15	17	18	20	22	24	25	27	29
	0.025		2	4	6	7	9	11	13	15	17	19	21	23	25	27	29	31	33	35
	0.05	1	3	5	7	9	12	14	16	18	20	22	25	27	29	31	34	36	38	40
	0.10	2	5	7	9	12	14	17	19	22	24	27	29	32	34	37	39	42	44	47
8	0.001				1	2	3	5	6	7	9	10	12	13	15	16	18	19	21	22
	0.005			2	3	5	7	8	10	12	14	16	18	19	21	23	25	27	29	31
	0.01		1	3	5	7	8	10	12	14	16	18	21	23	25	27	29	31	33	35
	0.025	1	3	5	7	9	11	14	16	18	20	23	25	27	30	32	35	37	39	42
	0.05	2	4	6	9	11	14	16	19	21	24	27	29	32	34	37	40	42	45	48
	0.10	3	6	8	11	14	17	20	23	25	28	31	34	37	40	43	46	49	52	55

Table A.7
Quantiles of the Mann-Whitney Test Statistic (Continued)

n_A	p	$n_B=2$	3	4	5	6	7	8	9	10	11	12	13	14	15	16	17	18	19	20
9	0.001				2	3	4	6	8	9	11	13	15	16	18	20	22	24	26	27
	0.005		1	2	4	6	8	10	12	14	17	19	21	23	25	28	30	32	34	37
	0.01		2	4	6	8	10	12	15	17	19	22	24	27	29	32	34	37	39	41
	0.025	1	3	5	8	11	13	16	18	21	24	27	29	32	35	38	40	43	46	49
	0.05	2	5	7	10	13	16	19	22	25	28	31	34	37	40	43	46	49	52	55
	0.10	3	6	10	13	16	19	23	26	29	32	36	39	42	46	49	53	56	59	63
10	0.001			1	2	4	6	7	9	11	13	15	18	20	22	24	26	28	30	33
	0.005		1	3	5	7	10	12	14	17	19	22	25	27	30	32	35	38	40	43
	0.01		2	4	7	9	12	14	17	20	23	25	28	31	34	37	39	42	45	48
	0.025	1	4	6	9	12	15	18	21	24	27	30	34	37	40	43	46	49	53	56
	0.05	2	5	8	12	15	18	21	25	28	32	35	38	42	45	49	52	56	59	63
	0.10	4	7	11	14	18	22	25	29	33	37	40	44	48	52	55	59	63	67	71
11	0.001			1	3	5	7	9	11	13	16	18	21	23	25	28	30	33	35	38
	0.005		1	3	6	8	11	14	17	19	22	25	28	31	34	37	40	43	46	49
	0.01		2	5	8	10	13	16	19	23	26	29	32	35	38	42	45	48	51	54
	0.025	1	4	7	10	14	17	20	24	27	31	34	38	41	45	48	52	56	59	63
	0.05	2	6	9	13	17	20	24	28	32	35	39	43	47	51	55	58	62	66	70
	0.10	4	8	12	16	20	24	28	32	37	41	45	49	53	58	62	66	70	74	79
12	0.001			1	3	5	8	10	13	15	18	21	24	26	29	32	35	38	41	43
	0.005		2	4	7	10	13	16	19	22	25	28	32	35	38	42	45	48	52	55
	0.01		3	6	9	12	15	18	22	25	29	32	36	39	43	47	50	54	57	61
	0.025	2	5	8	12	15	19	23	27	30	34	38	42	46	50	54	58	62	66	70
	0.05	3	6	10	14	18	22	27	31	35	39	43	48	52	56	61	65	69	73	78
	0.10	5	9	13	18	22	27	31	36	40	45	50	54	59	64	68	73	78	82	87

Table A.7
Quantiles of the Mann-Whitney Test Statistic (Continued)

n_A	p	$n_B=2$	3	4	5	6	7	8	9	10	11	12	13	14	15	16	17	18	19	20
13	0.001			2	4	6	9	12	15	18	21	24	27	30	33	36	39	43	46	49
	0.005		2	4	8	11	14	18	21	25	28	32	35	39	43	46	50	54	58	61
	0.01	1	3	6	10	13	17	21	24	28	32	36	40	44	48	52	56	60	64	68
	0.025	2	5	9	13	17	21	25	29	34	38	42	46	51	55	60	64	68	73	77
	0.05	3	7	11	16	20	25	29	34	38	43	48	52	57	62	66	71	76	81	85
	0.10	5	10	14	19	24	29	34	39	44	49	54	59	64	69	75	80	85	90	95
14	0.001			2	4	7	10	13	16	20	23	26	30	33	37	40	44	47	51	55
	0.005		2	5	8	12	16	19	23	27	31	35	39	43	47	51	55	59	64	68
	0.01	1	3	7	11	14	18	23	27	31	35	39	44	48	52	57	61	66	70	74
	0.025	2	6	10	14	18	23	27	32	37	41	46	51	56	60	65	70	75	79	84
	0.05	4	8	12	17	22	27	32	37	42	47	52	57	62	67	72	78	83	88	93
	0.10	5	11	16	21	26	32	37	42	48	53	59	64	70	75	81	86	92	98	103
15	0.001			2	5	8	11	15	18	22	25	29	33	37	41	44	48	52	56	60
	0.005		3	6	9	13	17	21	25	30	34	38	43	47	52	56	61	65	70	74
	0.01	1	4	8	12	16	20	25	29	34	38	43	48	52	57	62	67	71	76	81
	0.025	2	6	11	15	20	25	30	35	40	45	50	55	60	65	71	76	81	86	91
	0.05	4	8	13	19	24	29	34	40	45	51	56	62	67	73	78	84	89	95	101
	0.10	6	11	17	23	28	34	40	46	52	58	64	69	75	81	87	93	99	105	111
16	0.001			3	6	9	12	16	20	24	28	32	36	40	44	49	53	57	61	66
	0.005		3	6	10	14	19	23	28	32	37	42	46	51	56	61	66	71	75	80
	0.01	1	4	8	13	17	22	27	32	37	42	47	52	57	62	67	72	77	83	88
	0.025	2	7	12	16	22	27	32	38	43	48	54	60	65	71	76	82	87	93	99
	0.05	4	9	15	20	26	31	37	43	49	55	61	66	72	78	84	90	96	102	108
	0.10	6	12	18	24	30	37	43	49	55	62	68	75	81	87	94	100	107	113	120

Table A.7
Quantiles of the Mann-Whitney Test Statistic (Continued)

n_A	p	$n_B=2$	3	4	5	6	7	8	9	10	11	12	13	14	15	16	17	18	19	20
17	0.001		1	3	6	10	14	18	22	26	30	35	39	44	48	53	58	62	67	71
	0.005		3	7	11	16	20	25	30	35	40	45	50	55	61	66	71	76	82	87
	0.01	1	5	9	14	19	24	29	34	39	45	50	56	61	67	72	78	83	89	94
	0.025	3	7	12	18	23	29	35	40	46	52	58	64	70	76	82	88	94	100	106
	0.05	4	10	16	21	27	34	40	46	52	58	65	71	78	84	90	97	103	110	116
	0.10	7	13	19	26	32	39	46	53	59	66	73	80	86	93	100	107	114	121	128
18	0.001		1	4	7	11	15	19	24	28	33	38	43	47	52	57	62	67	72	77
	0.005		3	7	12	17	22	27	32	38	43	48	54	59	65	71	76	82	88	93
	0.01	1	5	10	15	20	25	31	37	42	48	54	60	66	71	77	83	89	95	101
	0.025	3	8	13	19	25	31	37	43	49	56	62	68	75	81	87	94	100	107	113
	0.05	5	10	17	23	29	36	42	49	56	62	69	76	83	89	96	103	110	117	124
	0.10	7	14	21	28	35	42	49	56	63	70	78	85	92	99	107	114	121	129	136
19	0.001		1	4	8	12	16	21	26	30	35	41	46	51	56	61	67	72	78	83
	0.005	1	4	8	13	18	23	29	34	40	46	52	58	64	70	75	82	88	94	100
	0.01	2	5	10	16	21	27	33	39	45	51	57	64	70	76	83	89	95	102	108
	0.025	3	8	14	20	26	33	39	46	53	59	66	73	79	86	93	100	107	114	120
	0.05	5	11	18	24	31	38	45	52	59	66	73	81	88	95	102	110	117	124	131
	0.10	8	15	22	29	37	44	52	59	67	74	82	90	98	105	113	121	129	136	144
20	0.001		1	4	8	13	17	22	27	33	38	43	49	55	60	66	71	77	83	89
	0.005	1	4	9	14	19	25	31	37	43	49	55	61	68	74	80	87	93	100	106
	0.01	2	6	11	17	23	29	35	41	48	54	61	68	74	81	88	94	101	108	115
	0.025	3	9	15	21	28	35	42	49	56	63	70	77	84	91	99	106	113	120	128
	0.05	5	12	19	26	33	40	48	55	63	70	78	85	93	101	108	116	124	131	139
	0.10	8	16	23	31	39	47	55	63	71	79	87	95	103	111	120	128	136	144	152

Table A.8
Critical Values of the Kruskal-Wallis Test Statistic for Three Samples and Small Sample Sizes[a]

Sample sizes			Critical Value	α	Sample sizes			Critical Value	α
n_1	n_2	n_3			n_1	n_2	n_3		
2	1	1	2.7000	0.500	4	2	2	6.0000	0.014
2	2	1	3.6000	0.200				5.3333	0.033
2	2	2	4.5714	0.067				5.1250	0.052
			3.7143	0.200				4.4583	0.100
								4.1667	0.105
3	1	1	3.2000	0.300					
3	2	1	4.2857	0.100	4	3	1	5.8333	0.021
			3.8571	0.133				5.2083	0.050
								5.0000	0.057
3	3	2	5.3572	0.029				4.0556	0.093
			4.7143	0.048				3.8889	0.129
			4.5000	0.067					
			4.4643	0.105	4	3	2	6.4444	0.008
3	3	1	5.1429	0.043				6.3000	0.011
			4.5714	0.100				5.4444	0.046
			4.0000	0.129				5.4000	0.051
								4.5111	0.098
3	3	2	6.2500	0.011				4.4444	0.102
			5.3611	0.032					
			5.1389	0.061					
			4.5556	0.100	4	3	3	6.7455	0.010
			4.2500	0.121				6.7091	0.013
								5.7909	0.046
3	3	3	7.2000	0.004				5.7273	0.050
			6.4889	0.001					
			5.6889	0.029				4.7091	0.092
			5.6000	0.050				4.7000	0.101
			5.0667	0.086					
			4.6222	0.100	4	4	1	6.6667	0.010
4	1	1	3.5714	0.200				6.1667	0.022
								4.9667	0.048
4	2	1	4.8214	0.057				4.8667	0.054
			4.5000	0.076				4.1667	0.082
			4.0179	0.114				4.0667	0.102

Source: Obtained from Kruskal and Wallis (1952).

[a] The null hypothesis may be rejected at the level α if the Kruskal-Wallis test statistic (H), given by Equation 18.8, is *equal to or greater than* the critical value given in the table.

Table A.8
Critical Values of the Kruskal-Wallis Test Statistic for Three
Samples and Small Sample Sizes[a] (Continued)

n_1	n_2	n_3	Critical Value	α	n_1	n_2	n_3	Critical Value	α
4	4	2	7.0364	0.006	5	3	2	6.9091	0.009
			6.8727	0.011				6.8281	0.010
			5.4545	0.046				5.2509	0.049
			5.2364	0.052				5.1055	0.052
			4.5545	0.098				4.6509	0.091
			4.4455	0.103				4.4121	0.101
4	4	3	7.1439	0.010	5	3	3	7.0788	0.009
			7.1364	0.011				6.9818	0.011
			5.5985	0.049				5.6485	0.049
			5.5758	0.051				5.5152	0.051
			4.5455	0.099				4.5333	0.097
			4.4773	0.102				4.4121	0.109
4	4	4	7.6538	0.008	5	4	1	6.9545	0.008
			7.5385	0.011				6.8400	0.011
			5.6923	0.049				4.9855	0.044
			5.6538	0.054				4.8600	0.056
			4.6539	0.097				3.9873	0.098
			4.5001	0.104				3.9600	0.102
5	1	1	3.8571	0.143	5	4	2	7.2045	0.009
								7.1182	0.010
5	2	1	5.2500	0.036				5.2727	0.049
			5.0000	0.048				5.2682	0.050
			4.4500	0.071				4.5409	0.098
			4.2000	0.095				4.5182	0.101
			4.0500	0.119	5	4	3	7.4449	0.010
5	2	2	6.5333	0.005				7.3949	0.011
			6.1333	0.013				5.6564	0.049
			5.1600	0.034				5.6308	0.050
			5.0400	0.056				4.5487	0.099
			4.3733	0.090				4.5231	0.103
			4.2933	0.112	5	4	4	7.7604	0.009
5	3	1	6.4000	0.012				7.7440	0.011
			4.9600	0.048				5.6571	0.049
			4.8711	0.052				5.6176	0.050
			4.0178	0.095				4.6187	0.100
			3.8400	0.123				4.5527	0.102

Table A.8
Critical Values of the Kruskal-Wallis Test Statistic for Three Samples and Small Sample Sizes [a] (Continued)

n_1	n_2	n_3	Critical Value	α	n_1	n_2	n_3	Critical Value	α
5	5	1	7.3091	0.009				5.6264	0.051
			6.8364	0.011				4.5451	0.100
			5.1273	0.046				4.5363	0.102
			4.9091	0.053					
			4.1091	0.086	5	5	4	7.8229	0.010
			4.0364	0.105				7.7914	0.010
								5.6657	0.049
5	5	2	7.3385	0.010				5.6429	0.050
			7.2692	0.010				4.5229	0.100
			5.3385	0.047				4.5200	0.101
			5.2462	0.051					
			4.6231	0.097	5	5	5	8.0000	0.009
			4.5077	0.100				7.9800	0.010
								5.7800	0.049
5	5	3	7.5780	0.010				5.6600	0.051
			7.5429	0.010				4.5600	0.100
			5.7055	0.046				4.5000	0.102

Table A.9
Critical Values of r_S, the Spearman Rank
Correlation Coefficient

N	Significance level (one-tailed test)	
	0.05	0.01
4	1.000	
5	0.900	1.000
6	0.829	0.943
7	0.714	0.893
8	0.643	0.833
9	0.600	0.783
10	0.564	0.746
12	0.506	0.712
14	0.456	0.645
16	0.425	0.601
18	0.399	0.564
20	0.377	0.534
22	0.359	0.508
24	0.343	0.485
26	0.329	0.465
28	0.317	0.448
30	0.306	0.432

Source: Adapted from Olds, E.G. 1938. Distributions of sums of squares of rank differences for small numbers of individuals. *Ann Math. Statist.*, *9*, *133–148*, and from Olds, E.G. 1949. The 5% significance levels for sums of squares of rank differences and a correction. *Ann. Math. Statist.*, *20*, *117–118*, with the kind permission of the author and the publisher.

Table A.10
Probabilities Associated with Values as Large as Observed Values of S in
the Kendall Rank Correlation Coefficient

S	Values of N				S	Values of N		
	4	5	8	9		6	7	10
0	0.625	0.592	0.548	0.540	1	0.500	0.500	0.500
2	0.375	0.408	0.452	0.460	3	0.360	0.386	0.431
4	0.167	0.242	0.360	0.381	5	0.235	0.281	0.364
6	0.042	0.117	0.274	0.306	7	0.136	0.191	0.300
8		0.042	0.199	0.238	9	0.068	0.119	0.242
10		0.0083	0.138	0.179	11	0.028	0.068	0.190
12			0.089	0.130	13	0.0083	0.035	0.146
14			0.054	0.090	15	0.0014	0.015	0.108
16			0.031	0.060	17		0.0054	0.078
18			0.016	0.038	19		0.0014	0.054
20			0.0071	0.022	21		0.00020	0.036
22			0.0028	0.012	23			0.023
24			0.00087	0.0063	25			0.014
26			0.00019	0.0029	27			0.0083
28			0.000025	0.0012	29			0.0046
30				0.00043	31			0.0023
32				0.00012	33			0.0011
34				0.000025	35			0.00047
36				0.0000028	37			0.00018
					39			0.000058
					41			0.000015
					43			0.0000028
					45			0.00000028

Source: Reproduced by permission of the publishers, Charles Griffin & Company Ltd. of London and High Wycombe, from *Kendall Rank Correlation Methods*, 4th Edition, 1970.

Table A.11
Random Numbers

61285	07068	22452	57376	21242	17862	34455	87861	68267	73886
56842	75893	24824	11508	73241	61735	87493	91590	58567	21433
83379	07294	96793	71360	35503	38350	55267	71611	67475	15985
70032	23033	67592	17276	58125	68188	34370	66462	27282	29982
12462	17135	82285	95753	72772	20736	10242	29611	67957	00426
82634	46763	92838	21357	72444	13543	15809	56684	68346	57939
96288	58891	62386	17345	21011	75315	12239	95406	34908	29806
60788	40846	31794	36868	86804	76726	53447	78068	94493	06503
56104	81174	85519	17681	36376	28255	87115	44834	20653	42302
60416	33391	95645	14311	01915	44145	37830	26807	53745	57189
82946	86707	38368	45776	80965	25234	11764	19408	92702	58544
31581	58877	45303	90248	29117	39095	63392	21665	26225	34986
63073	35656	22682	11393	59638	27235	74410	71774	29982	64961
98016	77985	94501	74547	81681	36775	77601	39152	21773	74123
78197	70345	88237	39805	89978	61843	27307	95173	74710	56784
23827	06325	63920	70424	80786	59019	28567	29707	89598	36356
92164	82637	83241	24578	15728	84939	57621	36562	41567	21006
61030	72793	25436	97741	62807	55664	40248	52078	74119	68021
58039	86465	59843	86135	05306	89891	93785	41738	82244	90937
67543	48098	97777	38627	11806	69406	14576	84619	96532	09634
13044	94442	12523	69018	31939	19070	15634	62396	81360	02132
44931	41325	96536	87367	54417	92886	39762	75261	13206	41835
82238	34257	62862	15504	29917	85453	62881	84922	87751	14752
93964	88645	95125	61660	62121	93252	28715	66944	13011	83788
19042	12154	23520	92775	47127	26218	48990	67822	90293	60806
74721	79323	63876	56214	61009	33510	71061	31849	86786	11658
30346	21063	65983	84323	52471	66339	57997	57181	45319	65037
30147	13958	68591	57428	51905	69148	63821	39120	92991	58551
84647	96966	49332	20037	42826	78226	52437	10882	19960	17084
98085	72782	76664	90210	76685	37996	95198	94087	25471	86138

Table A.11
Random Numbers (Continued)

95361	29804	78834	55956	60645	62333	35475	05122	96533	36429
10726	35647	15378	59696	17342	61950	91228	53597	14148	92233
58455	47498	93707	29871	34042	48985	90319	57889	52422	54195
81925	54845	82789	29488	97063	30296	87476	60331	24536	69650
31181	50909	26202	78877	77833	32687	77073	32507	57137	36272
82466	98292	90654	23205	10717	13602	14000	48249	82350	53061
25289	11437	26827	81063	33205	76649	34206	70213	85665	65923
54646	82902	31923	78934	45737	33020	49601	83344	53476	67911
33657	91329	71616	49971	56344	79052	26477	32117	37135	41311
06141	63062	83870	76303	42135	68886	71377	51283	65810	70902
75006	79591	93584	62368	25549	87338	88724	04188	19523	16771
97147	40230	45367	87124	98812	34559	28094	70708	35852	64364
41327	40009	25422	66272	37604	72743	58694	05026	96955	18881
30104	73933	59772	85998	63715	65077	51710	69888	20129	15195
53875	76952	35241	97504	46780	39117	43916	54159	61474	13169
02373	36459	69553	57019	24324	19078	52219	06724	29883	38368
15484	51475	14474	66519	87360	59837	79668	57847	37330	85020
79348	49057	94798	63480	11508	94978	29766	13627	49224	94136
13921	66981	57202	89998	91302	17460	56816	30661	16412	29734
46484	53042	68542	92935	62473	80166	46917	27163	87342	51503
61731	46087	89812	76551	98075	41721	97320	98835	24249	54851
96838	98073	18700	92070	21344	28927	73299	80284	21039	60219
96219	15660	95916	61982	38819	30633	53969	39474	96862	52471
73188	29062	37888	80712	33576	10956	37179	14887	23532	15233
40206	51561	93265	78655	95645	11411	88897	12059	84779	87462
00507	88479	71233	15485	58749	87625	10726	42247	98119	93685
96386	98240	19885	10512	82524	87139	51675	61092	56010	53684
25806	38611	40534	48032	80928	52516	44576	12788	11578	58797
84269	55398	72668	28861	74374	98534	59813	83317	56962	55638
10811	94720	56727	10097	61554	34539	93525	11807	63933	29875

APPENDIX B
*Answers to Selected
Odd-Numbered
Problems*

APPENDIX B
ANSWERS TO SELECTED
ODD-NUMBERED PROBLEMS

Chapter 4

4.3   Median = $91,428.6
4.11   a. Median = 28.6   b. $\mu = 29.5$ c. $\mu_3 = 199.4375$, coefficient of skewness = 0.4534
4.13 $10,748.00
4.15   a. $500   b. first quartile = $440.5, second quartile = $606
        c. $\sigma = \$114.0769$ d. $\mu_3 = 955,660.016$, $\mu_3 = 0.6437$
4.19   a. 7.5, 9.58   b. 20.83, 31.08
4.21   8.14%
4.25 $\mu = 0.535$, $\sigma^2 = 0.332$
4.31   a. 0.963   b. total value = $233.60
4.33   5.74%
4.37   a. -1.46875 b. 2,800   c. median = -1.50

Chapter 5

5.11   b. $\dfrac{3}{4}$ c. $\dfrac{5}{8}$
5.13   a. 20   b. 36   c. ∞
5.15   a. $\dfrac{1}{3}, \dfrac{1}{4}$ b. $\dfrac{14}{15}$ c. $\dfrac{1}{2}$
5.19   a. 0.034   b. 0.6025
5.21   0.335

5.23　a. 0.56　　b. 0.857
5.25　a. 0.04　　b. 0.042　　c. 0.0365　　d. 0.0714
5.29　a. 9,600　　b. 91,390
5.31　440
5.33　a. 1,679,616　　b. 1,413,720

Chapter 6

6.7　$E(X_1) = 25$, var $(X_1) = 125$, $\mu_3 = 0$
　　$E(X_2) = 31.25$, var $(X_2) = 285.9375$, $\mu_3 = -1,698.2422$
6.11　$E(X) = \$17,000$, $SD(X) = \$20,000$, $\mu_3 = 0$
6.13　$E(X) = 286.92$DM, $SD(X) = 8,449.6364$DM, $\mu_3 = 7.0061898 \cdot 10^{13}$,
　　coefficient of skewness = 116.1363
6.19　a. $\mu_3 = 0.432$　　b. $\mu_3 = 4.32 \cdot 10^{11}$　　c. $\mu_3 = -4.32 \cdot 10^{11}$, coefficient
　　of skewness = -0.5145

Chapter 7

7.3　0.4511, 0.7218, 0.5414, 0.2707
7.5　a. 0.5665,　　b. 0.0733,　　c. \$9.656,　　d. 0.000335
7.7　0.7061, 0.6988
7.9　$n = 10$; $p = 0.8$
7.11　a. 0.0054　　b. 0.000097
7.13　0.8226, 0.0169
7.15　a. 0.1680　　b. 0.5768

Chapter 8

8.1　a. 0.3085　　b. 0.9893　　c. 0.9332　　d. 0.0165
　　e. 0.1357　　f. 0.0062　　g. 0.9999
8.3　a. 70　　b. 63.28　　c. 80.24
8.5　\$4,684
8.7　10.05
8.9　Stock *B* should be chosen.
8.11　0.9177
8.13　0.8502

Chapter 9

9.7　0.2857
9.11　For $n = 9$ the probability is 0.8413.
　　For $n = 64$ the probability is 0.9962.
　　For $n = 81$ the probability is 0.9987.
　　For $n = 100$ the probability is 0.9996.
9.17　a. 0.9999366　　b. from 100.715 to 103.285
9.19　from \$153.218 to \$186.782
9.21　0.1359

9.23   point estimate = \$65.143
        confidence interval is from \$54.738 to \$75.548
9.25   Interval estimate is from 0.2081 to 0.6491
9.27   a. 0.45   b. from 0.44025 to 0.45975   c. point estimate = 0.45;
        interval estimate is from 0.3525 to 0.5475
9.29   9,604
9.31   a. 27.93%   b. from 21.73% to 34.13%   c. from 18.62% to 37.24%

Chapter 10

10.9   a. all values greater than 623.67   b. all values less than 571.80 or
        greater than 628.20   c. 0.7852   d. 0.8708
10.11   Test statistic is −3.015
10.13   a. Test statistic is −4.81   b. Test statistic is −8.10
10.15   Test statistic is −2.928
10.17   Test statistic is −7.396
10.19   Test statistic is 0.3310

Chapter 11

11.5   Test statistic is 32.0920
11.7   b. Test statistic is 775.41
11.9   Test statistic is 16.67
11.11   Test statistic is 22.22
11.15   Test statistic is 29.437
11.17   Test statistic is 18.169
11.23   Test statistic is 3.112
11.25   b. Test statistic is 2.964
11.27   Test statistics are: a. 3,540.77,   b. 197.67

Chapter 12

12.13   Test statistics are: a. 27.0   b. 2.175
12.15   Test statistics are: a. 2.84   b. 0.51   c. 0.61
12.17   b. Test statistic is 1.74

Chapter 13

13.7   a. 0, 0.5   b. 0, 2   e. 0, 1.42857
13.9   a. 1,200, 0.28   b. 3,000, $\frac{7}{9}$ c. 1,200, $\frac{28}{90}$
13.11   b. −840, 220
13.13   a. $\Sigma(Y - \hat{Y})^2 = 108,001.1$ b. 511.34
13.17 $\hat{Y} = 10$
13.19   b. $\hat{Y} = 0.162 + 0.732X$ c. $\hat{Y} = 0.017 + 0.920X$

Chapter 14

14.17   a. $\hat{Y} = 15.4937 + 0.0942X$ c. From 46.98 to 115.88
            d. Test statistic is 5.12

14.19   a. $\hat{Y} = 4.8501 + 1355X$; slope is not significantly different from zero.
            b. $\hat{Y} = 6.8120 - 0.7947X$; slope is significantly different from zero.
            c. $\hat{Y} = 4.4683 + 0.1910X$; slope is not significantly different from zero.

14.21   a. $b_{GM} = 1.313$, $b_{IBM} = 0.862$, $b_{AMC} = 0.864$

14.23   118

14.27   a. \$8,900   b. An increase in X_1 will increase $\hat{Y}$ more.

14.29   a. $\widehat{CR} = 4{,}692.1004 + 0.019946631C$
            b. $\widehat{CR} = -3{,}936.6150 + 0.078995147C + 321.60592V + 435.90042T$

Chapter 15

15.1   a. possible   b. 100

15.3   1

15.9   0.50, 0.50

15.11   a. 0.9413   b. 0.0479   c. 0.0553

15.13   1.00

Chapter 16

16.9   a. winter 1983: \$1.5375 million, spring: \$1.5759 million, summer: \$1.6153 million, fall: \$1.6557 million

16.11   a. $b_A = 4.952$, $b_B = 12.370$

16.13   b. The seasonal index is: winter = 99.81, spring = 97.52 summer = 95.34, fall = 107.34

16.15   d. 258.91

Chapter 17

17.3   7.2%

17.13   a. 105.6, 120.7   b. *Paasche:* 111.1, 125.2; *Laspeyres:* 111.1, 125.3
            c. *Paasche:* 104.8, 118.2; *Laspeyres:* 104.8, 118.3

Chapter 18

18.11 $h = 2.467$

18.13 $r_s = -0.90$

18.15   a. $\tau = 0.542$ b. $r_s = 0.683$, $R = 0.7133$

18.23 $H = 0$

18.25 $r_s = -0.489$, $\tau = -0.333$

Chapter 19

19.1   a. A_3 b. A_1

19.3   b. A_3

19.5   a. maximax: A_4; maximin: A_1; minimax regret: A_2

19.7   a. expected value: ∞ b. expected value: ∞

19.11   a. B   b. A; risk averter   c. A

19.13   A, B, C and F

19.15   a. yes   b. yes

INDEX

Table of Areas for Standard Normal Probability Distribution

z	0.00	0.01	0.02	0.03	0.04	0.05	0.06	0.07	0.08	0.09
0.0	0.0000	0.0040	0.0080	0.0120	0.0160	0.0199	0.0239	0.0279	0.0319	0.0359
0.1	0.0398	0.0438	0.0478	0.0517	0.0557	0.0596	0.0636	0.0675	0.0714	0.0753
0.2	0.0793	0.0832	0.0871	0.0910	0.0948	0.0987	0.1026	0.1064	0.1103	0.1141
0.3	0.1179	0.1217	0.1255	0.1293	0.1331	0.1368	0.1406	0.1443	0.1480	0.1517
0.4	0.1554	0.1591	0.1628	0.1664	0.1700	0.1736	0.1772	0.1808	0.1844	0.1879
0.5	0.1915	0.1950	0.1985	0.2019	0.2054	0.2088	0.2123	0.2157	0.2190	0.2224
0.6	0.2257	0.2291	0.2324	0.2357	0.2389	0.2422	0.2454	0.2486	0.2518	0.2549
0.7	0.2580	0.2612	0.2642	0.2673	0.2704	0.2734	0.2764	0.2794	0.2823	0.2852
0.8	0.2881	0.2910	0.2939	0.2967	0.2995	0.3023	0.3051	0.3078	0.3106	0.3133
0.9	0.3159	0.3186	0.3212	0.3238	0.3264	0.3289	0.3315	0.3340	0.3365	0.3389
1.0	0.3413	0.3438	0.3461	0.3485	0.3508	0.3531	0.3554	0.3577	0.3599	0.3621
1.1	0.3643	0.3665	0.3686	0.3708	0.3729	0.3749	0.3770	0.3790	0.3810	0.3830
1.2	0.3849	0.3869	0.3888	0.3907	0.3925	0.3944	0.3962	0.3980	0.3997	0.4015
1.3	0.4032	0.4049	0.4066	0.4082	0.4099	0.4115	0.4131	0.4147	0.4162	0.4177
1.4	0.4192	0.4207	0.4222	0.4236	0.4251	0.4265	0.4279	0.4292	0.4306	0.4319
1.5	0.4332	0.4345	0.4357	0.4370	0.4382	0.4394	0.4406	0.4418	0.4429	0.4441
1.6	0.4452	0.4463	0.4474	0.4484	0.4495	0.4505	0.4515	0.4525	0.4535	0.4545
1.7	0.4554	0.4564	0.4573	0.4582	0.4591	0.4599	0.4608	0.4616	0.4625	0.4633
1.8	0.4641	0.4649	0.4656	0.4664	0.4671	0.4678	0.4686	0.4693	0.4699	0.4706
1.9	0.4713	0.4719	0.4726	0.4732	0.4738	0.4744	0.4750	0.4756	0.4761	0.4767
2.0	0.4772	0.4778	0.4783	0.4788	0.4793	0.4798	0.4803	0.4808	0.4812	0.4817
2.1	0.4821	0.4826	0.4830	0.4834	0.4838	0.4842	0.4846	0.4850	0.4854	0.4857
2.2	0.4861	0.4864	0.4868	0.4871	0.4875	0.4878	0.4881	0.4884	0.4887	0.4890
2.3	0.4893	0.4896	0.4898	0.4901	0.4904	0.4906	0.4909	0.4911	0.4913	0.4916
2.4	0.4918	0.4920	0.4922	0.4925	0.4927	0.4929	0.4931	0.4932	0.4934	0.4936
2.5	0.4938	0.4940	0.4941	0.4943	0.4945	0.4946	0.4948	0.4949	0.4951	0.4952
2.6	0.4953	0.4955	0.4956	0.4957	0.4959	0.4960	0.4961	0.4962	0.4963	0.4964
2.7	0.4965	0.4966	0.4967	0.4968	0.4969	0.4970	0.4971	0.4972	0.4973	0.4974
2.8	0.4974	0.4975	0.4976	0.4977	0.4977	0.4978	0.4979	0.4979	0.4980	0.4981
2.9	0.4981	0.4982	0.4982	0.4983	0.4984	0.4984	0.4985	0.4985	0.4986	0.4986
3.0	0.49865	0.4987	0.4987	0.4988	0.4988	0.4989	0.4989	0.4989	0.4990	0.4990
4.0	0.4999683									

Illustration: For z = 1.93, shaded area is 0.4732 out of total area of 1.

Source: From John Neter, William Wasserman, and George A. Whitmore, *Fundamental Statistics for Business and Economics*, 4th ed. Copyright © 1973 by Allyn and Bacon, Inc., Boston. Reprinted by permission.